Marine Pollution and International Law

Marine Pollution and International Law

Principles and Practice

Douglas Brubaker

Belhaven Press
London and Florida

Belhaven Press
(a division of Pinter Publishers)
25 Floral Street, Covent Garden, London, WC2E 9DS, United Kingdom

First published in 1993

British Library Cataloguing in Publication Data
A CIP catalogue record for this book is available from the British Library

ISBN 1 85293 273 2

Distributed in North America by
CRC Press Inc., 2000 Corporate Blvd., N.W.,
Boca Raton, Florida, 33431

Library of Congress Cataloging-in-Publication Data
A CIP catalog record for this book is available from the Libarary of congress

Printed and bound in Great Britain by Biddles Ltd, Guildford and Kings Lynn

Table of Contents

Acknowledgements

The author has made all reasonable efforts to obtain permission for the use of copyright material. The following individuals and publishers have kindly granted permission for the reproduction of figures and tables.

Roald Saetre at the Havforskningsinstituttet, Bergen, for Figs. 1.1, 1.6 and 1.7 (Figs. 11, 27 and 28, Final Report from the Norwegian Marine Pollution Research and Monitoring Programme, The Fate and Significance of Oil in the Sea, 1976-1984); Rolf Lange, Oslo, for Figs. 1.2, 1.3, 1.4 and 1.5 (Figs. 14, 15, 16 and 17, Final Report from the Norwegian Marine Pollution Research and Monitoring Programme, The Fate and Significance of Oil in the Sea, 1976-1984); National Academy Press, Washington D.C. for Figs. 1.7, 1.8 and 1.10 (Tables 5-7, 5-8 and 5-9, Oil in the Sea: Inputs, Fates and Effects, National Research Council et al., 1985) and 1.14 (Table 1, Pollution in the North Sea from Oil Related Industry: An Overview, 1987 [based upon a study conducted by the US Academy of Science 1985]); Elsevier Applied Science Publishers, Barking, Essex, for Fig. 1.9 (Fig. 2, Petroleum Effects in the Arctic Environment, Ottawa, Canada, 1985); Tyncho Anker-Nilssen for Figs. 1.11, 8.18 (Appendix 2, Table 7.1, Anker-Nilssen, 1987); Veritec for Fig. 8.19 (Table 4.1 [based upon data from Veritec], Anker-Nilssen, 1987); Ian Dundas for Fig. 1.12 (Fig. 21, Final Report from the Norwegian Marine Pollution Research and Monitoring Programme, The Fate and Significance of Oil in the Sea, 1976-1984); Edwin Gold for Fig. 1.13 (Gold, 1985, p.133); Richard Westergaard at Senter for Industrieforskning, Oslo, for Fig. 1.15 (Table D.1, Westergaard, 1987); Statensforurensningstilsyn, Oslo, for Figs. 1.16, 1.18, 1.19 and 1.20 (SFT [Norwegian State Pollution Control Agency], 1987); The Paris Commission, London for Fig. 1.17; International Hydrographic Bureau, Monaco, for Figs. 8.1 and 8.2 (From maps on pp.200 and 201 of Limits of Oceans and Seas, International Hydrographic Organization, 1986); Pinter Publishers Ltd, London, for Figs. 8.3, 8.5 and 8.12 (Fig. 2.1, Table 2.1 and Fig. 3.5, Ostreng, 1987); Noroil Publishing House Ltd, Stavanger, Norway, for Fig. 8.4 (Euroil, Issue 5, August 1990, p.30); Pinter Publishers Ltd, London, for Fig. 8.6 (Fig. 1.1, Brigham, 1991); Lawson Brigham for Fig. 8.7; Norwegian Ministry of Oil and Energy, Oslo, for Figs. 8.8 and 8.9 (Maps 1 and 2 from St.meld.nr.40, 1988-89, p.8) and 8.10 (Table 6.1 from St.meld.nr.40, 1988-89, p.42); Norwegian Bureau of Statistics for Fig. 8.11 (Fig. 1 from Oil and Gas Activity); Butterworth Heinemann Ltd, Oxford, for Fig. 9.1 (Fig. 1 [modified from Griffiths: A Northern Foreign Policy, Canadian Institute of International Affairs, 1979], Marine Policy, Volume 11, No. 4, 1987, p.289); Fisheries and Oceans Canada, Canada Communications Group, Crown Copyright Administration, for Fig. 10.1 (Map 1, Ocean Development and International Law, Vol.18, No.1, 1987, p.52).

Preface

Recommendation for Study of National Environmental Law

It must be noted that chiefly the *international* legal aspects of environmental protection and prevention are addressed here. However an integral part of the regimes set forth by the international law is the *national* legislation, Court decisions and executive declarations implementing these provisions. It would be far beyond the scope of this project to carry out any analysis of the national environmental law of the countries which are Parties to the Conventions and Agreements discussed below. This also holds true concerning even the national environmental law of the countries involved in the Barents Sea Case Study, the States bordering the Arctic.[1] Such a study however would be of great benefit in clarifying the law applicable to the Barents Sea as well as providing a better basis for establishing new international provisions where required.[2] For the first an analysis of Arctic rim State legislation, Court decisions and executive declarations would indicate any legal areas from which it could be presumed environmental principles have entered into customary law though *State practice* and *opinio juris*. This is very important since as will be seen below in PART I Section 2.1. customary law *could* provide legal coverage in areas lacking conventional coverage, a situation not too uncommon.[3] The present state of international customary law however is that only a *few* general principles appear to have become customary law, and they do not provide the specificity required to deal actively with environmental problems. An analysis of the national anti pollution law of the Arctic rim States, especially that of the Soviet Union and the United States, both superpowers and sea powers, might add specificity to this vague yet important area of international law.

For the second, upon such an analysis a comparison could be made of national law, and the international provisions as set forth in the environmental conventions ratified by each Arctic State, as well as the United Nations Convention on the Law of the Sea (LOSC).[4] This would indicate any areas of *non compliance* by the Arctic rim States. Additionally in conjunction with the recommendation above, it might be argued that certain provisions of the international environmental conventions have passed into international customary law for the Arctic if all the Arctic rim States have complied with them over time. This comparison is especially important as regards the LOSC, which is not in force yet which often is referred to as a codification of customary law, but under which legal authors vary to a great extent as to which provisions represent customary law. Additionally since industrial States seem reluctant to ratify the LOSC, parallel Arctic rim State national law might assist in firming up the customary law, and hencc negate the need for State ratification of this Convention. A very brief outline of Arctic State national implementation of LOSC is presented in PART I Section 9.1.[5]

For the third an analysis of national Arctic State law would give an idea of the starting point for any negotiations involving new environmental provisions such as included in the Arctic Convention and Protocols proposed below in PART II Chapters 10 - 17. If as is noted by one author[6] that the Soviet legislation regarding dumping and vessel-source pollution generally follows LOSC provisions, then their starting point regarding negotiations in these areas can be presumed. If a similar comparison and analysis could be done for all the Arctic States and covering the pollution sources together with liability provisions and contingency plans, then drafts could be formulated even more concisely than has been done with the Arctic Convention and Protocols. Both environmental provisions and political realities could be much more in balance with a consequent more probable Arctic rim State interest in ratification of the proposed environmental regime. Additionally it would be more evident where stronger environmental provisions should be proposed.

Finally as part of any analysis suggested above there should as well be included an examination of the national law in *practice*. Are these rules *effective* or are they merely paper without any teeth similar to some of the international agreements mentioned below? This information would as above contribute to more concise and substantial proposals being made providing for Arctic environmental protection.

Saying this I would like to thank Associate Professor Geir Ulfstein, University of Oslo, for the interesting and illuminating discussions giving direction in this complicated area of law. Additionally I would like to thank Senior Lecturer in Law Robin Churchill, University of Wales, for taking the time to give illuminating comments as well as co-authoring an incredibly concise book covering the law of the sea, without which this work would have been much more difficult to conduct. As well thank you should be addressed to Research Scholar Olav Schram Stokke, Fridtjof Nansen Institute (FNI) for his ready assistance in providing up to date information in a legal area characterized by rapid developments, Office Manager Grete F. Haram for her essential aid in preparing this work for publication, Computer Consultant Ivar

M. Liseter for his crucial contribution, Consultant Kjersti Amundsen, the Norwegian Ministry of Oil and Ministry for her invaluable help in tracking down information concerning Soviet environmental standards, and the University of Tromsø for the funding and time to carry out such a project. The sole responsibility for the views expressed in this work as well as for any mistakes is mine alone. Finally I would like to express my gratitude to my family for their assistance and their all too frequent sharing of their husband, father and son with this project.

Tromsø, January 8, 1992
Douglas Brubaker

Notes - Preface

1 Canada, Denmark/Greenland, Norway, the Soviet Union, and the United States. These will hereafter be referred to as the 'Arctic rim States.' Note that due to the dissolution of the Soviet Union only some time ago it is impossible to state whether Russia will follow the international environmental policies established by the Soviet Union. For this reason any reference to the Soviet Union in this work is left standing until such a time as Russian policies especially in the Arctic become clear.

2 There do exist a few works of this nature. Erik Franckx with the Free University of Brussels has reportedly written a doctorate thesis concerning national legislation covering shipping in the Arctic, conversation Professor David Caron, Boalt Law School, University of California at Berkeley, July 24, 1990. In addition a study of indefinite scope involving Arctic environmental legislation of the Soviet Union, Norway, Denmark/Greenland, Canada and the US - Alaska is currently being carried out by J. Enno Harders, Institutt fur Internationales Recht an Der Universitat Kiel, correspondence July 1, 1988.

3 This is especially evident in the area of State liability for transboundary environmental damage.

4 This latter Convention is generally known as the 'Constitution of the Law of the Sea.'

5 Along these lines but on a basis broader than just that related to the Arctic, study also needs be done establishing concretely which LOSC environmental provisions have passed into international customary law.

6 Franckx, Erik, 'The New USSR Legislation on Pollution Prevention in the Exclusive Economic Zone,' *International Journal of Estuarine and Coastal Law*, Vol. 1, No. 2, 1986 (Franckx) pp. 156-159.

Abbreviations

ACOPS: Advisory Committee on Pollution of the Sea (ACOPS), Survey of Oil Pollution around the Coasts of the United Kingdom, 1986-87. 3 Endsleigh Street, London WC1H ODD.

Aftenposten: Norwegian newspaper published in Oslo, Norway.

CBT's: Clean ballast tanks, on ships under MARPOL 73/78 Annex I.

CFC's: Chlorofluorocarbons, ozone depletion substances.

COW's: Crude oil washings, on ships under MARPOL 73/78 Annex I.

dwt: Dead weight ton, used in shipping conventions.

ECA: Economic Commission of Africa.

ECLA: Economic Commission for Latin America under the United Nations.

ECWA: Economic Commission for Western Asia.

ECOSOC: Economic and Social Council of the United Nations.

ECU: European Currency Unit.

EEC: European Economic Community.

EMEP: Cooperative Programme for Monitoring and Evaluation of Long-range Transmission of Air Pollutants in Europe, under the ECE Convention.

EPA: United States Environmental Protection Agency.

ESCAP: Economic and Social Commission for Asia and the Pacific under the United Nations.

FAO: Food and Agricultural Organization of the United Nations.

FNI: Fridtjof Nansen Institute, Fridtjof Nansens vei 17, P.O. Box 326, N-1324 Lysaker, Norway.

GOP: Group on Oil Pollution under Paris Convention.

GRT: Gross tonnage of vessels.

IAEA: International Atomic Energy Agency.

ICJ: International Court of Justice.

ICJ Report: International Court of Justice Reports

IGS's: Inert Gas Systems, on ships under MARPOL 73/78 Annex I.

IJO: International Juridical Organization of the United Nations Environmental Programme.

ILC: International Law Commission.

ILO: International Labour Organization.

IMO: International Maritime Organization.

IMO News: Newsletter of International Maritime Organization.

INTERTANKO: International Association of Independent Tankers Owners.

IOPC Fund: International Oil Pollution Compensation Fund.

ITOPF: International Tanker Owners Pollution Federation Limited Administration of TOVALOP.

IUCN: International Union for the Conservation of Nature, Gland, Switzerland.

JMP: Joint Monitoring Program of the Paris and Oslo Commissions under the Paris Convention and the ODC.

LOT: Load-On-Top system of loading oil on ships under the OILPOL and MARPOL 73/78 Conventions.

MEPC: Marine Environment Protection Committee of the IMO - 25 Session Agenda Item 19, MEPC 25/19, July 1, 1987.

NATO: North Atlantic Treaty Organization, signed in Washington, April 4, 1949, 34 *United Nations Treaty Series* 243, entered into force August 24, 1949.

Nordlys: Norwegian newspaper published in Tromsø, Norway.

OCIMF: Oil Companies International Marine Forum - the organization which formed the PLATO/CRISTAL Liability proposals in 1985.

OECD: Council of the Organization for Economic and Cooperation and Development.

OPA: United States Oil Pollution Act for ships signed into law August 18, 1990, obtained from Burlingham, Underwood & Lord, *Report on United States Oil Pollution Laws*, October 1990, One Battery Park Plaza, New York, New York 10004.

OPEC: Oil Producing and Exporting Countries.

PAH: Distillate factions of crude oils with boiling point above 350 degrees C, associated with carcinogenic potential.

P. & I. Club: Shipping abbreviation for Protection and Indemnity Club.

ppm: Parts per million, used in shipping and oil platform conventions.

SBT's: Segregated Ballast Tanks, on ships under MARPOL 73/78 Annex I.

SDR's: Special Drawing Rights. IMO unit of payment under the CLC. 1 SDR equals roughly 1.3528 U.S. dollars (1991).

SFT: Norwegian State Pollution Control Authorities (Statens forurensningstilsyn).

Sovjet Nytt: Periodical distributed by the Soviet Embassy, Oslo, Norway.

toe: Tons oil equivalent, used in shipping conventions.

TWG: Technical Working Group under the Paris Convention.

UES: Uniform Emission Standard under the Paris Convention.

UOA's: Units of Account. SDR's defined by the International Monetary Fund as units of payment under the 1984 Protocol to the CLC.

UNDP: United Nations Development Programme.

UNEP: United Nations Environment Programme, P.O. Box 30552, Nairobi, Kenya; and 16, Avenue Jean-Trembley, Petit-Saconnex, 1209 Geneve, Switzerland.

UNESCO: United Nations Educational, Scientific and Cultural Organization.

UNU: United Nations University.

USCG: United States Coast Guard.

WARSAW PACT: Treaty of Friendship, Cooperation and Mutual Assistance, signed in Warsaw May 14, 1955.

WHO: World Health Organization.

WQO: Water Quality Objective under the Paris Convention.

WSF: Prudhoe Bay Crude Oil.

List of Figures

Part I - International Environmental Law

1. Introduction

1.1. Background

In the past few years there has occurred a marked increase of public interest in the effects human activities are having on the natural surroundings. Although there are undoubtedly many reasons for this, one of the most significant are the ship and oil platform catastrophes, dumping thousands of tons of oil into the sea.[1] Examples of these include the 1989 EXXON VALDEZ accident in Alaska (an estimated 30,000 to 40,000 tons of oil),[2] the AMOCO CADIZ accident off the coast of France in 1977 (230,000 tons)[3] and the 1979 IXTOC I blowout in the Gulf of Mexico (500,000 tons).[4] Another type of incident raising public anxiety though much less covered by the news media is the dumping of atomic wastes at sea,[5] and the sinking of submarines with nuclear reactors and weapons on board.[6]

Although smaller environmental accidents occur at regular intervals, a major accident, especially followed by charges of technical defects and operator incompetence, usually result in public outcries for better governmental control. Although these incidents help, albeit somewhat cynically, to raise public awareness and push through control measures for cleaner oceans, the picture given by these events is distorted. There actually is marine pollution quietly occurring every day with a higher cumulative affect than the individual catastrophes. At the same time there exists on a world wide basis extensive, functioning environmental control measures, as evidenced among others by the conventions established under the International Maritime Organization (IMO) and the United Nations Environmental Programme (UNEP). The picture in reality is complicated and represented by these two seemingly opposite situations.

The purpose of this book is to present as clear a picture as possible of marine pollution today and its attempted legal control[7]. This is to be achieved by dividing the book into two parts. The first deals generally with pollution itself, and the governing legal principles. This part attempts to give the reader a broad overview. The second part concentrates on the Barents Sea, north of Norway and the Soviet Union, as a case study, the specific problems occurring there and subsequent proposals forwarded for dealing with these problems.[8]

In Part I first is presented a general background into the effects of *oil* in the sea. Oil is chosen since it has been the most prevalent pollutant, although as can be noted radioactive materials and hazardous chemicals also present an increasing problem.[9] The effects of oil upon fish, animals and the environment has been studied somewhat, but there remains much to be done. Perhaps this is why so many answers are controversial. Then presented are definitions of marine pollution, statistics and background information. Following this a brief overview of legal principles are presented. For each defined sources of pollution,[10] relevant international customary law, general principles, conventions and bilateral and regional agreements are examined. For the pollution classifications presented, relevant questions involving liability, compensation, and intervention in case of mishaps are discussed. In addition measures involving preparedness, consultation and dispute settlement are reviewed. Sections of this law are complicated. Part I is meant as a *handbook* in which it is hoped that ready information can be gained concerning international environmental law.

In Part II environmental, political, fishery and ecological considerations affecting the Barents Sea will be addressed. Specific *problems* with the present international environmental coverage or lack of such in the Barents Sea represented by the conventions and principles discussed in Part I will be noted. Such discussion is placed in Part II rather than Part I for easier reading in relation to the specific proposals forwarded. Discussion of LOSC provisions are also included in Part II rather than Part I due to the Convention's *not* being in force, its *lack* of ratification by any Arctic rim State, as well as its *indefinite* legal status concerning many environmental provisions under international customary law. This was considered best so as to prevent conveying the impression that these environmentally progressive provisions are governing law.

In making recommendations a proposal is made that for *optimal* environmental protection of the Barents Sea, the Arctic rim States enter into a *regional* sea, pollution control agreement covering the Arctic. This is in line with recommendations made by international legal experts.[11] As part of the Arctic regional agreement, proposals are forwarded for individual Protocols covering vessel-source, land-based, oil platform and dumping pollution as well as State liability and contingency plans. Provisions are discussed for all these areas including the legal basis for such a formation. Full documents with comments are presented covering oil platform pollution and contingency plans due to the general paucity of legal documents covering offshore platform pollution as well as the present lack of a joint contingency plan in the Barents Sea.

1.2. Characteristics and Effects of Marine Oil Pollution[12]

1.2.1. Oil in the Sea[13]

In order to better understand the effects of marine oil pollution, an idea of the consistency of oil is needed. The Norwegian Marine Pollution Research and Monitoring Programme (FOH) defines oil as, ‘a naturally occurring, complex mixture of organic components, resulting from the transformation of plants and animal remains under special geological conditions, e.g. high temperature and pressure.’[14] Further, ‘(C)rude oil contains many thousands of different chemical components. The components in the largest group are called ’hydrocarbons’, since they are composed entirely of the elements carbon and hydrogen. Other groups of components also contain sulphur, nitrogen and oxygen. Most crude oils also contain traces of various metals.’[15]

Not all oil compounds in the seas comes from human related activities. A small amount has seeped naturally for thousands and possibly millions of years, and much of it has been assimilated without problem into the ocean environment. Other sources though human related are not those commonly thought associated with oil pollution. The combustion of coal for example gives off aromatic hydrocarbons similar to or the same as found in oil. Additionally though oil pollution may easily be seen to be the pollutant in cases of ship or oil platform catastrophes, it may well form part of a more general pollution threat especially in urban and industrial regions.

Oil discharged into the sea immediately is exposed to weathering which causes physical and chemical changes. The processes involved include spreading on the water surface, evaporation of the lighter components, release into the sea water, formation of emulsions, natural dispersal of droplets in the water, adsorption into particles, formation of tar balls, uptake and elimination by organisms, microbial degradation and photochemical conversion. Of these processes, evaporation and solubility in water are the most important, causing immediate chemical changes in the oil. Processes such as photochemical oxidation and microbial degradation affect the oil chemically several hours or even days later. Evaporation accounts for the loss to the atmosphere of approximately 1/3 to 2/3 of the oil spilled on a sea surface, and nearly all of evaporated oil becomes photochemically oxidized in the atmosphere. Photochemical oxidation is also a significant process acting on oil on the ocean surface, whereas dissolution into the water column is of considerably less importance because of the low solubility in water of most compounds.

The rate of oil spreading on the water surface is determined by the wind, the waves and currents. For an idea for some of the currents along the Norwegian coast see Figure 1.1. In an experimental discharge performed off the coast of mid Norway, it was discovered that in spite of constant monitoring by boats and aircraft and use of statistical oil drift models, 'determination of the drift and spreading of oil was (however) no easy matter.'[16] See Figures 1.2, 1.3, 1.4, and 1.5.

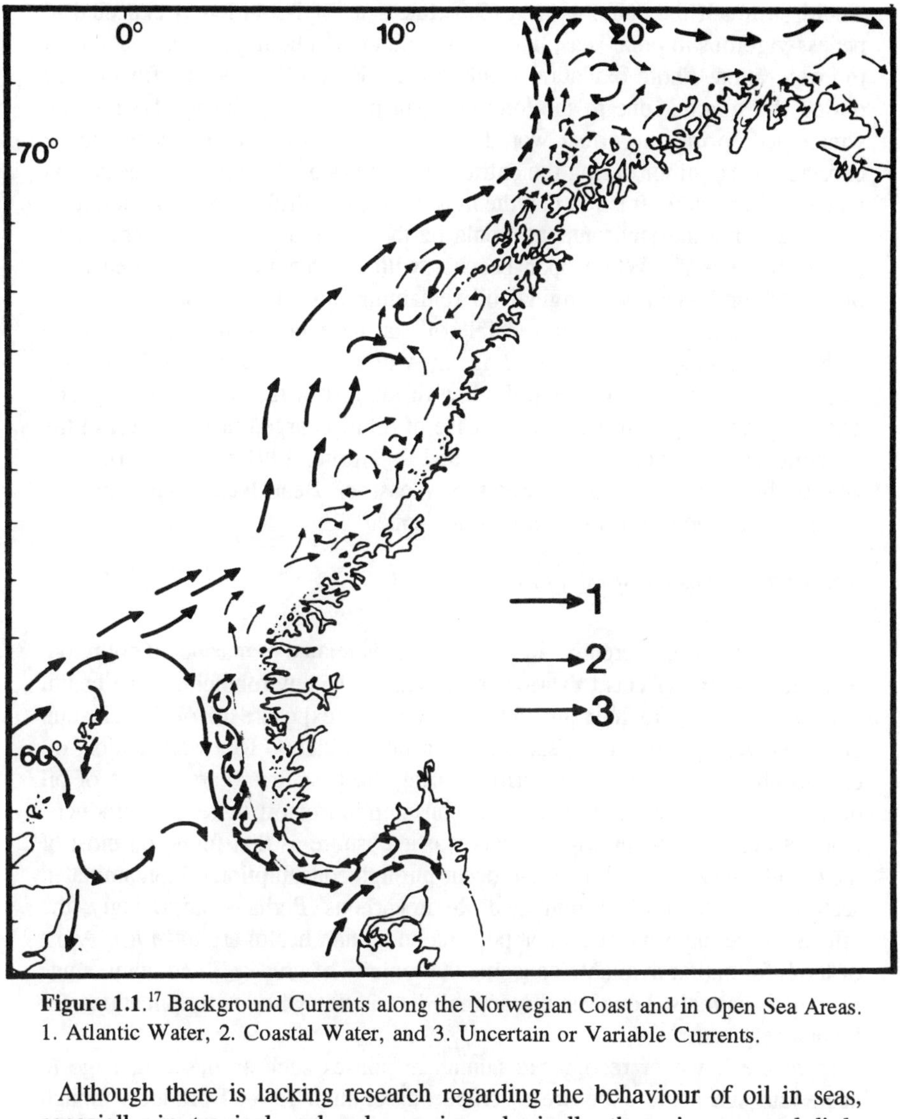

Figure 1.1.[17] Background Currents along the Norwegian Coast and in Open Sea Areas. 1. Atlantic Water, 2. Coastal Water, and 3. Uncertain or Variable Currents.

Although there is lacking research regarding the behaviour of oil in seas, especially in tropical and polar regions, basically there is expected little difference between these seas and the temperate ones, at least in ice free zones.[18] For the polar seas evaporation of crude oil is expected to proceed fairly rapidly since the main factors governing evaporation are the wind and the sea, rather than temperature. The same applies to emulsification due to the wave action which mixes water into the oil. Sedimentation and adsorption would be much the same as in a temperate sea.[19] Dissolution is expected to be less vigorous in polar seas, but this is a minor weathering process compared to evaporation. Photo-oxidation would have a large effect on the floating oil during the summer due to the long daylight periods, but during the autumn, winter and spring the effect would be much less. Auto-oxidation would be expected to be minor during the entire year. The rate of microbial degradation would be very slow. It is felt that the most noticeable difference of oil pollution in the cold marine environment would be the oil stranding on 'sloping, low-energy beaches.'[20] While open beaches with much wave action often have only 10% or less of the original oil remaining after 1 year, on low energy beaches with little wave action visible oil can be expected to last for at least 1 year and up to several *decades*. Lagoons with low wave action will act as oil traps and as reservoirs for the re-fouling of the rest of the coast in later years. The same trapping will occur in the case of oil discharged farther north in the ice zone. At least one study was planned for Spring 1991 whereby crude oil was to be released in the Barents Sea east of Bear Island[21] in order to examine the reaction in the marginal ice zone.[22]

1.2.2. The Effects of Oil upon Humans[23]

One of the questions foremost in people's minds related to marine oil pollution, in addition to the effects to various segments of the environment, is the health hazard it presents to humans. Acute or chronic exposure to oil by cleanup crews, investigating scientists and coastal residents followed major oil catastrophes such as those occurring during the past few years. Intake of oil occurs by inhalation, skin contact or consumption of oil or refined products. Contact may also occur with the tars washed ashore as that found on most of the world's beaches, and may also occur through consumption of contaminated seafoods, which have accumulated hydrocarbons. Perhaps surprisingly, the effects of the majority of oil components on human health are *unknown*. Most of the information available concerns the toxicity of crude oil and compounds related to animals, however the animal data is not without 'implications for human responses.'[24]

There are however recognized human responses such as tissue damage to acute and sub acute exposure to crude oils or their refined products though these responses appear to be of short duration unless the levels are very high. Protective equipment should be used however. The ingestion of hydrocarbons through consuming seafoods is generally thought not to be so serious. The quantities of PAH's[25] found in seafoods generally are equal to or less than levels found in other foods. Exceptions include however seafoods gathered near city outfalls, creosote pilings and local oil industry sources.[26] Reports indicate that, 'at present there is no demonstrated relationship that chronic exposures through eating petroleum-derived PAH-contaminated seafood are related to the incidence of cancer or other diseases in humans,' however exceptions may exist in local areas where the population's major diet is seafood.[27] As yet there exists no studies covering this situation.

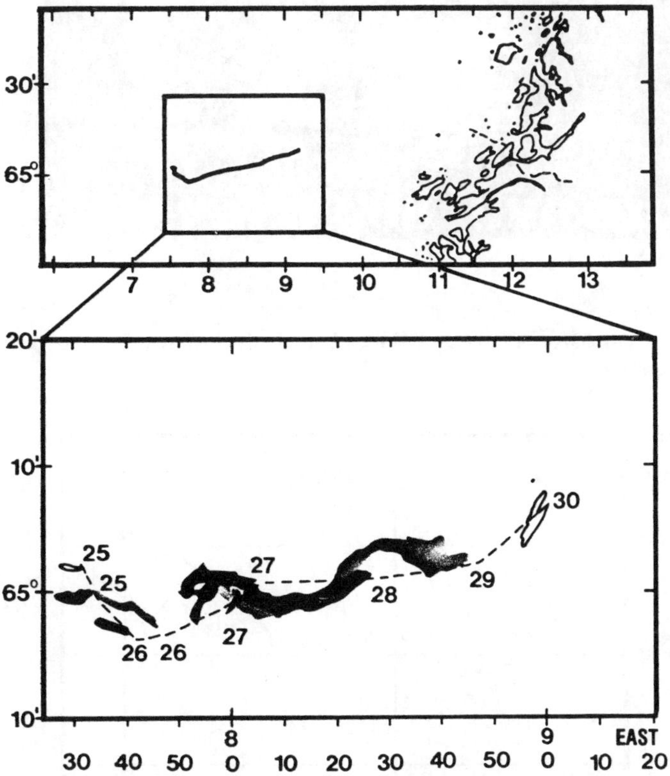

DAY	25	25	26	26	27	27	29	29	30
AREA, km²	1,4	8,1	5,7	10,8	2,9	17,4	63,7	86,1	9,7

Figure 1.2.[28] Spread of Oil Slicks After the Experimental Discharge of Oil on the Halten Bank.

Figure 1.3. Discharge of Oil on the Halten Bank Radar Picture (SLAR) of the Oil Slick and Its Position.

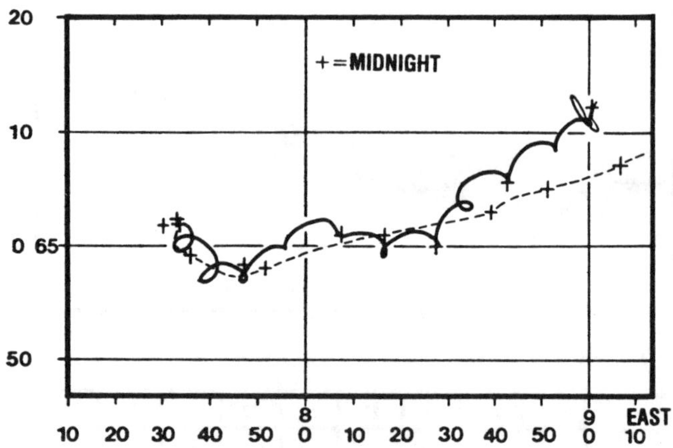

Figure 1.4. Position of the Satellite Positioning Buoy (solid line) and Position of the Central Part of the Oil Slick (dotted line).

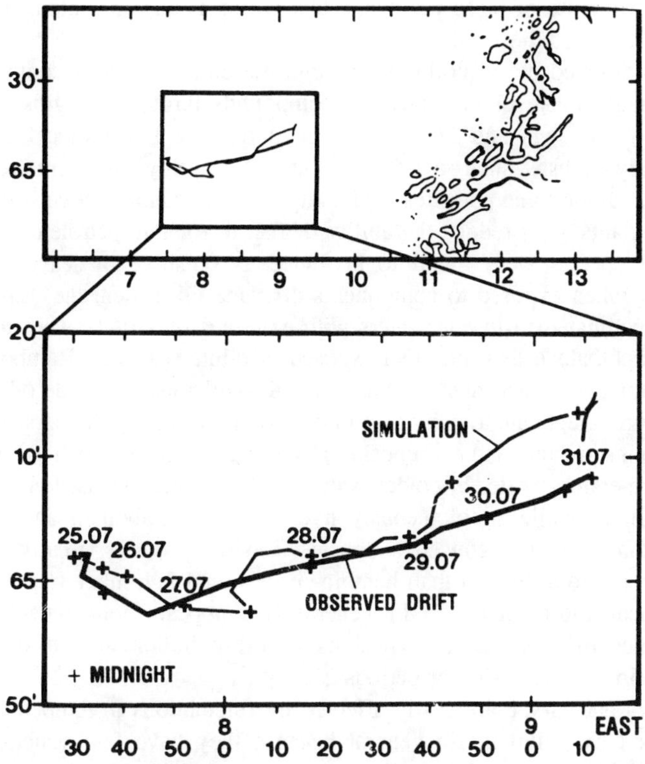

Figure 1.5. Results of the Numerical Oil Drift Model (broken line) Compared with the Observed Drift of the Slick (solid line).

1.2.3. The Effects of Oil upon Fish[29]

Generally and perhaps unexpectedly, it is somewhat *difficult* to say anything definite about the effects of oil upon fish in the sea. This is due to there being so many variables involved, including crude oil's thousand or more differently reacting chemical compounds, the varied types of fish present which may react differently, non corresponding and contradictory laboratory results, and generally field studies *failing* to document widespread effects of oil spills on fish. Additionally even if the difficulties and expenses of conducting a large scale field study are surmounted and the effects to fish populations measured, cause and effect relationships are seldom understood due to other factors being present such as overfishing, food abundance, natural fluctuations and other pollutants. In spite of these problems, that which is known, obtained mostly from laboratory studies, will be presented.[30]

Fish are affected by oil pollution through the intake of oil or contaminated prey, through intake of dissolved oil compounds through the gills, through effects on fish eggs and larval survival, and through ecological changes. Additionally stress, such as fluctuations in salinity, temperature, food abundance, disease and parasites, and competitive or augmentative stress from other pollutants, may reduce the ability of fish to tolerate petroleum. Several salmonoid species were found to be twice as sensitive in seawater as in freshwater when exposed to components of crude oil or benzene. Euryhaline killifish acclimated to live in waters with extremes of salinity were found to tolerate naphthalene less than fish exposed in dilute seawater. Rainbow trout when acclimated to seawater were also found to tolerate less crude oily water. Temperature affects the toxicity of oil pollution by affecting the persistence of hydrocarbons in water and by imposing physiological stress of fish at both ends of the temperature scale. In colder waters hydrocarbons persist longer in the water column, while simultaneously hydrocarbon metabolism and rates of clearing may become reduced.[31] Species sensitivity varies greatly. Pelagic species seem more tolerant than benthic species, and fish inhabiting intertidal areas appear most tolerant.[32] In general fish appear more sensitive than invertebrates to short, acute exposures of oil pollution and absorb lethal quantities in relatively shorter periods.[33]

Fish eggs and larvae are easily affected by fluctuations in temperature and salinity and by pollutants in general because they have few structures and organs capable of detoxifying oil, they are not mobile enough to avoid oil exposures and many of them develop at or near the surface where oil exposure can be expected. See Figure 1.6 for a picture of normal cod development. It appears that some stages of eggs and most stages of larvae probably are more vulnerable to oil than juveniles and adults. There are however nuances, and salmon eggs were found to be 'extremely tolerant to benzene and oil WSF.'[34] Damage to an embryo may not be apparent until after hatching of the eggs into larvae, with several studies indicating larval deformities following egg survival. Though cod eggs survived exposure to oil, the larvae were deformed and could not swim.[35] Crude oil is most toxic to cod eggs at the beginning of gastrulation, and although cod eggs develop considerable resistance to oil toxicity, their resistance decreases again when the larvae hatch and continues to fall as the yolk is used up. Typical larval responses to toxic concentrations include a brief increase of activity, reduced activity, sporadic twitching, narcosis and ultimately death. Since natural mortality in fish eggs and larvae is however very high, it might be possible that even larval losses of 50% or more may have little effect on an adult fish population.

Even though fish exposed to oil discharged into the sea might not be killed, the *sublethal* effects might be adverse. Studies indicate effects of such exposure include deformed spinal columns, tissue destruction especially of fins, abnormal energy consumption and damage to cells, tissues or organs, which may ultimately result in the reduction of the fish population in a given area. As might be expected these effects are more difficult to detect than death, and it is nearly impossible to determine a direct relationship between sublethal effects and the oil. There have been however some observations of this nature on fish in the field. It is known that oil components are taken up into the fish tissues after exposure in the water, in their food or the sediment. However once petroleum is accumulated in fish, many compounds are naturally metabolized and excreted.[36] As noted above in relation to humans there is no evidence of food web magnification. The excretion occurs chiefly from the liver into the bile, but also from the gills, the mucous secretions on the skin and the gonadal tissues. In spite of the metabolization and excretion, tissue damage and physiological disturbances may occur, and the balance between organisms and their environment changed through disease.[37] Though fish may recover from the sublethal effects of oil, because recovery requires energy, the fish may become more vulnerable to disease, including increased fin rot, reduction in external bacterial flora, reduction in rate of tissue repair or regeneration, and increase parasitism. Growth rates may decrease and have been observed in herring embryos, killifish larvae, cutthroat trout juveniles, pink salmon alevins, English sole, and coho salmon.[38] This may have significance for fisheries since the entry of fish into the fishery may be delayed.

Low levels of oil pollution comprising less than 1 part per million (ppm) may affect fish reproduction through the malformation of gonads or gametes or by decreasing the energy required for reproduction.[39] This damage has *significance* since such concentrations are present under actual spill conditions. Generally however, oil concentrations in the sea are low, especially in deep water, and though some sublethal effects could damage a population by affecting growth or reproduction, 'these effects would be more likely in estuarine populations that suffer long-term exposures to many pollutants.'[40]

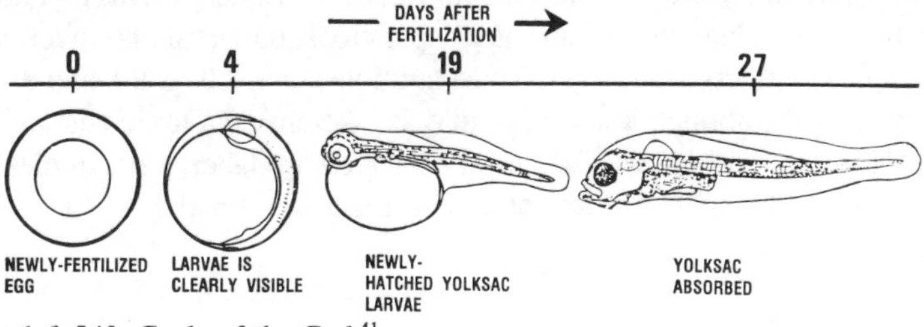

Figure 1.6. Life Cycle of the Cod.[41]

Fish have an excellent sense of smell, which could be used to avoid spills. However short term exposures could damage this sense, and one species has been observed to lose its ability to school after exposure to crude oil, 'presumably because olfactory and lateral-line organs were damaged.'[42] Some species have been observed avoiding oil spills, while others were observed in the close vicinity of such; so the conclusions are contradictory. Fish may in fact be attracted to oiled areas due to the presence of easy prey.[43]

Another effect of oil pollution involves the commercial and sport fisheries. The foremost fear in these industries is the destruction of stocks, but this may be overrated.[44] Some fish kills have been reported after localized oil spills, '(H)owever large numbers of dead fish have *not* been found after many of the large crude oil spills, such as ARGO MERCHANT, EKOFISK or IXTOC I. A decrease in fishery stocks has never been attributed to oil spills, not even to the largest spills.'[45] If large kills do occur, this occurs as seen above to the sensitive egg and larval stages found on the surface. This has been difficult to document however since these stages are so fragile that once killed they decompose within hours and cannot be collected. Available evidence from the ARGO MERCHANT accident indicates however that some kill occurred of cod and pollock. Another factor involved in the fisheries is the economic damage caused due to the consumer *fear* of buying and eating tainted fish, even when the fish are not tainted.[46] Tainting (defined as a change in appearance, texture, odour or taste) can however be a very real problem, and experienced tasters can detect 10-30 ppm of crude oil in cooked or raw fillets.[47] Tainting has been frequently reported for spill accidents including the TORREY CANYON, the EKOFISK blowout and the AMOCO CADIZ. It is unlikely however that tainting will occur, except in the case of major catastrophes, lessor spills occurring in restricted areas such as bays, or under chronic oiling conditions.

In summary, as can be seen experience dealing with the effects of oil on fish is limited. There is very little evidence of increased accumulation of oil pollutants in the higher members of the food web, and for humans, consumption of oil polluted seafood is not a widespread problem, though in individual areas it may be. More experiments need be done in the *field* to coordinate laboratory and field work and substantiate laboratory findings. This includes the impact of oil on pelagic organisms and populations, on larval and juvenile fish (including the effects of low oil concentrations) and on polar and tropical organisms.[48] Even though knowledge may be sketchy, the toxic and sublethal effects of oil to individual fish is *documented* in laboratory studies with concentrations of less than a few ppm. See Figures 1.7 and 1.8.

1.2.4. The Effects of Oil upon Mammals[49]

As is the case with fish, and perhaps even more so, there do not exist comprehensive studies showing the damaging effects of oil upon marine mammals, although there are some documented or inferred fouling from oil spills.[50] In many cases however clear effects are expected, due to mammals' grooming habits and a *large* proportion of mammals living in *ice-covered* areas. In ice-covered waters oil will be concentrated in and under leads, polynyas (open water areas) and breathing holes, and wind will move the oil between ice floes and up against ice edges. The open areas are where the seals, whales and polar bears surface for air, as well as represent routes and areas for feeding and reproduction for many mammals.[51] Oil that is discharged near coastal areas and washed upon low energy beaches can effect the seals and the otters.[52]

Almost *all* polar bears, sea otters and otters which come into contact with oil will die due to the fur coat losing its insulating quality and the adverse effects upon the digestive, the nerve and the circulatory systems. Studies indicate that following fouling body metabolism in seals and polar bears nearly doubled which can eventually lead to hypothermia, especially for the bears in the presence of wind. One study notes, '(M)aximum responses were latent, requiring 4-5 weeks to develop, culminating in the unexpected deaths of two of the three exposed bears.'[53] Highest hydrocarbon levels in the bears were found in the kidney, brain, and bone marrow. Severe pathological abnormalities including extensive anaemia and renal failures were noted following oiling. Polar bears have also been observed eating small quantities of oil in dumps. Seals with oil on their skin appear to be able to tolerate a temporary contact without damaging results.[54] Eye irritation occurs, but if the oil concentration is heavier, lethal damage to the intestines, the liver and the lungs is possible.[55] Several observations indicate that seals do not actively avoid oil covered waters. Knowledge however regarding extended seal contact with oil is incomplete, and the relationship between skin inflammation, energy balance and seals' ability to survive need be studied.

In the case of whales, various types have been observed eating in an oil spill, indicating they either have the inability to recognize oil or are indifferent to it.[56] Temporary contact with oil appears to have little effect, however prolonged contact can cause skin damage.[57] Little is known of the effects to the whales' inner organs, and this needs study, especially since some species can swallow large amounts of oil by skimming the surface for food with an open mouth.[58]

Besides damage from the oil itself, oil related *activities* in general may have damaging effects upon marine mammals. These include underwater explosions,

Type of Response	Threshold Concentrations for Response (mg/L)	Reference
Lethal		
Increased mortality	5.1(TH) (1-d LC$_{50}$)	
	1.3-2.2(TH), 0.20-0.47(TN)	15
	<0.68(TH)	2
	0.1->44(TH) -2(4-d LC$_{50}$)	9
	0.02-0.2(TH)(6-d LC$_{50}$)	
	0.01-0.5(TH)	17
	1.5(TH), 0.4(TN)(4-d LC$_{50}$)	21
		17,18
Sublethal		
Fertilization decrease	>14.1(TH)	14
Uptake, accumulation of hydrocarbons		
Contamination chorion	<0.25(TH)	17
Uptake, metabolism, and retention	0.01-2.1(B)	5
Physiology		
Heartbeat changes	0.01-0.1(TH)	10
	<3.1-11.9(TH)	14
	5.0-6.0(TH), 1-1,2(TN)	1
	<2.1(TH), 0.54(TN)	21
Respiration affected	<2.1(B)	4
	2.4-8.0(TH)	3
Behaviour		
Embryonic movements	<3.1-11.9(TH)	14
Development		
Pathological tissues	1.1-2.2(TH), 0.6-1.1(TN)	6
Morphological anomalies	<0.1(TH)	11
Rate changes	0.0001(B(a)P)	29
Growth		
Length, height changes	<1.7(TH), 0.38(TN)	16
Tissues	2.1(B)	4
Hatching		
Occurrence affected	<1.9(TH)	26
	2.2-4.4(YH), 1.1-1.3(TN)	6
Premature or delayed	<0.1(TH)	11
	<3.1-11.9(TH)	14
Success decreases	<2.1(TH), 0.54(TN)	21
	<0.8(B)	24
	0.01-0.1(TH)	11
	5-6(TH), 1.0-1.2(TN)	1
	0.0001(B(a)P)	29
Duration changes	<1.3(TH), 0.28(TN)	15
	>0.68(TH)	22

Figure 1.7.[59] Effects of Oil on Eggs and Embryos of Marine Fish. TH, total hydrocarbons; TN, total naphthalenes; B, benzene; B(a)P, Benzo (a) pyrene.

Type of Response	Threshold Concentrations for Response (mg/L)	Reference
Lethal		
Increased mortality	43(TH)(4-d LC_{50})	Vourinen & Axell (1980)
	<0.8(B)	Struhsaker (1977)
	1.5(TH, 0.4(TN) 4dLC_{50}	Sharp et al. (1979)
Sublethal		
Uptake, accumulation of hydrocarbons	0.01-2.1(B)	Eldridge et al. (1978)
Physiology		
Respiration changes	2.4-8.0(TH)	Davenport et al (1979)
	<0.885(TH), 0.21(TN)	Sharp et al. (1979)
Development		
Delays	<0.68(TH)	Smith & Cameron (1979)
Tissue pathologies	<o.1-2.2(TH)	Vuorinen & Axell (1980)
Morphological anomalies	<3.1-11.9(TH)	Linden (1978)
	<0.68(TH)	Smith & Cameron (1979)
	<8.0(TH)	Davenport et al. (1979)
	<3.0(TH)	
Eye pigmentation		
Irregular cytolysis		
Growth		
Length decreases	<5.4-5.8(TH)	Linden (1978)
	0.1-2.2(TH)	Vuorinen & Axell (1980)
Length increases	<0.68	Smith & Cameron (1979)
	<2.1(B)	Eldridge et al. (1977)

Figure 1.8.[60] Summary of Effects of Oil on Marine Fish Larvae. Threshold concentrations are based on direct hydrocarbon measurements. TH, total hydrocarbons; B, Benzene; TN, total naphthalenes.

noise from drilling, discharge of poisonous chemicals, and noise from ships and aircraft. Small dolphins and seals have been killed by underwater explosions, however it is expected that the larger whales, with the exception of the humpback, will not be especially bothered by noise since they can easily change course.[61]

In summary, it is difficult to generalize on the effects of oil on marine mammals since the levels and duration of exposure and type of petroleum tested have not been comparable. Much more investigation needs be done.[62] However, '(A)lthough information of oil effects on marine mammals is scanty at best, the sum of the results indicates that both habit and habitat make the various species highly vulnerable to oil pollution at sea.'[63] See Figure 1.9.

1.2.5. Oil's Effects on Marine Birds[64]

Marine birds are one of the most vulnerable animals to oil spills because they generally live most of their lives on the sea and obtain all of their food from the sea.[65] Both marine birds and oil spills are sea surface phenomenon, which means that the probability for birds coming into contact with a spill is quite large if the spill is in one of the heavily populated areas. Some geographical areas are so important to marine bird populations that if a single large oil spill occurred, it could have disastrous results.

If birds come into contact with oil, the effects of the exposure depend upon the oil type, the species of bird, the circumstances of the encounter in terms of weather or season, and the dosage. If the waterproofing and insulating qualities of the feathers are damaged by contact with the oil, the result can be an increased energy need at the same time as an impaired ability for food gathering. The oil clogs the fine structure of the bird's feathers, which is primary in maintaining water repellency and heat insulation. The final result of loss of repellency is that the feathers absorb water and the bird sinks and drowns. The loss of the insulation results in greatly increased metabolism to maintain body temperature, exhausting the fat and muscular energy reserves leading to death.[66]

ANIMAL	OPEN WATER	ICE EDGE	POLYNYA/ LEADS	FAST ICE
HARP SEAL	○	◒	—	—
HOOD SEAL	○	◒	◒	—
RINGED SEAL	○	◒	◒	●
BEARDED SEAL	○	◒	◒	—
WALRUS	○	◒	◒	—
BELUGA	○	◒	◒	—
NARWHAL	○	◒	◒	—
BOWHEAD	○	○	◒	—
POLAR BEAR	—	◒	◒	◒
SEA BIRDS	◒	●	●	—

○ UNLIKELY TO BE HARMFUL

◒ POSSIBLY HARMFUL

● PROBABLY HARMFUL

— ANIMALS NOT PRESENT

Figure 1.9. Anticipated Vulnerability of Marine Mammals (and Sea Birds) to Oil Spilled in Various Ice-associated Habitats.[67]

While oil can prove fatal to large numbers of marine birds, additional *sublethal* effects have also been shown. Oil is consumed from preening fouled feathers, resulting in abnormalities in the lungs, adrenals, kidneys, liver, nasal salt gland, gastrointestinal tract, and a reduction in white blood cell count.[68] Physiological changes occur as well including retardation of weight gain in young birds, induction of hepatic enzymes, a temporary depression of egg laying and reduction of hatching success of laid eggs.[69]

Due to the very uncertain factors involved it is next to impossible to obtain accurate figures showing casualties of birds following an oil spill. The greatest source of error is the unknown percentage of oiled birds that die at sea, therefore estimates of actual losses following major catastrophes are, 'little more than informed guesses.'[70] Other uncertain factors include the location of bird communities and the differences in oil spills. Generally there is little relationship between the size of a spill and the number of bird casualties.[71] The estimated risk is greater of course in heavily travelled sea routes and near oil operations. For an idea of seabird casualties following a number of known pollution incidents, see Figure 1.10.

In terms of vulnerability those birds which are 'gregarious, spend most of their time on the water, and dive rather than fly up when disturbed,'[72] are those most susceptible. These include the auks, especially murres and dovekies in the northern hemisphere, and the penguins especially the jackass penguin in the southern.[73] Susceptible ducks include scoters, oldsquaw, scaup and canvasback, mergansers, common eiders, grebes and loons, and phalaropes. Waterfowl and shorebirds include the greater snow goose and the semipalmated sandpiper.[74] As can be seen, those birds noted come from temperate and cold regions. Generally seabirds of tropical and subtropical waters, with the exception of the cormorant,[75] are less affected since they usually do not purse their prey under water to feed. For a further idea of birds which may be threatened by an oil spill associated with the Norwegian waters, see Figure 1.11.

Remedial and preventive measures, including rescuing and cleaning oiled birds, chemical dispersion of slicks, scaring birds from threatening areas, and protection of colonies by use of booms, have for the most part proved *ineffective*. One positive measure however is the restocking of colonies which was successfully carried out on the east coast of Maine in the US, however for polar birds the logistics involved would probably be too great.

In summary as was the case with other animals, surprisingly little is known on a worldwide basis of the effect of oil on birds, the number of seabird mortalities due to oiling, species vulnerability and the impact on populations. More research needs to be carried out in order to provide more realistic remedial and preventive measures. One thing can be said with certainty, the placement of shipping lanes or offshore oil platforms away from known nesting or feeding grounds is crucial.

Incident	Tons Spilled	Dead Birds Found	Dead Birds per km
Waddensee, Netherlands, Feb. 1969	<1.000	14.564	40
Poole, UK, Jan. 1961	300	487	?
Seestern, Medway, UK, Sept. 1966	1.700	2.772	?
Torrey Canyon, English Channel, April 1967	119.328	7.815	?
Tank Duchess, Tay, UK, Feb.-March 1968	87	1.368	?
Loch Indaal, UK, Oct. 1969	115	449	25
Hamilton Trader, Irish Sea, May 1969	700	4.092	?
San Francisco, Jan. 1971	2.700	7.380	?
Arrow, Cape Breton, Feb.-March 1970	10.400	567	26
Irving Whale, SE Newfoundland, Feb. 1970	<30	625	47
Kurdistan, Cape Breton, March-April 1979	7.900	1.697	26
Amoco Cadiz, Brittany, March 1978	200.000	4.572	?

Figure 1.10.[76] Relationship between Amount of Oil Spilled in an Incident and Number of Dead Birds found.

The *actual* numbers killed, allowing for extrapolation for known length of contaminated shoreline and for the sinking of dead birds before they can come ashore, would be at least an order of magnitude greater in each of these cases. 'Dead birds per kilometre' is a rather more meaningful figure than 'dead birds found' for comparisons between spills.

Norwegian, English and Scientific names of seabird species considered

Smålom	Red-throated Diver	*Gavia stellata*
Storlom	Black-throated Diver	*Gavia arctica*
Islom	Great Northern Diver	*Gavia immer*
Gulnebblom	White-billed Diver	*Gavia adamsii*
Toppdykker	Great Crested Grebe	*Poldiceps cristatus*
Gråstrupedykker	Red-necked Grebe	*Podiceps griseigena*
Horndykker	Slavonian Grebe	*Podiceps auritus*
Havhest	Fulmar	*Fulmarus glacialis*
Grålire	Sooty Shearwater	*Puffinus griserus*
Havlire	Manx Shearwater	*Puffinus puffinus*
Havsvale	Storm Petrel	*Hydrobates pelagicus*
Stormsvale	Leach's Storm-petrel	*Oceanodrama leucorhoa*
Havsule	Gannet	*Sula bassana*
Storskarv	Cormorant	*Phalacrocorax carbo*
Toppskarv	Shag	*Phalacrocorax aristotelis*
Knoppsvane	Mute Swan	*Cygnus olor*
Sangsvane	Whooper Swan	*Cygnus cygnus*
Kortnebbgås	Pink-footed Goose	*Anser brachyrhynchus*
Grågås	Greylag Goose	*Anser anser*
Kanadagås	Canada Goose	*Branta canadensis*
Hvitkinngås	Barnacle Goose	*Branta leucopsis*
Gravand	Shelduck	*Tadorna tadorna*
Stokkand	Mallard	*Anas platyrhynchos*
Ærfugl	Common Eider	*Somateria mollissima*
Praktærfugl	King Eider	*Somateria spectabilis*
Stellerand	Steller's Eider	*Polystcta stelleri*
Havelle	Long-tailed Duck	*Clangula hyemalis*
Svartand	Common Scooter	*Melanitta nigra*
Sjøorre	Velvet Scooter	*Melanitta fusca*
Kvinand	Goldeneye	*Bucephala clangula*
Siland	Red-breasted Merganser	*Mergus serrator*
Laksand	Goosander	*Mergus merganser*
Svømmesnipe	Red-necked Phalarope	*Phalaropus lobatus*
Polarsvømmesnipe	Grey Phalarope	*Phalaropus fulcarius*
Polarjo	Pomarine Skua	*Stercorarius pomarinus*
Tyvjo	Arctic Skua	*Stercorarius parasiticus*
Fjelljo	Long-tailed Skua	*Stercorarius longicaudus*
Storjo	Great Skua	*Stercorarius skua*
Hettemåke	Black-headed Gull	*Larus ridibundus*
Fiskemåke	Common Gull	*Larus canus*
Sildemåke	Lesser Black-backed Gull	*Larus fuscus*
Gråmåke	Herring Gull	*Larus argentatus*
Svartbak	Great Black-backed Gull	*Larus marinus*
Krykkje	Kittiwake	*Rissa tridactyla*
Ismåke	Ivory Gull	*Pagophilia eburnea*
Splitterne	Sandwich Tern	*Sterna sandvicensis*
Makrellterne	Common Tern	*Sterna hirundo*
Rødnebbterne	Arctic Tern	*Sterna paradisaea*
Lomvi	Guillemot	*Uria aalge*
Polarlomvi	Brûnnich's Guillemot	*Uria lomvia*
Alke	Razorbill	*Alca torda*
Teist	Black Guillemot	*Cepphus grylle*
Alkekonge	Little Auk	*Alle alle*
Lunde	Puffin	*Fratercula arctica*

Figure 1.11.[77] Systematic List of Species.

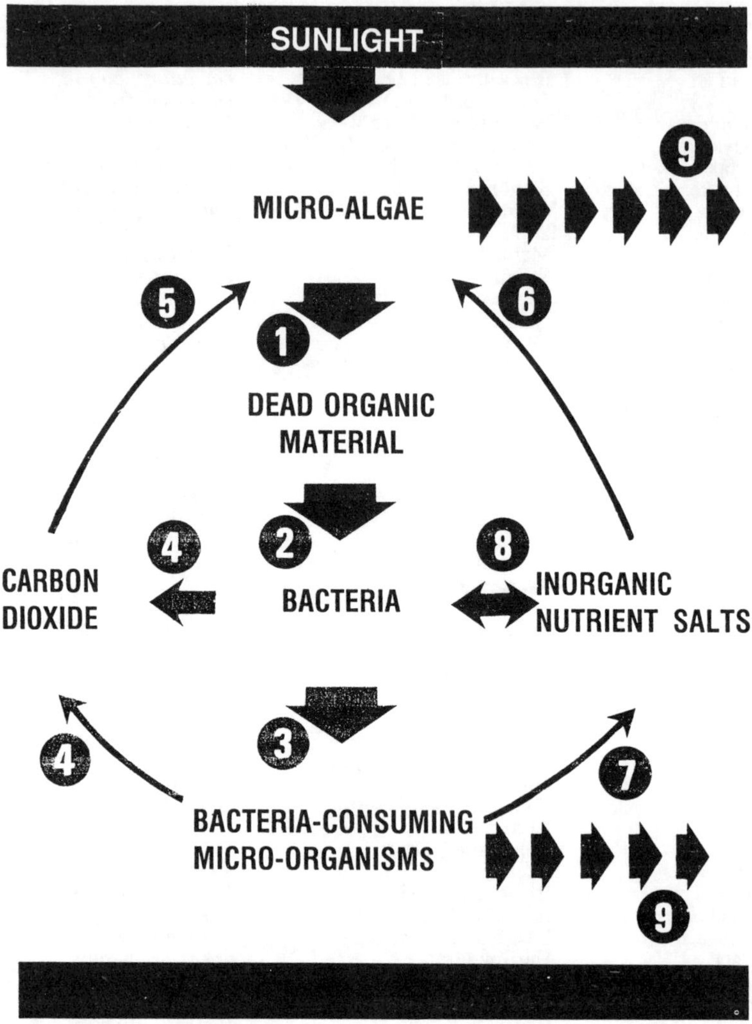

1. Algae release organic material
2. Bacteria convert dead organic material. This includes soluble released from living microalgae and animals, particulated fecal material, and the remains of dead plants and animals.
3. Bacteria form a food source for bacteria-consuming microbes.
4. Bacteria and bacteria-consuming microbes convert organic material and produce carbon dioxide as they grow.
5. Algae consume carbon dioxide as they grow.
6. Algae also require inorganic nutrient salts as they grow.
7. Bacteria-consuming microbes release inorganic nutrient salts as they grow.
8. Inorganic nutrient salts are released by bacteria growing in phosphate- and nitrogen-rich food.
9. Microalgae and bacteria form the basis for the food webs which terminate in larger animals, e.g. fish or jellyfish.

Figure 1.12. The Interactions between Populations of Micro-Organisms.[78]

1.2.6. Oil's Effects upon Marine Bacteria, Phytoplankton and Invertebrates[79]

Though unpolluted sea water appears relatively 'clean', actually one litre contains up to a thousand million living bacterial cells and a million bacterial-consuming micro-organisms. These different populations of micro organisms transfer energy, sunlight or organic material to food chains of higher marine animals. For an outline of the interrelations between populations of micro-organisms in the sea see Figure 1.12.

Following an oil spill there is a large increase to the 'dead organic material' appearing in Figure 1.12. As mentioned, chemical and physical processes lead to the conversion and spread of oil in the sea with the more volatile oil components evaporating and soluble components dissolving and becoming diluted in sea water. The percent of oil-degrading bacteria increases, feeding on the dissolved oil and on oil droplets. The rate of bacteriological degradation of oil in the marine environment provides a means by which the environment can self purify itself, however it is limited by several factors, including the original concentration of oil and the external supply and regeneration of inorganic nutrient salts.[80]

Another micro-organism, phytoplankton, is at the bottom of the marine food chain and creates the basis of all life in the sea by its photosynthesis. Data indicates that these plants are less sensitive to oil pollution than are other micro-organisms and small animals. It was found that of the different species of algae studied, all were very different in their ability to tolerate oil, however the active hydrocarbon concentrations were almost 50% before any significant reduction of growth rate was noted. Light was found to be of importance for the toxic effect. It was also found that oil pollution which was too weak to affect algae directly may nevertheless inhibit the regeneration of nitrogen salts, because it is capable of affecting the more sensitive animal plankton, and thereby reduce the growth of the algae through lack of these salts. Further research is necessary in order to understand the reactions of phytoplankton to light, temperature and nutrient supply and changes in these conditions, to better evaluate how oil affects these primary marine organisms.

There is evidence that oil can upset entire *invertebrate* populations. Patterns of long term impact on snails, crabs and soft shell clams 6 to 8 years after oil spills have been shown.[81] Recruitment problems, variation in isozyme patterns, imbalance in genetic patterns and low population densities are some of the problems involved. Generalizations concerning the effects on individual invertebrates are difficult because of the great variation occurring among the species and even among the various life cycles of any one species. Intertidal invertebrates, though very susceptible to oiling, may be slightly more tolerant than offshore benthic or pelagic species. Juvenile and moulting stages are more sensitive than adult stages.[82] Those species most susceptible are those living in the intertidal zone such as lagoons, bays, estuaries, marshes and tidal flats, due to high oil concentrations and shallow depth of the water. Intertidal species are as well more vulnerable to smothering, narcotization and immobilization with the risk of being washed out to sea or becoming easy prey. Benthic invertebrates are also susceptible due to the vertical mixing of hydrocarbons into the water column with a sub tidal impact. Though there exists speculation regarding the particular sensitivity of polar species to oiling, generally these exhibit the same variability in sensitivity over the same concentration range as other found at other latitudes. One relevant factor however is that reduced evaporation at lower polar temperatures with a corresponding lower loss of the toxic lower molecular weight compounds may make a difference related to the acute impact of oil spilled in the Arctic or Antarctic. Sublethal effects include alterations in respiration, growth, reproduction, and behaviour related to calcification, moulting, ion transport, and enzyme function. Most studies however have been made in the laboratory.

1.2.7. Special Areas - Tropical and Polar Regions[83]

From above it can be seen that especially little is known and few studies conducted concerning the impact of oil in the tropical and polar regions, though more information is beginning to be available. In addition to that noted more specifics regarding these regions' vulnerability, sensitivity and recoverability include the following.

For the tropics there lacks information especially concerning *coral* reefs and *mangroves* and their related flora and fauna. In spite of lack of information these ecosystems are characteristic of most tropical and sub tropic coastlines and often lie very close to heavy tanker traffic or oil activities. They are additionally of great economic and cultural importance and characterized by a rich biological life including fish and other marine resources. Those studies which do exist indicate that these ecosystems are at least as sensitive, if not more so, than those found on temperate coasts.

Responses of coral to oil pollution include decreases in reproduction and colonization, and variations in feeding and behaviour. Premature expulsion of the coral larvae seems to be a common reaction. Other reported reactions of different species include a decrease in number of tentacles expanded during feeding, changes in the pulsation rate of the polyps, and production of excessive mucus. In summary, '(D)espite their awesome mass and apparent robustness, reef corals are very sensitive organisms, forming the matrix for an extensive and highly intricate food web. ... (C)onsiderations of evidence to date rank this marine tropical system high in terms of potential impact of oil and the need for further research.'[84]

Mangroves occur along approximately 75% of the coastlines between 25 degrees North and 25 degrees South.[85] They act to protect coastal systems against storm and current erosion and provide food and shelter for a large number of species both invertebrates and vertebrates. Many commercial species, including the spiny lobster, snapper, drum, sea trout, crabs, shrimp, mullet and menhaden depend upon mangroves for part of their life cycles, and at least one study finds mangrove fringed bays the *most* important nursery grounds in the tropical marine environment. The mangrove system is vulnerable through the aerial root system and the permeability of the swamp to tidal water and consequently oil. The aerial roots are very susceptible to oiling which clogs the lenticels and inner air passages and eventually cuts off the respiratory system. The permeability of the mangrove system due to low wave energy, large numbers of small channels, fine sediments, and highly organic sediments makes it especially susceptible to oil entrapment and hence highly sensitive to oiling. Few observations have been made, but known effects include reduced rooting rates, especially in young trees, and leaf loss. Areas severely oiled can be completely defoliated, though partial defoliation usually occurs in every spill area. Totally defoliated trees do not recover. In partially defoliated trees recovery usually begins in 4 months but has been known to be continuing 7 years later. On the positive side recovery has been observed following most spills. Microbial degradation of oil in tropical sediments seems more rapid than in temperate climates, but may be stopped if buried below the surface. Generally very little is known about the degradation of the oil in mangroves. The mangroves recover functionally with canopy cover within 6 - 7 years and rootmat cover after 10 years. Artificially reseeding areas damaged have been successful. In summary, '(M)angrove regions represent a large unknown for the tropical areas of the world's ocean in terms of petroleum impact.'[86] Further research is recommended.

For the *Arctic* an increase in both offshore oil activities and shipping activities has shown the need for more knowledge concerning the effects of oil in the polar marine environment and corresponding protective measures. Due to lack of knowledge predicted effects range from 'virtually negligible to the cataclysmic.'[87] Recently however there has occurred both basic and applied research allowing identification of areas of concern and resulting in some progress in developing counter measures. In *Antarctica* offshore oil development is not planned in the near future, and therefore little research has been carried out studying the effects of oil on the fauna and flora of the southern polar ocean.[88]

Though many small spills have occurred in the Arctic little study has been made from these due to logistic difficulties, interference from cleanup activities and lack of pre spill comparative information. This is in contrast to temperate areas where much useful information has been gained from such spills. Several small experimental spills have been initiated in order to study the behaviour and fate of oil on or under ice, but these have rarely been used for study of the biological effects. Though natural seeps occur in the Arctic, the total quantity is low and little use made of studying the effects on surrounding flora and fauna. Except with few local exceptions concentrations of oil hydrocarbons in seawater and flora and fauna presently are low.[89]

Generally there is little evidence that polar species of animals are more sensitive to oil hydrocarbons than comparable temperate species, information however is too meagre to make generalizations. As noted it is thought that lower temperatures may amplify the toxicity of oil by altering the component's chemical characteristic or by lengthening the residence time of the toxics in the water. As also seen certain types of physiological disturbances including interference with energy storage and carbon flux could have more serious consequences for polar species since many must deal with extremes in food supply, dependent upon summer or winter, and any interference could be critical for overwintering. Recovery rates are slower than in temperate or tropical zones, with the oil remaining unweathered for extended periods, and biological recovery delayed due to reduced fecundity, dispersal, and growth rates of many polar species. Of special importance for the Arctic are the characteristic shallow coastal lagoons, which are heavily used at certain times of the year by birds and fish populations. Though natural cleaning can be very rapid, depending upon degree of shoreline exposure and amount of fetch offshore, in very low energy areas stranded oil can remain for longer than 10 years, similar to that observed in lower latitudes.

In summary there lacks information about the basic biology of many polar species, especially their ecological relationships and, 'the precise ecological significance of such aspects as the several unique polar habitats - leads, polynyas, ice edge, and the under ice.'[90] '(I)nformation on oil impact on polar environments is still fragmentary, with large knowledge gaps making spill impact assessment more *guesswork* than sound appraisal.'[91]

In concluding this section it can be said that with certain exceptions our knowledge of the effects of oil in the seas are *rudimentary* and *fragmentary*. Present fish census techniques are too crude to give us clear knowledge of standing fish stocks. Evidence on the effects of oil on some larval fish and fish eggs indicate that impact is possible, this has not been strictly examined. A major problem involves the inability to transfer information obtained from laboratory studies to field conditions. Oil can have very adverse effects on local environments sometimes for years, and some oil compounds are carcinogens and mutagens. Products of oil degradation can be hazardous. Oil pollution has the greatest effect in coastal areas where shallow water and currents trap the pollution. Areas characterized by low level continuous exposure such as waters near industrial or municipal regions are especially impacted, not only from oil pollution itself, but also possibly from a synergistic impact with other pollutants.[92] Finally the lack of solid information concerning oil's effects in the polar and tropic regions is even more pronounced, yet these areas have unique characteristics, both physical and biological, which may make them highly vulnerable. At the same time main shipping routes pass close to many sensitive tropical regions, and northern shipping and offshore oil activities are on the upswing in the Arctic.

Upon this somewhat sombre note, *statistics* of polluting substances entering the oceans, definitions and background information concerning pollution sources will be presented.

1.3. Marine Pollution: Definitions, Statistics and Background Information

As noted there are recognized 6 sources of global marine pollution, land-based, vessel-source, ocean dumping, through and from the atmosphere, and exploitation of both the seabed of the continental margin and of the international Area. For simplicity the pollution source 'through and from the atmosphere' will be combined with 'land-based pollution,' which was the practice prior to the LOSC PART XII classifications.

Many differing and often confusing statistics are published regarding quantities of polluting substances emanating from these sources, and some will be presented here in order to give an approximate idea of quantities of marine pollution occurring on a world wide basis. Statistics will additionally be briefly presented covering chiefly Norway and the Norwegian North Sea in order to lend 'reality' to the general figures.[93]

1.3.1. Land-based Pollution

Land-based pollution, including emissions carried from and through the atmosphere, is by far the most common form of marine pollution, accounting for 75% or more.[94] Another estimate gives figures ranging from 50% to 90% on a world wide basis,[95] but despite the variance of the statistics the point is that this source contributes at least over 50%.[96] This type of pollution includes in addition to oil sewage and industrial wastes, chemical fertilizer and pesticides, warm water from power stations, atmospheric discharges from vehicles, chimney fumes and sprayed agricultural chemicals.[97]

1.3.2. Vessel-source Pollution

Vessel-source pollution consists mainly of pollution from ship cargoes. This arises when seawater has been pumped into an empty ship to clean out the tanks or to serve as ballast, and is later pumped out again. Most often this occurs with oil or fuel tanks.[98] Additionally, oil and chemicals, liquid gas and radioactive materials are discharged into the sea as a result of ship accidents noted above including collisions, strandings and explosions. One author notes, 'globally 3.5 million tons of *oil* per annum are deliberately discharged or result

Accidental Tanker Oil Spills Total Losses and Deaths

Year	Oil spills		Total losses		Deaths	Tankers at risk	
	No	Tons Spilled	No	Millions DWT		No	Millions DWT
1973	36	84.458	12	0.682	70	3.750	218.9
1974	48	67.115	14	0.536	94	3.928	253.6
1975	45	188.042	22	0.815	90	4.140	296.2
1976	29	204.235	20	1.172	226	4.237	336.7
1977	49	213.080	20	1.000	113	4.229	369.0
1978	35	260.488	17	0.913	148	4.137	380.4
1979	65	723.533	26	2.501	306	3.945	376.0
1980	32	135.635	15	1.703	132	3.898	375.7
1981	33	45.285	21	1.166	73	3.937	372.5
1982[2]	7	1.491	16	0.818	72	3.950	364.7
Totals	379	1923.362	183	11.306	1.324	-	-

1

[2] To September 1982 only
Source: Tanker Advisory Centre, New York, 1983

Serious Casualties to Oil/Chemical Tankers, 1968-1983		
Year	**Tankers at risk**	**Serious casualty rate per 100 tankers**
1968	3.071	2.54
1969	3.126	2.37
1970	3.169	1.89
1971	3.260	1.96
1972	3.300	2.27
1973	3.361	1.96
1974	3.490	1.89
1975	3.659	2.41
1976	3.725	2.60
1977	3.593	2.39
1978	3.440	2.44
1979	3.346	3.20
1980	3.362	2.14
1981	3.274	2.84
1982	3.215	1.84
1983	3.100	1.87
	Total: 53.491	**Average: 2.29**
Source: IMO News, No. 3, 1984, p.3		

Figure 1.13.[99] Accidental Tanker Oil Spills Total Losses and Deaths. Serious Casualties to Oil/Chemical Tankers, 1968-1983.

from accidental spills - half from land-based sources; half from marine-based, the main source of the latter being *shipping*, not offshore oil exploitation - production platforms contributed very little, though blow-outs could have a serious impact in bays, estuaries and semi-enclosed seas like the North Sea if and when they occur.'[100] Of this *1,750,000 tons* of *oil* is deliberately or accidentally discharged yearly from ships.[101] See Figure 1.13 for accidental oil spills from tankers. Another author argues that in spite of the publicity vessel-source pollution receives, ship-source pollution amounts to less than 5% of the total marine pollution and is now a very minimal part of the total.[102] This statistic's variance from that above is difficult to explain but possibly due to not being limited to just oil but rather to include all types of marine pollution. Another author indicates that shipping accounts for 47% of the total *oil* marine pollution.[103] As is seen in Figure 1.14 below, shipping (46.8%) and land-based pollution including atmospheric (43.8%), times a total of 3.2 million tons of oil annually, roughly equal the former figures noted above.[104]

Source	**Million tonnes annually**	
Municipal and industrial waste water discharges and run offs	1.0	(31.3%)
Refinery waste water discharges	0.1	(3.1%)
Offshore oil production	0.05	(1.6%)
(Tanker operations Marine (Other shipping activities Transportation (Accidental spills	0.7) 0.4) 0.4	(34.3%) (12.5%)
Natural seeps and erosion	0.25	(7.0%)
Atmospheric fall-out	0.3	(9.4%)
Total	**3.2**	

Figure 1.14.[105] Petroleum Inputs to the Sea (NAS 1985).

1.3.3. Dumping

Dumping is defined as the deliberate disposal into the sea from ship or aircraft of waste loaded on board for this purpose and used to be a legitimate means of disposal in the past because of the low cost. It is estimated that this form for pollution contributes approximately 10% of the total marine waste disposed, whether land or sea based.[106] The materials dumped include radioactive materials, weapons (including mustard and nerve gas, explosives and obsolete munitions), highly toxic waste, dredge spoils, sewage sludge, construction and demolition debris, and garbage including paper, food, metals, glass, plastics and textiles.[107] Dredge spoils, sediments dredged from harbours and bays, are estimated to make up about 80% of the total dumping waste, while sewage (from large metropolitan areas) and industrial waste are the other major contributors. The disposal of especially radioactive materials has caused much public concern due to the long 'half-life' (the time required for the initial activity of an isotope to be reduced by one half) of some of the isotopes dumped.[108] Possible seawater corrosion of the containers holding these isotopes, which might cause their release, is as well a major cause of concern.[109] Another type of dumping which has appeared recently is the burning of land-based waste at sea on special incinerator ships. The idea behind burning wastes is that they be changed chemically into more environmentally acceptable substances, however these forms as well may cause marine pollution.[110] The popularity of marine dumping was due partly to the inexpensiveness and the ease of disposal and partly due to the tightening up of pollution controls on land.[111] In one author's opinion, 'the hazardous - sometimes ultra-hazardous - nature of some of these wastes certainly necessitates the strictest controls, or even outright prohibition.'[112] Dumping has been a problem in the North Sea with oil equipment being dumped and left to lie on the ocean floor, which in turn has caused damage to fishing gear,[113] and this problem is probably the most controversial issue between the oil and fishing interests.[114] A dumping problem just beginning to emerge relates to the disposal of oil platforms after the offshore oil fields dry up.[115] The importance of the issue was moreover illustrated by the recent, March 6, 1988, underwater collision of a small German submarine with the Norwegian 'OSEBERG B' oil platform in the North Sea.[116] It can only be surmised what the results would have been had one of the large Soviet Typhoon submarines collided with an offshore platform. Generally, although dumping at sea has been reduced, some Western European States, including the United Kingdom still conduct these activities.[117]

1.3.4. Pollution from Seabed Activities (Continental Margin)

As seen, internationally pollution from platforms is but a small percentage of the total marine pollution.[118] As a quick look at Figure 1.14 will indicate related to the other pollution sources, this source contributes in fact less than the individual amounts of refinery waste water discharges, atmospheric fallout and even natural seeps and erosion. In spite of the low figures however there is always the chance for a catastrophic blowout.[119] Other than from blowouts this type of pollution arises from the exploration and exploitation of the continental shelf for oil or natural gas and consists of two types, that associated with offshore platforms, and that associated with the transmission by pipeline to reception facilities.[120] From platforms pollution can occur through accidental spills and flaring operations, through the drilling cuttings and through production water. A problem that is currently being addressed by the Norwegian Pollution Authorities (SFT)[121] involves pollution occurring from offshore platforms using oil based drilling mud. See Part I Section 1.3.6. for statistics of oil pollution and others discharged by offshore platforms 1981-1986 chiefly in the North Sea.

1.3.4.1. Background Information on Blowouts

Concise background information on blowouts is given by thc book, *All About Blowout.*[122] See Figure 1.15 for a list of oil rig catastrophes which have occurred internationally during the period 1955 to 1981.

Generally, the blowout frequency is *5 times* higher offshore, than on shore. There is a risk of secondary blowout on platforms because platforms contain clusters of wells. Human lives are more endangered offshore since more people are concentrated on platforms, and it is more difficult to escape in the event of unsafe conditions. There are different types of blowouts. A cratered blowout, occurring outside the pipe casing, often leads to the formation of a crater on the surface and can occur both on land and on the sea floor. In this case gas and liquid can *liquify* the soil, 'and a whole rig on land or a platform resting on the sea floor may sink in and disappear.'[123] An underwater blowout, the blowout orifice located below the sea level, tends to last longer, is more difficult to stop, and pollution control is more difficult. Offshore blowouts are more likely to ignite in fire than blowouts on shore, because there is much more equipment and platform activity that can be sources of ignition.

Year	Owner/rig name	Type	Damage loss (mill USD)	Damage/unit returned to service (yes, no)
1955	Chevron S-44	Sub Recessed pontoons	1.8	Blowout and fire in Gulf Mexico. Retired./yes
1959	Reading & Bates C.T. Thornton	JU	1.0	Damage by blowout and fire in Persian Gulf - repaired and returned to service
1964	Reading & Bates C.P. Baker	Drilling barge	2.3	Capsized, turned over end-for-end during blowout and fire in Gulf og Mexico. 22 killed./no
1965	Royal Dutch Shell Trion	JU	1.5	Destroyed by blowout and fire in Nigeria./no
1965	SNAM-SAPEM Paguro	JU	6.0	Destroyed by blowout and fire in Adriatic Sea - not salvaged
1968	Coral (now: Fluor Drilling) Little Bob	JU	2.0	Blowout and fire in Gulf of Mexico (Gulf Derrick) col-lapsed - 7 killed./no
1969	Floor Drilling Wodeco III	Drilling barge	0.5	Blowout in the Red Sea./yes
1969	Sedco, Inc. Sedco 135G	Semi	3.5	Blowout caused severe dam-age while in Timo Sea off Australia./yes

1969	ODECO (formerly Rimrock Tidelands)	Sub	0.1	Blowout in Gulf of Mexico./yes
1970	Storm Drilling Co. Stormdrill III	JU	3.5	Blowout caused severe fire damage off Texas. Repaired and returned to service.
1970	Offshore Co. Discover III	Ship	0.6	Blowout in South China Sea./yes
1970	Offshore Co. Discover II	Ship	0.1	Blowout off Malaysia./yes
1971	Atwood Oceanics Big John	Drilling barge	4.3	Hull sank after blowout and fire off Brunei - 9 killed./yes
1971	Floor Drilling	Drilling barge	4.5	While off Peru had blowout and fire - 7 killed./no
1972	Marine Drilling Co. J. Storm II	JU	8.0	Blowout in Gulf of Mexico. Not salvaged.
1972	Reading & Bates M.G. Hulme	JU	7.5	Rig capsized in Java Sea because of blowout (no fire), crater./no
1972	Transworld Drilling Rig 60	Ju	10.0	While off Burma, had blowout in Gulf of Martaban./no
1973	Santa Fe Drilling Mariner I	Semi	0.1	Blowout off Trinidad 3 killed./no
1974	Offshore Co. Meteorite	JU	3.0	Blowout off Nigeria./no
1975	Zapata Off-Shore Topper III	JU	10.0	Sank following blowout in Gulf of Mexico. Later dynamite used to sink unit.
1975	Santa Fe Drilling Mariner II	Semi	0.3	Lost BOP during Blowout in Gulf of Mexico./no
1975*	Marine Drilling J. Storm II	JU	N/A	Blowout in Gulf of Mexico. Partially salvaged.
1976	Petrobas Petrobas III (formerly Penrod 63)	JU	N/A	No Information
1976*	Reading & Bates W.D. Kent	JU	5.9	Damage when drilling relied well off Dubai. Hit by Western Offshore III during storm. It sank - 1 killed./no
1977	Maersk Drilling Maersk Explorer	JU	6.5	Blowout and fire in the North Sea./yes
1978	Scan Drilling Scan Bay	JU	2.5	Blowout and fire in the Persian Gulf./yes

1979	Salen Offshore Salenergy II	JU	28.0	Blowout in Gulf of Mexico./no
1979	Sedco Drilling Sedco 135	Semi	22.0	Blowout and fire in the Bay of Kampuchea./no
1980	Sedco Drilling Sedco 135 C	Semi	18.0	Blowout and fire off Nigeria./no
1980	Offshore Co. Discoverer 534	Ship	Claims pending	Escaping gas caught fire off Australia./yes
1980	Reading & Bates Ron Tappmeyer	JU	1.1	Blowout in Persian Gulf - 5 killed./yes
1980	Peoples Republic of China Nanhai II	JU		Blowout off Hainan Island./yes
1980	Maersk Drilling Maersk Endurer	JU	10.0	Blowout in Red Sea - 2 killed./Will be salvaged
1980	ODECO Ocean King	JU	25.0	Blowout and fire in Gulf of Mexico - 5 killed./Salvage under way
1980	Marlin Drilling Marlin 14	JU	-	Blowout in Gulf of Mexico./yes
1981	Penrod Drilling Penrod 50	Sub	-	Blowout and fire in Gulf of Mexico./no

* Blowout indication.

Comments. It is interesting to note that in 18 of the 35 blowouts fire is reported. 13 do not tell, but probably did not have fire. The high incidence of fire is due to the fact that relatively few blowouts without fire fall in the category 'Major rig mishaps'. The highest damage loss reported is 28 million USD. Better fire control would in principle reduce the drilling blowout damage. A preparedness based in the TOTAL water injection concept (or similar concepts) may be the answer. Also carbon dioxide can be used to stop well fire.

Figure 1.15.[124] Major Rig Mishaps.

Though one report considers 30% to 50% of all blowouts to result in fire,[125] 15% to 20% may be more realistic. Few accidents are 'Acts of God', *human error* is involved in an estimated 92% of all blowouts.

Sub sea wells, where the wellhead is located on or just below the sea floor, are in use presently and are expected to a greater degree in the future, possibly in deep and very deep waters.[126] Up to now there have not been any blowouts from sub sea wells. Underwater completion operations are assumed to have about 4 to 5 times the chance of blowout than for surface completion. Sub sea wells are protected from fire and explosions and from the weather, but are exposed to fishing gear, dragging anchors, dropped objects and icebergs. It is believed that overall a sub sea well is more protected from external forces than a platform well, and with regular production sub sea wells have a smaller blowout probability than platforms wells by a factor of 2 to 3. However if there is a blowout from a completed sub sea well, the *majority* of such blowouts will cause *catastrophic* oil spills, easily resulting in spills of 500,000 tons. This is because the chances of the blowout self-stopping are small, and relief wells will have to be drilled to achieve control. It is estimated that 'each sub sea-completed oil well contributes at least *10 to 20 times* more to the pollution risk than each platform well does.'[127] Yet, there presently exists no special underwater blowout control preparedness in existence, and the blowout risks of subsea wells have hardly been addressed.

Generally, existing contingency plans for oil spill preparedness based on booms and skimmers cannot contain and collect a significant portion of oil from a large blowout. Oil collection is possible only in very good weather, and even then the efficiency is somewhat questionable. The *dispersion* method with well-injected chemicals is another possible solution where there is a high probability that oil will drift to the shore and do significant harm and containment is not possible.[128] Better training is definitely needed for offshore personnel to reduce the human error factor noted above associated with blowouts, and anti-pollution measures need be made as cost effective as possible for the oil companies to achieve compliance.

1.3.5. Pollution from Deep-Ocean Mining (International Area)

This type of pollution is presently inconsequential since deep sea mining has yet to become established, but when mining is initiated, it is likely to cause pollution problems.[129] The problems associated with these activities will not be dealt with here.

1.3.6. Pollution Statistics concerning Norway and the North Sea

This Section is a composite of statistics reported for Norway for roughly the period 1974-1986 dealing with the Norwegian mainland and the North Sea. There are presented to give an even more concrete and close-to-home example of what goes on in practice on land, in shipping lanes and offshore oil fields possibly not evident in the general statistics presented above.

To get an idea of the extensive *variability* of the pollution substances, reference is made to the section of the book, *The First Decade*, dealing with just Norway.[130] Larger industrial countries would of course produce even *more* varieties and quantities of polluting substances as a quick look at the same reference will indicate.[131] Norway produces approximately 1.7 million tons of *municipal* waste per year, consisting of refuse and objects from households, solid waste from shops, offices and industries, demolition waste and various others.[132] Generally knowledge is *lacking* of the exact amounts and ultimate fates of industrial waste, and this seems to be a common problem for all countries.[133] Norway's industrial waste has been classified into 4 groups including, waste containing heavy metals, persistent or slowly degradable organic compounds, nutrients and readily degradable organic compounds, and chiefly inert matter. The industries chiefly involved in the generation of the above are industries producing pulp and paper, iron alloys, aluminum, chemical, electroplating, herring-oil and meal, slaughterhouses and dairies.[134]

An idea of oil spill reports directed to SFT can be obtained from Figure 1.16 covering the years 1980 through 1986. The Norwegian spills consisting of oil or oil mixtures are required reported to SFT in accordance with the Norwegian Pollution Act.[135] Though there is an increase of reported spills, the increase is due to small discharges (0-5 tons), and comes primarily from *shipping*.[136] There was only one large spill of significance (greater than 50 tons) which drifted close to the Norwegian coast, but did minimal damage.[137] In addition an oil spill along the Finnmark coast caused extensive damage to seabirds.[138]

In 1984 *cuttings* were the largest source of oil discharged from platforms. The use of diesel dropped from 40% of the total oil in cuttings in 1983 to 10% in 1984. The total amount of oil discharged *increased* by 36% from 1983 to 1984. The quantity of oil discharged in production water decreased in the Norwegian and British sectors though the volume of production water increased, due to improved effluent treatment systems.[139] In 1985 the largest source of oil discharged was in connection with exploration and exploitation activities, and, '*no* diesel oil was discharged via cuttings.'[140] The total amount of oil discharge increased slightly, 6% since 1984, however, 'important uncontrolled oil spills around platforms', detected by Dutch surveillance flights between 1984-1986 on the Norwegian, Danish and United Kingdom offshore sectors, 'were *far in excess* of the amounts reported to the Paris Commission.'[141] Norway in 1985 had 14 oil and gas platforms in operation, and drilled 88, 34 with oil-based muds and 54 with water-based muds.[142] The three Norwegian platforms, which discharged effluents containing more than 40 mg/l oil in 1985, produce respectively:[143] gas (FRIGG) (1,495 mg/l); gas (ODIN) (134 mg/l); gas and oil (VALHALLA) (145 mg/l).[144] As can be seen from Figure 1.17 Norway's worst year in terms of total discharges in the Paris Convention area was 1984 while totally the worst year was 1985. 1986 showed approximately an 8% reduction in total quantity of oil discharge by offshore platforms 1985 to 1986 with significant decreases for all Parties but the Netherlands.

Year/Source	1980	1981	1982	1983	1984	1985	1986
Land-based	15	10	19	22	21	44	41
Ship	90	89	65	57	57	46	90
Offshore	57	23	10	6	12	2	17
Total	162	122	94	85	90	92	148

Figure 1.16.[145] *Number* of Reported Discharges and Related Sources, 1980-1986.

For an idea of quantity of oil released on the Norwegian continental shelf from offshore oil activities in 1986 and by what means, see Figures 1.18 and 1.19. It can be seen that 2472 tons of oil were released from oil platforms in 1986. Of this *82%* was released in connection with oil based mud, *9%* production water and displacement water and *9%* by accidents.[146] SFT notes that this amount was only 70% of the total quantity released in 1985, and the reduction is the result of a reduction of oil released when drilling wells with oil based mud.[147]

In 1986 the monitoring around the platforms on the Norwegian continental shelf showed *increased* values of hydrocarbons at 5,000 meters from the platforms, and SFT estimated that the total contaminated area in the offshore sector to be around 100 square kilometres.[148] Significant changes in the diversity of benthic communities was seen 5,000 meters from one of the platforms, STATFJORD C, and the change from diesel to low aromatic oils did not make any difference. SFT expected the affected area around the platforms where oil based mud was used to increase even more in the future.[149] Where diesel based mud had been used, no degradation in the sediment was found 5 years after discharge. Base oils were found in fish, but the complete effects are not known, and was planned discussed with fishery authorities. In cod the mean value of concentrations was found to be 10 times that of the control fish. In pelagic fish such as pollack, no changes were observed. No changes were observed in fish caught near platforms using water based muds. The concentration of heavy metals (Pb, Cd, Cr, Cu, Sr, Ba, V and Ni) in sediment was noted to increase near the platforms, originating from drilling mud and other discharges.[150] Figure 1.19 shows how many wells were drilled with oil based mud in 1986 and how many of these were discovery wells. SFT reports there were drilled 103 wells on the Norwegian shelf in 1986, and of these 32 were drilled with oil based mud. In connection with drilling these 32 wells, as seen in Figure 1.18, 2030 tons of oil were released. This is a decrease of 40% from

1985 when 34 wells were drilled with oil based muds and 3331 tons released.[151]

	PRODUCTION WATER					
	1981	**1982**	**1983**	**1984**	**1985**	**1986**
Denmark	<6	10	32,5	57	87	43,3
Netherlands	<2,5	<3,7	6,8	76	116	209
Norway	177	239	203	154	151	123
United Kingdom	526	927	1628	1430	2152	2622
Total	712	1180	1870	1717	2506	2997

cont....

ACCIDENTAL SPILLS AND FLARING OPERATIONS					
1981	**1982**	**1983**	**1984**	**1985**	**1986**
60	20	35	30	0	1
-	-	-	59	159	12,9
14	165	95,5	280	8	225
104	162	186	130	310	3540
178	347	317	499	477	3739

cont....

CUTTINGS									
1981	**1982**	**1983**		**1984**		**1985**		**1986**	
		O	**D**	**O**	**D**	**O**	**D**	**O**	**D**
-	900	1755	0	676	0	264	0	178	0
125,7	281,9	399,1	123,5	969	48	1932	0	4553	0
400	950	727,5	743,5	2819	647	3331	0	2030	0
6900	10600	<7000	<7000	18400	1400	20230	0	13000	0
7426	12732	9882	7867	22864	2095	25757	0	19761	0

O = Others
D = Diesel

cont....

	TOTAL					
	1981	**1982**	**1983**	**1984**	**1985**	**1986**
Denmark	<66	930	1822	763	351	142
Netherlands	<128	<285	529	1153	2207	4775
Norway	591	1354	1770	3900	3490	2378
United Kingdom	7530	11689	<15814	21360	22692	19162
Total	8315	14258	19936	27176	28740	26537

Figure 1.17.[152] Quantity of Oil Discharged by Offshore Installations, Comparative Data for the Years 1981-1986 (tons).

Source	**Discharge**
Discharges from drilling	2030 tons oil
Discharges from production water	123 tons oil
Discharges from displacement water	94 tons oil
Other accident/slop	225 tons oil
Total	**2472 tons oil**

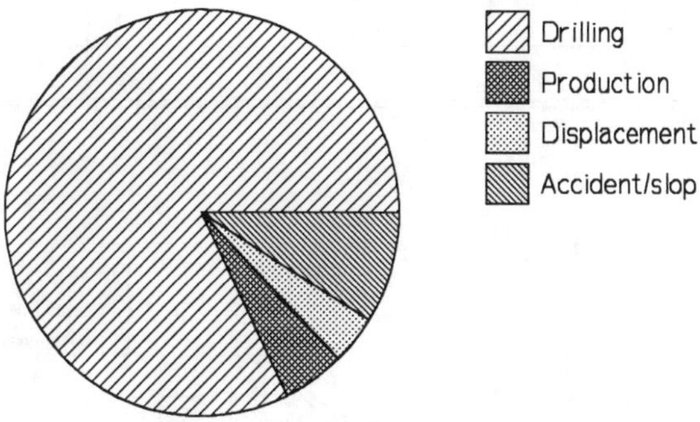

Figure 1.18.[153] Summary of Oil Discharges from Offshore Platforms on the Norwegian Continental Shelf in 1986. The circle figure shows how large a percentage of discharges come from the various sources, drilling, production, displacement water and accidents/other.

	Waterbased mud	Oilbased mud	Total
Production drilling	30	30	60
Discovery drilling	41	2	43
Total	71	32	103

Figure 1.19.[154] Summary of number of wells drilled in 1986.

A summary of quantity of oil discharges from *Norwegian* continental shelf activities given in tons through drilling, production and displacement water and accidents for the years 1981-1986 is given in Figure 1.20. Regarding production and displacement water and other oil discharges, of the 9 discharges in 1986, only 5 were reported to SFT by the offshore industry.[155] The 4 others were seen and reported by ships or airplanes. It is estimated that the total discharge ranged in size from 35 tons to 160 tons, which means that approximately from 13% to 42% of oil discharged into the sea in this 'other' group was not reported. Also of interest are the 8.8 tons of *chemicals* released from pipelines in 1986.[156]

Year	**1981**	**1982**	**1983**	**1984**	**1985**	**1986**
Drilling	400	950	1471	3466	3331	2030
Production			74	75	67	123
Displacement	177	239	129	79	84	94
Accident/slop	14	165	95	280	8	225
Total	**591**	**1354**	**1769**	**3900**	**3490**	**2472**

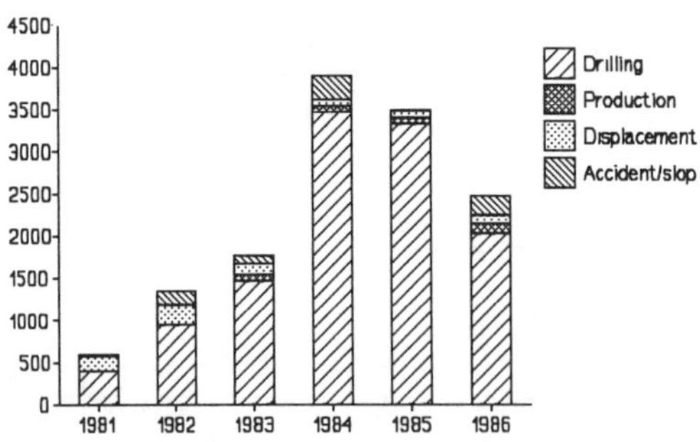

Figure 1.20.[157] Summary of Oil Discharges Given in Tons.

1.4. Conclusion

From the above it may be seen that there are substantial quantities of oil and other substances being spilled accidentally or intentionally in the seas. It may seem that these sources of pollution are uncontrolled. However this is not completely the case. As noted there already exist international regimes regulating marine pollution, most predominantly in shipping. Conventions and agreements as well as international customary and general principles of law will now be presented in Part I Chapters 2 - 7. From this can be gained an idea of the governing law for each of the sources of pollution, including the areas of contingency plans and State liability.

Notes - Chapter 1

1 As a rule these are followed closely by television networks. Other significant reasons include highly visible air-borne pollution present in most major cities.

2 Aftenposten 28. mars 1989, Uke 13. Nr. 139. 130. årg s. 3. 'Aftenposten' is a newspaper based in Oslo, Norway.

3 Gold, Edgar, *Handbook on Marine Pollution*, Halifax, 1985, (Gold), p. 29.

4 Johnston, Douglas M., (ed.) *The Environmental Law of the Sea*, International Union for Conservation of Nature and Natural Resources, (IUCN) Switzerland, 1981, (Johnston), pp. 197 and 248.

5 Aftenposten 24. desember 1990, Uke 52., Nr. 510., 131. årg., s. 10.

6 Nordlys, 3. februar 1990, Nr. 29, 89. årg., s. 13. 'Nordlys' is a newspaper based in Tromsø, Norway.

7 For a summary of international environmental law of wider scope than marine pollution see Adede, A.O., 'International Environmental Law from Stockholm to Rio - An Overview of Past Lessons and Further Challenges, *Environmental Policy and Law*, 22/2 (1992), (Adede), p. 88.

8 See Part II Section 8.2.1. Figure 1 for map with coordinates of the Barents Sea.

9 Churchill, R.R., and A.V. Lowe, *The Law of the Sea*, 2nd Edition, 1988, (Churchill) p. 244 notes oil is the pollutant the marine effects of which have received most attention through legislation or from the public. Though hazardous chemicals can be as toxic or more so, little research presently has been completed.

10 The Law of the Sea Convention (1982) (LOSC), 21 *International Legal Materials* 1261, Part XII identifies 6 sources of global marine pollution: land-based, vessel-source, ocean dumping, through and from the atmosphere, and exploitation of both the seabed of the continental margin and of the international Area (area beyond the continental margin).

11 See for example Harders, J. Enno, 'In quest of an Arctic legal regime, Marine regionalism - a concept of international law evaluated', *Marine Policy*, October 1987, (Harders Marine) pp. 785-298; and Lamson, Cynthia and David Vander Zwaag, 'Arctic Waters: Needs and Options for Canadian-American Cooperation', *Ocean Development and International Law*, 1984, vol 18, No. 1, (Lamson Arctic), pp. 49-99.

12 See *Oil in the Sea, Inputs, Fates and Effects*, National Research Council, et al., National Academy Press, (Oil Sea), 1985, pp. 369-582 for extensive coverage of the effects of marine oil pollution. Further topics covered include effects of oil on communities and ecosystems, effects of natural seeps, and the case histories of six major oil spills, including the AMOCO CADIZ and IXTOC I.

13 The following information is obtained from Oil Sea pp. 3, 472, 477, 481 and 482; Final Report from The Norwegian Marine Pollution Research and Monitoring Programme, 'The Fate and Significance of Oil in the Sea', Sponsored by The Ministry of the Environment, 1976-1984, (FOH), pp. 19-27; and Griffiths, D.J., N.A. Øritsland, T. Øritsland, 'Marine Mammals and Petroleum Activities in Norwegian Waters, A Review of the Literature on the Effects of Petroleum on Marine Mammals and Recommendations for Further Research,' The Fishing Directorate's Institute of Marine Research, Bergen, 1987 (Griffiths), p. 67-69, unless noted otherwise.

14 FOH p. 19.

15 *Ibid.*

16 *Ibid.* (parentheses added). Oil Sea p. 10 notes that modelling the drift of spilled oil is a complex and difficult problem partly because of the many oil types and the constantly changing environmental conditions. An estimated considered best for drift velocity is between 3% and 4% of the wind speed.

17 Obtained from Figure 11, FOH p. 31.

18 Existing research concerning the input of petroleum is lacking particularly in the southern hemisphere and the tropics.

19 Oil Sea p. 10 notes however for these processes, 'the relationship between chemical composition and the formation and stability of oil-water emulsions and of the sorption characteristics between oil and organic particulate matter are only poorly known.'

20 Griffiths p. 68.

21 Bjørnøya.

22 Aftenposten 9. november 1990, Uke 45., Nr. 514., 131. årg., s. 4.

23 The following information is obtained from Oil Sea pp. 472, 477, 481 and 482 unless otherwise noted.

24 Oil Sea p. 472.

25 Specific hydrocarbon compounds found in crude oils, refined products and other related fossil fuels can be carcinogenic or mutagenic to humans. *Ibid.* p. 477 notes that carcinogenic potential is 'invariably associated with the PAH fraction, the distillate fractions of crude oils with boiling point above 350 degrees C.' PAH content varies from 0.2 - 7.4% in crude oils to 7.7 - 14.6% in three synthetic crude oil samples.

26 *Ibid.* p. 482 notes that, 'in most instances the natural intake, depuration, and metabolic processes are such that the PAH tissue levels will drop rapidly if the source is removed.'

27 *Ibid.* This information is based upon 3 reports.

28 Figures 2, 3, 4, and 5 were obtained from Figures 14, 15, 16, and 17 FOH p. 34.

29 The following information is obtained from Oil Sea pp. 10-15, 417-424, and Engelhardt, F.R. (ed.), *Petroleum Effects in the Arctic Environment*, Ottawa, Canada, (1985), (Engelhardt) pp. 157-173 unless noted otherwise.

30 For the polar and tropical regions especially, extrapolations are carried out from studies of fish in temperate environments since there exist few laboratory or field studies.

31 Pink salmon exposed to toluene appear to be more sensitive at low temperatures than higher temperatures. Arctic fish *may* be more vulnerable than other fish stocks due to the coldness which increases the persistence of aromatic hydrocarbons that enter the waters. Generally the Arctic is a stressful environment that could increase the sensitivity of some species to spilled oil, yet the sensitivities of several Arctic fish appear similar to the sensitivities of similar species from temperate waters.

32 Oil Sea p. 418 notes, '(B)enthic species are of special concern because of their lifelong association with benthic sediments, which are known to become hydrocarbon traps in the event of oil spillage.'

33 The fish include the Arctic cod and pink salmon.

34 *Ibid.* p. 418. WSF refers to Prudhoe Bay crude oil.

35 The University of Bergen found that although there were lack of effects on cod eggs and larvae including the 'osmotic parameters studied, the cellular and sub-cellular structures, the oxygen uptake of the embryo and the larval body dry weight', there were *definite* and conclusive effects of the oil exposure on the 'oxygen uptake peak of the late yolk sack larva, and on larval body length.' Fyhn, Hans Jørgen (ed.) Fish Larval Physiology and Anatomy, Basic Research and Effects of Oil, Final Report 1983-1985, Zoological Laboratory, University of Bergen, supported by A/S Norske Shell, (Fyhn). Oil Sea p. 420 notes that many oil-exposed Baltic herring eggs developed normally but failed to hatch.

36 Oil Sea pp. 420-421 notes that there is some evidence that metabolites persist in the tissues longer than parent hydrocarbons and while as their toxicity is not known, '(T)here are indications, however, that some metabolites or intermediates may be toxic or even mutagenic.'

37 This includes specifically metabolic perturbations; altered reproductive potential; histological damage including damage to liver, gut, vertebrae, eye lens, stomach, brain and olfactory organs; and physiological changes including increased heart beat, increased coughing, ionic and osmotic imbalances, changes in respiration, changes in blood parameters, decreased energy reserves and changes in gill enzymes.

38 *Ibid.* p. 421 notes that several of these studies were carried out over considerable time and involved 'flow-through systems where exposures may be comparable to field exposure conditions.'

39 One study indicates the starry flounder exposed to low levels of monoaromatic hydrocarbons showed changes in gonadal maturation, and another study indicated that adult fish exposed to low concentrations of WSF during gonad maturation resulted in reduced survival of larvae hatched from eggs laid by these exposed fish.

40 Engelhardt p. 169.

41 Obtained from Figures 27 and 28, FOH p. 51.

42 *Ibid.* Oil Sea p. 422 notes that Atlantic silverside 'lost their normal schooling behaviour when exposed to crude oil, possibly because the olfactory organs and lateral lines of the fish were blocked or damaged.'

43 '(O)bservations are few and circumstantial,' Oil Sea p. 422. Studies indicate that a high percentage of salmon will avoid a contaminated fish ladder containing 3.2. mg/l of monoaromatic hydrocarbons, but an equally high percentage will pass though higher concentrations approaching toxic levels due to other factors including food.

44 Pollution from oil spills also interferes indirectly through the fouling of gear and the closing of harbours and fishing seasons.

45 Engelhardt p. 171, emphasis added.

46 As an example the price of fish in Paris dropped dramatically after the 1967 TORREY CANYON spill of 117,000 tons of crude oil off the coast of south-west England, even though few fish sold came from the vicinity of the spill.

47 Though little is known of the actual substances responsible for tainting, Oil Sea p. 423 lists 'aromatic hydrocarbons, substituted benzenes and naphthalenes, naphtenic acids, organosulphur compounds, and olefins.'

48 *Ibid.* pp. 13-14 lists the following areas as needing further research, 'mutagenicity/tumorigenicity, alteration of behaviour, mechanisms of toxicity, polar and tropical environments, synergistic toxicity, and ecosystem effects.'

49 The following information is obtained from *ibid.* pp. 424-430 and Griffiths pp. 8-14 unless otherwise noted.

50 This is especially so of those mammals found in the polar seas with only some of the species being definitely studied and found to be sensitive. Generally some deaths of grey whales, a dolphin, northern fur seals, California sea lions and northern elephant seas are thought associated (though somewhat tenuously) with oil pollution from a 1969 blowout in the Santa Barbara Channel in southern California. Further the deaths of seals are thought associated with the TORREY CANYON and ARROW spills, deaths of harp seals with a spill in the Gulf of Saint Lawrence, Canada, and deaths of grey and harbour seals with the KURDISTAN spill. Further descriptions of oil fouled animals are reported from the Arctic, the Antarctic and Europe.

51 Specific examples of mammals inhabiting polynyas include the narwhal, the ringed seals, the harp seals, the bearded seals, and the hooded seals. The bowhead whale could be especially threatened by oil spills since there exist few animals, and these migrate through the ice leads of the Bering and Beaufort Seas.

52 Specific examples include the harbour seal and the grey seal.

53 Engelhardt p. 235.

54 Oil Sea pp. 425-426 notes that the extent of oil adhesion to the skin or pelage depends at least on the, 'texture of the exposed body surface, frequency and duration of exposure, and characteristics of the oil.' Ringed seas and older harp seal pups showed no temperature response to crude oil, probably because of the thick blubber layer serving as sufficient insulation.

55 Such heavy concentrations are unlikely in the open sea, but are possible in ice filled areas. *Ibid.* p. 430 notes that though it appears that seals are not likely to show toxic effects from short term exposures to high doses of crude oil, long term exposures may lead to 'adrenal steroid exhaustion.'

56 Observations indicate that humpback whales, fin whales, right whales, grey whales and white-sided dolphins either ate or travelled through polluted waters. Under experimental conditions however bottlenose dolphins detected, visually and by echolocation, an oil spill and chose to avoid it.

57 There have been no substantiated reports of oil fouling of whale skins which generally is quite smooth, except for the skins of the grey whale, the right whale and the bowhead whale.

58 *Ibid.* p. 427 notes that spilled oil interferes with baleen functioning, '(T)he filtering efficiency of bowhead whale baleen (was) reduced by about 10% when coated with Prudhoe Bay crude oil and by 85% when coated with oil with a higher wax content', parentheses added. Another study indicated however that 'the temporary inhibition of baleen function would not be important in the long term feeding strategies of baleen whales.'

59 Reprinted with permission from *Oil in the Sea: Inputs, Fates, and Effects.* Published by the National Academy Press, Washington, D.C., 1985. Table 5-7, p. 419.

60 Reprinted with permission from *Oil in the Sea: Inputs, Fates, and Effects.* Published by the National Academy Press, Washington, D.C., 1985. Table 4-8, p. 420.

61 The humpback whale is effected by explosions and other noise when it is on its usual migration pattern along the coast of Finnmark in northern Norway, and this interference could have serious consequences since the stock in the northeastern Atlantic is reduced.

62 Some authors recommend that stranded animals, those influenced by oil spills and those hunted as a renewable resource be used since the experimental use of marine mammals for toxicological research is often neither desirable nor feasible, Engelhardt p. 238. See Griffiths p. 14 for recommendation of use of marine mammals found dead in connection with oil spills.

63 Oil Sea p. 430.

64 The following information is obtained from *ibid.* pp. 430-436 and Engelhardt pp. 206-208 unless noted otherwise. Much of the information related to the short term and long term effects of oil on seabirds concerns west European waters only, though recent data is obtained from western Atlantic waters of which, 'interpretations of can be applied to the North American waters with some degree of confidence,' Oil Sea p. 431. It is obvious that though extrapolation is possible, concerning species in the rest of the world this cannot be done with a high degree of confidence.

65 Anker-Nilssen, Tyncho, Metoder til Konsekvensanalyser Olje/Sjøfugl, Direktoratet for naturforvaltning 1987 (Anker-Nilssen), p. 79. Additionally birds are the most *conspicuous* casualties of oil pollution, receiving much publicity and evoking great emotional public reaction, 'such as the death of few other animals does,' Oil Sea pp. 430-431.

66 Because of this birds are more likely to perish from exposure to oil pollution in colder climates and following stormy weather when feeding has been low and energy reserves low.

67 Obtained from Fig. 2. Engelhardt p. 237. Copyright owner, Elsevier Applied Science Publishers Ltd, Crown House, Linton Road, Barking, Essex, IG11 8JU England.

68 Oil Sea p. 431 notes however that there is not evidence, 'that any of the observed (pathological) abnormalities were a primary cause of death, which in the majority of cases, is likely to have been drowning or hypothermia' (parentheses added).

69 There are conflicting findings however which may be attributable to different oils and differences in species and dosing. Oil Sea pp. 431-432 notes, '(T)he importance of all these biochemical and physiological change in the wild are unknown.'

70 *Ibid.*

71 *Ibid.* Notes that one of the spills resulting in the most casualties (30,000 birds appearing on the beaches) was in the Skagerrak, Denmark, January 1981 which was caused by, 'relatively small amounts of oil discharged by perhaps two ships.' Additionally noted is that figures of casualties from regions such as the North Sea with frequent *small* oil slicks from oily bilge and ballast water from ships are even less certain. However since pollution in the North Sea 'approaches the chronic', estimates run between *150,000 - 450,000* seabirds killed *annually*, accounting for at least as many deaths as associated with the large catastrophes. (The North Atlantic area is also included in the calculation).

72 *Ibid.* p. 434.

73 The latter is regularly affected by oil pollution associated with the shipping routes along the coast of South Africa.

74 Generally in spite of the large losses of seabirds noted, it appears that this may not have a major impact on *some* species, including the auks of the British Isles and possible the diving ducks. The Arctic and Atlantic auk populations however are in trouble, and oil pollution will affect them seriously.

75 *Ibid* p. 434 notes, 'several hundred of these birds were killed by a leakage of fish oil off Namibia.'

76 Reprinted with permission from *Oil in the Sea: Inputs, Fates, and Effects.* Published by the National Academy Press, Washington, D.C., 1985. Table 5-9, p. 433.

77 Obtained from Appendix 2, Anker-Nilssen s. 90.

78 Obtained from Fig. 21. FOH p. 39.

1. algae release organic material
2. Bacteria convert dead organic material. This includes soluble material released from living microalgae and animals, particulated fecal material, and the remains of dead plants and animals.
3. Bacteria form a food source for bacteria-consuming microbes.
4. Bacteria and bacteria-consuming microbes convert organic material and produce carbon dioxide as they grow.
5. Algae consume carbon dioxide as they grow.
6. Algae also require inorganic nutrient salts as they grow.
7. Bacteria-consuming microbes release inorganic nutrient salts as they grow.
8. Inorganic nutrient salts are released by bacteria growing in phosphate- and nitrogen-rich food sources, and are taken up by bacteria growing in nitrogen- and phosphate-poor organic material.
9. Microalgae and bacteria form the basis for the food webs which terminate in larger animals, e.g. fish or jellyfish.

79 The following information is obtained from *ibid.* pp. 412-417 and FOH pp. 38-48 unless noted otherwise.

80 Other factors include the competition between oil-degrading bacteria and micro-algae for nutrient salts; grazing, and consequent reduction of bacteria populations by bacteria-consuming micro-organisms; the availability of oil to oil-degrading bacteria; the oil components and the toxicity of their derivatives; and the influence of environmental conditions on biotic and abiotic/physical degradation mechanisms.

81 Seen from a commercial viewpoint lobsters, mussels, clams and oysters are also liable to tainting.

82 Oil Sea p. 412 notes however, 'these two statements must be made with considerable care, and each instance should be interpreted in its own context.'

83 The following information is obtained from *ibid.* pp. 13-14 and 449-468 unless otherwise noted.

84 *Ibid.* p. 453.

85 In Australia they extend to 39 degrees South and in Japan and Bermuda to 32 degrees North.

86 *Ibid.* p. 458.

87 *Ibid.*

88 *Ibid.* p. 458 notes additionally that, '(T)he many pronounced physical and biological differences between the two polar marine areas will make it *difficult* to utilize the information gathered in the Arctic for anticipating any ecological impact of oil in the Antarctic,' emphasis added.

89 *Ibid.* p. 464 gives a figure of less than 50 ug/l in both the Arctic and the Antarctic. See however Part II Section 10.2.1. for pollutants associated with the Soviet Arctic rivers.

90 Oil Sea p. 468.

91 *Ibid.*, emphasis added.

92 The synergistic impact is only poorly understood.

93 See Part II Sections 8.5. and 10.2.1. for more information concerning marine pollution related to the Barents Sea and the Arctic generally.

94 Johnston p. 197. See also Churchill R.R. and A.V. Lowe, (eds.) *The Law of the Sea*, 1983 (Churchill and Lowe), p. 213.

95 Hakapää, Kari, *Marine Pollution in International Law, Material Obligations and Jurisdiction, with Special Reference to the Third United Nations Conference on the Law of the Sea*, Dissertationes Humanarum Litterarum 28, (1981), (Hakapää), p. 45, citing such sources as, 'Competence to Establish Standards for the Control of Vessel Source Pollution', Working paper presented by the US in the U.N. Sea-Bed Committee, April 1973, UN Doc. A/AC.138/SC. III/L. 36, p. 1; and The National Council on Marine Resources and Engineering Development, *Annual Report* 20, p. 41, April 1970.

96 Even if limited to marine oil pollution specifically, land-based sources contribute some *54%*. *Ibid.*, citing National Petroleum Council, 'Protection of the Marine Environment' 8 *Natural Resources Lawyer*, 1975, pp. 511-543 at 512. The same study estimates that 35% marine oil pollution comes from shipping, 10% from natural seeps and 1% from oil platforms.

97 Churchill and Lowe p. 214. See Part II Section 10.2.1. for more information concerning specifically land-based pollution occurring in the Arctic.

98 Churchill and Lowe p. 213.

99 Obtained from Gold p. 133. Sources of information are Tanker Advisory Centre, New York, 1983 and *IMO News*, No. 3, 1984, p. 3. In first table only tankers over 6,000 dead weight ton (dwt) are considered and the information current only to September 1982.

100 Birnie, Pat. W., 'Pollution and Fisheries in the North Sea,' *The regulation of fisheries: legal, economic and social aspects*, Proceedings of a European workshop, University of Tromsø, Norway, (eds.) Ulfstein, Andersen and Churchill, June 2-4, 1986, (Birnie) p. 220, emphasis added. Figures derived from 'The State of the Environment' 1985, Organization for Economy Co-operation and Development, (OECD), Paris, 1985.

101 This approximate figure is arrived by the author based upon figures presented in Birnie p. 220. Another author reports that 80% to 85% of oil spillage comes from *intentional* releases resulting from tank cleaning, ballasting and discharges, and that as of 1975 approximately 1,000,000 tons of oil per year were discharged in standard operations, while 200,000 tons were spilled through accidents such as tanker collisions or running aground, Dempsey Paul Stephen and Lisa L. Helling, 'Oil Pollution by Ocean Vessels - An Environmental Tragedy: The Legal Regime of Flags of Convenience, Multilateral Conventions, and Coastal States,' *Denver Journal of International Policy*, Vol. 1, (1980) (Dempsey), p. 49.

102 Gold, pp. 18 and 142.

103 McIntyre, A.D., 'Pollution in the North Sea from Oil Related Industry: An Overview', paper presented at the International Conference on Environmental Protection of the North Sea, March 1987 (McIntyre).

104 Yet another author puts marine oil pollution from shipping at 35%, from land-based sources at 54%, from oil platforms at 1% and from natural seeps 10%, Hakapää p. 45. A Norwegian study evaluating oil spill protection on the Tromsø continental shelf considers the probability of an oil spill from a *tanker* to be 20 times more likely than an oil discharge from an oil platform blowout. Hustoft, A. G., M. Østby, E. Næess-Ulseth, 'Vurdering av oljevernberedskapen på Tromsøflaket, Rapport nr. 83 02 15 - 2, 9 mai 1984, (Hustoft), p. 7-1. See also Part II Section 10.2.1. for more information concerning specific vessel-source pollution related to the Arctic. To indicate the surrounding circumstances and magnitude of pollutants released in a ship accident the recent EXXON VALDEZ incident can be considered. The super tanker EXXON VALDEZ ran aground in the Prince William Sound off the coast of Alaska, March 1989. The captain, the third steersman and the rudderman were required to take blood tests following the accident, examining whether they had used stimulants or depressants. The third steersman did not have the proper certificate for sailing in the channel. Fishermen had frequently protested that tankers were allowed by the US Coast Guard to sail out of the established shipping zones (because of an iceberg in this case) and directed to this channel full of small islands and reefs, as well as large quantities of fish, whales and seals. The oil spill consisting of at least 250,000 barrels spread to an area of approximately 40 square kilometres, and oil was found in large quantities in an adjacent area consisting of 125 square kilometres. Seals were photographed with clear signs of respiratory difficulties, and fishermen had demanded compensation for lost wages. The estimated amount spill of 30,000 to 40,000 tons comprises only approximately *2%* of the total yearly shipping discharge figure discussed above. Experts believe that this catastrophe will effect he environment for many years to come. See Aftenposten 28. mars 1989. Uke 13. Nr. 139. 130. årg. s. 3.

105 Obtained from Table 1. McIntyre based upon a study conducted by the US Academy of Science 1985.

106 Hakapää p. 52.

107 *Ibid.* and Churchill p. 243.

108 Plutonium - 24,000 years; Strontium 90 and Cesium 137 - 30 years.

109 Hakapää, p. 53 notes 'one percent of the drums dumped are broken on landing on the sea bottom while *99 per cent* are broken after some ten years from the dumping', (emphasis added).

110 Timagenis, Gr. J. *International Control of Marine Pollution*, Oceana Publication, Inc. Dobbs Ferry, New York, Sitjhoff & Noordhoff International Publishers BV; Alphen aan den Rijn, The Netherlands, Vols. I-II, 1980, (Timagenis) p. 227.

111 Churchill and Lowe p. 214.

112 Johnston p. 197.

113 Churchill, Robin, 'The Conflict between Oil and Fisheries: A Survey of Norwegian and United Kingdom Law and Practice, ' The regulation of fisheries: legal, economic and social aspects, Proceedings of a European workshop, University of Tromsø, Norway, (eds.) Ulfstein, Andersen and Churchill, June 2-4, 1986 (Churchill Conflict) pp. 204-205 and McIntyre p. 16.

114 Miles, Edward, Laura Geselbracht, 'Fisheries and Conflicting Uses of the Sea', The regulation of fisheries: legal, economic and social aspects, Proceedings of a European workshop, University of Tromsø, Norway, (eds.) Ulfstein, Andersen and Churchill, June 2-4, 1986 pp. 146.

115 See Birnie pp. 19-21 and Ulfstein, Geir, 'The Conflict between Petroleum Production, Navigation and Fisheries in International Law,' *Ocean Development and International law*, Vol. 19, (Ulfstein Conflict) pp. 248-251. See Part II Section NEED for more information concerning dumping as related to the Arctic.

116 Aftenposten, 7. mars, 1988, Uke 10. Nr. 111., 129. årg., s. 1. See also Aftenposten 6. april 1988, Uke 14, Nr. 156, 129. årg. s. 11, 'Oljeinstallasjoner og sikkerhet, Grunnlag for mer fleksible regler?' Ulfstein Conflict p. 236 notes that during the period 1980-1984 there were *55 collisions* between ships and oil installations in the Gulf of Mexico, emphasis added.

117 See Aftenposten 17. januar 1990, Uke 3, Nr. 27. 131. årg. s. 52. Hakapää pp. 53-54 notes Belgium, the Netherlands and Switzerland still dumped as late as 1981. See also generally Oslo-Paris Comm.

118 See Part II Section 10.2.1. for more information concerning specific pollution from offshore activities in the Arctic.

119 The extent of off shore activity on a world basis is indicated by Draft Convention on Offshore Mobile Craft, International Maritime Organization, LEG XXXXIV/66, (IMO Mobile), 19 December 1977 p. 2. At the end of 1976 there were 374 mobile drilling rigs in existence, excluding tenders. They were registered in 23 countries, though approximately 2/3 (306) were registered in 7 countries. 46 additionally drilling unites were on order. The 7 countries having the largest number of rigs were the US (139), Panama (79), Norway (27), Liberia (19), Venezuela (16), the UK (13), and Canada (13). Other countries had less than 10 each. The drilling rigs were placed in approximately 20 geographical areas comprising about 25 countries. The most important areas included the Gulf of Mexico (104), the North Sea (44), Arabian Persian Gulf (36), Venezuela (26), Brazil (21), Mediterranean (18), West Africa (17), South east Asia (14) and Red Sea/Gulf of Suez (12).

120 Johnston p. 197.

121 Statens forurensningstilsyn.

122 Westergaard, Rich. H. 'All About Blowout', Senter for Industriforskning, Oslo, Norway, 1987, (Westergaard), pp. 26-31, 41, 103, 125, 138, 172, 222, 242, 250, 269-277, 289-290. The following information is obtained from Westergaard unless noted otherwise.

123 *Ibid.* p. 27. A crater is reported which was 99 meters deep and 500 meters wide.

124 Obtained from Table D.1., *ibid.* pp. 35-37.

125 Mundheim, Ø., Tørhaug, M., Helgaker, P., Ose, T., Westergaard, Rich., H., Sjøen K., Vesterhaug, O., Olshausen, K., Wold, M. (1981), 'Offshore Blowout Control'. OTTER Group Report No. STF 88 A81004, Trondheim, Norway, (Mundheim).

126 Future development in the Barents Sea will probably include sub sea wells due to the depth and the climate.

127 *Ibid.* p. 153, emphasis added.

128 An additional solution noted if the risk for shore contamination is small, and the cost of offshore oil combat high and its efficiency low, is doing nothing in spite of public criticism.

129 Johnston p. 198.

130 The Oslo-Paris Commission, *The Fist Decade, International Cooperation in Protecting the Marine Environment*, 1984, (Oslo-Paris Comm.) pp. 235-260.

131 See for example *ibid.* pp. 313-359 (UK) and pp. 108-137 (France).

132 *Ibid.* pp. 239-241.

133 *Ibid.* pp. 368-369. It is noted that the disposal of municipal waste seems to be under control in all or most of the countries.

134 Ibid. p. 241.

135 Forurensningsloven.

136 Statens forurensningstilsyn, 'Rapport om akutt forurensning i 1986. Registrerte utslipp til land, kyst og hav,' SFT mai 1987, (SFT Akutt), s. 2.

137 *Ibid.* and s. 8. The spill occurred when a pipeline broke on the British side of the oilfields.

138 *Ibid.*

139 Eighth Annual Report on the Activities of the Paris Commission, Paris Commission 1986, New Court, 48 Carey Street, London WC2A 2JE, (Par. Comm. 8) pp. 22, 86-89. The Par. Comm. reports cover activities dated almost 6 to 18 months previously and are made public even later.

140 Ninth Annual Report on the Activities of the Paris Commission, Paris Commission 1987, (as above) (Par.Comm.9), p. 170.

141 Par. Comm. 8 p. 12 (emphasis added). These Dutch sightings were discussed at meetings taking place under the Agreement of 9 June 1969 for Cooperation in Dealing with Pollution of the North Sea by Oil, (Bonn Agreement), 704 *United Nations Treaty Series* 3 and the Convention for the Prevention of Marine Pollution from Land Based Sources, (Paris Convention) 13 *International Legal Materials* 353 (1974). See Part I Sections 4.2.2. and 7.4. Two conclusions reached were that *most* offshore operators do *not* report oil discharges, and the UK part of the continental shelf was the most *insufficiently* inspected, with 70% of the observed uncontrolled discharges occurring in this area, Advisory Committee on Pollution of the Sea, Survey of Oil Pollution around the Coasts of the United Kingdom, 1986 (ACOPS), p. 1. Surprise was noted by some State Parties to the Paris Convention that the oil slicks were from the oil platforms and *not* ships as previously thought. See Par. Comm. 9.

142 Par. Comm. 9. pp. 1, 170-175 and Par. Comm. 10, pp. 1, 189-198 covering the periods 1985 to 1986.

143 The Paris Convention 40 mg/l *recommended* discharge limit is discussed in Part I Section 4.2.2.4.

144 Par. Comm. 9. p. 172. Concerning FRIGG (1,495 mg/l), the operator had a dispensation from the 40 mg/l standard until December 31, 1987. During this period the operator was to evaluate, test and install new cleaning equipment. It was stated however that in spite of this effort, it was not possible to reach the standard because of problems with very small condensate droplets in the production water. Details about the most probable discharge concentration level following December 31, 1987 were to be submitted as soon as the information was available. The standard was reached for ODIN (134 mg/l) in 1986. For VALHALLA (145 mg/l), improvements were made and the standard was to be reached by December 31, 1987.

145 Obtained from Tab. 1. SFT Akutt, s. 5.

146 Statens forurensningstilsyn, 'Utslipp av olje fra plattformene på norsk kontinentalsokkel i 1986,' SFT, mai 1987, (SFT Utslipp 1986) p. 1.

147 *Ibid.* p. 1.

148 Miljøundersøkelser Rundt Oljeplattformer på Norsk Sektor, Årsrapport for 1986, SFT, januar 1988, (SFT Miljø 86), s. 4.

149 Conversation Chief Engineer Jens Koefoed SFT October 7, 1988. Various effects to the marine environment as far out as 10,000 meters from drilling platforms as well as effects to fish were also noted.

150 Par. Comm. 9, pp. 127-148 notes that the biological effects of contaminants such as heavy metals in fish such as cod, plaice and flounder are also under study by the Paris Commission. See Part I Section 4.2.2.

151 SFT Utslipp 1986 p. 2.

152 Obtained from Table 10, Par. Comm. 10, p. 199.

153 Obtained from Tabell 1, SFT Utslipp 1986.

154 Obtained from Tabell 2, SFT Utslipp 1986.

155 *Ibid.* p. 6.

156 *Ibid.* p. 7. See Tabell 7, s. 7 for a list chemicals including biocides, corrosion inhibitors, oxidizing agents and colouring agents, discharged into Norwegian waters.

157 Obtained from Tabell 8 and Fig. 8, *ibid.* p. 8.

2. International Customary and General Principles of Law

Important legal sources in international law besides conventions are customary law and general principles of law. Due to the frequent ambiguity found in the environmental conventions or the non existence of any conventions whatsoever, these 2 sources should also be examined when considering a legal problem.[1] As noted in the Preface *customary* law is what needs examined extensively in order to fill out the conventional provisions applying to international environmental law. The problem is made more complex in that most legal authors note that provisions *may* have passed into international customary law but rarely state definitely what provisions are *considered* customary law.[2] Consequently in what follows, both in Parts I and II, where legal authors have come out definitely and declared that certain provisions can be considered customary law, then this is noted. Otherwise a generally conservative view is taken that it is unclear and doubtful whether such a transition has occurred. This is done to avoid false expectations concerning environmental legal coverage. *General* principles of international environmental law are even more vague, ambiguous and complicated and are presented here for the sake of background.

2.1. Customary Law

2.1.1. Introduction

Customary law is usually considered to consist of two elements, State practice and its acceptance as law.[3] Due to the relative short time in which marine environmental law has been in existence, it can be questioned whether there has been any customary law formed based upon these elements, and in spite of the progress made both in national and international marine environmental policies, even today, 'the measures taken have not resulted in the creation of a detailed body of customary law covering, in comprehensive terms, the whole area of marine pollution.'[4] However regarding certain aspects of environmental law, '(S)tate practice may have emerged which is to the degree convincing and supported by an *opinio juris* to the extent firm that one might speak of generally binding international law'.[5]

2.1.2. Customary Law - Vessel-source, Dumping, Offshore Platforms, Contingency Plans

The most extensive source of customary marine environmental law is found in the conventions.[6] The conventional provisions must have a 'norm-creating character' and as such establish, 'rules of general applicability, rather than merely settling issues between particular State parties on the basis of expediency.'[7] The relevance of provisions passing into customary law lies in the fact that States *not* Parties to the conventions would still be bound by these norm creating rules. Other sources consist of national legislation and resolutions by the United Nations General Assembly.[8]

The environmental conventional rules most in force are those covering *vessel-source* pollution. One author considers, 'the various documents regulating vessel-source pollution, as well as the work of the IMCO (International Maritime Organization (IMO)) in general, appear to offer a solid basis for the verification of the existence of a fairly strict customary regime regarding vessel-based oil discharges.'[9] Areas specifically covered are the discharge provisions established in the 1969 Amendments to OILPOL,[10] including the total discharge prohibition for tankers within the 50 mile limit and the 'load-on-top' system.[11] Another author notes that presumably generally accepted international rules include OILPOL but leaves it an open question whether MARPOL 73/78[12] is included because the Convention has not been widely ratified, especially the non compulsory Annexes III, IV, and V.[13] Still another author notes generally that at a certain point in time, 'a series of State practices, codified in *treaties* or working agreements, and supported by the writings of legal scholars and the acknowledgement or acceptance of the world community, passes from mere examples of national action to a customary norm of international law.'[14]

Pollution from *dumping* is also considered covered by customary rules stemming from international conventions.[15] Where a number of conventions prohibit the dumping of specific substances, (organohalogen compounds, mercury and cadmium), it can be said that, 'application as a norm of customary law is justified.'[16] Additionally there is a possibility that, 'a customary rule has emerged...that harmful substances not included in such a *customary* 'black list' may be dumped only with national or international approval.'[17] Another author mentions customary law associated with dumping only in relation to a State's enforcement jurisdiction under which a coastal State can take enforcement action against ships illegally dumping in its territorial sea and against ships about to leave its ports to dump without having the proper permit.[18]

As regards land-based and offshore platform pollution and contingency plans none of the conventions and agreements dealing with these topics and appearing in Part I are global conventions, and hence are ratified by only a few States. For these areas of international law there exist few if any provisions under customary law. The authors noted above do not even address these areas specifically in their discussions of international customary law.[19]

An argument might be made that *regional* customary law has been formed covering land-based and offshore platform pollution as well as contingency plans in the North Sea, stemming from provisions of the Paris Convention, the Bonn Agreement and the Copenhagen Agreement.[20] This is mentioned since it might have relevance to the Barents Sea case study presented in Part II. These are all agreements formed in the period late 1960's to mid 1970's. However, due to the inconsistency of State practice especially under the Paris Convention, this argument seems of little use. The same can probably be said of the different Arctic State contingency plans.[21] Besides being characterized by inconsistent State practice, most of the dates of formation are quite recent thereby putting into doubt whether any definite provisions have norm creating value and have passed into regional customary law.

2.1.3. Customary Law - Liability and Compensation

The area of liability and compensation *is* characterized by customary law provisions. It is generally acknowledged that a State has responsibility for an act or omission which results in a breach of a legal obligation resting upon treaty, custom or other, such as unilateral acts.[22] This responsibility can be based upon principles of *objective* responsibility and *culpa*, although objective responsibility seems to be more the general principle and a, 'better basis for maintaining good standards in international relations and for effectively upholding the principle of reparation.'[23] State practice, arbitral tribunals and the International Court of Justice (ICJ) have followed the principle of *objective* responsibility, while international law presently lacks the principle of strict liability for ultra-hazardous activities.[24] Objective responsibility seems at least similar to the strict liability known under national legal systems, being based upon a voluntary act and a breach of duty occurring by result alone.[25] Another author notes in the area of State liability and compensation under customary law there are *trends*, but they are exclusively included within the relevant conventions and are, 'far from becoming part of customary law.'[26]

Although various international cases support the objective and culpa principles, regarding transboundary pollution 2 seem to be central, the *Corfu Channel Case*[27] and the *Trail Smelter Case*.[28] Under the *Corfu Channel Case*, the ICJ stated a State's obligation, 'not to allow knowingly its territory to be used for acts contrary to the rights of other States', and found Albania responsible for damage and loss of human life which resulted from explosions of mines with a duty to pay compensation to the UK[29] In the *Trail Smelter Arbitration*, the Tribunal stated, '(U)nder the principles of international law, as well as of the law of the United States, no State has the right to use or permit the use of its territory in such a manner as to cause injury by fumes in or to the territory of another or the properties or persons therein, when the case is of serious consequence and the injury is established by clear and convincing evidence.'[30]

Two additional cases, the *Nuclear Tests Cases*,[31] and a Canadian claim against the USSR for damages caused by a nuclear satellite,[32] have relevance to transboundary pollution. These can be partly distinguished however in that they involve activities occurring outside the territory of the State, whereas the offshore polluting event originally occurs within State territory.[33] Under the *Nuclear Tests Cases* Australia and New Zealand served complaints on the French government stating the undesirability and unlawfulness of atmospheric nuclear testing in the Pacific. The French were requested to stop conducting such tests. The dispute was held by the ICJ to have disappeared when the French discontinued the tests. Under the Canadian claim against the USSR for damages caused by the Cosmos 954 nuclear satellite, CAD 6,041,174.70 was claimed based upon, 'the relevant international agreements'[34] and on general principles of international law.[35] The general principles of international law was stated to include a violation of sovereignty giving rise to an obligation to pay compensation, absolute liability for space activities, limited damages to those costs reasonable and proximately caused by the intrusion and capable of being calculated with a reasonable degree of certainty, and a Canadian duty to take steps to mitigate damages.[36] The Soviets limited any possible liability to that allowable under the Convention,[37] and the claim was settled on April 2, 1981 which provided for Soviet payment of CAD 3,000,000 'in full and final settlement.'[38]

Generally, awards available for damages include monetary compensation for actual injury or damage suffered, restitution, monetary compensation for non-material damage, and satisfaction, which includes apologies, ceremonial honours, or a Court declaration.[39] Possible defenses include extinctive prescription, acquiescence and waiver (usually considered as issues of admissibility of claims), assumption of risk and contributory negligence, *force majeure* applying to acts of war and to insurrection and civil war, military necessity at times in particular contexts, and self defence including collective and defense of third States.[40]

2.1.4. Customary Law - National Legislation and Declarations

As noted in addition to international conventions and resolutions an important source of customary law is that of national legislation and executive declarations. State practice is defined broadly as, 'any act or statement by a State from which views about customary law can be inferred', and national legislation and declarations can easily be included.[41] Besides indicating *implementation* of the environmental conventions indicated in Part I, national legislation and declarations also *can* indicate implementation of provisions from conventions *not in force* including the LOSC.

Since as noted it is the desire that chiefly black letter law be covered in Part I as a handbook, all discussion and analysis of the environmentally progressive yet legally indefinite LOSC is restricted to Part II.[42] However brief mention will be made here since State *practice* through national legislation and declarations may be of most influence in implementing LOSC environmental provisions through customary law. As noted the LOSC already represents a *partial* codification of international customary law though to what extent is indefinite. One author notes that LOSC codifies pollution control and protection of the ocean environment including principles that the oceans are the unique responsibility of the world community; all nations share the obligation to assure their continued survival as an international resource; States must take appropriate steps to protect the ocean areas under their sovereignty and cooperate with one another to *protect ocean areas under multiple jurisdiction*; States have both a joint and individual obligation to safeguard the high seas; all States have the obligation to control their citizens to assure protection of the marine environment; and a State can be held internationally responsible for violations by its citizens.[43] Another author argues that *all* the LOSC transnational pollution Articles 192 to 237 are reflections of established principles of customary law.[44] See Part II Section 11.2. for a brief summary of these Articles as well as individual Chapters 11 - 17 for more specific discussion. Another author notes that while the establishment of a State's exclusive economic zone has undoubtedly passed into customary law, it is doubtful LOSC provisions related to jurisdiction over *pollution*, fisheries and research have passed or are likely quickly to pass into customary law.[45] Still another author differentiates between international customary law and practice, a draft treaty and an international treaty and concludes that LOSC including Part XII, the environmental protection section, cannot be said not to be international law merely because it has not come into force. The different provisions might be either international law, customary international law, customary international practice or merely hollow words.[46] In spite of the controversy most legal writers address all of the LOSC pollution provisions in their works.[47]

2.1.5. Conclusions

Although the liability area of international customary law seems to offer possibilities for demanding compensation from a State for transboundary pollution, in *reality* it appears this approach is little used based upon the apparent paucity of case law. The exact reasons are not known and can only be speculated upon here. Perhaps States are waiting for the proposals of the International Law Commission (ILC) regarding liability schemes,[48] especially given the diversity of legal opinion noted above. Perhaps there are political reasons. Perhaps it is difficult to find proper causes of action as occurred under the *Nuclear Tests Cases*.[49] In addition there exist problems with the cases noted. The *Trail Smelter Case* was arbitrated by an international tribunal and is of questionable precedence value. Under the *Nuclear Tests Cases*, a final decision by the ICJ was *not* arrived at, the dispute disappearing. The Canadian claim was settled by a Protocol, though compensation was received.

Concerning coverage by international customary law of vessel-source, dumping, land-based and offshore platform pollution as well as contingency plans even if it is formed through State practice and *opinio juris*, it seems more beneficial as noted to take a sober view and follow the author who states, '(C)ustomary international law contains few rules relevant to the question of marine pollution.'[50] Another author notes that States have increasingly resorted to conventions not through a failure to recognize a duty to prevent significant marine pollution under a prior duty of customary law, but rather through a, 'need to flesh out and expand upon the fundamental customary obligation with a view to rendering it more readily effective.'[51] Thus due to the vagueness and complexity of the customary law provisions, when they do exist, all the legal areas associated with transboundary pollution control will have to be achieved through *conventional* coverage to be effective.

2.2. General Principles of Law[52]

'General principles of law recognized by civilized nations' is also a source of international law in accordance with Article 38 of the ICJ Statute. Some writers however including the Soviets do not recognize this as a separate source. Although it appears to be unsettled what exactly is meant by 'general principles' and the role of national law in its formation, the doctrines, 'use your own property in such a manner as not to injure that of another',[53] 'abuse of rights', 'good neighbourliness' and 'custodianship' appear to be central. These will be examined briefly in relation to international environmental law. This is done for background and also because as one author notes, environmental threats are constantly increasing, the law is 'defective and fragmentary', and '(U)ntil more specific norms materialize, general principles of law emerge as the *last resort* to avoid a blank 'license to pollute.''[54]

2.2.1. 'Sic Utere Tuo Ut Alienum Non Laedas'

The principle of 'use your own property in such a manner as not to injure that of another' has been referred to as noted previously in such cases as the *Corfu Channel* case, and the *Trial Smelter Arbitration* and in such documents as Principle 21 of the Stockholm Declaration, Article 30 of the Charter of Economic Rights and Duties of States[55] and LOSC Article 194(2).[56] In practice however due to the phrase's vagueness, various interpretations are used and it is difficult to establish any specific State obligations. The phrases 'abuse of rights' and 'good neighbourliness' are often resorted to in an attempt to provide direction.[57]

2.2.2. 'Abuse of Rights'

'Abuse of rights' is defined 'to consist of the prohibition of the exercise of a right for an end different from that for which the right was created, to the injury of another person or the community.'[58] There are difficulties, however, in that 'the doctrine appears superfluous since adequate basis for the negation of the 'abuses' may be found in other, material, rules of law.'[59] Additionally there are problems concerning its scope, with some writers requiring 'intentional infliction of damage', and others 'the mere occurrence of excessive use of rights, where resulting in external damage.'[60] Apart from the theoretical problems exists the problem involving the doctrine's acceptance as a general principle of law. It has been applied in State practice and in the international courts, but writers differ as to the sufficiency of its application 'to indicate customary consolidation of an international prohibition of abuse of rights.'[61] Additionally, even if it is not a customary principle of international law, it possibly has been included in national legislation and forms a general principle 'recognized by civilized nations.'[62] Though writers give conflicting results regarding State legislation, it seems that a prohibition of *deliberate* and *malicious* 'abuse of rights' is found widely and hence would appear to form a general principle of law under Article 38(1) of the ICJ Statute. Thus, in order for this doctrine to apply to international environmental law, there must probably be a *deliberate* intent to pollute and cause damage to another country. This intent in most cases is not likely, and any attempt to modify the malicious intent element through using mere *knowledge* of likely environmental damage, 'branches away from the theoretical understructure of the doctrine.'[63]

2.2.3. 'Good Neighbourliness

'Good neighbourliness' is defined as, '(N)o State may conduct, promote or sustain in its territory activities, which cause other than inconsiderable and usual damage in the territory of a neighbouring State.'[64] 'Good neighbourliness' is distinguished from 'abuse of rights' as being somewhat more concrete, and '(A)lthough it does not, in specific terms, distinguish what is right from what is wrong, it does appear to aim at some identification of State rights and duties rather than merely 'interpreting' the mode of their exercise.'[65] It also encompasses intentional and unintentional conduct thus avoiding the problems noted above under 'abuse of rights.' The doctrine is accepted by authors and is incorporated into the national legislation of Western, Socialist and Third World countries. Specific elements are lacking however with the exception of a development involving international river pollution and the concept '*equitable utilization.*'[66] It can be questioned, however, whether this concept can be transferred to that of marine pollution, especially since there are difficulties in determining the 'equitable shares' of 160 States to the many various ocean uses. Thus under present circumstances, 'activities liable to cause *marine pollution* are not subject to any *specific* rule of 'equitable utilization', as distinct from a general postulate of reasonable behaviour.'[67]

2.2.4. 'Custodianship

The principle of 'custodianship' gives the coastal State a right and a duty to prevent pollution within its coastal waters and is said to have originated in a speech made by Prime Minister Trudeau of *Canada* in 1969,[68] in which an extension of coastal jurisdiction beyond the traditional limits was justified, as well an obligation stated for coastal States to preserve the marine environment. As 'custodian' of its part of the marine environment the coastal State is responsible for pollution even in its own territory since the effects 'derogate from the total value of the world's amenities.'[69] Several international documents contain statements implying a similar idea, including LOSC Article 192.[70]

2.2.5. Conclusions

As can be seen from the brief explanations given, the main common characteristic of the general principles is *vagueness*, even more so than customary international law. Even 'good neighbourliness', the one perhaps coming closest to being practical, is itself, 'imprecise and offers little definite guidance for State behaviour.'[71] As such it can perhaps be seen how important international environmental conventions are in controlling marine pollution. The general principles being the last resort to avoid a blank license to pollute, unfortunately are a weak last line of defense. They do however provide basic obligations for States to protect the marine environment, which can act as legal support for the foundation of conventions with specific provisions needed for pollution control.

This background being given the legal coverage represented by conventions will be presented.

Notes - Chapter 2

1 Fleischer, C., 'Et folkerettslig perspektiv på global ressursforvaltning. Behov for forskning?' NAVF's forskningspolitiske konferanse, 22-23 mars 1988 for oppfølging av Brundtland-kommisjonens rapport, (Fleischer NAVF), s. 17, considers State practice, an element of customary law, often the *most* important legal source when an international legal problem is examined.

2 See for example Belsky , Martin, 'Management of Large Marine Ecosystems: Developing a New Rule of Customary International Law,' *San Diego Law Review*, Vol. 22, No. 4, 1985, (Belsky) p. 753, 'even if the provisions of the Treaty (LOSC) are not automatically accepted as customary international law, they are recognized as the best available evidence of customary law,' (emphasis added).

3 Hakapää p. 130, 'the existence of a usage *(consuetudo)* embodying a rule of conduct,' citing Brownlie, Ian, *Principles of Public International Law*, 3 ed. Oxford, (1979) (Brownlie Principles), pp. 4-12; and 'the *opinio juris* of States.' The author also notes the unorthodox argument that customary law can be founded upon only the *opinio juris* element and thus be formed instantaneously and independent of the usage over time element, which might be of some benefit to international environmental law. Churchill p. 5 notes 'international custom, as evidenced by a practice generally accepted as law' is a source of international law. See generally Henkin, Pugh, Schachter, Smit, *International Law, Cases and Materials*, American Casebook Series, West Publishing Co. (1980), pp. 36-69.

4 Hakapää p. 131.

5 *Ibid.*

6 *Ibid.* and Timagenis pp. 21-39. See Article 38 of the Vienna Convention on the Law of Treaties (Vienna Convention) 8 *International Legal Materials* 679.

7 Churchill p. 6, citing *North Sea Continental Shelf* Case, 1969 ICJ Rep. 3 at 39.

8 See Hakapää pp. 131-132, Timagenis pp. 21-39 and Franckx pp. 156-159. See Part I Section 2.1.4. for discussion of national legislation. The Stockholm Declaration and Action Plan on the Human Environment, Report of the United Nations Conference on the Human Environment U.N. Doc. A/CONF. 48/14 and Cor. I, (Stockholm Declaration), 11 *International Legal Materials* 1416 (1972), should be mentioned generally, since its principles reoccur in subsequent environmental treaties and have had a generally influence on the treaties, see Johnston pp. 46-48. Specifically, Principle 7 states, '(S)States shall take all possible steps to prevent pollution of the seas by substances that are liable to create hazards to human health, to harm living resources and marine life, damage amenities or to interfere with other legitimate uses of the sea;' Principle 21, '(S)tates have, in accordance with the Charter of the United Nations and the principles of international law, the sovereign right to exploit their own resources pursuant to their own environmental policies, and the responsibility to ensure that activities within their jurisdiction or control do not cause damage to the environment of other States or of areas beyond the limits of national jurisdiction;' and Principle 22, '(S)tates shall co-operate to develop further the international law regarding liability and compensation for the victims of pollution and other environmental damage caused by activities within the jurisdiction or control of such States to areas beyond their jurisdiction.' Johnston pp. 46-48 notes that following this Declaration the U.N. General Assembly adopted 11 environmental resolutions, but that the legal impact of the Declaration and of U.N. resolutions in general depends upon a 'somewhat subjective interpretation.' Hakapää pp. 132-133 notes that the U.N. resolutions generally seem too broad to contribute to specific rules. Professor Richard Bilder, University of Wisconsin, Madison, Wisconsin, US believes, based upon the US Restatement of Foreign Law that Principle 21 is considered to have become international customary law, conversation September 10, 1989.

9 Hakapää p. 132, parentheses added.

10 International Convention for the Prevention of Pollution of the Sea by Oil (OILPOL), 9 *International Legal Materials* 1. See Part I Section 5.2.2. for specific OILPOL provisions.

11 Hakapää p. 132.

12 1973 International Convention for the Prevention of Pollution from Ships, 12 *International Legal Materials* 1319 and the 1978 Protocol to the 1973 International Convention for the Prevention of Pollution from Ships, 17 *International Legal Materials* 546 (MARPOL 73/78). See Part I Section 5.2.3. for specific MARPOL 73/78 provisions.

13 Churchill pp. 255-256. The author notes on p. 262 however a strong possibility that the International Convention relating to Intervention on the High Seas in Cases of Oil Pollution Casualties (Intervention Convention), 9 *International Legal Materials* 25, 'crystallized and clarified' international customary law. See Part I Sections 5.3.1. and NEED for specific Intervention Convention and Protocol provisions.

14 Belsky p. 743 citing MacRae, 'Customary International Law and the United Nations Law of the Sea Treaty,' 13 *Cal.W.Int'l.L.J.* 181, 200-04 and Burke, Book Review, 78 *Am.J.Int'l.J.* 270, 271 (1984) quoting D. O'Connell, *The International Law of the Sea* 38 (1982), emphasis added. Under this view arguments could be made that provisions from those conventions which are widely ratified and are somewhat dated have become international customary law. This would probably include in addition to those already mentioned, provisions from the Convention on the Continental Shelf (1958 Continental Shelf Convention) 499 *United Nations Treaty Series* 311, the Convention on the Territorial Sea and the Contiguous Zone (1958 Territorial Sea Convention) I *New Directions in the Law of the Sea* 1, and the Convention on the High Seas (1958 Convention on the High Seas) I *New Directions in the Law of the Sea* 257, (see Part I Sections 4.2.1. and 5.2.1.) (Fleischer s. 88 and 103 notes that the latter 2 Conventions can be assumed to represent international customary law) and provisions in force from the International Convention for the Safety of Life at Sea (SOLAS), 14 *International Legal Materials* 959 and 17 *International Legal Materials* 579; the International

Convention on Load Lines (Load Lines Convention) IMO Document, 701 81.01.E. (1989) p. 33.; and the Convention on the International Regulations for Preventing Collisions at Sea (Collisions Convention) IMO Document, 904 85.01.E, (1989) p. 46; where ratification comprises over 90% of the world's merchant fleets gross tonnage (see Part I Sections 5.3. and 5.4.).

15 See Part I Section 3.2. for the Convention for the Prevention of Marine Pollution by Dumping from Ships and Aircraft (ODC) 11 *International Legal Materials* 262, Intergovernmental Conference on the Convention on the Dumping of Wastes at Sea (LDC) 11 *International Legal Materials* 1294, Convention on the Protection of the Marine Environment of the Baltic Sea Area (Helsinki Convention) 13 *International Legal Materials* 544, and the Protocol to the Convention for the Protection of the Mediterranean Sea against Pollution (Barcelona Convention) 15 *International Legal Materials* 285.

16 Hakapää p. 132. The prohibition of dumping oil, high-level radioactive material, and chemical and biological warfare agents is also arguably included since they are subject to strict prohibitions under all the above Conventions but the ODC.

17 *Ibid.*

18 Churchill pp. 269-270.

19 See for example *ibid.* pp. 245, 262, 272-282 and Hakapää pp. 130-133.

20 Respectively Convention for the Prevention of Marine Pollution from Land Based Sources 13 *International Legal Materials* 353; Agreement of 9 June 1969 for Cooperation in Dealing with Pollution of the North Sea by Oil 704 *United Nations Treaty Series* 3; and Agreement between Denmark, Finland, Norway and Sweden concerning Cooperation in Measures to Deal with Pollution of the Sea by Oil 822 *United Nations Treaty Series* 324. See Part I Sections 4.2.2., 7.4. and 7.5. See generally Henkin pp. 53-54.

21 See Part I Section 7.8.

22 Brownlie Principles pp. 435. See also Churchill p. 245.

23 Brownlie Principles p. 438.

24 *Ibid.* pp. 435 and 477. Generally it can be said that under these concepts fault need not be shown which has important consequences for international environmental law. Simply this means that negligence for example need not be shown on the part of the polluter.

25 *Ibid.* pp. 436 and 476-477. The author calls the national rules 'absolute or relatively strict liability for failure to control operations which create a serious or unusual risk of harm to others' and in fact questions the non-existence of strict liability under international law believing that objective responsibility provides some protection in respect to dangerous activities. Handl, G., 'International Liability of States for Marine Pollution,' *Canadian Yearbook of International Law*, Vol 211. (Handl) pp. 100, 115-116 believes that present State practice is, 'insufficient to suggest a foundation in customary international law of a strict liability approach to transnational hazardous activities in general.' The author concludes, however, based upon justifications including similarities in all major domestic legal systems, works of various legal scholars, work of the International Law Commission (ILC), negotiations preceding Article 235 of the LOSC, and a principle of sovereign equality of States; where States or private persons under State authorization engage in activities that carry a recognizable *significant risk* of transnational harm, the States will be held *strictly liable* if harm typical of the risk occurs.

26 Timagenis pp. 20, 63-64. As regards the trends it is noted, 'a general undertaking is included of other development of rules and procedures for the assessment of liability' (Helsinki Convention and the Convention for the Protection of the Mediterranean Sea against Pollution, 15 *International Legal Materials* 285 (1986) (Barcelona Convention)),

see Part I Section 3.2.; 'recognition by implication of the existence of relevant rules and principles in international law' (LDC and the Convention for the Protection and Development of the Marine Environment and the Coastal Areas (Kuwait Convention) 2 *Ocean Year Book* 516), see *ibid*; with this trend commencing 'with the Stockholm Conference on Human Environmental and with its most concrete expression in the UN Conference on the Law of the Sea,' see Section 2.1.2. Generally, developments also include absolute liability, (not based on fault) coupled with the channelling of liability (concentration of obligation of compensation in a particular person such as the owner), the limitation of liability (fixed ceiling amount) and the establishment of compensation funds, (these are elements found in the shipping liability conventions, see Part I Section 6.4.). It is noted that the rules that have developed seem to cover mainly the areas of oil pollution from ships and radiation from nuclear plants. However even if customary law has developed regarding State liability, it has developed from conventions whose provisions, 'are at a very high degree of generality' and traditional concepts of responsibility and liability do not work, 'because the causal connection between pollution and damage, the extent of knowledge required, the foreseeability of harm and the standard of proof required are matters which lie beyond the traditional legal standards.'

27 1949 ICJ Rep. 4.

28 1941 III R.I.A.A. 1905 245.

29 Brownlie Principles p. 434, citing 1949 ICJ Rep. 4, pp. 22-23.

30 *Ibid*. p. 434, citing R.I.A.A. 1905, at 1965-66.

31 1974 ICJ Rep. 253.

32 18 *International Legal Materials* 899, (1979).

33 Brownlie Ian, *System of the Law of Nations, State Responsibility*, Part I, (Brownlie System), pp. 182, 183. The transboundary pollution of course drifts into another State's territory, which is the core of the problem.

34 Specifically the 1972 Convention on International Liability for Damage caused by Space Objects, March 29, 1972, 24 *United States Treaties and Other International Agreements* 2389.

35 Canadian note of 23 January 1979 of the Canadian Secretary of State for External Affairs to the Soviet Ambassador in Ottawa, 18 *International Legal Materials* 899.

36 *Ibid*. at 905 and 907. Specifically the Soviet obligation to pay compensation was claimed by Canada based upon the trespass of the satellite, the harmful consequences of the intrusion, the damage caused by hazardous radioactive debris and the interference with the sovereign right of Canada to determine the acts performed on its territory which based upon international precedents gave rise to an obligation to pay compensation. The absolute liability was further claimed to apply to activities having in common a high degree of risk and based upon, 'the general principles of law recognized by civilized nations' (Article 38 of the ICJ Statute) accepted as a general principle of international law as well as a principle accepted in numerous international agreements.

37 *Ibid*. at 923.

38 Brownlie System p. 97. Several legal authors oppose such settlements. Brownlie, 'Causes of Action in the Law of Nations,' 50 *Brit. Y.B. Int'l L.* 13, (1979) (Brownlie Causes) p. 40, notes, 'A major role of State responsibility and the law of claims is psychological and moral: that of putting a hard edge on legal rights and duties,' something which is not achieved through settlements. Handl p. 88 notes, '(I)nternational negotiated allocation of transnational losses has to be recognized as an undesirable development' because a *clear* State liability is a key element in pollution prevention.

39 Henkin pp. 568-570.

40 Brownlie Principles pp. 464-465.

41 Franckx, p. 157, citing the ICJ in the *Nottebohm Case*, 1955 I.C.J. 4, and the *North Sea Continental Shelf Cases*, 1969 I.C.J. 4. The *North Sea Continental Shelf Case* is additionally used to explain the *opinio juris* element, 'State conduct must be performed in such a manner... 'as to be evidence of a belief that this practice is rendered obligatory by the existence of the rule of law requiring it.'' For a contrary view regarding the role of national legislation see Hakapää p. 131-132, who notes that in 1981 despite, 'a clear trend in national legislation towards more comprehensive regimes of marine pollution control...(a)part from 'mere' implementation of conventional regimes...such legislation has, as yet, hardly had sufficient uniformity effectively to contribute to the process of international custom creation.' See Part II Sections 9.1. and 9.2.3.1. for brief summary of Arctic State national legislation and declarations.

42 The LOSC provisions especially involving prescriptive and enforcement jurisdiction can be said to be more progressive than those of the Part I environmental conventions. Generally jurisdiction is defined as, 'the power of a State to affect under international law the conduct of other subjects (private or public) as well as the status of their property, by measures of regulation adjudication or (executive) enforcement.' Prescriptive jurisdiction allows a State to impose an obligation or a right upon a natural or juridical person through legislation, executive act or order, administrative rules or regulations, or court judgment. Enforcement jurisdiction gives a State the competence to adopt reasonable measures to compel, induce compliance, or to impose sanctions for non compliance with applicable laws, regulations, or enforceable judgments by means of administrative or executive action, or judicial proceedings. Wang, Cheng-Pang, 'A Review of the Enforcement Regime for Vessel-Source Oil Pollution Control,' *Ocean Development and International Law*, Vol. 16, No. 4, (Wang) p. 309, citing Restatement (T.D. No. 3) of the Law of Foreign Relations (Revised) (1982) paragraphs 419 and 431. See also Henkin pp. 420-425 and Churchill pp. 9-10. Though these two jurisdictions often go together, this is not always the case, and the separation can be important as related to anti pollution provisions. A State has jurisdiction over persons and activities within its territory. Outside its territory a State may not adjudicate or enforce its regulations within another State without the second State's consent. Beyond any State's territory or maritime zones, jurisdiction is limited. A State's prescriptive jurisdiction applies to its nationals beyond the State's territory or zones, but the nationals' violations of State regulations can only be adjudicated or enforced when the national returns to the territory or zone of the State. Churchill p. 10 gives an example of separation when a flag State under international law may legislate for its ships wherever they are, but may not enforce its laws when its ships are in a foreign State's internal or territorial seas. Wang p. 310 notes when States enact statutes which are not enforceable, '(A)ccession to the Convention becomes a public relations ploy, window dressing to pacify domestic groups or foreign States when governments and their industries have really only wanted to continue business as usual.'

43 Belsky pp. 751-753, emphasis added. The author notes additionally that these provisions are generally accepted by all nations, but there is much controversy as to whether LOSC non sea bed provisions have become part of customary international law.

44 Kindt, 'International Environmental Law and Policy; An Overview of Transboundary Pollution' (Law of Sea Symposium), 23 *San Diego Law Review* 583 (1986) (Kindt) pp. 600-601.

45 Churchill p. 146.

46 Wang pp. 306-307.

47 See for example Churchill pp. 241-282 and Harders Seminar pp. 519-521.

48 See Part I Section 6.5.

49 Brownlie Principles p. 474 notes that the, '(A)pplicant States had some difficulty in relating the deposit of radioactive fall-out to existing legal categories.'

50 Churchill p. 245. The author cites the *Corfu channel* case, the *Trial Smelter Arbitration* and Article 2 of the 1958 High Seas Convention (see Part I Section 5.2.1.2.) to establish a general rule of customary law, 'that States must not permit their nationals to discharge into the sea matter that could cause harm to the nationals of other States', but notes this is too vague to be effective and incapable of being developed into detailed emission standards or liability regimes. Noted as well is that States' legislative and enforcement jurisdiction are defined in customary international law, but are considered by many States as 'both inadequate and incapable of sufficiently speedy or extensive development to sustain effective action against pollution.'

51 Handl p. 92.

52 The following is obtained from Hakapää pp. 134-147 unless noted otherwise.

53 *Sic utere tuo ut alienum non laedas.*

54 *Ibid.* p. 136.

55 GA Res. 3201 (S-VI), UN GAOR Supp. 1, p. 3, UN Doc. A/9559 (1974).

56 See Part II Section 11.2. for LOSC provisions. It is also noted that the International Law Commission (ILC) in its draft Articles of State responsibility includes a similar provision, see Part I Section 6.5.

57 It is noted, however, that it is not completely settled what the relation is between these three concepts.

58 Hakapää p. 138.

59 *Ibid.* This idea is made somewhat clearer by the example given, 'ill-treatment of foreigners amounting to excessive use of a State's sovereign discretion within its territory would not represent an abuse of rights but simply a violation of the prohibitory rules by which the minimum standards of alien treatment are determined in international law.'

60 *Ibid.* p. 139.

61 *Ibid.* p. 140.

62 *Ibid.*

63 *Ibid.* p. 141.

64 *Ibid.*

65 *Ibid.* p. 142.

66 *Ibid.* 'Equitable utilization' is described as, '(E)ach basin State is entitled, within its territory, to a reasonable and equitable share' in beneficial international uses of international drainage basins.

67 *Ibid.* pp. 144-145. The author notes another possible element includes the gravity of the damage caused which implies the exclusion of State responsibility for damages not amounting to 'substantial' or 'serious' pollution. Though LOSC seems to encompass all cases of 'extraterritorial pollution', in practice State responsibility for pollution will probably vary according to the seriousness. Additionally it is noted that the ILC in its work on State responsibility graduates pollution incidence into *massive* pollution, international *crimes*; and *other* environmental wrongs, not so grave international *delicts*, citing draft Article 19, *ILC Yearbook* 1976 - II(2), p. 75.

68 The speech includes the statement, '(C)anada regards herself as responsible to all mankind for the peculiar ecological balance that now exists so precariously in the water, ice and land areas of the Arctic archipelago', Canadian House of Commons Deb., p. 39 (October 24, 1969).

69 Hakapää p. 146.

70 It is noted that such provisions reflect a concept of 'collective custodianship', a term not identifying prohibited activities but rather stating a positive duty for States to establish relevant norms.

71 *Ibid.* p. 147.

3. Pollution by Dumping

3.1. Introduction

As noted in Part I Section 1.3.3. dumping is the deliberate disposal into the sea from ships or aircraft of waste loaded on board for this purpose and used to be a legitimate means of disposal in the past. It is treated as a separate source of pollution, since it usually is the *reason* for the ship's journey and in addition is usually an extension of the practice on land.[1]

3.2. Conventional Law[2]

The earliest conventions established covering dumping are the *global* London Dumping Convention[3] and the *regional* Oslo Dumping Convention[4] from which many of the later conventions are based.[5] See Part II Section 9.3.1. for discussion of the *problems* associated with these Conventions.

3.2.1. The LDC and the ODC, Definition of Dumping

The main definition of dumping in use is provided by the LDC, Article III (1) which reads as follows:

'For the purposes of this Convention:

(a) 'Dumping' means:
- (i) any deliberate disposal at sea of wastes or other matter from vessels, aircraft, platforms or other man-made structures at sea;
- (ii) any deliberate disposal at sea of vessels, aircraft, platforms or other man-made structures at sea;

(b) 'Dumping' does not include:
- (i) the disposal at sea of wastes or other matter incidental to, or derived form the normal operations of vessels, aircraft, platforms or other man-made structures at sea and their equipment, other than wastes or other matter transported by or to vessels, aircraft, platforms or other man-

made structures at sea, operating for the purpose of disposal of such matter or derived from the treatment of such wastes or other matter on such vessels, aircraft, platforms or structures.

(ii) placement of matter for a purpose other than the mere disposal thereof, provided that such placement is not contrary to the aims of this convention.

(c) The disposal of wastes or other matter directly arising from, or related to the exploration, exploitation and associated off-shore processing of seabed mineral resources will not be covered by the provisions of this Convention.'

The definition used in the ODC, Article 19 (1) is similar however does not exclude specifically waste dumped related to offshore activities. The ODC definition does not include as dumping similar to the LDC the disposal of wastes 'incidental to, or derived from the *normal* operations' of vessels, aircraft, platforms etc., and *this* may restrict the application concerning offshore platforms.[6] An official Recommendation was adopted in 1977 emphasizing control and cleaning the seabed of oil drilling debris, and noted that since the number of requests for compensation submitted by professional fishermen to the Parties had decreased, the Commission 'considered the implementation of the Recommendation was being respected,' though vigilance was necessary.[7] With the exception of disposal directly related to sea-bed activities, it may be difficult to think of any deliberate, 'non-operational' waste disposal which does not amount to dumping as defined in the LDC and the ODC, and therefore these Conventions can probably be considered 'fairly comprehensive'.[8]

3.2.1.1. Jurisdiction to Legislate and Enforce

The LDC in Articles I and II and the ODC in Articles 1 and 4 requires the Parties to take all possible steps and measures (and to harmonize these) consistent with the respective Conventions to prevent marine pollution by dumping. The LDC in Annex I prohibits, 'black lists', completely except under emergencies, the dumping of organohalogen compounds, mercury, cadmium, high-level radioactive materials, biological and chemical warfare materials, persistent plastics, and various oils including crude, fuel, heavy diesel, lubricating, hydraulic fluids, and mixtures of these.[9] Annex II, lists substances the dumping of which requires a prior special permit except in emergencies in accordance with Article IV (1)(b).[10] This 'grey list' contains arsenic, lead, copper, zinc, organosilicon compounds, cyanides, fluorides, pesticides, beryllium, chromium, nickel, vanadium, mixtures of these, radioactive wastes not included in Annex I, and containers, scrap metal and other bulky wastes liable to present a serious obstacle to fishing or navigation.[11] The dumping of all other wastes not appearing in either Annex I or II, requires a prior general permit in accordance with Article IV (1)(c) and (2).[12]

The ODC is similarly structured and lists in its three Annexes much the same substances and provisions to be considered as in the LDC. Exceptions include however that the ODC does not prohibit in any of the Annexes the dumping of the oils or oily mixtures, high-level radioactive wastes, or materials produced for biological and chemical warfare, but does include organosilicon compounds on its 'black list' rather than its 'grey list'.[13] The ODC also makes additional requirements in Annex II regarding dumping in deep water and 150 miles from land of certain substances and quantities, 'likely to seriously reduce amenities.'[14]

In both the LDC and the ODC jurisdiction rests upon the flag States, loading States and coastal States.[15] Thus for flag States each Party is obliged to enforce convention measures against vessels and aircraft registered in that State or flying its flag. For loading States each Party is obliged to enforce Conventional measures against vessels and aircraft loading in its territory or territorial sea waste for dumping, though loading State enforcement apparently can also be taken against vessels and aircraft flying the flag of a non-party.

The obligation of *coastal* State enforcement is not clear. The ODC states the obligation to enforce the dumping regulations applies only to the State's territorial sea.[16] The LDC states the obligation extends over 'vessels and aircraft and fixed or floating platforms under its jurisdiction believed to be engaged in dumping.'[17] The term 'under its jurisdiction' was deliberately left vague and did not define any coastal State area of jurisdiction, because of a compromise between coastal States and the major maritime States to leave it to the LOSC to define the extent of coastal State jurisdiction over dumping.[18]

Additionally both the LDC and the ODC state that the Parties, 'shall take in its territories appropriate measures to prevent and punish conduct in contravention of the provisions of this Convention[19] which seems to be a catch-all for States to take actions against all activities which might result in dumping, yet slip through the system outlined above in LDC Article VII (1) and ODC Article 15 (1).[20]

3.2.1.2. Dispute Settlement

The ODC contains no provisions covering the settlement of disputes.[21] Under the LDC the settlement of disputes is provided for under Articles XI and X, which respectively provide, '(T)he Contracting Parties shall at their first consultative meeting consider procedures for the settlement of disputes concerning the interpretation and application of this Convention,' and 'the

Contracting Parties undertake to develop procedures for the assessment of liability and the settlement of disputes regarding dumping'.[22]

3.2.1.3. Oil Installation Removal

Though proposals for oil platform dumping or removal after abandonment or disuse are under discussion in several different forums, including the Oslo-Paris Commissions of the ODC and Paris Conventions[23] and the IMO,[24] there are presently no definite provisions under the dumping Conventions. The basic law governing platform removal is Article 5 (5) of the 1958 Continental Shelf Convention, which requires that, '(A)ny installations which are abandoned or disused must be *entirely* removed.'[25] However the legal status of platform removal is unclear partly due to LOSC's uncertain legal status and the major offshore oil producing States non ratification of it.[26] Some North Sea States which have ratified the 1958 Continental Shelf Convention consider that Article 5 (5) is already modified by State practice, though others follow it and require removal.[27] Different options involved for platform removal include dumping them on site, loading them on barges for deep water dumping, dismantling and moving them to land, or dumping with the intention of using them as artificial fishing reefs.[28]

3.2.1.4. Monitoring, Contingency Plans

Monitoring of dumping is provided for in both the LDC, Article VI (1)(c) and (d) and Article VIII; and the ODC, Articles 11 and 13. The LDC requires States to keep records showing the nature and quantities of permitted dumpings as well as the location, time and method used; and requires States to monitor individually with other parties or international organizations the conditions of the seas. Under this program however there exists some dissatisfaction, due to among other things the continued dumping of low-level radioactive waste in an ocean area where no monitoring studies had taken place as late as 1979.[29] Under the ODC Articles 13 and 17 a Joint Monitoring Group was formed combining with the Paris Convention, where the monitoring program was to be based upon existing national arrangements, but 'harmonized by intercalibration of sampling, treatment of samples, methods of analysis and standard reporting procedures.'[30] There have been difficulties with this program due to the data obtained by different States not being comparable.[31] On the other hand the Oslo-Paris Commission's publication seems to contain a fair amount of data regarding wastes disposal and in addition *believable* since they criticize their own incomplete knowledge.[32]

Abatement rules concerning pollution *clean up* are provided for in the ODC Article 15 (2) in which the Parties are to establish an information network for the timely reporting of dumping incidents, and in Article 15 (4) in which the

Parties are to assist one another in dealing with dumping incidents.[33] Article VII (3) and Article IX of the LDC provides for much of the same regarding cooperation in reporting and the development of procedures for the effective application of the LDC, as well as collaboration to prevent or mitigate pollution caused by dumping. It is unclear how these measures function in practice, however from the number of contingency plans in existence in the ODC and LDC areas of application indicated in Part I Chapter 7, this signifies that the ODC and LDC leave contingency plans for pollution abatement to other agreements between the countries.[34]

3.2.1.5. Sovereign Immunity and Other Exceptions

Under the LDC, Article VII (4), vessels and aircraft entitled to sovereign immunity under international law are exempt from dumping provisions. Thus *government* ships and aircraft not used for commercial service are excluded from LDC provisions.[35] Under the same Article, however, each party must ensure that such vessels and aircraft are operated in a manner consistent with the object and purpose of the Convention. Under the ODC, Article 15 (6), '(N)othing in this Convention shall abridge sovereign immunity to which certain vessels are entitled under international law.'

These two Conventions seem generally to narrow the sovereign immunity exception when compared to various conventions on marine pollution.[36] The LDC allows the exclusion of government ships and aircraft not used for commercial service, however, the Parties are obligated to ensure that such vessels and aircraft exempted are operated in a manner consistent with the Convention's provisions. This obligation is *not* qualified as is done in many other Conventions by words such as 'so far as is reasonable and practicable' or 'not impairing the operations or operational capabilities of such ships'.[37] The ODC may vaguely apply to *any* vessels, warships included through the wording used.[38] In spite of the narrower interpretations however the sovereign immunity exception is broad enough and seems unfortunate when it is estimated that substantial dumping pollution occurs from military vessels.

Besides sovereign immunity, other important exceptions to dumping provisions include dumping necessary for the safety of human life or vessels,[39] dumping of 'black list' substances by special permit where emergencies pose an unacceptable risk to human health,[40] dumping of 'black list' substances if rapidly rendered harmless by the sea,[41] or if they appear as trace contaminant in wastes.[42]

3.2.1.6. Incineration at Sea

Another form of dumping involves incinerating wastes at sea. In 1981 the ODC added Annex IV as a Protocol addressing incineration at sea, which was signed by 11 Contracting Parties and ratified by 4 as of 1984. Most of the noxious substances incinerated are organohalogen compounds, i.e. Annex I substances.[43] The principle provisions include a definition of incineration, the incinerator ship and the incineration system; a decision of the Commission on a date to put an *end* to incineration at sea; and a strict limitation of the substances which may be incinerated.[44]

In 1978 the LDC adopted regulations covering incineration on the global level which are legally binding. One paragraph was added to each of LDC's Annexes I and II, which *allowed* incineration at sea of these substances subject to prior special permit, thus permitting an *exception* to the dumping of Annex I substances.[45] One author considers the criteria allowing incineration at sea to be similar in both Conventions, and notes that, '(T)he activity continues to be opposed by some LDC parties, and its prohibition sought'.[46]

3.2.1.7. ODC Reports - Practical Information

As noted previously the ODC is one of the few dumping conventions where fairly comprehensive information is published concerning its functioning. Though it is beyond the scope of this work to do an in depth coverage of these reports, it is considered beneficial to present an overview in order to show briefly current topics.

A quick look at the Oslo Commission Thirteenth Annual Report[47] shows reports of conferences including the International Conference on the Protection of the North Sea, progress and policy covering Amendments to Annexes I and II, and entry into force of the Annex on rules on incineration at sea. Concerning the latter specifically noted are a termination date, wastes containing PCB's and scenarios for the handling of organochlorine waste. Topics such as the disposal of sewage sludge and dredged material, the dumping of industrial wastes, methods of control of dumping and incineration operations, and the monitoring of dumping grounds are taken up as are the removal of offshore platforms, the transfrontier movement of industrial waste, the establishment of a North Sea Task Force, and monitoring.[48]

3.2.2. The Helsinki Convention[49]

The Helsinki Convention was formed in 1974 to control marine *pollution* in the Baltic Sea, the Gulf of Bothnia, the Gulf of Finland and the entrance to the Baltic.[50] One Article and 5 Annexes address *dumping* under the Helsinki Convention. The structure is different from the LDC and the ODC, with emphasis being placed in the Annexes rather than the Article, hence the

Annexes are complicated. Thus in addition to the Article the relevant Annexes will be briefly discussed to give an idea of their contents.

Article 9 covers specifically dumping and was *modelled* upon the LDC.[51] It is however much *stricter* and gives better protection than the LDC.[52] The definition of dumping in Article 2 for example *includes* exploration and exploitation of the seabed. In addition *all* substances are 'black listed' and thus prohibited from dumping in the convention area, with the exception of dredged spoils which can be dumped subject to a prior special authorization given by the appropriate national authority in accordance with Annex V. Each Contracting Party undertakes to ensure compliance with Article 9 by vessels and aircraft; registered in its territory or flying its flag, loading, within its territory or territorial sea, matter which is to be dumped, or believed to be engaged in dumping within its territorial sea. The provisions do not apply when the safety of human life or of a vessel or aircraft at sea is threatened similar to Article V of the LDC and Article 8(1) of the ODC. In cases of dumping which is suspected to be in violation of this Article, the State Parties must cooperate in investigating the matter in accordance with Annex IV.

Annex V, the Annex dealing exclusively with dumping, consists of 4 regulations. Regulation 1 deals with exceptions to the prohibition of dumping and provides that prohibition of dumping does not apply to the disposal at sea of dredged spoils *provided that* they do not contain significant quantities and concentrations of substances to be defined by the Organization and listed in Annexes I and II;[53] and provided that the dumping is carried out under a prior special permit issued by the proper national authority, either within the area of the territorial sea of the Contracting Party or outside the area of the territorial sea, whenever necessary, after prior consultations with the Organization. Regulation 2 deals with the issuance of permits and is essentially the same as appears in Article VI of the LDC and Article 6 of the ODC but is limited to permits for dumping dredge spoils.[54] Regulation 3 deals with factors the national authorities must take into account when issuing Regulation 1 permits and include much the same as noted above in Annex III of the LDC and ODC, but as related to *only* dredged spoils. Regulation 4 deals with information which must be included in the reports of dumpings required in accordance with the Article and include much the same as noted above under the LDC Article VI and ODC Article 11.

Annex I[55] lists the hazardous substances consisting of DDT (1,1,1-trichloro-2,2-bis-(chlorophenyl)-ethane) and its derivatives DDE and DDD; and PCB's (polychlorinated biphenyls).

Annex II lists noxious substances and materials including mercury, cadmium, and their compounds; antimony, arsenic, beryllium, chromium, copper, lead, molybdenum, nickel, selenium, tin, vanadium, zinc, and their compounds, as well as elemental phosphorus.[56]

Annex IV provides for appropriate State assistance in investigating violations of existing legislation on anti pollution measures which have occurred or are suspected to have occurred within the Convention area. This assistance includes the inspection by authorities of oil record books, cargo record books, log books and engine log books and taking oil samples for analytical identification purposes and in respect of the system of tagging oil residues.

Annex VI deals with State cooperation in combatting marine pollution and covers State contingency plans and bilateral or multilateral agreements, surveillance, salvage and recovery.

The Helsinki Convention also covers pollution from land-based sources, offshore installations and ships and aircraft, and the relevant provisions are presented in the respective Chapters.[57] For simplicity the general provisions of this Convention will be presented here since this is the fist Section dealing with the Helsinki Convention. Though the presentation will be brief, it will be somewhat involved, since this Convention is one of the first environmental conventions which is comprehensive and consequently has been used as a model. In later discussions of the Helsinki Convention reference can be made to this Section for these general rules.

As noted definitions are covered in Article 2. Besides defining 'pollution'[58] and the meaning of pollution from the various sources, definitions are also presented for 'oil,'[59] 'harmful substances'[60] and 'incident.'[61]

Article 3 sets forth fundamental principles and obligations whereby the State Parties must take all appropriate measures including legislative and administrative to prevent and abate pollution and to protect and enhance the marine environment of the Baltic. They must use their best endeavours to ensure that implementation of this Convention will not cause increases in pollution in areas outside the Baltic.

Article 4 specifies areas of application of the Convention and includes the water-body and sea-bed, living resources and other forms of marine life. It is restricted to each State's territorial sea, although provision is made for the States in their respective internal waters to undertake to ensure that the purposes of the Convention are also obtained, however without prejudicing their sovereign rights. Government non commercial ships including the military are exempted from the Convention although States must ensure that such ships and aircraft act consistently so far as is reasonable and practicable with Convention provisions.

Article 12 deals with the Convention's institutional and organization framework and sets up a Commission to carry out conventional objectives. The main duties of the Commission are set forth in Article 13 and include checking the implementation of the Convention; making recommendations; reviewing the contents of the Convention including the Annexes and recommending to the States necessary amendments; defining pollution control criteria and objectives of reducing pollution; and promoting research, informational dissemination and international collaboration in scientific and technological research. Various administrative, voting and financial rules are provided for in Articles 12, 14 and 15 while Article 16 deals with State scientific and technological cooperation.

The settlement of disputes if provided for by Article 18 and states that in case of dispute in the interpretation or application of the Convention the Parties should negotiate. If this fails they should seek the good offices or jointly request the mediation by a third State Party, a qualified international organization or a qualified person. If this fails, the dispute must upon common agreement be submitted, by common agreement, to an ad-hoc arbitration tribunal, to a permanent arbitration tribunal, or the ICJ.

Article 19 sets forth guarantees that the Convention does not infringe upon the freedoms of navigation, fishing, marine scientific research and other legitimate uses of the high seas, as well as innocent passage through the territorial sea.

The remaining Articles are for the most part general and deal with among other things amendment procedures for the Articles and Annexes, reservations, signature, ratification entry into force and withdrawal.

3.2.3. United Nations Environmental Programme (UNEP) Conventions - Barcelona Protocol II, South Pacific Protocol and Others

The UNEP Barcelona Convention for the Protection of the Mediterranean Sea against Pollution[62] (Barcelona Convention) was adopted in 1976 and is a comprehensive though very *general* Convention covering all pollution sources, liability and joint contingency plans. It is used as a model for the Arctic Convention appearing in *Section 11.4.*, where an idea can be obtained of its contents. Though the proposed Arctic Convention is more environmentally progressive, note has been made where the variations have been made. The UNEP Regional Seas Programme in addition to the Barcelona Convention encompasses 9 other regions, the Kuwait Action Plan Region, West and Central Africa, Wider Caribbean, East Asian Seas, Southeast Pacific, Southwest Pacific, Red Sea and Gulf of Aden, East Africa and Southwest Atlantic.[63] These Conventions are generally the same as the Barcelona Convention.

The Barcelona Convention has a sub agreement dealing with dumping, the Barcelona Protocol for the Prevention of Pollution of the Mediterranean Sea by Dumping from Ships and Aircraft[64] (Barcelona Protocol II). As with the

Helsinki Convention the Barcelona Protocol is modelled upon the LDC. The definition of dumping however resembles the ODC definition in that dumping associated with offshore oil activities is included, and the same exception exists concerning 'normal operations'.[65] The use of a 'black list'[66] and a 'grey list'[67] with the traditional jurisdictional system, prohibitions, permits and exceptions also resembles the LDC and the ODC with only minor differences.[68]

Of the other UNEP Conventions only one will addressed in relation to dumping, because as one author notes, '(T)he other regional conventions are much more *general*, and simply call on contracting parties to take all appropriate measures to prevent and reduce pollution caused by dumping and ensure effective compliance with the internationally agreed rules on dumping.'[69] Under these agreements should dumping be felt to be a problem, specific Protocols can be established.

Mention should be made however of the South Pacific Convention and Protocol. The South Pacific Convention prohibits the dumping of all radioactive waste including sub-seabed disposal which as seen in Section 3.2.1.1. is still under discussion in the LDC.[70] The South Pacific Protocol follows the progressive LOSC jurisdictional provisions on dumping rather than the traditional rules noted above in Section 3.2.1.1.[71] Thus both agreements, rather than being more general in controlling dumping than the LDC, are in fact more progressive.

In addition to the UNEP Conventions mention should be made of the Multilateral Consultation and Surveillance Mechanism for Sea Dumping of Radioactive Waste established by the Council of the Organization for Economic Co operation and Development (OECD) in 1977.[72] This Decision covers the creation of guidelines and recommendations concerning dumping of radioactive waste by OECD States, procedures for notification and consultation concerning proposed dumping, international surveillance, and recording and reporting of dumping.

3.3. Conclusion

As perhaps noted that which is interesting with the present world wide dumping regime is the *duplicity*. The LDC is global and yet has been used as a model for many of the regional conventions. One might ask what is the point in having parallel systems in different regional areas unless the regional agreements establish stricter provisions than the LDC. This is what has occurred in the cases of the Helsinki Convention and the South Pacific Convention and Protocol and to a lesser extent the ODC, but not at all in the cases of the remaining UNEP Conventions. As noted the way is open for developing UNEP Protocols with stricter dumping provisions, however at present most States seem generally content with the dumping regime established by the LDC. The parallel system can perhaps be explained by this 'open endedness.' States not really interested in strict regimes can nevertheless put forth a good face by being a Party to the LDC, a moderately successful environmental convention. Should a specific dumping problem develop, resort can then be made to a regional agreement, either under the UNEP system or independently as in the cases of the Helsinki Convention and the ODC. This type of State practice while very typical in international law is unfortunate in environmental law since it can allow questionable dumping practices to continue, albeit on a reduced scale, through the special and general permits system covering 'grey list' and other substances, through incineration at sea and through the trace contaminants exception. It is only *after* the problems become too large as is what happened in the Baltic Sea that really environmentally strict regimes are established.

Notes - Chapter 3

1 Churchill and Lowe p. 214.

2 For some of the regional conventions with general provisions which follow, the specific ratifying States are not presented due both to the lack of space and the information not being readily available. See Johnston pp. 337-353 for information regarding regional *programs* and arrangements dealing with conservation and environmental protection including the Food and Agricultural Organization (FAO); the Economic and Social Council (ECOSOC), including the Economic and Social Commission for Asia and the Pacific (ESCAP), the Economic Commission for Latin America (ECLA), the Economic Commission of Africa (ECA) and the Economic Commission for Western Asia (ECWA); the United Nations Educational, Scientific and Cultural Organization (UNESCO); the United Nations Development Programme (UNDP); the United Nations Regional Training Programmes including the United Nations University (UNU); the World Health Organization (WHO); and the International Atomic Energy Agency (IAEA). Regional programs for Asia and the Pacific are as well dealt with. These will not be covered here due to the level of *generality* involved. Churchill for example does not even mention these arrangements in his book.

3 Intergovernmental Conference on the Convention on the Dumping of Wastes at Sea (LDC), 11 *International Legal Materials* 1294, (1972). Amendments to the Convention and to the Annexes, October 12, 1978. Amendments to the Annexes, September 24, 1980. Ratifying Parties include as of December 1988 Afghanistan, Argentina, Australia, Belgium, Byelorussia, Brazil, Canada, Cape Verde, Chile, China, Cuba, Denmark, Dominican Republic, Democratic Republic of Germany, Federal Republic of Germany, Finland, France, Gabon, Greece, Guatemala, Haiti, Honduras, Hungary, Iceland, Ireland, Italy, Japan, Jordan, Kenya, Kiribati, Libya, Mexico, Monaco, Morocco, Nauru, Netherlands, New Zealand, Nigeria, Norway, Panama, Papua New Guinea, Philippines, Poland, Portugal, South Africa, Spain, Suriname, Sweden, Switzerland, Tunisia, Ukraine, USSR, United Arab Emirates, United Kingdom, United States, Yugoslavia, and Zaire. International Maritime Organization, 'Status of Multilateral Conventions and Instruments in Respect of which the International Maritime Organization or its Secretary General Performs Depositary or Other Functions as of 31 December 1988,' J/2735/Rev.3, (IMO

Status) pp. 353-364; and De Cesari, P. et.al. (eds.) *Index of Multilateral Treaties on the Law of the Sea*, Milano, Dott. a. Giuffrè Editore 1985. (De Cesari) pp. 281-292. Though the Convention is not normally applicable to the internal waters of the Parties, Hakapää p. 99 notes Articles I and XII, state that Parties shall 'promote the effective control of all sources of pollution of the marine environment', which would seem to include internal waters. Additionally noted is that Article I requires Parties to act in internal waters in such a way that possible dumping will not spread pollution to waters beyond.

4 Convention for the Prevention of Marine Pollution by Dumping from Ships and Aircraft (ODC), 11 *International Legal Materials*, 262 (1972). This Convention covers according to Article 2 the Parties' territorial seas and the high seas situated north of 36 degrees latitude, and between 42 degrees west longitude, and 51 degrees east longitude, (excluding generally the Baltic, the Mediterranean and the north Atlantic between 59 degrees north and 44 degrees west and 42 degrees west longitude). Ratifying Parties include Norway, Denmark, Sweden, Portugal, Spain, Iceland, France, the United Kingdom, the Netherlands, the Federal Republic of Germany, Belgium, Finland, Ireland and EEC, Oslo-Paris Comm., p. 6. The Soviet Union was invited to join but declined, 11 *International Legal Materials* 262, (1972). The *Soviets* however were again invited by Norway in 1989 and are expected to *accede* soon, thus bringing the Soviet Barents Sea east to 51 degrees east longitude under ODC auspices, conversation Chief Consultant Per Antonsen Norwegian Department of the Environment November 1989. A press release of 24 November 1989 from the Norwegian Ministry of Foreign Affairs notes that an amendment will be open for signature the 5 December 1989 for including internal waters within the geographical coverage of the ODC. This is because it is mostly in internal waters pollution from the dumping of dredged spoils occurs.

5 Johnston p. 225 notes that the LDC and the ODC were formed almost simultaneously with the same governments interested in the ODC also participating in developing the LDC, and thus, '(T)he ODC was heavily influenced by earlier drafts of the LDC and the LDC borrowed its black list/grey list format from the ODC.' The Paris Convention, 13 *International Legal Materials* 353, (1974), covering land-based pollution as will be seen in Part I Section 4, in Article 3 (c) also controls pollution from off-shore platforms involved in the exploration and exploitation of sea-bed resources, and thus might have some application to dumping from these platforms. It does not however include any specific dumping provisions, and in addition, basically applies to the same geographic area as the ODC which was established especially to cover dumping. As such the Paris Convention will only be mentioned in this connection generally.

6 See ODC, Article 19 (1)(a) and (2) emphasis added. Hakapää p. 98 notes that ODC Article 19 (1) also seems to exclude from its definition the disposal of ships or aircraft where not also loaded with substances and materials to be disposed of, through use of the words 'deliberate disposal of substances and materials into the sea *by* or *from* ships or aircraft,' (emphasis added).

7 The Oslo and Paris Commissions, *The First Decade*, International Cooperation in Protecting the Marine Environment 1984 (Oslo-Paris Comm.) p. 18.

8 Hakapää p. 98. The author was referring to only the LDC, however due to the 2 Conventions' similarities the ODC can probably be included as well in this characterization.

9 Churchill p. 268 notes that the 1958 High Seas Convention Article 25 (1) even before the LDC came into effect required States to take measures to prevent dumping pollution from radioactive waste. The LDC considers *seabed burial* of high level radioactive waste separately because some countries including the US do not consider it as a form for dumping. *Council on Ocean Law*, Ocean Policy News, The Newsletter of International Ocean Law, 1709 New York Ave., N.W. Washington D.C. 20006 USA. January/February 1988 (OPN 1988) p. 6. See Calmet, D.P. and J.M. Bewers, 'Radioactive waste and ocean dumping, The role of the IAEA', *Marine Policy*, Vol 15, Number 6, November 1991, (Calmet), p. 413 who argue that the present trend through the LDC and other regional conventions is towards the prohibition of the dumping of any radioactive waste into the marine environment. This however is in opposition to IAEA

guidelines for dumping which complies with radiological protection principles and could result in the transfer of the problem from one sector to another without ensuring overall benefits to the environment and man.

10 Article IV (1)(b) and (2) allows the dumping of Annex II substances only upon prior special permit issued only upon careful consideration of factors appearing in Annex III including the total amount, toxicity, location of dumping site, method of packaging, water characteristics, existence and effects of other dumpings, possible effects and the practical availability of alternatives, as well as prior studies of the characteristics of the dumping site.

11 There is at present a voluntary moratorium in effect among LDC Parties covering the disposal of low-level radioactive waste, which will remain in effect pending results of studies covering scientific and technical questions and socio economic risks, OPN 1988 pp. 5-6.

12 Article IV(1)(c) and (2) allows the dumping of all other wastes or matter only upon a prior general permit issued only upon careful consideration of the Annex III factors listed above, as well as prior studies of the characteristics of the dumping site.

13 See Oslo-Paris Comm. p. 23, where it is noted that a recommendation was made to include organosilicon compounds in the List of Hazardous Substances and not in Annex I.

14 ODC, Annex II. Note that all parties to the ODC, with the exception of Ireland, are also parties to the LDC, which would indicate that the *stricter* provisions of *either* Convention would apply in the ODC areas of application for all the Parties but Ireland.

15 Under LDC, Article VII (1), the State obligation is denoted by the phrase, 'shall apply the measures required to implement the present Convention'. Under ODC, Article 15 (1), the State obligation is stated, 'undertakes to ensure compliance with the provisions of this Convention.'

16 Article 15 (1).

17 Article VII (1).

18 Hakapää p. 114. See Article XIII and VII(3). As noted in Part I Section 2.1.2. Churchill pp. 269-270 notes that the LDC is somewhat imprecise regarding enforcement, but the general rules of customary law would apply under which a coastal State can take action against a ship illegally dumping in its territorial sea and against any ship about to leave its port for dumping without the proper permit. As will be seen in Part II Section 15.1.2. LOSC Article 210 (5) defines coastal State jurisdiction covering dumping in that State's territorial sea, economic zone or onto its continental shelf, which would seem to allow under LDC Articles XIII and VII (3) the use of the same jurisdictional system. See also OPN 1988 p. 6, where at a meeting of legal experts associated with LDC it was agreed, 'that the geographic scope of coastal State authorities should now be assumed to comport with the jurisdictional concepts set forth in the LOSC treaty.'

19 Article VII (2) and Article 15 (3) respectively.

20 Hakapää p. 115, who notes as an example industrial and domestic dumping.

21 This seems somewhat odd considering the ODC's sister convention, the Paris Convention Article 21 Annex B, makes provisions for compulsory arbitration in addition to negotiation. See Part I Section 4.2.2.5.

22 In 1979 Article XI was amended to read, '(A)ny dispute between two or more Contracting Parties concerning the interpretation or application of this Convention shall, if settlement by negotiation or other means has not been possible, be submitted by agreement between the parties to the dispute to the International Court of Justice or upon the request of one of them to arbitration. Arbitration procedures, unless the parties to the dispute decide otherwise, shall be in accordance with the rules set out in the Appendix to this Convention', 18 *International Legal Materials* 516, (1979). The rules

in the Appendix are patterned after Protocol II of MARPOL 73, but improve upon it according to Timagenis p. 282 by placing emphasis on the possibility of a sole arbitrator procedure and allowing for a 30 day period for appointing arbitrators.

23 Birnie p. 232.

24 Marine Environmental Protection Committee of IMO 25 Session Agenda Item 19, MEPC 25/19, July 1, 1987 (MEPC) pp. 1-2. The Eleventh Consultative Meeting (October 3-7 1988) began discussion covering the dumping of oil platforms at sea, correspondence with IMO, February 29, 1988. Recommendations have been submitted by Friends of the Earth International and ACOPS; and the FAO and UNEP will possibly be contacted in the future. The Draft Guidelines being discussed at the IMO brought forth by Norway, the U.K. and the US require installation removal except where non or partial removal is consistent within certain standards and guidelines, based upon factors including effects on navigation, rate of material deterioration, effects on the environment, costs and risks of injury associated with removal and the determination of a new use. Additionally proposed is that all abandoned installations standing in less than 75 meters of water and weighing less than 4000 tons, excluding deck and superstructure, should be entirely removed. By 1998 all abandoned installations standing in less than 100 meters and weighing less than 4000 tons, excluding deck and super structure, should additionally be removed.

25 Convention on the Continental Shelf (1958 Continental Shelf Convention) I *New Directions in the Law of the Sea* 101, emphasis added. See Part I Section 4.2.1.

26 LOSC Article 60 (3) provides that, '(A)ny installations or structures which are abandoned or disused must be removed to ensure safety of navigation, taking into account any generally accepted international standards established in this regard by the competent international organization. Such removal must also have due regard to fishing, the protection of the marine environment and the rights and duties of other States. Appropriate publicity must be given to the depth, position and dimension of any installations or structures not entirely removed.' See Part I Section 2.1.4. for discussion of LOSC provisions and their relation to international law. Birnie p. 232 believes that this provision may 'offer a somewhat vague possibility of modification' of the 1958 Continental Shelf Convention provision, though which international organization is the competent international body is unclear, though the IMO is preferred over the Oslo-Paris Commission since it is global.

27 Birnie p. 232.

28 *Ibid.* pp. 232-233 notes however that the latter use raises questions of non interference with other uses of the sea as required under the 1958 Continental Shelf Convention Article 5 (1) and LOSC Article 58 (3).

29 Johnston p. 223.

30 *Ibid.* p. 226. See Part I Section 4.2.2.4.1. for more information regarding the ODC-Paris Convention Joint Monitoring Program.

31 Johnston p. 226. See also Birnie p. 233, who states additionally that publication of the results of the monitoring programs is slow and that there is room for improvement. Churchill p. 272 and Roginko, Alexei, 'Arctic Environmental Cooperation: Prospects and Possibilities,' Institute of World Economy and International Relations, USSR Academy of Sciences, Prepared for the International Studies Association Annual Convention, London, March 1989, (Roginko) pp. 26-27 equally note problems obtaining information.

32 Oslo-Paris Comm., p. 368.

33 Timagenis p. 161 notes the assistance refers to investigation of and violation proceedings, but also includes clean-up and mitigating the effects of the pollution.

34 Under ODC Article 9 the Oslo Commission must make certain recommendations in cases of emergency. Both the Oslo Commission and the IMO were contacted regarding contingency plans under the ODC and the LDC respectively. No reply was received from the Oslo Commission, but Mr. Manfred Nauke, Chief of the Marine Science Section of the IMO in a correspondence of February 25, 1988 replied that the LDC Articles, 'have so far not been interpreted by Contracting Parties as setting up contingency plans with a view to clearing up pollution from dumping'. Noted however was that in the event of a major pollution incident due to dumping, 'support would be provided through this Organization (IMO) in cleaning up the respective pollution', parentheses added.

35 Under sovereign immunity although foreign warships cannot be arrested, they can be expelled from coastal waters due to environmental violations.

36 Hakapää p. 160.

37 See for example MARPOL 73/78 Article 3 (3) and LOSC Article 236.

38 See Hakapää p. 160.

39 LDC, Article V (1); ODC Article 8 (1).

40 LDC, Article V (2); ODC, Article 9.

41 LDC, Annex I (8).

42 LDC, Annex I, (9); ODC, Article 8 (2).

43 Oslo-Paris Comm. pp. 20-21.

44 *Ibid.* p. 22. Churchill p. 271 notes the ODC as the only regional convention to deal with incineration through its Protocol.

45 See Timagenis p. 227. The resolution appears at 18 *International Legal Materials* 508 (1979).

46 Birnie p. 234.

47 Oslo Commission Thirteenth Annual Report on the Activities of the Oslo Commission, (1989), Oslo Commission, New Court 48 Carey Street, London, WC2A 2JE, (Oslo Comm.).

48 The latter specifically consists of an assessment of the Joint Monitoring Programme for 1985 and 1986, monitoring of seawater, biological effects of monitoring, future developments, analytical problems and intercalibration exercises and monitoring of nutrients.

49 The Convention on the Protection of the Marine Environment of the Baltic Sea Area, 13 *International Legal Materials* 546. The Convention entered into force May 1980 with Denmark, Finland, German Democratic Republic, Federal Republic of Germany, Poland, Sweden and the USSR as ratifying Parties, Johnston p. 226.

50 Article 1.

51 Johnston p. 226.

52 Churchill p. 271 notes that the Helsinki Convention is in fact much stricter than *any* of the dumping agreements. Johnston p. 227 notes that the higher level of protection offered by this Convention was necessary due to the semi-enclosed Baltic Sea's past vulnerability to dumping.

53 Thus as Churchill p. 271 notes even the dumping dredged spoils is prohibited if they contain significant Annex I and II trace contaminants.

54 See Section 3.2.1.

55 Annexes I and II under the Helsinki Convention are lists as well for the other Articles dealing with other pollution sources.

56 Further Annex II noxious materials include phenols and their derivatives; phthalic acid and its derivatives; cyanides; persistent halogenated hydrocarbons; polycyclic aromatic hydrocarbons and their derivatives; persistent toxic organosilicic compounds; persistent pesticides, including organophosphoric and organostannic pesticides, herbicides, slimicides and chemicals used for the preservation of wood, time, wood pulp, cellulose, paper, hides and textiles, not covered by the provisions of Annex I; radioactive materials; acids, alkalis and surface active agents in high concentrations or big quantities; and oil and wastes of petrochemical and other industries containing lipid-soluble substances, substances having adverse effects on the taste and/or smell of products for human consumption from the sea, or effects on taste, smell, colour, transparency or other characteristics of the water seriously reducing its amenity values; materials and substances which may float, remain in suspension or sink, and which may seriously interfere with any legitimate use of the sea; lignin substances contained in industrial waste waters; the chelators EDTA (ethylenedinitrilotetraacetic acid or ethylenediaminetetraacetic acid) and DTPA (diethylenetri - aminopentaacetic acid).

57 See Section 4.2.3.1. for land-based pollution, Section 4.2.4.1. for offshore installation pollution, Section 5.5. for vessel-source pollution, Section 6.2 for liability, and Section 7.6. for contingency plans under the Helsinki Convention.

58 'Pollution' under the Helsinki convention means the introduction by man, directly or indirectly, of substances or energy into the marine environment, including estuaries, resulting in such deleterious effects as hazard to human health, harm to living resources and marine life, hindrance to legitimate uses of the sea including fishing, impairment of the quality for use of sea water, and reduction of amenities.

59 'Oil' is defined as petroleum in any form including crude and fuel oil, sludge, oil refuse and refined products.

60 'Harmful substance' is defined as hazardous, noxious, or other substance, which if introduced into the sea is liable to cause pollution.

61 'Incident' is defined as an event involving the actual or probable discharge into the sea of a harmful substance, or effluents containing such a substance.

62 15 *International Legal Materials* 290. State Parties include Algeria, Cyprus, Egypt, the European Economic Community (EEC), France, Greece, Israel, Italy, Lebanon, Libya, Malta, Monaco, Morocco, Spain, Syria, Tunisia, Turkey, and Yugoslavia. De Cesari pp. 325-326.

63 UNEP, 'Guidelines and principles for the preparation and implementation of comprehensive action plans for the protection and development of marine and coastal areas of regional seas, *UNEP Regional Seas Reports and Studies* No. 15 (1982), p. 1. The Conventions and Protocols can be ordered from UNEP Headquarters, United Nations Environment Programme, P.O. Box 30552 Nairobi, Kenya. See Johnston pp. 324-337 and ACOPS Yearbooks 1986/87 pp. 20-25 and 1987/88 pp. 21-25, for information on developments in specific UNEP areas.

64 15 *International Legal Materials* 300. State Parties to the Protocol include all State Parties to the Barcelona Convention with the exception of Israel. De Cesari pp. 325-326.

65 See Section 3.2.1.

66 The 'black list' Annex I of the Barcelona Convention consists of organohalogen compounds and organosilicon compounds (except those which are non toxic or rapidly converted which are biologically harmless), mercury and mercury compounds, cadmium and cadmium compounds, persistent plastic and other synthetic materials, crude oil and its derivatives, high, medium and low level radioactive wastes and matter, acid and alkaline compounds which may seriously impair the quality of sea water, and biological and chemical warfare materials in whatever form (other than those rapidly rendered

harmless). Trace contaminants of the above are not meant to be included as 'black list' substances.

67 The 'grey list' Annex II includes arsenic, lead, copper, zinc, beryllium, chromium, nickel, vanadium, selenium, antimony and their compounds, cyanides and fluorides, pesticides and their by products not appearing in Annex I, synthetic organic chemicals other than those covered in Annex I, acid and alkaline compounds the composition and quantity of which have not yet been determined in accordance with Annex I, acid and alkaline compounds not covered by Annex I, containers, scrap metal and other bulky wastes which may impair fishing or navigation, nontoxic substances which may become harmful due to the quantities dumped, and radioactive waste or matter not included in Annex I. Concerning the latter in issuing permits for such dumping, full account should be made of Atomic Energy Agency recommendations.

68 See Part I Section 3.2.1. Churchill p. 271 additionally notes that institutional arrangements differ with each agreement.

69 Churchill p. 271 emphasis added.

70 *Ibid.* notes that the 1985 Treaty of Raratonga also contains a similar prohibition.

71 See Part II Section 11.2. and 15.1.2. for LOSC dumping provisions.

72 17 *International Legal Materials* 445-452. See Churchill p. 271.

4. Land-based Pollution - Offshore Oil and Gas Installations - Airborne

4.1. Introduction

Pollution from land-based sources refers to contamination including energy introduced into the seas through water or through the atmosphere from land-based activities.[1] As noted in Part I Section 1.3.1. this source is the *largest* source of marine pollution. Due to pollution from *offshore* oil and gas exploration and exploitation being joined with the land-based pollution regime through the Convention most relevant to the Barents Sea Case study in Part II, the Paris Convention,[2] and airborne pollution addressed in a Protocol (not in force), for clarity all three will be considered together in this Section. This is in spite of the LOSC procedure for separating these regimes as noted in Part I Section 1.1.[3]

4.2. Conventional Law[4]

Pollution from land-based sources percentage wise is the most extensive, yet is covered by the least amount of international agreements.[5] There is *no* global convention covering specifically this pollution source. The 1958 Continental Shelf Convention[6] is global but covers on offshore installations and has as a rule general provisions. The Paris Convention covers pollution from land-based sources and offshore platforms in the North Sea and Barents Sea areas. See Part II Section 9.3.1. for problems associated with application of this Convention. The Helsinki Convention covers pollution from land-based sources and offshore installations in the Baltic. The UNEP Barcelona Protocol III[7] and Southeast Pacific Protocol[8] cover specifically land-based pollution in their respective regions while the other UNEP Conventions noted in Part I Section 3.2.3. contain only general provisions. In Europe there also exist some regulation of land-based sources by the EEC and some river agreements as well as the Nordic Convention.[9] There exist several bilateral agreements covering pollution from offshore installations including Norwegian and British agreements[10] and a 1983 Canadian - Danish Agreement.[11] Finally airborne pollution is covered specifically by the Helsinki Convention and Barcelona

Protocol III as a part of land-based pollution. In addition the ECE Convention[12] entered by European States and Canada and the US apply as do the LDC and the ODC cover incineration at sea. Various bilateral agreements are in existence between States touching upon airborne pollution including agreements between the Soviet Union and Finland, Norway, Sweden, Canada and the US.[13]

4.2.1. The 1958 Continental Shelf Convention[14]

Important provisions from the 1958 Continental Shelf Convention relevant to pollution from offshore oil activities include the following. The coastal State has exclusive sovereignty under Article 2 (1), (2), and (4) over its continental shelf for the purpose of exploring it and exploiting its natural resources, including mineral and non-living, and living sedentary species. The continental shelf is defined under Article 1 to extend, 'outside the area of the territorial sea[15] to a depth of 200 meters or, beyond that limit to where the depth of the superjacent waters admits of exploration of the natural resources of the said areas.'[16] Under Article 4 the coastal State may not impede the laying or maintenance of submarine cables or pipe lines, subject to reasonable measures of exploration and exploitation on its continental shelf.[17]

Article 5 (1) and (6) requires that exploration and exploitation and consequent installations on a coastal State's continental shelf may not result in unjustifiable interference to navigation. Neither the installations nor safety zones around them may be established where interference may be caused to recognized international sea lanes. Article 5 (2) and (3) entitles the coastal State to construct and operate installations on its continental shelf for exploration and exploitation of the natural resources, complete with safety zones up to 500 meters around the installations.[18] Ships of all nationalities must respect these zones. Under Article 5 (5) due notice must be given of installation construction and as well as a permanent means of warning.[19] Any installations which are abandoned or disused must be *entirely* removed.[20] Article 5 (7) requires the coastal State in the safety zones to take appropriate measures to protect living resources from harmful agents. Under Article 5 (8) other non resource oriented research related to the continental shelf may be controlled.

Under Article 6 (1) and (2) in determining boundaries of the continental shelf, in the absence of an agreement, the equidistance principle is to be used unless special circumstances dictate otherwise.[21]

4.2.2. The Paris Convention[22]

The convention governing land-based pollution and offshore platforms in the North Sea and the Barents Sea is the *regional* Paris Convention (in addition to the 1958 Continental Shelf Convention). The Paris Convention is covered in most detail in this Chapter because it was one of the earliest international convention formed covering land-based pollution and later agreements contain similar provisions, and because it is the most relevant Convention in relation to the case study area.

4.2.2.1. Definition of Marine Pollution from Land-based Sources

'Pollution from land-based sources' is defined in Article 3 (c) to mean pollution of the maritime area through watercourses, from the coast, including through underwater or other pipelines, and from man-made structures placed under the jurisdiction of a Contracting Party within the area of application of the Convention. As seen the Paris Convention somewhat paradoxically through this Article includes pollution from man-made structures under State Party jurisdiction, within its definition of 'pollution from land-based sources', and thus pollution from offshore oil and gas platforms. Noting this one author comments, 'as far as sea-bed resource activities are concerned, the Convention's provisions are *not* particularly well-adapted for preventing 'platform-oriented' releases, often accidental in character.'[23]

4.2.2.2. Jurisdiction

The Paris Convention applies to the 'maritime areas' within the coordinates noted above. 'Maritime areas' are defined by the Convention to include the high seas, the territorial seas of Contracting Parties and waters on the landward side of the base lines from which the territorial seas are measured, extending in watercourses up to the freshwater limit.[24] 'Freshwater limit' is defined, 'as the place in the water-course at low tide and in a period of low freshwater flow, there is an appreciable increase in salinity due to the presence of sea-water'.[25]

The Contracting Parties must adopt individually and jointly measures to combat marine pollution from land-based sources and to harmonize their policies.[26] Further each Party undertakes to ensure compliance with the provisions of the Convention and to take in its *territory* appropriate measures to prevent and punish violative conduct.[27] In order to further define '*territory*' resort must be made to the 1958 Continental Shelf Convention above Articles 1, 2 (1) and 5 (7).[28]

In the areas noted generally the Contracting Parties *undertake* to eliminate by stages if necessary, and to establish programs and measures for the elimination as a matter of urgency, pollution of 'black list' substances.[29] These substances include organohalogen compounds,[30] mercury and mercury compounds, cadmium and cadmium compounds, persistent synthetic materials[31] and persistent oils and hydrocarbons of petroleum origin.[32]

The Parties likewise undertake to limit strictly, and to establish programs and measures for the reduction or as appropriate, elimination of pollution by 'greylist' substances.[33] The discharge of such substances is to be regulated by authorities of the Contracting State, and such approval must be periodically reviewed.[34] 'Greylist' substances include organic compounds of phosphorous, silicon and tin,[35] elemental phosphorus, non-persistent oils and hydrocarbons of petroleum origin, elements and compounds of arsenic, chromium, copper, lead, nickel, and zinc, and substances agreed by the Commission as having a deleterious effect on the taste and/or smell of products derived from the marine environment for human consumption. Licensing is required only for the discharge of these substances.

Programs or measures to prevent, reduce or eliminate land-based pollution from substances not listed in Annex A may be implemented if scientific evidence establishes a serious hazard may be created and urgent action is needed.[36] States *undertake* to consult affected States concerning proposed discharges of 'prejudicing' substances not appearing on the 'blacklist' with the object of negotiating a cooperation agreement,[37] however as of 1983 no agreements had been concluded.[38] Additionally, the Parties undertake to adopt measures to forestall and as appropriate eliminate pollution by radioactive substances, including wastes.[39]

4.2.2.3. General Provisions

Besides the principle of eliminating, reducing or preventing pollution from land-based sources, the Paris Convention also includes the principle of joint action.[40] Under Articles 15 and 16 a Commission is formed with duties including the supervision of overall Convention implementation and the establishment of the Convention's joint measures.[41] The Commission is required to adopt programs and measures for the reduction or elimination of pollution from land-based sources and for advancing scientific research and monitoring.[42] This is to be adopted by unanimous vote, and if not possible, 'the Commission may nonetheless adopt a program or measures by a three quarters majority vote of its members.'[43] These programs and measures apply for those Parties which voted for them. Similarly the Commission has a duty to adopt recommendations for amendments to Annex A and submit them to the Governments of the Contracting Parties.[44] Any Government unable to approve an amendment must notify the Depositary Government in writing, and should no such notifications be received, the amendment will enter into force for *all* Contracting Parties.[45]

Other duties of the Commission include reviewing the condition of the seas within the scope of the Convention, reviewing the effectiveness of the control measures and the need for any additional or different measures;[46] fixing limits of maritime areas;[47] making recommendations if requested in cases of pollution from 'prejudicing' non 'blacklist' substances to reach agreement between the Parties;[48] and receiving, reviewing and distributing information covering monitoring, pollution technology and legislative and administrative measures taken by Parties to enforce the Convention.[49]

Under Article 24 the Paris Convention is open for signature to *upstream* States in addition to coastal States whose waters flow into protected areas in order to attempt to solve the problem of polluting upstream neighbours.[50]

4.2.2.4. Provisions covering Oil and Gas

Although various mainly 'blacklist' substances including mercury, PCB's, PCT's, cadmium, aldrin, dieldrin, endrin, persistent synthetic floating materials and titanium dioxide, have been under study and review by the Paris Commission,[51] the work carried out concerning pollution from offshore oil and gas activities in the North Sea will be of focus here due to its relevance to the Barents Sea case study. This work is also presented to give an example of the complexity of pollution control.

In 1978 the Paris Commission recommended that only waste or used oils appear under Part I of Annex A 'blacklist' and classified the other hydrocarbons as being 'greylist' substances.[52] Further it was decided to classify the pollution into 4 groups according to sources, offshore oil and gas production *platforms*, oil *refineries*, oil waste *reception* facilities and *other* sources of oil pollution from land-based sources. It was decided that the Working Group on Oil Pollution (GOP)[53] should initially consider only the first three sources, with the exception of the regeneration and disposal of waste oils.[54] Following these classifications, each category will be considered briefly below.

4.2.2.4.1. Offshore Oil and Gas Platforms

The Paris Commission agreed soon after its formation in 1978 upon a target standard of *40 mg/l* for oil discharges from offshore platforms.[55] This is a *goal* and not a maximum value, and the Commission recommended that national authorities fix a limit of allowable quantity of discharge from each platform. Note the non binding character of the State obligations. The Commission additionally recommended to Contracting Parties that the 40 mg/l standard of hydrocarbons in platform discharges should be applied to all platforms constructed after January 1, 1988, and if the national standard is not the same or less lenient, then the Party setting the standard must inform the Commission of the reason for its decision.[56] A proposal for a *binding* limit of 40 mg/l for new platforms built after January 1, 1988 was not accepted. The Commission did decide however it was 'no longer appropriate to refer to a 'provisional' target standard for discharges from existing platforms'. It was made clear that this standard covers effluents not covered by MARPOL 73/78, formation water, ballast water and drainage water from the production unit.[57] Comprehensive reports covering rig discharges were to be prepared every three years, as well as brief annual reports which are prepared by the Secretariat. Exploratory activities were also considered and annual reports required regarding oil spills occurring during these activities.

Studies were also conducted covering pollution from drilling fluids including oil-based drilling muds. It was initially decided not to undertake environmental monitoring routinely in the proximity of each rig, but rather to take measurements annually in 'major input zones'.[58] An Ad Hoc Working Group on Test Methods for Oil Based Muds was formed in 1986-87 with the aim of preparing test methods for oil-based muds which could be accepted by all Contracting Parties.[59] The Commission agreed that GOP should continue reviewing the use of chemicals in drilling operations and took note of GOP's conclusions that surfactants as used in 1986 were not a serious environmental problem.[60] The work has centred around studying the polluting effects of discharges resulting from exploration activities and oil based drilling muds and has resulted in the adoption of several 'Par.Comm. Decisions', which involve prohibiting the use of diesel based drilling muds and restricting the use of oil based muds generally.[61] The Commission also agreed upon methods of sampling and analysis to apply the target standard.

As noted in Part I Section 3.2.1.4. under both the Paris Convention and the ODC there is a duty for both the Commissions and the Parties to monitor the seas within their respective jurisdictions.[62] Joint contingency plans are not covered however. The two Commissions in 1978 set up a Joint Monitoring Program (JMP), and initially the activities were limited to examining mercury, cadmium and the PCB's in organisms and to mercury and cadmium in seawater. These continue to be the main areas of interest.[63] Additionally reports have been prepared covering such contaminants as nutrients, nuclear discharges, titanium dioxide, mercury from the chloralkali industry and platform discharges.[64]

Norway is the *lead* country under the Paris Convention in examining methods for monitoring contaminants in the vicinity of offshore platforms.[65] The methods of platform monitoring as well as results are found in reports published by SFT, and will briefly be outlined here since these represent the 'lead' in platform monitoring.[66]

Continuous discharges from platforms occur through drilling mud or production water, and in Norway oil based mud must be tested and approved for use. Oil must be cleaned from the drilling mud and production water before they are discharged in accordance with established limits. For oil based mud and the cuttings, the limit is 10% in dry weight. Water based mud and the cuttings can be discharged uncleaned. *Daily* testing is required of the mud and production water (40mg/l) to ensure compliance, and SFT additionally requires periodic biological and chemical examinations to be carried out studying the effects these discharges have upon the marine environment to ensure there is no unnecessary damage.[67] The operator of each oil field is responsible that the programs be carried out, but the studies themselves must be conducted by *independent* institutions. Initially an environmental impact statement is required covering the drilling area previous to even exploratory drilling and at least one year before production starts subject to approval by SFT.[68] After production starts and as long as the oil field is in operation, periodically monitoring studies must be carried out. These are to be based upon the original environmental impact statement and be of such a format that results from the different oil fields can be compared.

The standard frequencies for monitoring programs suggested are as follows:

Oil and condensate field

Drilling phase	Chemistry	Once a year
	Biology	Once every 3rd year
Production phase	Chemistry	Once every 3rd year
	Biology	Once every 3rd year

Gas field[69]

Drilling phase	Chemistry	Once a year
	Biology	Once every 3rd year
Production phase	Chemistry	Once every 6th year
	Biology	Once every 6th year

The sampling stations are to be arranged in a *grid* orientated in the direction of the residual current or if no current is established, in the most likely current direction. The monitoring stations are to be placed on perpendicular axes with the platform in the centre, and one axis parallel to the current. If possible one station should be 10,000 meters but not less than 6,000 meters from the platform in the opposite direction of the residual current. Where there are several installations, another grid system determined by the State authorities may be more suitable. The precision of locating a station is to be within limits of 50 to 100 meters. Sampling is to be performed in the northern areas between

April 1 and June 15th with SFT receiving the report prior to February the following year.[70]

The standard monitoring program must include the following.[71]

Chemical surveys

Descriptive Sedimentology:	Visual description of the sediment surface, odour, colour, total organic carbon content (TOC), water content, Eh.
Hydrocarbons:	Total hydrocarbon content (THC), naphthalene, phenanthrene/ anthracene, dibenzothiofene and their C3 alkyl homologues (NPD), bicyclic aliphatic hydrocarbons, e.g. decalins, n-C17/pristane to n-C18/phytane ratio, 2-6 ring aromatics.
Metals:	Barium (Ba), Cadmium (Cd), Copper (Cu), Lead (Pb), Zinc (Zn), Mercury (Hg), Iron (Fe); If desirable; Cobalt (Co), Nickel (Ni), Vanadium (V), Chrome (Cr), Strontium (Sr), Aluminum (Al).

Biological surveys

Descriptive Sedimentology:	Particle size distribution.
Zoobenthos:	Number of species per sediment volume or surface area, number of individuals per species.

Additional parameters

At the discretion of the State authorities the following parameters may be added but are not recommended for use in the standard monitoring program because of the large variation and interpretation difficulties related to results obtained.

Fish:	THC, NPD, bicyclic hydrocarbons in muscle and liver, individual age and size, sample size and statistical analysis.
Crustaceans:	THC, NPD, bicyclic hydrocarbons in tissue, individual age and size.

Mussels:	THC, NPD, bicyclic hydrocarbons in tissue, individual age and size, used for monitoring effects of processed water.
Degradation of hydrocarbons:	Biodegradation capacity.

4.2.2.4.2. Refineries

Target guidelines were also approved by the Commission covering discharges from *new* refineries, including *5 mg/l* of oil levels in process waters and other polluted effluents, and *20 mg/l* for ballast water.[72] Regarding existing refineries it was agreed that Parties should establish measures to reduce discharges which, 'they consider might be harmful to the environment'.[73] Parties were asked to 'pay special attention to' refineries discharging over 100 tons of oil per year, over 30 grams of oil per ton refined crude or of effluents with oil concentrations over 20 mg/l measured annually.[74] It was noted that most Parties had *not* filed environmental impact reports on the consequences of refinery discharges, and consequently the Commission emphasized the importance of such reports.[75] The State Parties did agree to file such reports to GOP in 1987. In 1984 a general reduction was noted of oil concentrations in effluents and in total quantity of hydrocarbons discharged.[76] The Paris Commission concluded in 1985 that, 'it appeared that refineries do not cause serious environmental damage,' although industrial pollutants are difficult to differentiate from one another, and further research is needed.[77]

4.2.2.4.3. Reception Facilities

Discharges from reception facilities were studied by the Commission, but no definite results established.[78] In 1986 a Recommendation was adopted setting a standard of *15 mg/l* for discharges from reception facilities and oil terminals with a review to be made of its State implementation by GOP in 1989. In contrast to platform and refinery discharges, triennial reports were not required until following the 1989 study.[79] Additionally Environmental Impact Assessments were not required due to an EEC Directive and relevant Scandinavian national legislation.

4.2.2.4.4. Other Land-based Sources of Oil and Gas Pollution

Recommendations were adopted by the Paris Commission for a registration system for used oils, for public information campaigns, for an increase in the number of reception facilities available and for the reuse of oils, and it was agreed that the problem of used oils would again be reviewed in 1988.[80] A discussion was planned for the TWG in 1987 concerning PCB's and PCT's in waste oils and estimates were to be made of land-based sources of hydrocarbons.[81]

4.2.2.5. Dispute Settlement

The Paris Convention contains fairly extensive provisions covering dispute settlement. Under Article 21 any dispute between Contracting Parties regarding the interpretation or application of the Convention may be settled by the Parties themselves, including by inquiry or conciliation within the Commission; or upon request by any Party, by mandatory submission to arbitration under the conditions set forth in Annex B.

Annex B allows the Parties to the dispute to decide otherwise, but lacking this, provides that the arbitral procedure must be conducted in accordance with the Annex's provisions which include the following.[82] The request for arbitration under Article 21, must be addressed by one Contracting Party to another and must state the disputed interpretation or application of the Convention Articles.[83] An arbitral tribunal must be established, and the tribunal itself made up of three members, two appointed by the Parties and the third, who will be chairman, by the two arbitrators.[84] The Secretary General of the United Nations must designate a arbitral chairman or an arbitrator to the tribunal in the event neither has been designated.[85] The dispute is to be decided according to international law and according to the rules of the Paris Convention, the tribunal making its own procedural rules.[86] Decisions are to be by majority voting,[87] and the absence of a Party does not procedurally present an obstacle.[88] The Tribunal may recommend interim measures of protection. Awards are to be final and binding, and any further disputes concerning award interpretation or execution may be submitted by either Party to the same tribunal if available or to another tribunal established in the same manner as the first.[89]

A quick review of Par.Comm. 7, 8, 9 and 10 reveals *no* arbitral proceedings taken between any State Parties. One author reports (in 1987) that in the 11 years the Paris Convention had been in existence, there had yet to be brought any proceedings before an arbitral tribunal due to the use of the 'rotating' veto in all important cases which exempts Parties from implementing measures with which they have disagreed.[90] This presumably has totally avoided the occurrence of Party disputes.

4.2.2.6. Exceptions to Convention Provisions

In addition to the allowable discharges of 'black list' and 'grey list' substances noted in Part I Section 4.2.2.2. there is one exception to Paris Convention provisions listed, and it deals with pollution originating in the territory of a non-Contracting Party.[91] Under this exception the Paris Convention's provisions may not be invoked against a Contracting Party to the extent that this State is prevented from ensuring full application of the Convention. The Contracting Party must however co-operate with the non-Contracting State 'so as to make possible the full application' of the Paris Convention.[92]

It could be assumed that at least one other exception could be argued, that of *force majeure*, for at least the offshore platforms, under customary international environmental law since very similar provisions appear in the LDC,[93] ODC,[94] MARPOL 73/78,[95] and OILPOL[96] as well as other marine environmental conventions.[97] In addition many of the Paris Convention State Parties are also State Parties to these Conventions.[98]

Under the Paris Convention it might be questioned whether the Soviets if and when they become Parties could invoke the *sovereign* immunity exception in the case of offshore installation pollution. The former Socialist States, including the Soviet Union, have in the past claimed that a sovereign immunity exception should include *all* government ships irrespective of their character.[99] The answer would seem dependent upon whether an offshore platform could be considered a vessel, a questionable point considering the contact with the continental shelf. However UNEP notes that it is only just recently that the IMO has taken any action regarding the unclear application of MARPOL 73/78 to offshore platforms.[100] The MARPOL definition namely states that a ship includes fixed or floating platforms.[101] The same definition also occurs in the ODC,[102] the Helsinki Convention[103] and essentially the LDC,[104] though as noted in Part I Section 3.2.1. the LDC limits and the ODC partially limits their application to non offshore oil related activities. Arguments against the Soviets claiming this exception include the increasingly environmentally progressive attitudes evidenced by statements in official negotiations and the press,[105] and in addition Soviet ratifications without relevant reservation to the Helsinki Convention and the MARPOL 73/78, both of which include a sovereign immunity Article *excluding* commercial State owned vessels.[106] Sovereign immunity of the *military* could have application to any possible run off of atomic wastes from Novaya Zemlya into the Barents Sea mentioned in Part II Section 8.5.

4.2.3. Other Land-based Pollution Conventions

4.2.3.1. Helsinki Convention

As seen the Helsinki Convention is structured somewhat differently than the other instruments under discussion. The Articles dealing with land-based pollution are Articles 5 and 6, while the relevant Annexes are Annexes I, II, and III. Annexes I and II are the same as presented previously.[107]

Briefly, land-based pollution is defined in Article 2 to be marine pollution caused by discharges from land reaching the sea waterborne, airborne or directly, including outfalls from pipelines. Under Article 5 Parties generally undertake to counteract the introduction of Annex I hazardous substances in the Baltic through the air, through the water or otherwise.[108] Under Article 6 Parties must take all appropriate measures to control and minimize land-based pollution to the Baltic, specifically through the control and strict limitation of pollution by Annex II noxious substances. To this end programmes, guidelines, standards or rules concerning discharges, environmental quality and Annex II substances must be developed. Discharge of Annex II substances must not occur in significant quantities without a special permit given by the appropriate national authority and subject to periodic review. The Parties must endeavour to develop and use common criteria for the issuance of such permits. Should significant discharges of Annex II substances, occur the Helsinki Commission must be notified. Parties must endeavour to use the best practicable means to minimize airborne pollution of the Baltic by noxious substances. Finally in order to control and minimize pollution by hazardous substances, in addition to carrying out Article 5 noted above, Parties must attempt to attain the goals and apply the criteria covered in Annex III.

Briefly *Annex III* requires States to treat municipal sewage in order that the oxygen content in the Baltic not be harmed nor the nutrients eutrophicated. Sewage must be treated to ensure hygienic qualify and epidemiological and toxicological safety to human health and such that no significant amount of Annex I and II substances are formed. The amount of industrial wastes are to be minimized in order to reduce the amount of harmful substances, organic matter and nutrients, through processing techniques, re circulation and re use of processing water, water economy, and the application of mechanical, chemical, biological and other treatment measures. Discharges of cooling water from nuclear power plants or other industries using large quantities of water must be carried out so as to minimize marine pollution. Finally the Helsinki Commission is required to define pollution control criteria and measures more precisely.[109]

4.2.3.2. UNEP Barcelona Protocol III and other UNEP Agreements

The Barcelona Protocol III[110] was formed in 1980 as a Protocol to the Barcelona Convention, pursuant to Article 8 to cover pollution from land-based pollution in the Mediterranean area.[111] One author notes that the Mediterranean deserves special attention 'as a victim of land-based pollution' since its coastal areas are characterized by lack of adequate treatment of domestic sewage, and industrial discharges and extensive use of agricultural pesticides.[112]

Its structure is similar to the Paris Convention with the use of Annex I 'blacklist' substances and Annex II 'grey list' substances, with similar prohibition and conditions of discharge. Airborne pollution however is not yet covered under the Barcelona Protocol III which is to be dealt with under Article 4(1)(b) at a later date and defined in an Annex. Pollution from offshore platforms is also not dealt with by this Protocol, though pollution from fixed offshore platforms *other* than those used for exploration and exploitation of mineral resources is dealt with under Article 4(2). Parties *undertake* under Article 5 to eliminate pollution from 'blacklist' substances more extensive than the Paris Convention's 'blacklist',[113] and under Article 6 shall strictly limit pollution from 'greylist' substances much more extensive than the Paris Convention's 'greylist.'[114] Factors to be considered for the issuance of permits for 'greylist' discharges are also more extensive than factors of the Paris Convention.[115] To implement these State Parties under Article 7 must establish timetables, guidelines, common emission standards, and standards for use or criteria dealing in particular with the length, depth and position of pipelines for coastal outfall, special requirements for effluents necessitating separate treatment, quality of sea water necessary for protection of human health, and specific requirements concerning quantities of Annexes I and II substances.[116] Monitoring programs must be carried out under Article 8 to evaluate levels of pollution along State Party coasts and to evaluate the effects of Protocol measures. Assistance in implementing programs with developing States in the fields of science, education and technology to prevent land-based pollution is required in Article 10. Under Article 11 a State is not responsible for pollution emanating from a non contracting Party, however the former 'shall endeavour to cooperate with the said State so as to make possible full application of the Protocol' in order to deal with the problem of the *upstream* polluting neighbour. UNEP and other State Parties must be held informed of the results achieved and of difficulties encountered in applying the Protocol under Article 13. Finally, meetings must be held under Article 14 to review implementation of the Protocol, amend Annexes and adopt the timetables, measures, guidelines and standards noted.[117] One author concludes that in spite of the Protocol including no provisions concerning its enforcement, it should be regarded as a model Protocol for use by the States of other regions.[118]

Other UNEP regions developing land-based pollution regimes include the Red Sea and Gulf of Aden, the Kuwait Action Plan, the West and Central African, the Wider Caribbean, the East Asian Seas and the Southeast Pacific, although with the exception of the latter these areas lie *behind* the Mediterranean in terms of coverage. One author notes the Protocol to the Southeast Pacific Convention is similar to the Barcelona Protocol III, but that the other regional agreements simply provide that State Parties are to take all appropriate measures to prevent, reduce and control marine pollution from land-based sources, though a Protocol to the Kuwait Convention is being drafted.[119] Another author notes that the Kuwait plan and the West and Central African Region deal with land-based pollution in only 1 Article, and these two regions, with the exception of the Red Sea and Gulf of Aden, have carried the legal developments the furthest.[120] The Red Sea and Gulf of Aden region approved a *draft* convention in 1981 dealing with land-based sources requiring Contracting States under Article VI to, 'take all appropriate measures to prevent, abate and combat pollution caused by discharges from land reaching the Red Sea and Gulf of Aden, whether water-borne, air-borne, or directly from the coast including outfalls and pipelines.'[121] Additionally an environmental assessment programme exists consisting under Article 12(4) of the Action Plan, of surveys of land-based sources of industrial and municipal wastes; impact studies of industrial and municipal wastes on human health; research on effects of pollutants; research on other human activities, such as dredging and land-reclamation on important marine species, communities and ecosystems; and base-line studies and monitoring of the levels of particularly heavy metals in marine organisms.[122] As under the dumping regime, Protocols to give effect to the general obligations can be developed later, when pollution levels, economy and politics allow.

4.2.3.3. EEC and Others

The EEC has adopted environmental measures for Europe some of which deal with land-based pollution.[123] Briefly, the Council of Ministers has passed various environmental Directives. The Commission of European Communities has generally developed operational functions in pollution control including implementation and monitoring of existing Directives; information systems, exchanges and training courses; and participation in other international organizations to which the EEC is a member including the Paris Convention, Bonn Agreement, Barcelona Convention and the North Sea Conference.[124] The Commission issued charges against various Member States for non compliance with the 1979 Directive of Shellfish Waters.[125] The UK had to answer to the Commission with regard to 2 environmental Directives which could be taken to the European Court of Justice involving sewage along coasts of Blackpool and Southport and the non classification of a special protection area for birds for Duich Moss on the island of Islay in Scotland. Other Directives regulated by the Commission include the Drinking Water Directive and the Bathing Water Quality Directive. The latter was complied with by only one Member State in the proper time limit and reports by Member States had not been filed as required. The Commission planned to open infringement proceedings. In addition the Commission initiated a Fourth Action Programme to guide the EEC's environmental program between 1987 and 1991 dealing with such topics as encompassing more than 129 dangerous substances by Directives, forming water quality objectives for 'grey list' substances including lead, nickel, copper and zinc, and developing a more coherent approach towards controlling dangerous substances involving a discriminating use of discharge limits and water quality standards. The European Parliament created a new budgetary line for intervention in case of emergency action to safeguard species and to combat accidental pollution, though in 1987 the fund was reduced from 600,000 ECU's[126] to 581,000 ECU's and the use limited only to accidental marine pollution. The same amount was expected for 1988 however.

The Nordic Convention[127] between Denmark, Finland, Norway and Sweden adopted in 1974 deals with land-based pollution and pollution from continental sea bed activities in these States' territories and is characterized mainly by liability and compensation provisions.[128] One author also notes that there exist a number of agreements made to reduce the pollution of *rivers*, the most important of which are 2 formed in 1976 covering the Rhine.[129] The first agreement attempts to reduce the release of chlorides into the Rhine, while the second attempts to stop the release of the most noxious chemical pollutants and control the discharge of less noxious chemicals by means of permits.

In the coastal areas of the North Sea it was felt that the applicable environmental conventions including the Oslo and Paris Conventions did not give sufficient protection and as a result that these areas were in ecological unbalance. Three Ministerial Declarations, the Bremen Declaration (1984), the London Declaration (1987) and the Hague Declaration (1990), were proclaimed on the behalf of the 8 North Sea States in addition to the EEC in which the States are to intensify their efforts to reduce pollution in these areas. It is especially the Hague Declaration which concretizes what they have signed. The Hague Declaration covers chiefly the environmental toxins, nitrogen and phosphate salts, oil and oil products. The North Sea States are challenged to reduce discharges of 36 specific toxins at least 50% from 1985 to 1995, with the 4 most dangerous being 70% reduction. Nitrogen and phosphorus are to be reduced 50% during the same period in specially exposed coastal areas. In addition the States are challenged to follow up their obligations concerning dumping and incineration at sea, pollution from ships and oil platforms, discharge and storage of radioactive waste, salvage of ship and animal and habitat protection.

Finally, a Protocol on Environmental Protection to the Antarctic Treaty was signed recently in October 1991 but is not yet in force. This instrument establishes and advisory body and sets up a system of Annexes with mandatory provisions for environmental protection covering Antarctic fauna and flora, waste disposal, marine pollution and environmental impact procedures. Any activities other than scientific research related to mineral resources are prohibited, though this can be reviewed after 50 years following entry into force of the Protocol. The possibility for adding other Annexes is also open.[130]

4.2.4. Other Sea Bed Activities Pollution Conventions

4.2.4.1. Helsinki Convention

The Article dealing with seabed pollution under the Helsinki convention is Article 10, which *generally* requires State Parties to take all appropriate measures to prevent marine pollution of the Baltic from exploration or exploitation of the seabed, subsoil or any associated activities. Parties are additionally required to ensure that adequate equipment is available to abate pollution. No reference is made to the expansive Annexes mentioned in Part I Section 3.2.2. However one author notes that the Helsinki Convention after defining 'vessels and aircraft' as including 'fixed or floating platforms' for general purposes, does not exclude them when dealing specifically with land-based pollution under Article 6, indicating that platforms are covered under the land-based regime, similar to the Paris Convention.[131]

4.2.4.2. UNEP Draft, IMO Mobile and Others

The UNEP Draft[132] is one of the few international instruments dealing fully with pollution from seabed activities, however it is *not* in force.[133] An idea can be gained of this Protocol from Part II Chapter 12 where it is used as a model for forming a Protocol covering offshore activities in the Arctic. One author notes that this separate development of legal instruments covering offshore platform pollution is *fortunate* since, 'the potential threat they (platforms) represent for the marine environment can clearly not be treated nowadays in the same way as the threat from a sewer.'[134] Additionally '(A)s offshore technology advances in sophistication, it may become increasing artificial to draw analogies between shore-based and platform-based activities.'[135]

Briefly, the UNEP Draft contains comprehensive provisions covering such points as definitions, requirements for granting State authorization, sanctions, wastes and harmful or noxious substances and materials[136] and discharge standards and exceptions, reception facilities, safety measures, contingency planning, monitoring, removal of installations, specially protected areas, studies and research programs, State assistance, international rules, liability and compensation. Generally it can be said that should it come into force it would serve the Mediterranean well in addition to serving as a good model for other areas interested in forming comprehensive pollution instruments covering solely offshore activities.

The IMO has approached pollution from offshore platforms from the perspective of mobile platforms having many the same characteristics and being exposed to the same hazards as ships. It consequently has formed a draft Convention on Offshore Mobile Craft which is *not* in force.[137] This instrument focuses mainly on liability for pollution damage and will be presented in Part I Chapter 6.

In addition to the instruments noted above various others include an informal Diplomatic Conference on Safety and Pollution safeguards in the Development of North West European Offshore Mineral Resources, and the 1983 Canadian - Danish Agreement.[138] Under the former, installation safety and personnel and pollution safeguards have been studied but the practical outcome of such studies is still unclear. Under the latter a general requirement is made under Article V for the Parties to undertake to take measures to ensure offshore installations are designed, constructed, placed, equipped, marked, operated and maintained to minimize the risk of pollution. Additionally required is that the Parties notify and consult one another over the initiation of environmentally risky offshore projects. Other agreements include bilateral agreements between Norway and the UK for co operation over safety standards for North Sea installations and pipelines.[139]

Several of the Conventions appearing in Part I Chapters 3 and 5 cover intentional but relatively minor pollution from offshore installations, including the dumping and disposal of waste, sewage and garbage.[140] MARPOL 73/78 covers pollution from offshore installations other than that associated with the exploration, exploitation and associated offshore processing of seabed minerals.[141] Annex I regulates the amount of oil which can be discharged, Annex II chemicals in bulk, Annex III[142] harmful substances in packaged forms, Annex IV the amount of sewage and Annex V the amount of garbage. The LDC, ODC, Helsinki and Barcelona Convention limit with their provisions dumping from offshore platforms *not* associated with waste disposal incidental to or derived from the normal operation of the platforms.[143] In addition as noted for the LDC dumping is limited other than that associated with the exploration, exploitation and associated offshore processing of seabed minerals.

4.2.5. Airborne Pollution - Paris Convention, Helsinki Convention, UNEP Conventions, ECE and Others

As mentioned a Protocol to the Paris Convention covering airborne pollution from land-based sources has been adopted but is not yet in force.[144] As also mentioned the LDC and ODC somewhat weakly regulate incineration at sea.[145] The Helsinki Convention and the UNEP Barcelona Protocol III both treat marine pollution from or through the atmosphere as land-based and hence those regulations shown above apply in these areas.[146]

In addition, the European States together with Canada and the US signed in 1979 the ECE Convention,[147] which under its definition would also apply to the seas.[148] This Convention will not be elaborated upon extensively since it does not establish specific emission standards or other concrete rules.[149] One of its fundamental principles, Article 2, exhorts States to '*endeavour* to limit and as far as possible, *gradually* reduce and prevent air pollution including long-range transboundary air pollution', but does not require any definite anti pollution obligations. Generally its provision deal with co-operation between Parties in research projects, exchange of information and monitoring. There are also 2 Protocols, the Protocol to Transboundary Air Pollution Convention on Reduction of Sulphur Emissions of July 8, 1985,[150] and Protocol to the 1979 Convention on Long-Range Transboundary Air Pollution on Long-Term Financing of the Co-operative Programme for Monitoring and Evaluation of the Long-Range Transmission of Air Pollutants in Europe, September 28, 1984.[151] The former, which requires Parties to *specifically* reduce their national annual sulphur emissions or their transboundary fluxes by at least 30% (based upon quantities measured in 1980) latest by 1993, entered into force on September 2, 1987 with 18 ratifying States. To enter into force the latter Protocol required 19 States to ratify and in addition the required the aggregate of the United Nations assessment rates for such States and organizations exceeded 40%, which has been satisfied.

Various other bilateral agreements exist between States including the 1985 Soviet - Finnish Agreement concerning Cooperation in the Field of Environment Protection,[152] the Norwegian - Soviet Agreement,[153] the Soviet - Norwegian Agreement on Notification of Nuclear Accidents, the 1988 Soviet Swedish Agreement on Prompt Notification of Nuclear Accidents, the Soviet Finnish Program of Cooperation in Atomic Power Plant Engineering and Safety, the 1972 extended Soviet - US Agreement on Cooperation in Environmental Protection, and the Canadian - Soviet Technical Agreement.[154]

As an example the Norwegian - Soviet Agreement is taken. This agreement generally deals with the protection of airspace from long-range transboundary air pollution and the exchange of research results, methods and measures for the cleaning of industrial and community discharges as well as new technology. A percentage reduction of airborne emissions on the Kola Peninsula has been agreed upon and a project established for bettering the monitoring of pollutants by the USSR and Norway.[155] The Arctic environment is specifically mentioned as needing protection and preservation.[156] As in the ECE Convention however its provisions deal mainly with co-operation between the Parties in research projects and exchange of information, though a Commission has been established with the specific obligation to establish concrete measures and co-operative programs to realize the objectives mentioned.[157]

4.3. Conclusions

As can be noted with the exception of *Europe* both the sources of land-based pollution and continental shelf seabed pollution is lacking international control in most of the world. Though the UNEP Regional Seas Programme is making inroads in some of these areas, unfortunately most UNEP agreements contain only general promises. For sources producing so much world wide marine pollution, or capable of producing catastrophic blowouts, this is unfortunate.

One author notes that regarding land-based pollution this can be attributed to among other factors such pollution being the most 'national' source.[158] The pollution emanates solely in an area in which no other State has any rights, which is unlike the situation regarding the other forms of marine pollution.[159] Consequently there is only one State which can legislate and enforce these rules. Additionally, different national traditions of legislation and policy make it difficult for countries to agree on a uniform approach, including for example the uniform emission standards or water quality standards controversy under both the EEC and the Paris Convention noted above.[160] Other problems relate to the many varying pollutants produced, the wide variety of activities, the varying ways of entering the sea, and finally the difficulty of monitoring. The economic and political power many large offshore industries exert in most countries should also be mentioned. States may simply not be willing to put too many constraints upon the corporations providing them with both revenue and lower unemployment figures.

Thus perhaps a similar comment can be made as noted above in the Chapter covering dumping. It is only after the problems become too large, as happened in the Baltic and Europe in general, that environmental regimes will be formed covering land-based sources and offshore installations.[161]

Notes - Chapter 4

1 'Airborne' is differentiated from 'air-based' pollution, the former referring to industrial and automobile fumes, and the latter to aircraft emissions and radioactivity from atmospheric nuclear explosions. Despite the LOSC combining these two into one separate source of atmospheric pollution in Article 212, the previous practice has been to treat airborne pollution as a form of land-based pollution which will be followed here.

2 13 *International Legal Materials* 353 (1974).

3 For pollution statistics from offshore oil activities, see Part I Sections 1.3.4. and 1.3.6.

4 For some of the more general regional conventions which follow the specific ratifying States are not presented due to lack of space and to information not being readily available.

5 Churchill and Lowe p. 245.

6 499 *United Nations Treaty Series* 311, (1964).

7 19 *International Legal Materials* 869.

8 UNEP Headquarters, United Nations Environmental Programme, P.O. Box 30552, Nairobi, Kenya. See ACOPS Yearbooks 1986/87 pp. 20-25 and 1987/88 pp. 21-25 for information on developments in specific UNEP regions.

9 See Churchill p. 281. Information on EEC Directives can be obtained from ACOPS Yearbooks 1986/87 pp. 31-33 and 1987 88 pp. 38-39, 104-106. The Nordic Agreement is found at 4 *New Directions in the Law of the Sea* 514 (1975).

10 See Churchill pp. 273 and 284.

11 23 *International Legal Materials* 269.

12 18 *International Legal Materials* 1442.

13 See Osherenko, Gail, 'Environmental Security in the Arctic: Prospects for Soviet cooperation,' Dartmouth College, Prepared for the International Studies Association Annual Convention, London, March 1989, (Osherenko) for an overview. The agreements can be obtained from the respective Ministries of the Environment.

14 Convention on the Continental Shelf (1958 Continental Shelf Convention), 499 *United Nations Treaty Series* 311, (1964). The Convention entered into force June 10, 1964 and is ratified, etc. by the Yugoslavia, Venezuela, US, UK, USSR, Ukraine, Uganda, Trinidad and Tobago, Tonga, Thailand, Switzerland, Sweden, Swaziland, Spain, South Africa, Solomon Islands, Sierra Leone, Senegal, Rumania, Portugal, Poland, Norway, Nigeria, New Zealand, Netherlands, Mexico, Mauritius, Malta, Malaysia, Malawi, Madagascar, Lesotho, Kenya, Kampuchea, Jamaica, Israel, Haiti, Guatemala, Greece, German Democratic Republic, France, Finland, Fiji, Dominican Republic, Denmark, Czechoslovakia, Cyprus, Costa Rica, Colombia, Canada, Byelorussia, Bulgaria, Australia and Albania. De Cesari, P. et al. (eds.) *Index of Multilateral Treaties on the Law of the Sea*, Milano, Dott. A. Giuffrè Editore 1985 (De Cesari), pp. 155-157.

15 The breadth of the territorial sea most commonly practised by States is 12 miles, Brownlie Principles, pp. 191-194. Norway has a 4 mile territory sea, while the Soviets have adopted the 12 mile limit, Gold pp. 104 and 107.

16 Fleischer s. 97 notes that it must be presumed that international law gives coastal States authority of placing the continental shelf under their jurisdiction.

17 Article 24 of the 1958 High Seas Convention additionally requires every State to draw up regulations to prevent marine pollution from pipelines or from the exploitation and exploration of the sea bed and its subsoil. See Part I Section 5.2.1.2. for further provisions of the 1958 High Seas Convention.

18 Emphasis added.

19 Churchill pp. 274 and 284 footnotes 42 and 43 notes in addition that there are general rules of international law which are relevant regarding avoidance of ships and installations. Under these, offshore installations must not be placed where interference may be caused to essential and recognized international sea lanes. Installations must be properly marked and lighted. In addition the IMO has recommended routing systems for shipping through offshore areas, including the Gulf of Mexico, IMO Assembly Resolution A.379(X) of 1977. The IMO has also recommended that non offshore industry related ships should avoid areas including the Gulf of Kampuchea due to the high level of activity, IMO Assembly Resolution A. 527 (XIII) of 1983. See also Part I Section 5.4.3.

20 Emphasis added.

21 This Article is the legal basis for both the Norwegian and the Soviet claims to different boundaries in the Barents Sea. See Part II Section 9.2.2.

22 Convention for the Prevention of Marine Pollution from Land-based Sources (Paris Convention), 13 *International Legal Materials* 353, (1974). Entered into force May 6, 1978. Norway ratified April 6, 1977. Other ratifying parties include Denmark, Sweden, France, Netherlands, United Kingdom including Northern Ireland, Portugal, Spain, Iceland, Federal Republic of Germany, Belgium and Republic of Ireland. The EEC signed the Convention on June 23, 1975; this signature needs no ratification, Oslo-Paris Comm. p. 7. The Paris Convention is a *regional* convention with a geographic area of application of maritime areas north of 36 degrees north latitude, and between 42 degrees west longitude and 51 east degrees, excluding generally the Baltic, and the Mediterranean, and in the north Atlantic, between 59 degrees north latitude and 44 degrees west and 42 degrees west longitude. This is generally the same area as that of the ODC. The Paris Convention was amended on March 26, 1986 to include within its scope marine pollution through the *atmosphere*. Norway ratified this Protocol in Spring 1988, as did the UK, Iceland and Spain. However as of April 1988 this Protocol was not yet in force, correspondence with Paris Commission, April 13, 1988. Eastern block countries were not present at the Diplomatic Conference initiating the Paris Convention, and the Paris Commission indicates that it is not known whether they were invited. Norway has recently forwarded a proposal to suggest *membership* to the *Soviet Union* as a Party to the Convention, possible under Article 24 (2) which allows accession to any Party to the ODC, and the Soviets are expected to accede soon. Conversation Chief Consultant Per Antonsen, Norwegian Ministry of the Environment November 1989. In the Barents Sea upon Soviet accession the high seas and Norwegian and Soviet territorial seas east to 51 degrees east as well as waters generally on the land side of base lines up to the freshwater limit will be encompassed, Article 3 (a).

23 Hakapää pp. 94-95.

24 Article 3 (a). Johnston p. 235 notes that the Helsinki Convention Part I Section 4.2.4.1., does not expressly provide for the applicability of its provisions to the internal waters of the Parties, in contrast to both the Paris Convention and the Barcelona Protocol III. This is due to State reluctance in the former Convention to surrender sovereignty over its internal waters.

25 Article 3 (b). Johnston p. 235 notes that under this Article, as well as 16 (c), the Paris Commission is empowered to fix an inner limit different from that of the freshwater limit in the case of watercourses in order to take into account 'geographic reality', similar to the Barcelona Protocol III's inclusion of *salt marshes* in the applicable legally protected area.

26 Article 1.2. Contingency plans are not covered in the Paris Convention and must be supplemented from other sources. See Part I Chapter 7.

27 Article 12.1., emphasis added.

28 Hakapää p. 277 notes that there might occur jurisdictional problems over land-based pollution in ocean areas outside of territorial seas, exclusive economic zones or not in direct physical contact with the continental shelf. See Part I Section 5.2.1.1. for 1958 Territorial Sea Convention provisions related to State jurisdiction over territorial sea and Part II Sections 10.3.3. and 14.1.2.1. for LOSC provisions related to State jurisdiction over the territorial sea and the exclusive economic zone.

29 Article 4.1 (a) and 4.2 (a) emphasis added. As can be seen the State obligation is less than mandatory.

30 Substances forming these compounds are included as well, but the compounds and substances which are biologically harmless or which are rapidly converted in the sea into biological harmless substances are excluded, Annex A, Part I.1.

31 This includes those which may float, remain in suspension or sink, and which may seriously interfere with any legitimate uses of the sea, Annex A, Part I.4.

32 Annex A, Part I.1.-.5.

33 Article 4.1. (a) and (b). Johnston p. 232 notes that '(B)anning all discharges containing harmful substances, even if the toxic component of the waste is infinitely small, would unduly affect agricultural and industrial development of a coastal area,' and therefore the 'greylist' disposal is strictly regulated rather than banned.

34 Article 4.2. (b).

35 Substances forming these compounds are included as well, but the substances and compounds which are biologically harmless or which are rapidly converted in the sea into substances which are biologically harmless are excluded, Annex A, Part II.1.

36 Article 4.4.

37 Article 9, emphasis added.

38 Churchill and Lowe p. 246.

39 Article 5.1.

40 Oslo-Paris Comm. p. 9.

41 Churchill p. 276 notes that the Paris Commission and the Oslo Commission hold both joint and separate meetings, are served by the same Secretariat, and have 'as far as possible the same commissioners.'

42 Articles 4, 10, 11, 16 (c).

43 Article 18.

44 Articles 16 (g) and 18.

45 Article 18, emphasis added.

46 Article 16 (b).

47 Article 16 (d).

48 Articles 9 and 16 (e).

49 Articles 11, 12, 16 (f), and 17.

50 Emphasis added.

51 Paris-Oslo Comm. pp. 24-31. See generally Par.Comm.7 - 10.

52 Oslo-Paris Comm. p. 28.

53 *Ibid.* notes the GOP was established to study oil pollution specifically and to report to both the Technical Working Group (TWG) and the Paris Commission.

54 *Ibid.* notes that both GOP and the TWG study this waste oil problem.

55 *Ibid.* See Part I Section 1.3.6. for discharge statistics from Norwegian offshore oil and gas platforms.

56 Par.Comm.8, pp. 11-12.

57 *Ibid.*

58 Oslo-Paris Comm. p. 29.

59 *Ibid.* pp. 10, 44-45. Difficulties of arriving at methods and standards which can be accepted by all Parties should not be underestimated. Churchill pp. 278-279 notes that under the EEC and Paris Commission there has been an intense and so far *unresolved* debate over whether to use 2 different standards in studying the sources of the more noxious pollutants. Some countries follow the uniform emission standards (UES) and others water quality objectives (WQO). (The latter would allow more noxious discharges in areas of lessor polluted waters whereas the former would require uniform pollutant discharges over the entire convention area). It is noted further that substances with which discharge standards have been adopted, mercury, *hydrocarbons*, cadmium and organohalogen compounds, the Paris Commission has established *both* UES's and WQO's.

60 *Ibid.* p. 11.

61 See Par.Comm.10. Annex 12. As noted in Part I Section 1.3.6. the effects of oil-based drilling muds are more severe than previously thought. A target standard of *100 grams* of oil per kilogram dry cuttings applies to exploration and appraisal drilling operations after the end of 1990 and to platforms where the detailed design has started after the end of 1988. Before these dates every effort had to be made to reduce the discharge of oil on the cuttings in accordance with these principles, Par.Comm.10 Annex 12. Because of these more severe effects GOP proposed stronger regulation of oil-based muds, and SFT had as a goal for the Norwegian continental shelf that no environmental effects would occur from drilling activities further than 500 meters from platforms. Consequently SFT proposed to completely *prohibit* oil-based drilling muds by January 1, 1991, which gave the oil companies a 2 year moratorium. During this period however SFT strictly controlled the use of such muds, proposing fines of 100,000 Norwegian crowns for every *gram* oil in excess of legal allowable limits in dry cuttings, conversation Chief Engineer Jens Koefoed, SFT, October 7, 1988.

62 Paris Convention Articles 11 and 16; See also Oslo-Paris Comm. p. 34.

63 See generally Par.Comm.9, pp. 26-31, 79-147, covering reports on mercury, cadmium, and PCB's in the benthic community, contaminants in fish, cadmium and mercury in seawater and contaminants in sediments.

64 *Ibid.* pp. 33, 44-78, 149-175.

65 *Ibid.* p. 10, and Par.Comm.10, p. 9. Norway published in 1988 as a Par.Comm. document, 'Guidelines for Monitoring Methods in the Vicinity of Platforms.' In addition Aftenposten 13 oktober 1991, Uke 46., Nr.524., 132 årg., reports that a satellite used for environmental surveillance (ERS-1) is currently being tested and the results are very positive. The satellite pictures of marine oil pollution are capable in Tromsø of being processed and distributed in less than one half hour. It is expected that this will allow SFT to achieve more efficient surveillance of Norwegian waters and more effective use of the surveillance aircraft currently used.

66 See SFT Miljø 86 and Miljøundersøkelser Rundt Oljeplattformer på Norsk Sektor; Årsrapport for 1985, SFT, mai 1987, (SFT Miljø 85). The following information is obtained from SFT Miljø 86 pp. 2-24 unless noted otherwise.

67 See SFT Miljø 86, pp. 6 and 7.

68 This is with the exception as noted in Part II Section 8.2.2. of the controversial strategic areas opened for drilling in the Barents Sea prior to any environmental studies being conducted.

69 Underwater installations are to be considered the same as gas fields.

70 The guide lines presented, effective from 1988, were to standardize of the monitoring programs. See generally SFT Miljø 85 for previous less stringent requirements.

71 Documentation of company expertise in monitoring techniques is also required.

72 Oslo-Paris Comm. p. 29.

73 *Ibid.*

74 *Ibid.*

75 Par.Comm.8, p. 12.

76 *Ibid.* pp. 22, 90-99.

77 Par.Comm.9, p. 10. No standards appear in Par.Comm.10. regarding oil pollution from refineries, reception facilities or other.

78 Oslo-Paris Comm. p. 29.

79 Par.Comm.9 pp. 11 and 25.

80 Oslo-Paris Comm. p. 30.

81 Par.Comm.8, p. 13.

82 Annex B, Article 1. The EEC has the right to appear as a Party before the tribunal, *ibid.* Article 8.

83 *Ibid.* Article 2.

84 *Ibid.* Article 3.

85 *Ibid.* Article 4.

86 *Ibid.* Article 5.

87 *Ibid.* Article 6.

88 *Ibid.*

89 *Ibid.* Article 7

90 Sætevik, Sunneva, 'Miljøvernsamarbeid mellom Nordsjøstatene, virkelighets oppfatninger, handlingspreferanser og utfall. En studie av Pariskommisjonens arbeid for regulering av landbasert havforurensning.' FNI, April 1987 (Sætevik), p. 54.

91 Article 14.

92 Quotations added. This Article again relates to the polluting upstream neighbour noted in Part I Section 4.2.2.3.

93 See Article V.1. and V.2. The LDC specifically mentions 'platforms or other man-made structures at sea in cases of force majeure cause by stress of weather...'

94 Article 8(1).

95 See Regulation 11(a) of Annex I, Regulation 6(a) of Annex II, Regulation 7(a) of Annex III, Regulation 9(a) of Annex IV and Regulation 6(a) of Annex V.

96 See Article IV(a).

97 Article 9(4) of the Helsinki Convention.

98 See Part I Sections 2.1., 3.2. and 5.2. An argument could probably be made that under customary law, State practice supported by an *opinio juris* has emerged firm enough to follow these parallel exceptions and to allow it as generally binding international law as regards offshore platforms.

99 See Henkin pp. 496-498. Hakapää pp. 158-159 notes that the Socialist States made reservations to Article 9 of the 1958 High Seas Convention allowing sovereign immunity only to State non commercial ships.

100 'Report of the Technical Consultation on Pollution Resulting from Exploration and Exploitation of the Continental Shelf and the Sea-Bed and Its Sub-Soil', UNEP-/WG.155/1, (UNEP Report), Appendix. See Part I Section 5.2.3.1.2.

101 Article 2 (4).

102 Article 19 (2).

103 Regulation 3.1.

104 Article III.1.(a) and III.2.

105 See Protokoll fra Første Møte I den Blandete Norsk-Sovjetiske Kommisjon for Samarbeid på Miljøvernområdet i Oslo 23.-26. august 1988 (Joint Protocol 1). See also Part II Section 9.1.

106 Hakapää p. 159 additionally notes that the Soviets do not claim sovereign immunity in civil proceedings on the basis of *reciprocity*.

107 See Part I Section 3.2.2. The Helsinki Convention's 'grey list' appears more extensive than the Paris Convention but also contains substances appearing on the latter's prohibited 'black list'. Compare Helsinki Convention, Articles 5, 6 and Annexes I, II and III, and Paris Convention Annex A Part II. The DDT's and PCB's of the former are also better enumerated as DDT (1,1,1,-trichloro-2.2-bis(chlorophenyl)-ethane) and its derivatives DDE and DDD; and PCB's (polychlorinated biphenyls), but are however not strictly prohibited. Johnston p. 239 notes that the Helsinki Convention was the first land-based pollution instrument adopted covering an area, 'where the threat arising from land-based pollution is probably the most serious, due to high industrialization and an advanced level of development.'

108 Johnston p. 234 notes that whereas under most agreements only releases are subject to regulation Annex I's 'use of appropriate technical means, prohibitions and regulations of the transport, trade, handling, application and final disposal of products contains such substances' is extremely ambitious. However, '(G)iven the high technical nature of the regulations and means necessary to implement them, the innumerable loopholes they may permit, and the economic interest involved, it may be very difficult to apply such an ambitions policy.'

109 Churchill p. 280 notes that the Commission has done this so far for agricultural pollutants, sewage treatment, industrial waste water and stormwater.

110 Protocol for the Protection of the Mediterranean Sea against Pollution from Land-based Sources, (Barcelona Protocol III), 19 *International Legal Materials* 869. The Protocol came into force 17 June 1983 with ratifying Parties including Algeria, Egypt, France, Monaco, Tunisia and Turkey. Note the *absence* of other State Parties to the Barcelona Convention noted in Part I Section 3.2.3. including Cyprus, the EEC, Greece, Israel, Italy, Lebanon, Libya, Malta, Morocco and Spain. De Cesari pp. 329-330.

111 Johnston pp. 235 and 240 notes that, 'land-based marine pollution is the most conspicuous component both of the action plans and legal instruments' adopted under the UNEP Regional Seas Programme. Also noted is that the Protocol's application is broader than the other Conventions expressly including, besides discharges from rivers, coastal establishments or outfalls, land-based pollution emanating from any other land-based sources within their territories. Salt marshes are included as the Mediterranean Sea area, thus 'taking into account geographical and ecological realities.'

112 *Ibid.* Also noted is that more than 100 million tourists visit Mediterranean beaches yearly, 'thereby contributing to a general pollution load of toxic chemicals in biota and of nutrients and pathogenic organisms in waters,' the threat of which is increased due to the Mediterranean's almost complete enclosure.

113 In addition to organohalogen compounds, mercury and mercury compounds, cadmium and cadmium compounds, persistent synthetic materials, and oils appearing in both 'blacklists,' the Mediterranean Protocol III lists organophosphorus compounds and organotin compounds (excluding those biologically harmless or rapidly converted into biologically harmless), proven carcinogenic, teratogenic or mutagenic substances, and radioactive substances the discharges of which do not comply with principles of radiation protection put forth by the competent international organizations. If these substances are discharged in quantities below a certain limit however the prohibition does not apply.

114 In addition to zinc, copper, nickel, chromium, lead, arsenic, organosilicon compounds (excluding those biologically harmless or rapidly converted into biologically harmless), crude oils and hydrocarbons, and taste deleterious substances, the Mediterranean Protocol III's Annex II lists the following as noxious substances, selenium, antimony, molybdenum, titanium, tin, barium, beryllium, boron, uranium, vanadium, cobalt, thallium, tellurium, sliver, (and compounds of all of the above), biocides and derivatives not covered in Annex I, cyanides and fluorides, non biodegradable detergents and other surface active substances, inorganic compounds of phosphorus and elemental phosphorus, pathogenic micro organism. thermal discharges, oxygen reducing substances, acid or alkaline compounds which impair the quality of sea water, and harmless substances which may become harmful due to the quantities discharged.

115 See Barcelona Protocol III Annex III and Section 4.2.2.2. Generally the criteria include the characteristics and composition of the waste, characteristics of waste constituents with respect to their harmfulness, characteristics of discharge site and receiving marine environment, availability of waste technologies, and potential impairment of marine ecosystems and sea water uses. For each of these criteria are listed between 4 to 7 factors to be considered.

116 Johnston p. 234 notes however that the establishment of separate procedures for the treatment of discharges from existing sources and new installations is an innovation criticized for introducing too much complexity in the interpretation of the text.

117 Churchill p. 280 notes the State Parties have adopted measures concerning bathing water quality, shellfish water quality, shellfish quality for human consumption and mercury in marine products.

118 Johnston p. 241. The author notes additionally that enforcement is, 'a thorny issue, since land-based sources of pollution are subject to strong national sovereignty feelings.' See Section 4.3.

119 Churchill pp. 280-281.

120 Johnston p. 242. Article 8 of the 1978 Kuwait Convention and Article 7 of the 1981 Abidjan Convention deal with land-based pollution in their respective areas. *Ibid.* p. 283, footnote 252 notes that land-based pollution in the former region, bordered by Bahrain, Iran, Iraq, Kuwait, Oman, Qatar, Saudi Arabia, and the United Arab Emirates, 'deserves special attention, since it is undergoing the most rapid industrialization of any region in the world.' Also noted in footnote 263 is that in the Abidjan Convention region, 'implementation of Article 7 cannot be expected to proceed rapidly without substantial assistance from outside the region, where most of the countries suffer from serious development problems.'

121 *Ibid.* p. 242.

122 *Ibid.*

123 See ACOPS Yearbooks 1986/87 pp. 31-33, and 1987/88 pp. 38-39, 104-106.

124 For criticism of EEC environmental policy in the Paris Commission see Part II Section 9.3.1. See ACOPS 1987/88 p. 40 for more information concerning the North Sea Ministerial Conference attended by the eight North Sea States, Belgium, Denmark, FRG, France, the Netherlands, Norway, Sweden, and the UK

125 Greece was charged and Belgium, Italy and the Netherlands censured by the European Court of Justice, while the Federal Republic of Germany and France had cases outstanding.

126 European Currency Unit (ECU) which for example on October 23, 1991 was worth USD 1.2061 US International Herald Tribune No. 33,795, 43/91 Wednesday October 23, 1991, p. 15.

127 Convention on the Protection of the Environment and Protocol (Nordic Convention), 4 *New Directions in the Law of the Sea* 514 (1975).

128 See Part I Section 6.3.1. As will be seen in Chapter 7 and noted by Johnston p. 240 liability and compensation provisions, especially for land-based pollution are unusual.

129 Churchill p. 281.

130 See US Department of State Dispatch, October 14, 1991 p. 771.

131 Johnston p. 233.

132 UNEP Draft Protocol for the Protection of the Mediterranean Sea against Pollution Resulting from Exploration and Exploitation of the Continental Shelf and the Sea-bed and its Subsoil, UNEP/IG.74/Inf.9, Annex II, July 15, 1987, pp. 1-38.

133 Churchill p. 273 notes that a similar Protocol is under development to the UNEP Kuwait Convention. The other UNEP regional Conventions contain only general obligations to take appropriate measures to prevent pollution from sea bed activities.

134 Johnston p. 233, parentheses added. Note however the complexity involved in the Paris Convention's programs indicated in Section 4.2.2.4. above.

135 Johnston p. 282 footnote 229. Also noted on p. 233 is that, 'the industrial nature of such offshore activity calls for specific environmental treatment by lawyers as well as ecologists' and many States in fact deal with platforms under legislation govern vessel-source pollution.

136 These include oil and oil mixtures, drilling fluids and discharges, sewage, garbage, reception facilities, and exceptions.

137 Craft Convention on Offshore Mobile Craft (IMO Mobile), LEG XXXIV/6(b) 19 December 1977. Churchill p. 273 lists as well an IMO recommended Code for the Construction and Equipment of Mobile Offshore Drilling Units, IMO Assembly Resolution A. 414 (XI) of 1979. IMO Status p. 345 lists 'Protocol for the Suppression of Unlawful Acts against the Safety of Fixed Platforms Located on the Continental Shelf 1988 (SUA PROT 1988) which is not in force.

138 Churchill p. 273. Agreement between the Government of Canada and the Government of the Kingdom of Denmark for Cooperation Relating to the Marine Environment (1983) (Canadian - Danish Agreement), 23 *International Legal Materials* 269. See also Part I Section 7.8.3.

139 Churchill pp. 273 and 284 footnote 41 lists, Agreement relating to the Transmission of Petroleum by Pipeline from the Ekofisk Field and Neighbouring Areas to the United Kingdom, 1973, Article 8; and the Agreement relating to the Exploration of the Frigg Field Reservoir and the Transmission of Gas therefrom to the United Kingdom, 1976, Articles 7, 8, 17 and 18.

140 Churchill pp. 275-276 notes this 'cross over' by the various Conventions.

141 See Part I Section 5.2.3.1.

142 Neither Annex III nor Annex IV are yet in force. See Part I Section 5.2.3.2.4.

143 See Part I Section 3.2.

144 Par.Comm.9 does report however a pilot project was started in 1985 in which most member countries monitored airborne pollution into Convention waters.

145 See Part I Section 3.2.1.6. Churchill p. 282 notes that the Treaty Banning Nuclear Weapon Tests in the Atmosphere, in Outre Space and under Water, (Nuclear Test Ban Treaty), Moscow, 5 August 1963, 480 *United Nations Treaty Series* 43, in force October 1963, also resulted in ending radioactive marine pollution from testing nuclear weapons.

146 Johnston p. 233 notes that the 1981 Abidjan Convention deals with atmospheric pollution and land-based pollution in separate Articles. Additionally noted similar to his comments regarding the combination of land-based pollution and continental seabed pollution, '(D)ealing with both under a single legal instrument gives rise to unnecessary complexity. Normally separate texts are likely to deal more adequately with each issue.'

147 Economic Commission of Europe Convention on Long-range Transboundary Air Pollution, (ECE Convention) 18 *International Legal Materials* 1442, (1979). The Convention entered into force March 16, 1983, with ratifying States including Austria, Belgium, Bulgaria, Byelorussia, Canada, Czechoslovakia, Denmark, Finland, France, German Democratic Republic, Federal Republic of Germany, Greece, Hungary, Iceland, Ireland, Italy, Luxembourg, Netherlands, Norway, Poland, Portugal, Rumania, Spain, Sweden, Switzerland, Turkey, Ukraine, USSR, UK, US, and Yugoslavia and also including the Holy See, Liechtenstein, San Marino and the EEC. Johnston p. 378 footnote 317, and correspondence with United Nations Office in Norway, May 26, 1988. (Thus 31 of the 34 ECE members are Parties. Non Parties include Albania, Cyprus and Malta).

148 Article 1 (b) defines 'long range transboundary air pollution' as 'air pollution whose physical origin is situated wholly or in part within the area under the national jurisdiction of one State and which has adverse effects in the area under the jurisdiction of another State at such a distance that it is not generally possible to distinguish the contribution of individual emission sources or groups of sources.' This definition would cover the territorial sea under the jurisdiction of a State, but not the high seas, nor possibly the exclusive economic zone over which a State has only partial jurisdictional rights.

149 Churchill p. 282 notes that this Convention was formed to attempt to control 'acid rain' and is 'the most significant attempt so far to develop international rules dealing with atmospheric pollution.'

150 U.N. Doc. GE 86-42399/1799E

151 U.N. Doc. GE 86-42195/4033E.

152 The Soviets have obligated themselves to reduce the airborne pollution from the Kola industry by 50 % within 1993.

153 Overenskomst Mellom Kongeriket Norges Regjering og Unionen av Sovjetiske Sosialistiske Republikkers Regjering om Samarbeid på Miljøvernområdet (1988), (Norwegian - Soviet Agreement).

154 USSR - Canada Protocol on the Development of a Programme on Scientific and Technical Cooperation in the Arctic and in the North, (Canadian - Soviet Technical Agreement). See generally Osherenko Figure 4, p. 30 and pp. 2-3 Appendix. The various Agreements can be obtained from the respective States' Ministries of the Environment. Undoubtedly other bilateral agreements have recently come into existence as well between other States covering airborne pollution, however it is beyond the scope of this work to address these outside the case study area. This is both because of the normal generality of provisions, covering chiefly study programs, and lack of space. Bilateral control of airborne land-based pollution may however be under rapid development judging from the number of recent agreements in just the Arctic and the

fact that neither Churchill nor Johnston in their books address airborne marine pollution to a great extent.

155 Stokke, Olav, 'Environmental Threats in the Arctic: Ignorance, Poverty and Impotence, FNI, September 1989 (Stokke), pp. 13-14.

156 See Part I Section 7.7. for discussion of Agreement provisions.

157 Additionally recently the Joint Norwegian - Soviet Commission formed under the Norwegian - Soviet Agreement met and specified several points dealing with air pollution. See Joint Protocol 1 pp. 3-7 and Appendix 2. A proposal for a cooperative program regarding air pollution problems was discussed extensively with the result that a work group was formed, together with a program for the two years period 1989-1990. Both Norway and the Soviet Union presented measures to reduce air pollution in the border area Sør Varanger - Vest Finnmark, with the Soviets showing that sulphur dioxide is already reduced by 24% of the 1980 level and expected to be reduced by 50% by 1993. The Soviets have hopes of achieving a reduction of up to 80% if international technological assistance is obtained. An experts meeting was planned in February 1989 to form a proposal concerning the use of advance technology for the cleaning of gasses and drainage in these areas, with the goal of presenting the proposal to the second Commission meeting in April 1989.

158 Churchill p. 278.

159 In a historic perspective it must be noted that under international law it is only recently that complete State sovereignty over its territory and its subjects has begun to be challenged, albeit to a very small extent.

160 See above Section 4.2.1.5.1.

161 Johnston p. 236 believes however that, '(I)nternational efforts aimed at promoting uniformity of norms, standardized assessment of noxious wastes, and above all international control, are *not* likely to be efficacious in most cases,' and holds that the most import part of land-based pollution control regulation is to be found in *national* laws, emphasis added. This is due to national interests being affected most by noxious releases which in turn provides incentives for improvement. 'National controls are likely to be carried out, and national regulations are likely to be enforced.' The author does however recognize as an exception to this the case of the agreements formed in Northwest Europe in the 1970's noted above.

5. Vessel-Source Pollution

5.1. Introduction

As seen from Part I Section 1.3.2. most estimates place vessel-source pollution as contributing roughly from between 40% to 45% of the total marine pollution on a world wide basis. Luckily, perhaps, vessel-source pollution is one of the more internationally regulated areas of marine pollution.[1] A quick glance at the IMO publication, IMO Status, shows more conventions in force than is generally typical of the other pollution sources combined. While it may be questioned why some of the conventions are included since they seem of peripheral relevance to marine pollution and thus increase the complexity of an already complex area of law, consideration of the Alaskan EXXON VALDEZ example gives a clarification. As was true with offshore oil platforms, it is also true with ships that much of the problem with marine pollution lies in the element of *human* failure. Regulations which attempt to limit such human problems will also limit marine pollution. The relevant conventions will be examined here in order to give a general idea of how well this pollution source is legally covered.[2]

5.2. Principal Conventions[3]

5.2.1. The 1958 Territorial Sea Convention and the 1958 High Seas Convention[4]

5.2.1.1. The 1958 Territorial Sea Convention

This global Convention does not deal specifically with marine pollution and shipping but does give specific rules concerning a coastal State's *rights* governing marine pollution in their sea areas. Briefly coastal States may legislate anti pollution and navigation measures under Article 17 but under Articles 15(1) and 18(2) may not hamper innocent passage through its territorial sea or discriminate.[5] Innocent passage is described in Article 14 as navigation through the territorial sea which is not prejudicial to the peace, good order or security of the coastal State and in conformity with the Convention and other rules of international law.[6] Territorial sea is described in Article 1 as a belt of sea adjacent to a State's coast over which the sovereignty of a State extends beyond its land territory and its internal waters. Additionally under Article 17 all ships must comply with such laws. Under Article 19 a coastal State may enforce violations of its pollution legislation committed in its territorial sea.

5.2.1.2. The 1958 High Seas Convention

This global Convention requires ships to regulate their flag ships in relation to marine pollution. The flag State under Article 6 has exclusive jurisdiction over the ship on the high seas both to prescribe legislation and to enforce it. Article 10 requires States to take measures for flag ships as are necessary to ensure safety at sea with regard to construction, equipment seaworthiness and manning of ships, and in doing this to conform to generally accepted international standards and to take any steps which may be necessary to ensure their observance. Under Article 11 in the event of a ship collision or other navigation incident on the high seas, no penal or disciplinary proceedings may be brought except by the flag State or the State of which the master (or other involved employee) is a national. Additionally ships may not be arrested or detained even for investigation except by the flag State. Under Article 12 the flag State must require its ships to render assistance, without causing danger to the ship itself, crew or passengers, to persons lost at sea. Additionally every coastal State must promote the establishment of joint search and rescue arrangements. Article 26 entitles all States to lay submarine cables and pipelines on the sea-bed of the high seas, something the coastal State may not impede, but this right is subject to the coastal State's right to explore and exploit its natural resources. Articles 27 and 28 require States to provide that compensation is available for damage caused by their flag ships or owners of other pipelines or cables to pipe lines and cables under the high seas. Article 24 requires States to 'draw up regulations' to prevent marine pollution by the discharge of oil from ships or pipelines or resulting from sea-bed and subsoil exploration and exploitation taking into account existing treaty provisions. Article 29 requires States to provide that compensation is available to shipowners who sacrifice anchors, nets or other fishing gear in order to avoid damaging cables or pipelines. Article 23 covers hot pursuit of a ship suspected by a coastal State of violating the laws of that State in its internal waters, territorial sea or contiguous zone. The pursuit must be commenced when the foreign ship is within these sea zones and following an order to stop and be uninterrupted. Hot pursuit ceases when the suspected ship enters the territorial sea of its own State or another State. Compensation must be paid for unjustifiable hot pursuit and resulting stop and arrest. Article 25 requires States to cooperate with the competent international organizations in taking measures to prevent marine pollution resulting from any activities with radio-active materials or other harmful agents.

Generally the 1958 Territorial Sea Convention and the 1958 High Seas Convention have relevance in the jurisdictional rights given to States to prescribe and enforce environmental measures. They give a legal basis for carrying out specific MARPOL 73/78 as well as other measures to both coastal States and flag States.

5.2.2. OILPOL[7]

OILPOL is a global convention and is briefly considered here since its provisions are in force for some States pending ratification of MARPOL 73/78 (or denouncement of OILPOL).[8] Additionally OILPOL was the *first* convention which attempted to control marine oil pollution and served as a basis for establishing the succeeding conventions. However one author does not even address OILPOL since it is noted that several OILPOL discharge standards are 'over-rigid' and 'virtually unenforceable outside of ports.'[9] Most of the expected ship and tanker traffic would probably be covered by MARPOL 73/78.

OILPOL under Article II prohibits the intentional discharge of oil and oil mixtures by ships registered in the Contracting Parties or unregistered ships having the nationality of a Contracting Party. Excepted are tankers under 150 tons gross tonnage and other ships under 500 tons,[10] whaling ships, ships navigating the Great Lakes of North America and naval ships.[11]

Under Article III discharges of oil or oily mixtures are not prohibited when a ship, *other* than a tanker, is proceeding en route, the instantaneous rate of discharge does not exceed 60 litres per mile, the oil content of the discharge is less than 100 parts per million of the mixture, and the discharge is made as far as practicable from land.[12] Discharges from *tankers* of oil or oily mixtures are not prohibited when the tanker is proceeding en route, the instantaneous rate of discharge of oil content does not exceed 60 litres per mile, the total quantity of oil discharge on a ballast voyage does not exceed 1/15,000 of the total cargo-carrying capacity, and the tanker is more than 50 miles from the nearest land.[13] Certain exceptions are made for discharges following the flushing of tanks,[14] from necessity for the safety of the ship, cargo and life, and for damage to the ship.

Article VI provides that offenses punishable under Convention provisions both in the territorial sea and the high seas must be equal in severity. Under Article

VII carrying water ballast in oil fuel tanks must be avoided if possible. Article VIII requires Parties to take all appropriate steps to provide port and ship repair facilities to receive residues and oily mixtures from ships, and to provide oil loading terminals with facilities to receive residues and oily mixtures from tankers.

Under Article IX all loading and discharges of oil must be recorded in an 'oil record book' which is to be available for inspection at all reasonable times. For *tankers*, operations which must be registered in the book include the, 'loading of oil cargo, transfer of oil cargo during voyage, discharge of oil cargo, ballasting of cargo tanks, cleaning of cargo tanks, discharge of dirty ballast, discharge of water from slop-tanks, disposal of residues, the discharge overboard of bilge water containing oil which has accumulated in machinery spaces while in port, and the routine discharge at sea of bilge water containing oil unless the latter has been entered in the appropriate log book.' For ships other than tankers, operations which must be registered include 'ballasting or cleaning of bunker fuel tanks, discharge of dirty ballast or cleaning water from bunker fuel tanks, disposal of residues, discharge overboard of bilge water containing oil which has accumulated in machinery spaces while in port, and the routine discharge at sea of bilge water containing oil unless the latter has been entered in the appropriate log book.'

Discharges of oil or oil mixtures from new ships over 20,000 tons are prohibited in the 1962 Amendment, and the load-on-top (LOT) system adopted in the 1969 Amendment.[15] The LOT system attempts to reduce the amount of oily ballast water discharged and to retain on board slop from the separation process.[16] Essentially this system consists of washing cargo tanks with sea water, allowing the water to settle downward from the residue oil, and then pumping the lower layer of water into the sea. The process may be repeated once, and then new cargo loaded on top of the slop oil residues.[17]

Amendments to the 1954 Convention were also adopted in 1971 dealing with accidents and attempting to limit oil pollution from accidental strandings or collisions by restricting the size of the individual oil cargo tanks but are not yet in force.[18]

5.2.3. MARPOL 73/78[19]

The global MARPOL 73/78 is the *main* Convention controlling vessel-source pollution, hence the provisions briefly presented are what are controlling vessel-source pollution. See Part II Section 9.3.2. for discussion of *problems* associated with MARPOL 73/78 as well as several other conventions.

5.2.3.1. MARPOL 73

5.2.3.1.1. Introduction

As indicated OILPOL with Amendments proved unsatisfactory to deal with marine pollution, and an International Conference on Marine Pollution was held in 1973.[20] The result was the global MARPOL 73. This Convention was slow to enter into force and would have taken even longer were it not for the 1978 Protocol,[21] and only a brief mention will be given here to the original non incorporated MARPOL 73 provisions. Provisions comprising the combined final result, MARPOL 73/78, will be covered more extensively. The historical development is touched upon briefly however in order to give background into the problems and proposals noted in Part II Sections 9.3.2. and 14.2. This information as well might give an idea into the evolution involved in an anti pollution regime, points of which might generally be extrapolated to other pollution source legislation and practice. The following Articles and Annexes are common to both MARPOL 73 and the 1978 Protocol unless otherwise noted.

Articles 1 - 6 deal with prescriptive and enforcement jurisdiction.[22] Article 2 covers definitions and includes the following. *Harmful substance* includes any substance which if discharged into the sea is liable to create hazards to human health, to harm living resources and marine life and to damage amenities or to interfere with other legitimate uses of the sea. *Discharge* includes any release howsoever caused from a ship and includes any escape, disposal, spilling, leaking, pumping, emitting or emptying. It does *not* include dumping as defined in the LDC, nor the release of harmful substances directly arising from off shore exploration, exploitation or processing of sea-bed minerals, nor the release of harmful substances for scientific research involving pollution abatement or control. *Ship* is defined as any vessel of any type whatsoever operating in the marine environment and includes hydrofoil boats, air-cushion vehicles, submersibles, floating craft and fixed or floating platforms. *Incident* means an event involving the actual or probable discharge into the sea of a harmful substance, or effluents containing such a substance.

Under Article 3, Convention provisions do not apply to State ships such as warships or others on non commercial service. Each Party must however ensure by adopting appropriate measures not impairing the operation of such ships that they act as far as is reasonable and practicable in accordance with the Convention. Additionally application of the Convention, is to all ships under the flag of a State Party, and to ships, though not flying the State Party's flag, are operating under its authority.

Under Article 4 violations of Convention requirements must be prohibited and sanctions established under the law of the flag State for ships anywhere. If notice is given and the evidence sufficient that an alleged violation has occurred, the flag State must bring proceedings.[23] In addition any violation of Convention requirements within the jurisdiction of any Party must be prohibited and sanctions established under the national law. When a violation occurs, the State must bring proceedings or give to the flag State evidence that a violation has occurred, and the flag State must in turn inform the State and the IMO of action taken. Penalties under the law of a Party must be severe enough to discourage violations and must be equally severe regardless of where the violations occur.[24]

Article 5 requires that all ships hold a *certificate* of compliance with MARPOL 73/78 standards, both technical and operational. In the event of substantial non compliance evidenced by clear grounds, the port authority of the State Party carrying out the inspection must take such steps to ensure that the ship *not* sail until it can proceed 'without presenting an unreasonable threat of harm to the marine environment.' As recognized by one author this 'International Oil Pollution Prevention Certificate' may be the most important element in detecting violations in the MARPOL system, because the other checks, oil record books[25] and the oil discharge monitoring and control scheme[26] are flawed.[27]

Article 6 requires that State Parties 'cooperate in the detection of violations and the enforcement of the provisions.' A flag State is also required to initiate legal proceedings as soon as possible for discharge violations of the Convention upon the basis of sufficient evidence. A State must furnish to the flag State evidence of discharges violating the provisions of MARPOL 73/78.[28] A port State is allowed upon request to inspect a ship within its jurisdiction and if sufficient evidence exists regarding a violation which has occurred *anywhere*, issue a report to the flag State and the State Party requesting it. However, a port State may not bring legal action for violations outside its jurisdiction.

Articles 7 - 20 deal with more peripheral concerns and will be dealt with only briefly. Article 7 deals with undue delay to ships and requires that all possible efforts be made by a State to avoid unduly detention under Articles 4, 5 or 6 to ships, and when there has been such detention, the ship is entitled to compensation for damages suffered. Article 8 deals with reports on incidents involving harmful substances; Article 9 with other treaties and interpretation;[29] Article 10 with settlement of disputes;[30] Article 11 with communication of information;[31] and Article 12 with casualties to ships;[32] Articles 13 - 20 deal with subjects including signature, ratification, entry into force, amendments, promotion of technical cooperation and languages.

5.2.3.1.2. Annex I[33]

Annex I is mandatory and deals with operational oil pollution, which is permitted as long as the tanker is fifty nautical miles from land, not in a *special area*, the tanker is proceeding en route, the rate of discharge does not exceed 60 litres per nautical mile, the total quantity of oil discharged by existing tankers is not more than 1/15,000 of the particular cargo or for new tankers 1/30,000 of the particular cargo, and the tanker with certain exceptions has an oil discharge monitoring and control system operating and a slop tank in existence.[34] Oil covered by Annex I includes generally crude oil, fuel oil, sludge, oil refuse and refined products, and specifically includes approximately 45 varieties of asphalt solutions, oils, distillates, gas oil, gasoline blending stocks, gasolines, jet fuels and naphtha, though this is not comprehensive.[35]

Special areas are covered further in Annex I, Regulations 1(10) and 10, and special attention should be given, since they are of importance to the proposals made in Part II Section 14.2. A special area is defined as an area 'where for recognized technical reasons in relation to its oceanographical and ecological condition and to the particular character of its traffic the adoption of special mandatory methods for the prevention of sea pollution by oil is required ... and shall include the Mediterranean Sea, the Baltic Sea, the Black Sea, the Red Sea, and the Persian Gulf.'[36] All discharges except clean ballast and segregated ballast are prohibited, and pollutants must be retained on board and reception facilities provided to receive these.[37]

Due to the difficulties with enforcing the LOT system noted under OILPOL, a discharge *monitoring* and *control* system was established to attempt to replace reliance upon vessel operators.[38] A State is allowed under subsection 6 however to *avoid* this requirement if such equipment is not available, and the IMO reviews the unavailability of the equipment at least annually. In 1973 there were available *no* 'commercially viable monitoring systems.'[39] Other required equipment include slop tanks, oily-water separating or filtering systems, oil content meters, sludge tanks and appropriate pumping and piping systems.[40] Tank sizes are limited, taking into account factors as the arrangements of tanks, the fitting of double bottoms and the positioning of clean ballast tanks, with typical sizes for normal tankers being centre tanks limited to 30,000 cubic meters and wing tanks, 15,000 cubic meters.[41] New ships are designed according to subdivision and stability requirements to survive a specified hull damage.[42] Ship operators are required to maintain an oil record book, more comprehensive than under OILPOL, showing all loading, transferring and unloading of oil cargo, ballasting, cleaning and discharge of ballast[43] from cargo tanks and discharge of water and residues.[44] Segregated ballast tanks (SBT's) for all new tankers over 70,000 dwt are required[45] and must be large enough to give sufficient ballast on voyages in order that cargo tanks not be used for ballasting. An exception is made for severe weather conditions. Under Regulation 12 each Party *undertakes* to ensure that adequate reception facilities for residues and oil mixtures from tankers and other ships be provided, sufficient so as not to cause delay.

An interesting point is stated in Regulation 21 which deals with oil or oily mixture discharges from *drilling rigs* and other platforms. This regulation requires that fixed and floating rigs and platforms, 'when engaged in the exploration, exploitation and associated off-shore processing of sea-bed mineral resources' must comply with Annex I requirements associated with ships, other than oil tankers, of 400 tons or more. Further, platforms *shall* be equipped as far as practicable with oil discharge monitoring and control system, oily water separating equipment and tanks for oil residues (sludge); a record of all oil or oily mixture discharges must be kept in an officially approved form; and in any *special area* and subject to several conditions,[46] discharges into the sea are prohibited except when the oil content of the discharge not greater than 15 parts per million.[47] Recently the IMO took action to clear up the uncertainty regarding the application of the Convention to offshore platforms by declaring only *machinery* space drainage to be subject to Convention provisions.[48]

5.2.3.1.3. Annexes II, III, IV, and V

Annex II dealing with *noxious liquid substances* in bulk, as noted is probably the main reason that MARPOL 73 was not ratified by many States due to economic and technical problems meeting the required standards. This Annex is *mandatory* and covers approximately 200 chemicals of great variety and with classification levels denoting toxicity to aquatic life.[49] Annex II requires that noxious substances be discharged at a reception facility unless they can be *diluted* down to acceptable levels. The discharge of the most toxic chemicals (Category A) is prohibited. Following unloading, tank washings may be discharged if a set amount of residue[50] of Category A substances is diluted down by water of not less than 5% of the total volume of the tank, the ship is travelling at least 7 knots (self propelled), the discharge is made below the water line and the discharge is made at least 12 nautical miles from land and in a depth of water of not less than 25 meters.[51] The discharge of Categories B and C are somewhat less stringent, allowing residues of 1 cubic meter or 1/3,000 of the tank capacity for Category B substances and 3 cubic meters or 1/1,000 of the tank capacity in cubic meters for Category C substances, but still include the same ship movement and discharge requirements noted. Still the concentration of the substance in the wake astern cannot exceed 1 part per million for both categories.[52] The discharges standards to be used in special areas are similar to the above but with even *lower* allowable residue quantities.[53] Additionally Annex II contains provisions to minimize pollution in case of an accident.[54] Ship operators are required to maintain an cargo record book, similar to the oil record book of Annex I, showing all loading, transferring and unloading of chemical cargo, ballasting, cleaning transfer and discharge of ballast from cargo tanks, discharge of residues, and accidental or other discharges.[55] There are no provisions for independent monitoring and control systems as with Annex I. Additionally State Parties are required to control the design, construction and operation of chemical tankers which must be in compliance with the Code for the Construction and Equipment of Ships carrying Dangerous Chemicals in Bulk.[56]

Annexes III, IV and V are voluntary, and according to one author have not affected ratification.[57] They were incorporated without change into MARPOL 73/78. Annex III deals with the prevention of pollution by harmful substances carried by sea in packaged forms, freight containers, portable tanks or road and rail tank wagons, and consists of 8 Regulations.[58] Annex IV deals with the prevention of pollution by sewage by ships and consists of 11 Regulations and an Appendix.[59] Annex V deals with the prevention of pollution by garbage from ships and consists of 7 Regulations.[60] State Parties to Annexes II, IV and V are required to provide adequate port reception facilities.[61]

5.2.3.1.4. Summary

By 1978 opposition to MARPOL 73 was strong, and there was no uniform application of any international standards.[62] The LOT and crude oil washing regimes as well left much to be desired due to the random application by tanker and ship operators.[63] On top of this during a 2 month period from December 1976 to January 1977 a series of tanker disasters occurred, including ARGO MERCHANT, SANSINENA, OSWEGO PEACE, OLYMPIC GAMES, DAPHENE, GRAND ZENITH, BARCOLA, MARY ANN, UNIVERSE LEADER and IRENES CHALLENGE all of which were caused by human error or negligence.[64] These disasters lead to the IMO and the US and other coastal nations holding the Conference on Tanker Safety and Pollution Prevention, which resulted in the 1978 Protocol.[65]

5.2.3.2. MARPOL 73/78

5.2.3.2.1. Introduction

2 main changes were made in the 1978 Protocol to streamline State ratification. As noted Article I (2) provided that both MARPOL 73 and the 1978 Protocol were to be merged into one legal instrument, and the provisions of both were to be read and interpreted together. Ratification of one meant that a State was a party to both. In addition ratification of Annex II of MARPOL 73 was postponed for 3 years from the date of entry into force of the 1978 Protocol or even longer if so decided by a two-thirds majority of the Marine Environment Protection Committee (MEPC) of the IMO.[66]

5.2.3.2.2. Annex I[67]

Annex I was changed in the 1978 Protocol[68] to introduce Regulations attempting to deal with the problems noted above with the LOT system. Periodic surveys and inspections were made more stringent under *Regulation 4*, and the conditions for issuance of the International Oil Pollution Prevention Certificate were tightened up and certification inspections made more frequent under *Regulation 8*.[69] Briefly periodic and intermediate surveys are required of the ship's structure, equipment, systems, fittings, arrangements and material to ensure they comply with Annex I requirements. The ship's certificate is tested more frequently. The Article V requirement noted above is reiterated that if a Port State is aware of a ship which does not correspond substantially with the particulars of the Certificate or is not fit to proceed to sea, it must take such steps to ensure that the ship not sail until it can proceed 'without presenting an unreasonable threat of harm to the marine environment.' *Regulation 13*, concerning SBT's, dedicated clean ballast tanks (CBT's) and crude oil washing (COW) requires SBT's[70] and COW on all new oil tankers of 20,000 dwt and new product tankers of 30,000 dwt.[71] SBT's must be constructed such that they protect cargo tanks in cases of running aground or collisions.[72] Inert gas systems (IGS's)[73] in each cargo and slop tank are required to avoid explosions, and existing oil tankers of 40,000 dwt and larger are required to refit SBT's, or COW or CBT's.[74] Existing tankers engaged in specific trades between ports with adequate reception facilities are exempted from the SBT's, COW and CBT's requirements.[75] Improved stripping, drainage and discharge arrangements are provided for in Regulation 18. Regulations regarding discharge standards, reception facilities and oil monitoring systems were incorporated directly from MARPOL 73 to MARPOL 73/78 with no changes.[76]

As occurred previously and not unexpectedly, one of the largest problems connected with the new and revised Annex I is the lack of adequate and reasonably priced port reception facilities, as well as the lack a verification system consisting of oil monitoring devices.[77] One author notes that State Parties attempted in fact to amend regulations to allow a waiver of equipment requirements, something which was objected to by the Netherlands, Sweden and the US on the grounds that the amendments would, 'result in compliance deferral or avoidance'.[78]

5.2.3.2.3. Annex II[79]

As noted Annex II was incorporated in its entirety from the MARPOL 73 into the MARPOL 73/78 with the exception regarding its entry into force. This is in spite of the fact that an accident involving some of the chemicals appearing in Annex II's 'Category A' high hazard list might be more disastrous than an oil spill.[80]

5.2.3.2.4. Annexes III,[81] IV,[82] and V[83] [84]

As noted these Annexes were incorporated directly from MARPOL 73 into MARPOL 73/78 with no changes made. One author is pessimistic regarding their entry into force and states, 'there is little chance of international implementation in the near future.[85] Another is more optimistic stating that it is only a matter of time until Annex III enters into force and Annex IV is accepted by a number of States and many provisions required in others.[86]

5.3. Other IMO Conventions

5.3.1. Intervention Convention and Protocol[87]

The Intervention Convention is a global convention which in Article I gives State Parties the right after a marine casualty to, 'take such measures on the high seas as may be necessary to prevent, mitigate or eliminate' grave danger to their coastlines or related interests[88] from threats of or oil pollution itself.[89] Marine casualties of warships or other state owned non-commercial ships are excluded. Under Article III before taking any measures consultations must be made to other affected States, particularly the flag State, although in extreme emergencies measures may be taken without prior notification or consultation.[90] Under Article V measures taken by the coastal State must be *proportional* to the actual or threatened damage and must not unnecessarily interfere with the rights and interests of the flag State or other States.[91] Under Article VII compensation must be paid for excess measures taken under the Convention's provisions and which causes damage to others. According to Article VIII disputes involving measures taken in contravention of the provisions or compensation obligations upon request must be submitted to conciliation or to arbitration procedures set out in the Annex to the Convention. The 1983 Protocol was added to greatly broaden the scope of the polluting substances with which State Parties might take action following a marine casualty.[92] The Convention followed the TORREY CANYON disaster in which Great Britain *bombed* the wreck to try to prevent further pollution damage.[93] As one author notes the Convention permits for the first time, 'States other than flag States to take preventative and mitigating action against foreign vessels provided that there was a realistic concern that oil pollution might result in major harmful consequences.'[94] Additionally, '(T)he Convention has been criticized both for allowing too much discretion to coastal states and for limiting the rights of such states to take action.'[95]

5.3.2. SOLAS[96]

SOLAS 1960 came into force in 1965 and was amended in 1966, 1967, 1968, 1971 and 1973, none of which are in force.[97] SOLAS 1960 has been superseded by SOLAS 1974 which came into force in 1980 and which is cited above, complete with amendments and Protocol. One author notes that many of the SOLAS 1960 provisions will continue to be applicable for some time to come.[98] It must be noted that with the exception of the amendment procedure SOLAS 1974 is a very general Convention with few technical regulations. To obtain a clear picture of SOLAS it would be necessary to view all the SOLAS Conventions, Amendments, Protocol and Resolutions, which is beyond the scope of this work. Since SOLAS 1974 and the 1978 Protocol are those which are central,[99] these are what will be summarized to give a brief idea of the direction and scope.

The Contracting States generally undertake under SOLAS 1974 Articles I and II to give laws and all other necessary steps to give the Convention full effect to ensure that with respect to safety of life, a flagship is fit for the service which it is intended. Exceptions to Convention provisions are made for *force majeure* situations, shipwrecked persons or evacuation of persons in emergencies. Recommendations are made in Attachment II.1-8. for formulating international standards covering the intact stability of ships,[100] fire safety requirements,[101] life-saving appliances, main propulsion machinery,[102] periodically unattended machinery spaces,[103] nuclear ships[104] and safety measures for certain types of ships.[105] These general recommendations are supplemented in Resolution 4 with recommendations to State Parties that consideration be given to the Recommendations of the International Conference on Safety of Life at Sea, 1960, which have so far been only partially implemented as well as developments which have taken place since the 1960 Conference.

The Protocol which modifies and makes additions to the 1974 Convention gives specific rules. Chapter I, Part A, deals with Application and Definitions, Part B with Surveys and Certificates; Chapter II-1 with Construction, Subdivision and Stability, Machinery and Electrical Installations; Chapter II-2 with Construction, Fire Protection, Detection and Extinction; Chapter V with Safety of Navigation; and the Appendix with various forms to be used concerning the above. Specifically, unscheduled inspections and surveys are provided for to ensure that the ship and its equipment 'remain in all respects satisfactory for the service for which the ship is intended.'[106] In addition immediate repairs can be required or the relevant certificate may be withdrawn and Port Authorities ensure that the ship not sail until appropriate repairs are made.[107] Surveys are to be made in cargo ships of items including life-saving appliances, fire extinguishing appliances and the inert gas systems.[108] In new ships fire control plans, lights, shapes and means of making sound signals are to be surveyed to ensure that they comply with Convention and Protocol requirements as well as the International Regulations for Preventing Collisions at Sea.[109] Surveys of the hull, machinery and equipment of cargo ships are provided for at regular intervals not exceeding 5 years.[110] After such surveys are made, no changes are allowed made in the structural arrangement, machinery, equipment and other items surveyed.[111] Safety certificates issued are good for definite periods varying between 1 year to 5 years.[112] Control is provided for by Port Authority officers to check that the certifications issued are valid and that the condition of the ship or of its equipment corresponds substantially with the certificates.[113] Lacking this the ship can be prevented from sailing, with all possible efforts being taken to avoid unduly detention or delay, and the flag State required notified of particulars.[114]

Tankers of 10,000 tons and larger are required to have 2 remote steering gear control systems operable separately from the *navigating* bridge, and alarms are required in case of failure of either.[115] On new tankers of the same size the main steering gear must be capable of putting the rudder from 35 degrees on one side to 35 degrees on the other side when the ship is fully loaded and running ahead at maximum service speed.[116] For new tankers of 20,000 metric tons and larger fire extinguishing systems, including a fixed deck froth system and a fixed inert gas system, are required to protect the cargo tanks deck area and the cargo tanks.[117] If an equivalent system is used in place of the deck froth system, it must be capable of extinguishing spill fires and prevent ignition of spilled oil not yet ignited, and be capable of combating fires in ruptured tanks.[118] If an equivalent system is used in place of an inert gas system, it must be capable of preventing dangerous accumulations of explosive mixtures in intact cargo tanks and be so designed as to minimize the risk of ignition from the generation of static electricity by the system.[119] Radars are required on all ships greater than 1,600 tons with ships of 10,000 tons or greater requiring at least two independent radars.[120] Within 12 hours of departure the ships steering gear must be checked and tested by the ship's crew, and emergency steering drills must take place at least every three months to practice emergency steering procedures.[121]

5.3.3. INMARSAT[122] and Other Communication and Navigational Regimes

Basically INMARSAT deals with a space segment owned or leased by INMARSAT for satellite use. Its and other communication and navigation system's relevance lies in the fact that it contributes to improving distress and safety of life at sea communications, efficiency and management of ships, maritime public correspondence services and radio determination capabilities which would indirectly reduce pollution incidents.[123] Access is permitted for use by ships upon payment and without discrimination on the basis of nationality.[124] Various organs are provided for including the Assembly, the Council and the Directorate.[125] The Operating Agreement provides that a government may designate an entity to assume financial, technical and operational responsibilities within INMARSAT.[126] Managerial responsibility is vested in a strong governing body in which policy-making responsibilities are given to investors in direct proportion to their utilization of the INMARSAT system.[127] Primarily INMARSAT provides telephone and telex services between ship and shore and vice versa, first access for emergency and distress messages and information services. The system is not yet adequate to serve all potential users as regards access, coverage and costs, and these must rely on other communications systems.[128]

Other satellite systems in use which will be covered briefly here include the following.[129] SARSAT is a joint search and rescue satellite aided system between Canada, the US, France, Norway and Sweden to reduce detection time and location of ships and aircraft in difficulty.[130] This has recently been linked to the Soviet COSPAS system providing the possibility for a worldwide search and rescue system. Other systems available include the electronic position fixing system very low frequency (VLF) OMEGA[131] and 2 satellite navigational systems, TRANSIT and NAVISTAR.[132] Additional systems include ice and ocean condition surveillance and monitoring assisted by satellite imagery by Landsat, NOAA's National Environmental Satellite Services and the IMO Plan for the Establishment of a World-wide Navigational Warning Service (1977) for shipping information regarding safe navigation and meteorology.[133] Additionally the recent agreement signed between the Soviet Union and the US regarding sea and air transportation can be mentioned.[134] Under this agreement the American 'LORAN-C' radio navigation system and the Soviet 'TSJAJKA' will be standardized and the land based stations connected in the Arctic Pacific, with the common zone extended within 1992. Soviet experts expect that these navigation chains can also be established in the coastal areas of Europe or generally where there is found much ship or air traffic, and in order for a ship or an airplane to pinpoint their exact position, they merely have to make contact with three of the land based stations of either LORAN-C or TSJAJKA.

5.3.4. STCWS[135]

As noted almost all marine accidents have a human factor involved, and to combat this STCWS was formed after the IMO called an international conference to establish basic standards of *crew* training and ship operation. The Convention requires that the master, officer in charge of the watch, engineer officers and engine room ratings must, 'be aware of the serious effects of operational or accidental pollution of the marine environment and shall take all possible precautions to prevent such pollution, particularly within the framework of relevant international and port regulations.'[136] As well it is recommended that certification include testing concerning navigation and position determination in ice, ship manoeuvring and handling in ice or ice accumulation on board, and establishing adequate and safe watchkeeping arrangements under special operating conditions including ice.[137] Other minimum standards are set forth to reduce accidents, but as one author notes, 'it remains to be seen how it (the STCWS) will impact on the reduction of marine accidents'.[138]

5.4. Other General Maritime Safety Conventions[139]

Although the Conventions which follow may be somewhat peripheral, they do indicate important shipping problems with indirect application to marine pollution which must be dealt with.

5.4.1. International Convention for the Unification of Certain Rules of Law Relating to Assistance and Salvage at Sea (1910)[140]

This Convention deals with salvaging wrecked ships, saving life and property and rewarding those who salvage. One author notes that the Convention codified most of the ancient principles concerning *salvage* law, but that this did not cover marine pollution problems involved with salvage operations.[141] Because of this the Convention is out of date, and the rules often not followed. This has resulted in significant changes being made in an IMO draft Convention which was to be considered at a diplomatic conference in 1989.[142]

5.4.2. Load Lines Convention[143]

This Convention attempts to prevent the *overloading* of vessels, which causes ship instability and as well increases the danger of running aground, resulting in marine pollution from the cargo or fuel bunkers.[144]

5.4.3. Collisions Convention[145]

This Convention is a safety convention dealing with vessel traffic separation zones. Although presently ship traffic through many areas might not warrant such separation zones, it does not take much imagination to think of such problems arising should ship traffic increase or offshore oil production begin.[146]

5.4.4. Containers Convention[147]

This Convention contains provisions covering the safe handling and shipping of containers. Since container ships are presently in much use, and many varieties of dangerous substances are freighted, accidents involving containers are conceivable. One author notes that, '(A)ccidents involving such containers might cause serious and significant pollution damage - under certain circumstances more serious than oil pollution.'[148] Additionally noted is that liability questions of this type are presently unsettled.

Other agreements including IMO and ILO Conventions which can be mentioned with relevance to marine pollution include, the Memorandum of Understanding on Port State Control of 26 January 1982.[149] 14 States under this Agreement undertake to maintain effective port State control that ships in its ports comply with the main IMO safety conventions, ILO Convention No. 147[150] and MARPOL 73/78 dependent upon such Conventions being in force and the port State being a Party, but regardless of whether the flag State is a Party. Each authority must inspect a minimum of 25% of the ships in its ports and restrict the ship's sailing where deficiencies are discovered. The ILO Convention No. 147 of 1976 concerning Minimum (Crewing) Standards in Merchant Ships, ILO Convention No. 109 of 1958 concerning Wages, Hours of Work on Board Ship and Manning,[151] ILO Convention No. 92 of 1949 supplemented by No. 133 (1979) concerning Crew Accommodation on Board Ship,[152] and ILO Convention No. 145 of 1976 concerning Continuity of Employment of Seafarers,[153] IMO Convention on Facilitation of International Maritime Traffic of 1965,[154] IMO International Convention on Tonnage

Measurement of Ships of 1969,[155] IMO International Maritime Dangerous Goods Code of 1966,[156] IMO Convention on Maritime Search and Rescue of 1979,[157] and IMO Convention for the Suppression of Unlawful Acts against the Safety of Maritime Navigation of 1988[158] can also be mentioned.

5.5. Regional Conventions - Helsinki and UNEP

The Helsinki Convention deals extensively with vessel-source pollution through Articles 2, 4,[159] 7 and 8 and Annex IV. Under Article 2 'vessels and aircraft' are defined to mean waterborne or airborne craft of any type whatsoever, including hydrofoil boats, air-cushion vehicles, submersibles, floating craft, self-propelled or not, and fixed or floating platforms. Article 4 exempts application of Convention provisions to military ships and aircraft, however requires States to ensure, however without impairing the operational capabilities, that such ships and aircraft act in a manner consistent so far as is reasonable and practicable with Convention provisions. Article 7 requires States to take measures consistent with Annex IV to protect the Baltic from deliberate, negligent or accidental discharges from ships of oil, harmful substances, sewage and garbage. In addition States are required to develop and apply uniform requirements for capacity and location of reception facilities for residues of oil, harmful substances, sewage and garbage, paying attention to the special needs of passenger ships and combination carriers. Article 8 requires States to apply Convention provisions to pleasure craft as well as take any special measures to stop harmful effects from such craft, including providing adequate reception facilities.

Annex IV is complex consisting of 8 Regulations dealing with pollution from oil, noxious liquid substances in bulk, harmful substances in packaged forms, sewage and garbage; and 5 Appendices giving lists of oils, guidelines for categorization of noxious liquid substances, lists of noxious liquid substances carried in bulk, lists of other liquid substances carried in bulk, and a cargo record book for ships carrying noxious liquid substances carried in bulk. Though extensive they will not be covered here, since in 1987 the Helsinki Commission adopted Recommendations amending Annex IV and Annex VI (reporting provisions on incidents involving harmful substances) to allow supersession by MARPOL 73/78 provisions.[160] This was done both because Helsinki Convention Parties are also Parties to MARPOL 73/78 (and thus Parties to the mandatory Annexes I and II and amended Protocol I) and because of the result of the developments under MARPOL 73/78.[161] '(T)here was no need to retain in the Helsinki Convention texts which reproduce these provisions.'[162] The result of these changes is that all the regulations concerning oil, noxious liquid substances carried in bulk and definitions, as well as the respective lists were amended to those of MARPOL 73/78 noted above. As one author notes, though Helsinki Convention provisions are broadly similar to MARPOL 73/78 provisions, the Helsinki are in most cases stricter,[163] so the Recommendations may be not all that fortunate a development for the Baltic.

The UNEP Conventions noted above in Part I Section 3.2.3. deal only generally with vessel-source pollution,[164] requiring State Parties to take all measures in conformity with international law to prevent, abate and combat vessel-source pollution in the respective areas and to ensure effective implementation in those areas of generally recognized international rules relating to control such pollution.[165]

5.6. Conclusions

As can be seen in contrast to the other pollution sources vessel-source marine pollution is highly regulated especially at the global level. This is both because this source is much more international and because there have been as noted numerous catastrophes which have goaded States and international organizations into taking international action. This process is not static however and continuing development can be observed both in the IMO where Amendments and Protocols to the IMO conventions are under continual development, and in States where more stringent legislation often is passed following environmental shipping incidents.[166] In spite of the many developments achieved in this area however, the pollution statistics indicate that much still remains to be done. It is perhaps noticeable, even when action is finally taken by States and international organizations, unfortunately much marine pollution remains. This might indicate the scale of the problems involved.

Notes - Chapter 5

1 See generally Gold Chapter 6. Unluckily perhaps note the diversity of international interests involved in *one* ship as expressed by the legal advisor to Shell International Petroleum Co., '(A) ship may strand on the high seas and cause pollution in two neighbouring States, say, France and England (as with the TORREY CANYON in 1967). She may be owned say, by a Liberian company, bareboat chartered to a Bermudian company, managed by an English company, time chartered to a Greek company and voyage chartered to an American company. Her cargo may have been sold during the voyage by the American company to a Japanese one. The officers may be English and the crew, Indian. The international nature of shipping business creates such diversity of interests, with potential conflicts of law and jurisdiction, daily.' Curtis, Jeff B., 'Vessel Source Oil Pollution and MARPOL 73/78: An International Success Story?' *Environmental Law*, Vol. 15, 1985 (Curtis) p. 680.

2 Wang pp. 307 and 313 notes when mapping out the current international law for vessel-source pollution control that this includes relevant IMO conventions in force, especially MARPOL, customary law, and some provisions of LOSC that could be regarded as international law, though 'there is no guarantee that everything that has been said will be implemented in practice.' Further noted is that through experience from the IMO conventions, increased awareness regarding serious oil pollution, and conclusion of the more universal LOSC provisions, 'a relatively recognizable international law regime for vessel-source oil pollution control' has been developed.

3 For complete up-to-date lists of the IMO conventions and respective State ratifications thereof (December 31, 1988), see IMO Status. For the non IMO conventions and respective State ratifications thereof (1985), see De Cesari. Note that for most of the conventions mentioned below surveys of ratifying States are given.

4 Geneva Convention on the Territorial Sea and the Contiguous Zone (1958), (1958 Territorial Sea Convention), in force 10 September 1964, I *New Directions* 1; Ratifying, etc. States include Yugoslavia, Venezuela, US, UK, USSR, Ukraine, Uganda, Trinidad and Tobago, Tonga, Thailand, Switzerland, Swaziland, Spain, South Africa, Solomon Islands, Sierra Leone, Senegal, Rumania, Portugal, Nigeria, Netherlands, Mexico, Mauritius, Malta, Malaysia, Malawi, Madagascar, Lesotho, Kenya, Kampuchea, Japan, Jamaica, Italy, Israel, Hungary, Haiti, Democratic Republic of Germany, Finland, Fiji, Dominican Republic, Denmark, Czechoslovakia, Byelorussia, Bulgaria, Belgium, and Australia, De Cesari, pp. 119. Despite the lack of ratification by Norway, Fleischer s. 88 notes that this Convention can essentially be viewed as an expression of earlier valid customary law and therefore binding on *all* States. Geneva Convention on the High Seas (1958) (1958 High Seas Convention), in force 30 September 1963, I *New Directions* 257; Ratifying, etc. States include Yugoslavia, Venezuela, Upper Volta, US, UK, USSR, Ukraine, Uganda, Trinidad and Tobago, Tonga, Thailand, Switzerland, Swaziland, Spain, South Africa, Solomon Islands, Sierra Leone, Senegal, Rumania, Portugal, Poland, Nigeria, Netherlands, Nepal, Mongolia, Mexico, Mauritius, Malaysia, Malawi, Madagascar, Lesotho, Kenya, Kampuchea, Japan, Jamaica, Italy, Israel, Indonesia, Hungary, Haiti, Guatemala, Federal Republic of Germany, Democratic Republic of Germany, Finland, Fiji, Dominican Republic, Denmark, Czechoslovakia, Costa Rica, Central Africa, Byelorussia, Bulgaria, Belgium, Austria, Australia, Albania, and Afghanistan, De Cesari, pp. 133. Fleischer s. 103 again notes that this Convention can be considered an expression of generally valid law. See Part I Section 4.2.1. for the 1958 Convention on the Continental Shelf (1958). The Geneva Convention on Fishing and Conservation of the Living Resources of the High Seas (1958), in force 20 March 1966, I *New Directions* 353, does not contain any specific rules regarding marine pollution from shipping and will not be considered here.

5 See Churchill pp. 79-82 and 253 who adds that under customary law much the same right exists.

6 The journey must be for the purpose either of traversing the territorial sea without entering internal waters, of proceeding to internal waters or of proceeding to the high seas from internal waters.

7 International Convention for the Prevention of Oil Pollution (1954) (OILPOL), 9 *International Legal Materials* 1. The text presented here is a composite of the 1954 Convention, the Amendment of 1962 and the Amendment of 1969.

8 Ratifying States are not listed for OILPOL because upon a State's ratification of MARPOL 73/78, OILPOL lapses, and this has occurred for *most* States. See Section 5.2.3. for MARPOL 73/78. A brief comparison of IMO Status, pp. 69-72, 377-382 and 445-451 indicates only *Canada, Cyprus* and *Venezuela* to have not ratified MARPOL 73/78 after ratifying OILPOL. One author however considers while, 'OILPOL has now been superseded it will, nevertheless, continue to be a part of international marine pollution law for some time to come,' Gold p. 38.

9 Wang p. 313. Additionally noted is that there is no reason for States to prefer OILPOL over MARPOL 73/78 and MARPOL 73/78's rules and standards are those referred to in the LOSC.

10 For such ships each Party is required however so far as is reasonable and practicable to apply the Convention.

11 For naval ships each Party undertakes so far as is reasonable and practicable to adopt and apply measures parallel to the Convention.

12 Ships other than tankers are allowed to discharge if the ship is heading for a port not equipped with reception facilities.

13 Originally in the 1954 Convention discharges were as well prohibited in special zones including the North Sea, the Baltic and Black Seas. Special zones were abolished by the 1969 Amendments.

14 The discharge of ballast from a cargo tank must be cleaned so that if discharged from a stationary tanker into clean calm water on a clear day, it would produce no visible traces of oil on the surface of the water. The discharge of oil or oily mixtures from machinery space bilges are governed by the allowable discharge standards noted above.

15 Churchill p. 248 notes that the 1954 discharge standards were made more stringent by the 1962 Amendments, but they remained ineffective due both to the difficulty of detecting violations and to the reluctance of many flag States to prosecute their masters where violations were detected.

16 Curtis p. 690.

17 *Ibid.* Though the oil companies proclaimed LOT a success, Curtis notes that the system violated both the 1954 Convention and the 1962 Amendment, because LOT discharges exceeded the 100 ppm limitations required by the 1954 Convention in coastal zones and conflicted with the 1962 prohibition of any discharge by new tankers over 20,000 tons. Additional criticism generally includes that although the regulations protected coastal areas from immediate discharges, they did not prevent oily ballast waters and residues from drifting into these areas after a *legal* discharge 50 miles at sea, *ibid.* pp. 685-686. Problems also occurred with the enforceability of LOT as well as the lack of port State reception facilities in violation of Article VIII to handle the residue slop retained on board. *Ibid.* p. 693 notes that industry figures released to the public varied drastically from internal company reports. Public figures indicated an 80% use rate of LOT by oil tankers and an 90% effectiveness rate, while a major oil company found that 33% used LOT, 33% used it poorly and 33% did not use LOT at all. Additionally noted was that the impossibility for monitoring LOT compliance, was among other things, which lead to the formation of MARPOL 73. Churchill p. 249 is somewhat more optimistic stating that the LOT and the 1969 Amendments 'have made a significant contribution to the reduction of deliberate pollution of the sea by oil.' The same author cites a report by the IMO in September 1981 which concludes that a reduction of deliberate marine oil pollution by *30%* has taken place 1971 to 1980 while at the same time oil transport by sea during the same time period has increased *17%*. It is noted however that the 1969 Amendments are not wholly effective partly due to LOT not always being capable of being used (oil cargoes for example being incompatible) and partly because of many ports not having adequate reception facilities for oily wastes, in violation of Article VIII. State interest in 1954 in OILPOL provisions is presented on an interesting chart, Curtis p. 686, showing different States' positions on four proposals for general and zonal discharge prohibitions. Norway voted *against* all four, the US voted *against* three and abstained on one, the Soviet Union voted *for* two and abstained on two and the UK voted *for* all four. This voting denoting State environmental interest is almost opposite for the countries noted, 35 years later, see Aftenposten 30.mars 1989 Uke 13., Nr.143., 130.årg., s. 9.

18 De Cesari p. 101.

19 International Convention for the Prevention of Pollution from Ships (1973) (MARPOL 73), 12 *International Legal Materials* 1319. Ratifying, etc. States include Yugoslavia, Yemen, Uruguay, UK (with exception of Annexes III, IV, and V), Tunisia, Peru, Norway (with exception of Annex IV), Kenya, Jordan, Italy, Hungary, Federal

Republic of Germany, Colombia, Bulgaria (with exception of Annexes III, IV and V), Brunei Darussalam (with exception of Annexes III, IV and V), Benin, Belgium (with exception of Annexes III, IV and V) and Antigua and Barbuda. In total there were 18 Contracting States, the combined merchant fleets of which constituted approximately 11% gross tonnage of the world's merchant fleet. IMO Status pp. 443-451. The Protocol of 1978 Relating to the International Convention for the Prevention of Pollution from Ships, 1973 as Amended, (1978 Protocol) 17 *International Legal Materials* 546. Article I of the 1978 Protocol provided that the 1978 Protocol would merge with MARPOL 73 into one final instrument (MARPOL 73/78). Article V of the 1978 Protocol required that 'the Protocol shall enter into force twelve months after the date on which not less than fifteen States, the combined merchant fleets of which constitute not less than 50 percent of the gross tonnage of the world's merchant shipping, have become Parties.' This occurred October 1982 and MARPOL 73/78 entered into force 2 October 1983. Ratifying, etc. States include Yugoslavia, Uruguay, Tuvalu, Tunisia, Sweden, Suriname, Saint Vincent and the Grenadines, Portugal, Poland, Peru, Panama, Oman, Norway (with exception of Annex IV), Marshall Islands, Lebanon, Japan, Italy, Hungary, Greece, Federal Republic of Germany, German Democratic Republic, Gabon, France, Finland, Egypt, Denmark (accession subject to reservation with regard to obligations of Greenland), Democratic People's Republic of Korea, Czechoslovakia, Côte d'Ivoire, Colombia, Austria and Antigua and Barbuda. All of the following ratified, etc. with the exception of Annexes III, IV and V, US, UK, USSR, Syrian Arab Republic, Switzerland, Spain, South Africa, Republic of Korea, Netherlands (approval), Liberia, Israel, Indonesia, India, Iceland, China, Burma, Bulgaria, Brunei Darussalam, Brazil, Belgium, Bahamas and Australia. The number of Contracting States currently consists of 54, the combined merchant fleets of whom constitute approximately 81% of the world's merchant's fleet gross tonnage. Additionally, the 1984 Amendments entered into force 7 January 1986, and 1985 Amendments to Protocol I entered into force 6 April 1987. IMO Status pp. 63-98. States accepting the optional Annexes III-V at a later date will be shown below in Section 5.2.3.2.4.

20 Curtis p. 693.

21 Gold, p. 39. The 1978 Protocol allowed States to adopt MARPOL 73/78 while accepting only Annex I dealing with the Oil Pollution Regulations. Ratification of Annex II dealing with noxious chemicals was put on hold for a specified period of time due to State aversion. See Section 5.2.3.2.

22 Wang pp. 317-319 notes that MARPOL 73/78's strength lies in its consolidation of the construction, design, equipment and manning standards and the discharge standards of the OILPOL Convention, *not* in its provisions of enforcement jurisdiction.

23 Churchill p. 253 notes that under *customary* law a flag State can prescribe anti pollution provisions for its vessels anywhere. It can enforce them under customary law anywhere but in the territorial sea or port of another State, but even then the flag State can bring proceedings against the shipowner within its boundaries or against the ship when it returns.

24 Wang pp. 317-319 notes that under MARPOL 73/78 the content of coastal State jurisdiction and the geographical coverage was left to LOSC to decide and port State enforcement was treated as a part of coastal State jurisdiction. It is additionally noted coastal States have potential enforcement powers to inspect foreign ships in their ports, but coastal States' competence to inspect ships in out-of-port waters is left vague under Article 4 (2) and hence unconfirmed under MARPOL 73/78. Churchill p. 253 notes that under *customary law* a coastal State may prescribe anti pollution legislation for foreign ships in its territorial sea as long as it does not hamper innocent passage. Additionally a port State may adopt anti pollution legislation for foreign ships in its ports and make the observance of such legislation or particular international conventions a condition of entry. Normally though the provisions must be non discriminatory under bilateral treaties of friendship, commerce and navigation.

25 Annex I, Regulation 20 requires that all movements of oil be recorded, including times, dates, quantities and operations. The book may be inspected by port State authorities and certified copies are admissible in judicial proceedings as evidence of the facts stated in the entry.

26 *Ibid.* Regulation 15.

27 Curtis p. 706. Adequate monitoring system technology does not exist widely, and the oil record book is open to fraudulent entries by the ship operators and is impossible to verify. See Part II Section 9.3.2.1. Substantiating this view it can be noted that Regulation 4, dealing with surveys and inspections related to the International Oil Pollution Prevention Certificate, was substantially changed and much expands the list of flag State obligations, whereas Regulations 15 and 20 of Annex I, dealing with the monitoring system and the oil record book respectively, were incorporated from the 1973 Convention into MARPOL 73/78 with only minor alterations.

28 Wang p. 320 considers this provision in addition to Articles 7(2) and 8 (3) as giving a ship a sense of security against 'wanton port apprehensions' through violation reports to the flag States and compensation for loss or damage caused by undue detentions or delays.

29 It is under Article 9 that MARPOL 73/78 supersedes OILPOL.

30 Any dispute unable to be settled between the Parties and if not agreed otherwise, must upon request by any Party be sent to arbitration as outlined in Protocol II to MARPOL 73/78.

31 Article 11 (1)(b) was changed by the 1978 Protocol to require States to communicate to the IMO a list of nominated surveyors or recognized organizations authorized by the States to examine the design, construction, operation and equipment of ships carrying harmful substances in accordance with the Regulations.

32 Flag States undertake to investigate casualties to their ships subject to the Regulations if a major deleterious effect upon the marine environment has occurred and supply the IMO with their findings if the information may assist in determining what changes are desirable in the Convention.

33 Annex I Regulations 4, 8 and 13 were extensively changed by the 1978 Protocol. For these provisions see Section 5.2.3.2.2.

34 Annex I Regulation 9. Wang p. 313 notes that MARPOL 73 introduced the 'prohibited zone' concept in Regulation 9, but whether the IMO conventions 'have the authority to delineate universally binding national jurisdiction boundaries has been constantly under challenge.'

35 *Ibid.* Regulation 1, Appendix I.

36 Churchill p. 250 notes that special areas were *reintroduced* though different from those established under OILPOL and done away with by the 1969 Amendments. IMO News Nr. 1, 1991, pp. 5 and 16 note that the Antarctic is to be made a special area under Annexes I and V of MARPOL 73/78. Briefly no discharges of oil will be permitted at all, and Contracting Parties whose ports ships depart en route to or arrive from Antarctic will be required to ensure that reception facilities are provided as soon as practicable for sludge, dirty ballast, tank washing water and other oily residues. Garbage must be retained on board. Plans also exist for possibly making Australia's Great Barrier Reef and the Wider Caribbean a special area (under Annex V). Guidelines are also underway to deal with Particularly Sensitive Areas.

37 Annex I Regulation 10(4) and 10(2)(b).

38 *Ibid.* Regulations 15 and 16.

39 Curtis p. 695. Wang p. 315 writing in 1986 notes much the same.

40 Annex I Regulations 15 and 16.

41 *Ibid.* Regulation 24 and Gold p. 64.

42 Annex I Regulation 25.

43 *Ibid.* Regulation 20 (ix) excepts discharges from segregated ballast tanks.

44 *Ibid.* Regulation 20. This book can be inspected by any State Party to MARPOL 73/78. As noted this however implies the same weakness as under OILPOL, *compliance* is dependent upon the operators being conscientious.

45 *Ibid.* Regulation 13. Curtis p. 696 notes that major shipping states and France and Japan, the largest ship-building States, objected to this provision. It passed however since most states viewed the SBT's as a means of avoiding construction of the reception facilities.

46 These involve situations of *force majeure*, non intentional damage and officially approved pollution combating operations.

47 Emphasis added.

48 UNEP Report Appendix.

49 Annex II, Appendix I. Annex II itself is unchanged as incorporated into MARPOL 73/78. It is Article II of the 1978 Protocol which changes the *implementation* of this Annex. See Section 5.2.3.2.1.

50 This varies between 0.1 and 0.01 for residual concentration per cent by weight, dependent upon substance, Annex II Appendix II.

51 *Ibid.* Regulation 5 (1). A Category A substance for example is the element phosphorus, Appendix II.

52 *Ibid.* Regulation 5(2) and (3). Examples of Categories B and C include ammonia and potassium hydroxide respectively, Appendix II.

53 *Ibid.* Regulation 5(7), (8) and (9). Briefly the discharge of Category A substances and mixtures are prohibited unless diluted to specified levels noted in Appendix II and *added* to at least 5% of the total tank volume. The discharge of Category B and C substances are less stringent yet similar.

54 *Ibid.* Regulation 13.

55 *Ibid.* Regulations 8, 9 and Appendix IV.

56 Gold p. 69.

57 Curtis p. 697.

58 Annex III deals specifically with negligent handling and packaging during loading, carriage and discharge or accidents, and attempts to limit pollution caused thereby. The Regulations cover adequate packing, the carriage and prohibition thereof of listed harmful substances in accordance with Annex II, prohibition of empty containers, the proper marking and labelling of packages indicating contents, requirement of documentation on board, limits as to quantity, and proper notification to authorities in the loading and discharge ports. See also Gold pp. 69-70.

59 Annex IV specifically requires that ships not discharge sewage within 4 miles of the nearest land unless processed by an approved sewage treatment system meeting Convention standards. Disinfection systems, a holding tank, discharge lines, special certification and adequate reception facilities are additionally required. See also Gold pp. 70-71.

60 Annex V specifically encompasses all garbage of victual domestic and operational waste occurring during the operation of the ship, excluding fish. The disposal of all plastics is prohibited. Reception facilities must be provided. Disposal shall generally

take place at least 3 nautical miles from land for ground up garbage or 12 miles for non ground, and floating garbage at least 25 miles. See Gold pp. 71-72.

61 Annex II Regulation 7, Annex IV Regulation 10, Annex V Regulation 7.

62 Curtis pp. 698-699. Wang pp. 312-313 adds that vessel-source oil pollution control in its history of 3 decades, 'has been complicated by jurisdictional issues, technological immaturity, and the *passive* attitude of the maritime States'. Churchill p. 250 adds that ratification difficulties are due to 'the considerable economic cost and technical difficulties of complying with its provisions.' Gold p. 58 notes '(T)he original 1973 MARPOL Convention attempted to eliminate ship-source pollution from all sources almost at one stroke. Unfortunately a number of very complex technical problems, primarily related to measures designed to prevent pollution from chemicals, were such that many States could neither accept nor implement the new convention in the very ambitions time-frame originally planned.'

63 Curtis pp. 698-699.

64 Gold p. 29. Cutis p. 698 mentions 12 other accidents occurring between 1976 and 1977. Wang p. 312 notes that but for the momentum provided by some of the catastrophes such as the TORREY CANYON (see Section 5.3.1.) perhaps the IMO might have ceased to survive due to lack of State commitment. Additionally noted is that there has been little genuine State commitment to establishing workable pollution control standards, and some regard the IMO as a maritime States' club.

65 Gold p. 29. The author notes just a few weeks after this conference the 'total' tanker disaster took place off the coast of France, where VLCC AMOCO CADIZ lost all of its 230,000 ton crude oil cargo.

66 Article II (1). IMO Status p. 97 notes Annex II entered into force *April 6, 1987*. Article II (2) states that the Parties to the Protocol shall not be under any obligations (nor entitled to any privileges) under the Convention as related to Annex II during the period specified in Article II (1).

67 17 *International Legal Materials* 546. Entry into force 2 October 1983. Ratifying States shown above with 1978 Protocol. Entry into force of 1987 Amendments, 1 April 1989. IMO Status p. 63.

68 Churchill p. 250 writes that Annex I was further amended in 1984. Wang does not address this in his thorough analysis from 1986.

69 An initial survey is required, before the ship enters into service or an International Oil Pollution Certificate is issued, periodic surveys must occur not less often than every five years, a minimum of 1 survey is required during the certificate period, yearly surveys are mandatory and stricter action is required concerning ships not meeting required Protocol standards.

70 Curtis pp. 699-700 notes the use of SBT's (and COW) in addition to doing away with the human element involved in the LOT system, means that the volume of mixtures (oil or chemical) is reduced which implies reduced dependence on reception facilities, a main problem under MARPOL 73. Wang p. 319 considers this in terms of enforcement of construction, design equipment and manning standards to be an important contribution of MARPOL 73/78. These could in addition help to detect illegal discharges as well when used in connection with the Oil Record Book.

71 As noted in Section 5.2.3.1.2. the MARPOL 73 standard was 70,000 dwt.

72 Annex I Regulation 13(E).

73 Gold p. 66 notes that using the COW system necessarily results in a build up of explosive gases in cargo tanks, and the Protocol to SOLAS 1974 also adopted at the same 1978 Tanker Safety Conference and entered into force in May 1981, requires an IGS to be used during COW operations.

74 Annex I Regulations 13B(3), 13(7), 13(8), 13(9). CBT's, using only certain specified tanks for ballast purposes, were available for existing tankers for a period of 2-4 years from October 1983. This system was meant as a transitional system which allowed existing tanks, piping and pumps to be used. However after the legal period the other systems are required to be put into use or the ship removed from service, Regulation 13(9).

75 *Ibid.* Regulation 13C.

76 *Ibid.* Regulations 9, 12 and 16. See Section 5.2.3.1.2.

77 Gold pp. 60, 159-161.

78 Curtis p. 703-704. Also noted are the costs involved of converting existing tankers or building new tankers with the required equipment in a depressed market.

79 12 *International Legal Materials* 1319. Entry into force 6 April 1987. Ratifying States shown above with 1978 Protocol. 1985 Amendments entry into force 6 April 1987. IMO Status p. 63.

80 Gold p. 39.

81 12 *International Legal Materials* 1319. Not yet in force. Accepting States currently include Yugoslavia, Uruguay, US (OPN June 1991 p. 6), UK, USSR, Tuvalu, Tunisia, Sweden, Suriname, Saint Vincent and the Grenadines, Portugal, Poland, Peru, Panama, Oman, Norway, Netherlands, Marshall Islands, Lebanon, Japan, Italy, Hungary, Greece, Federal Republic of Germany, German Democratic Republic, Gabon, France, Finland, Egypt, Denmark, Democratic People's Republic of Korea, Czechoslovakia, Côte d'Ivoire, Colombia, Belgium, Austria and Antigua and Barbuda. The number of Contracting Parties currently includes 36, the combined merchant fleets of whom constitute approximately 48% of the world's merchant fleet's gross tonnage. IMO Status pp. 73-93.

82 12 *International Legal Materials* 1319. Not yet in force. Accepting States currently include Yugoslavia, Uruguay, USSR, Tuvalu, Tunisia, Sweden, Suriname, Saint Vincent and the Grenadines, Portugal, Poland, Peru, Panama, Oman, Marshall Islands, Lebanon, Japan, Italy, Hungary, Greece, Federal Republic of Germany, German Democratic Republic, Gabon, France, Finland, Egypt, Denmark, Democratic People's Republic of Korea, Czechoslovakia, Côte d'Ivoire, Colombia, Austria and Antigua and Barbuda. The number of Contracting States currently include 32, the combined merchant fleets of whom constitute approximately 40% of the world's merchant fleet's gross tonnage. IMO Status pp. 75-93.

83 12 *International Legal Materials* 1319. Annex V entered into force, 31 December 1988. Accepting States include Yugoslavia, Uruguay, US, UK, USSR, Tuvalu, Tunisia, Sweden, Suriname, Saint Vincent and the Grenadines, Portugal, Poland, Peru, Panama, Oman, Norway, Netherlands, Marshall Islands, Lebanon, Japan, Italy, Hungary, Greece, Federal Republic of Germany, German Democratic Republic, Gabon, France, Finland, Egypt, Denmark, Democratic People's Republic of Korea, Czechoslovakia, Côte d'Ivoire, Colombia, China, Belgium, Austria and Antigua and Barbuda. The number of Contracting Parties currently consists of 38 States, the combined merchant fleets constituting 56% of the world's merchant fleet's gross tonnage. IMO Status pp. 77-93.

84 IMO News Nr. 1:1991, p. 6 notes that plans exist for adding a new Annex to MARPOL 73/78 covering air pollution. Specific pollutants under consideration include chlorofluorocarbons (CFC's) (ozone depleters); exhaust gases which produce sulphur oxides and nitrogen oxides; volatile organic compounds produced from oil cargo; and incineration of garbage and ship generated wastes. See generally Churchill R. and D. Freestone (eds.) *International Law and Global Climate Change*, Chapter 9, 'Controlling Emissions of Greenhouses Gases' (Churchill Green), p. 147 for an extensive summary of CFC's and other gases and the present international control.

85 Curtis p. 705.

86 Gold pp. 69-71. Gold apparently was the more correct. As seen Annex V has already entered into force and IMO News Nr. 2:1991 p. 3 reports that Annex III is expected to enter into force shortly after July 1, 1992. The text is expected to be somewhat amended following entry into force to allow a more explicit connection to the International Maritime Dangerous Goods Code.

87 International Convention Relating to Intervention on the High Seas in Cases of Oil Pollution Casualties (1969) (Intervention Convention), 9 *International Legal Materials* 25. This Convention came in force 6 May 1975 with ratifying, etc. States include Yugoslavia, Yemen, US, UK, United Arab Emirates, USSR, Tunisia, Syrian Arab Republic, Switzerland, Sweden, Suriname, Sri Lanka, Spain, South Africa, Senegal, Qatar, Portugal, Poland, Papua New Guinea, Panama, Oman, Norway, New Zealand, Netherlands, Morocco, Monaco, Mexico, Liberia, Lebanon, Kuwait, Japan, Italy, Ireland, Iceland, Ghana, Federal Republic of Germany, German Democratic Republic, Gabon, France, Finland, Fiji, Ecuador, Dominican Republic, Denmark, Cuba, Côte d'Ivoire, Cameroon, Bulgaria, Benin, Belgium, Bangladesh, Bahamas, Australia and Argentina. The Protocol relating to Intervention on the High Seas in Cases of Pollution by Substances other than Oil, (1983) (Protocol), 16 *International Legal Materials* 1103, came in force 30 March 1983 with ratifying and acceding States including Yugoslavia, Yemen, US, UK, USSR, Tunisia, Switzerland, Sweden, Portugal, Poland, Oman, Norway, Netherlands, Mexico, Liberia, Italy, Federal Republic of Germany, France, Finland, Denmark, Belgium, Bahamas and Australia. IMO Status pp. 163-186.

88 Related interests are defined in Article II to include maritime coastal, port or estuarine activities, including fisheries; tourist attractions; and the health of the coastal population and the well-being of the area concerned, including conservation of living marine resources and of wildlife.

89 Oil is defined in Article II as crude oil, fuel oil, diesel oil and lubricating oil.

90 Additionally notification of measures taken must be made without delay to the concerned States as well as to the Secretary-General of the IMO.

91 Emphasis added.

92 The substances include asphalt solutions, oil, distillates, gas oil, gasoline blending stocks, gasolines, jet fuels, naphtha, noxious substances (approximately 100 substances including DDT, cyanide, arsenic and mercury compounds among other notables) and liquified bulk substances (about 20 substances including chlorine, methane (LNG) butane, propane and sulphur dioxide).

93 Gold p. 24 notes the oil tanker TORREY CANYON grounded off the coast of England (caused solely by human error) spilling the largest amount of oil into the seas up to 1967. 80,000 tons of crude oil washed upon the British and French coasts in a 200 mile arc. The oil discharge occurred both after the initial grounding and after the bombing of the vessel on orders of the British government when it was decided there was no other way to deal with the wreck. Over *2.5 million gallons* of chemical dispersants were additionally administered which by some were considered to cause more disastrous effects than the original oil.

94 *Ibid.* p. 38.

95 *Ibid.*

96 International Convention for the Safety of Life at Sea (1974), (SOLAS), 14 *International Legal Materials* 959, (Convention entered into force 25 May 1980, 1981 Amendments entered into force in 1 September 1984, 1983 Amendments entered into force 1 July 1986 and 1987 Amendments entered into force 30 October 1988). Protocol of 1978 Relating to the International Convention for the Safety of Life at Sea, 1974, As Amended, 17 *International Legal Materials* 579, 1981 Amendments *Gassetta Ufficiale della Repubblica Italiana* Suppl. 260, 20 September 1984, Protocol entered into force 1 May 1978, (1981 Amendments entered into force 1 September 1984). Ratifying, etc. States of Convention include Vanuatu, Uruguay, US, UK, United Arab Emirates, USSR, Ukrainian SSR, Tuvalu, Turkey, Tunisia, Trinidad and Tobago, Tonga, Thailand,

Switzerland, Sweden, Suriname, Sri Lanka, Spain, South Africa, Singapore, Saudi Arabia, Saint Vincent and the Grenadines, Rumania, Republic of Korea, Qatar, Portugal, Poland, Philippines, Peru, Papua New Guinea, Panama, Pakistan, Oman, Norway, Nigeria, Netherlands, Monaco, Mexico, Mauritius, Marshall Islands, Malta, Maldives, Malaysia, Libyan Arab Jamahiriya, Liberia, Lebanon, Kuwait, Jordan, Japan, Jamaica, Italy, Israel, Ireland, Indonesia, India, Iceland, Hungary, Honduras, Guinea, Guatemala, Greece, Ghana, Federal Republic of Germany, German Democratic Republic, Gabon, France, Finland, Fiji, Ethiopia, Egypt, Ecuador, Dominican Republic, Djibouti, Denmark, Democratic People's Republic of Korea, Czechoslovakia, Cyprus, Côte d'Ivoire, Congo, Colombia, China, Chile, Cape Verde, Canada, Cameroon, Burma, Bulgaria, Brunei Darussalam, Brazil, Benin, Belgium, Barbados, Bangladesh, Bahrain, Bahamas, Austria, Australia, Argentina, Antigua and Barbuda and Algeria. The number of Contracting Parties includes 104 States, the combined merchant fleets of whom constitute approximately 96% of the world's merchant fleets gross tonnage. The Protocol has been ratified, etc. by States including Yugoslavia, Vanuatu, Uruguay, US, UK, United Arab Emirates, USSR, Tunisia, Switzerland, Sweden, Spain, South Africa, Singapore, Seychelles, Saint Vincent and the Grenadines, Republic of Korea, Portugal, Poland, Peru, Panama, Pakistan, Oman, Norway, Nigeria, Netherlands, Mexico, Marshall Islands, Malta, Malaysia, Libyan Arab Jamahiriya, Liberia, Lebanon, Kuwait, Japan, Italy, Israel, Ireland, Indonesia, India, Iceland, Hungary, Honduras, Greece, Ghana, Federal Republic of Germany, German Democratic Republic, France, Finland, Ethiopia, Egypt, Denmark, Democratic People's Republic of Korea, Czechoslovakia, Cyprus, Côte d'Ivoire, Colombia, China, Burma, Bulgaria, Brunei Darussalam, Brazil, Belgium, Barbados, Bahamas, Austria, Australia, Argentina, Antigua and Barbuda and Algeria. The number of Contracting States is 69, the combined merchant fleets of whom constitute approximately 90% of the world's merchant fleet's gross tonnage. IMO Status pp. 9-42.

97 Gold p. 49. Lamson, Cynthia, 'Arctic shipping, marine safety and environmental protection,' *Marine Policy*, January 1987, (Lamson) notes on p. 6. that a *Document for Guidance* was issued in 1964 with amendments and supplements issued in 1975 and 1977 which followed a Recommendation from the 1969 International Conference on the Safety of Life at Sea urging all Parties to, 'take all practicable steps to ensure that the education and training of seafarers in the use of aids to navigation ship's equipment and devices was kept satisfactorily up to date' and encouraging the IMO and the ILO to cooperate to achieve these goals.

98 Gold p. 49.

99 These are the instruments listed in De Cesari pp. 309-317.

100 External forces affecting ships which might lead to capsizing or to unacceptable angles of heel are to be taken into account.

101 It is noted fire safety requirements for cargo ships, excepting tankers, and for passenger ships carrying not more than 36 passengers are still in many respects unsatisfactory.

102 It is noted that ship size and power have increased in complexity, and power rating of machinery and reliable main propulsion machinery is important to ensure safe navigation.

103 Note is made that due to automated and remotely controlled installations special attention must be given to the determination of the minimum number of crew needed to deal with emergency situations.

104 Note is made that due to nuclear engineering, experience in nuclear propulsion and the expected increase in the use of nuclear propulsion, Convention provisions with respect to nuclear ships must be revived.

105 It is noted that Convention provisions for cargo ships are not applicable or not adequate or suitable and recommends specific safety requirements for ships including ships carrying chemicals and liquefied gases in bulk.

106 Chapter I, Part B, Regulation 6.

107 *Ibid.*

108 *Ibid.* Regulation 8.

109 *Ibid.* See Section 5.4.3.

110 *Ibid.* Regulation 10.

111 *Ibid.* Regulation 11.

112 *Ibid.* Regulation 14.

113 *Ibid.* Regulation 19.

114 *Ibid.*

115 Chapter II-1, Regulation 29.

116 *Ibid.* Under the same conditions the rudder must be capable of being put over from 35 degrees on either side to 30 degrees on the other side in not more than *28 seconds*. An alternative power system is also required.

117 Chapter II-1, Part E, Regulation 60.

118 *Ibid.*

119 *Ibid.* Tankers of various sizes carrying crude oil or with tank washing machines are required to be fitted with an inert gas system.

120 Chapter V, Regulation 12.

121 Chapter V, Regulation 19. These drills must include direct control from within the steering gear compartment, the communications procedure with the navigation bridge and the operation of alternative power supplies if applicable.

122 Convention on the International Maritime Satellite Organization (1976), (INMARSAT) 15 *International Legal Materials* 1051. Entry into force 16 July 1979, 1985 Amendments, not in force. Operating Agreement on the International Maritime Satellite Organization, 15 *International Legal Materials* 233. Entry into force 16 July 1979, 1985 Amendments not yet in force. Ratifying, etc. States include US, UK, United Arab Emirates, USSR, Ukrainian SSR, Tunisia, Sweden, Sri Lanka, Spain, Singapore, Saudi Arabia, Republic of Korea, Qatar, Portugal, Poland, Philippines, Peru, Panama, Pakistan, Oman, Norway, New Zealand, Nigeria, Netherlands, Malaysia, Liberia, Kuwait, Japan, Italy, Israel, Iraq, Iran, Indonesia, India, Greece, Federal Republic of Germany, German Democratic Republic, Gabon, France, Finland, Egypt, Denmark, Czechoslovakia, Colombia, China, Chile, Canada, Byelorussian SSR, Bulgaria, Brazil, Belgium, Australia, Argentina and Algeria. IMO Status pp. 285-304.

123 Article 3. INMARSAT is considered by Lamson pp. 4-5 to have relevance for environment protection since it marks a step toward standardizing maritime communications. See also Churchill p. 216.

124 Articles 7 and 19.

125 Article 9.

126 In Norway for example this is the Norwegian Telecommunications Administration (Teledirektoratet).

127 Freeman, Arthur L., Introductory Note, 15 *International Legal Materials* 1051.

128 Lamson pp. 4-5 notes Arctic users are not covered.

129 This information is obtained from Lamson Arctic pp. 69-71 and Roginko, Alexei, 'Arctic Environmental Cooperation: Prospects and Possibilities,' Institute of World Economy and International Relations, USSR Academy of Sciences, Prepared for the International Studies Association Annual Convention, London, March 1989 (Roginko) p. 12 unless noted otherwise.

130 IMO Status p. 351.

131 This can provide continuous worldwide coverage with accuracy within 2 to 4 miles 95% of the time but is not yet fully operational.

132 The latter 2 have to be used together with other systems for continuous access and to obtain real-time information such as navigational fixes.

133 See Churchill p. 216.

134 See Sovjet-Nytt, 21 april 1989, (2619) 44 årgang s. 12. This is a Soviet periodical distributed by the Soviet Embassy in Oslo, Norway.

135 International Convention on the Standards of Training, Certification, and Watchkeeping for Seafarers (1978) (STCWS), London. IMO Document p. 49. The Convention entered into force in 28 April 1984. Ratifying, etc. States include the Yugoslavia, Venezuela, United Republic of Tanzania, US (OPN June 1991 p. 6.), UK, United Arab Emirates, USSR, Tuvalu, Switzerland, Sweden, Sri Lanka, Spain, South Africa, Singapore, Seychelles, Republic of Korea, Portugal, Poland, Philippines, Peru, Pakistan, Norway, Nigeria, New Zealand, Netherlands, Mozambique, Mexico, Maldives, Libyan Arab Jamahiriya, Liberia, Kiribati, Japan, Jamaica, Italy, Israel, Ireland, Indonesia, India, Hungary, Honduras, Greece, Federal Republic of Germany, German Democratic Republic, Gabon, France, Finland, Ethiopia, Egypt, Ecuador, Denmark, Democratic People's Republic of Korea, Czechoslovakia, Cyprus, Côte d'Ivoire, Colombia, China, Chile, Canada, Burma, Bulgaria, Brunei Darussalam, Brazil, Bolivia, Benin, Belgium, Bahamas, Australia, Argentina and Algeria. IMO Status pp. 319-330.

136 IMO Convention on the Standards of Training, Certification and Watchkeeping of Seafarers, London, 1978, pp. 22-26 and 43. See also Lamson p. 6.

137 *Ibid.* An interesting point considering the problems noted above with the EXXON VALDEZ in Alaska.

138 Gold p. 50, parentheses added.

139 These Agreements and Conventions were obtained from Gold pp. 48-52, Churchill pp. 216-220 and IMO Status. For some Conventions the specific ratifying countries will not be presented due both to lack of space and to information not being readily available.

140 CMI, International Conventions on Maritime Law, Antwerp, no date, p. 11. Not listed in IMO Status and De Cesari.

141 Gold p. 48.

142 *Ibid.* and Churchill p. 220.

143 International Convention on Load Lines, 1966, London, 1981 (Load Lines Convention), IMO Document p. 33. Entry into force, 21 July 1968. (1971 Amendments not yet in force, 1975 Amendments not yet in force, 1979 Amendments not yet in force, 1983 Amendments not yet in force). Protocol of 1988 relating to the International Convention on Load Lines, 1966, not yet in force. The Amendments of 1971, 1975 and 1979 are found in IMO Document p. 33. Acceding and accepting States of the Convention include Zambia, Zaire, Yugoslavia, Yemen, Viet Nam, Venezuela, Vanuatu, Uruguay, US, UK, United Arab Emirates, USSR, Tuvalu, Turkey, Tunisia, Trinidad and Tobago, Tonga, Syrian Arab Republic, Switzerland, Sweden, Suriname, Sri Lanka, Spain, South Africa, Somalia, Singapore, Seychelles, Senegal, Saudi Arabia, Samoa, Saint Vincent and the Grenadines, Rumania, Republic of Korea, Qatar, Portugal, Poland, Philippines, Peru, Papua New Guinea, Panama, Pakistan, Oman, Norway, Nigeria, New

Zealand, Netherlands, Morocco, Monaco, Mexico, Mauritius, Mauritania, Marshall Islands, Malta, Maldives, Malaysia, Madagascar, Libyan Arab Jamahiriya, Liberia, Lebanon, Kuwait, Kenya, Japan, Jamaica, Italy, Israel, Ireland, Iran, Indonesia, India, Iceland, Hungary, Honduras, Guinea, Greece, Ghana, Federal Republic of Germany, German Democratic Republic, Gabon, France, Finland, Fiji, Ethiopia, Egypt, Ecuador, Dominican Republic, Djibouti, Denmark, Democratic Yemen, Czechoslovakia, Cyprus, Cuba, Côte d'Ivoire, Congo, Colombia, China, Chile, Cape Verde, Canada, Cameroon, Burma, Bulgaria, Brunei Darussalam, Brazil, Benin, Belgium, Barbados, Bangladesh, Bahrain, Bahamas, Austria, Australia, Argentina, Antigua and Barbuda, and Algeria. Contracting Parties include 114 States the combined merchant fleets of whom constitute approximately *97%* of the world's merchant fleet's gross tonnage. IMO Status pp. 119-148.

144 Gold p. 49.

145 Convention on the International Regulations for Preventing Collisions at Sea, 1972, as amended 1981 and 1987 (Collisions Convention), IMO Document p. 46. Convention entered into force 15 July 1977, 1981 Amendments entered into force 1 June 1983 and 1987 Amendments entered into force 19 November 1989. Ratifying and acceding States of the Convention include Zaire, Yugoslavia, Yemen, Venezuela, Vanuatu, Uruguay, US, UK, United Arab Emirates, USSR, Tuvalu, Turkey, Tunisia, Trinidad and Tobago, Thailand, Tonga, Syrian Arab Republic, Switzerland, Sweden, Sri Lanka, Spain, South Africa, Solomon Islands, Singapore, Seychelles, Senegal, Saudi Arabia, Samoa, Saint Vincent and the Grenadines, Rumania, Republic of Korea, Qatar, Portugal, Poland, Peru, Papua New Guinea, Panama, Pakistan, Oman, Norway, Nigeria, New Zealand, Netherlands, Morocco, Monaco, Mexico, Marshall Islands, Maldives, Malaysia, Liberia, Kuwait, Japan, Jamaica, Italy, Israel, Ireland, Indonesia, India, Iceland, Hungary, Honduras, Guinea, Greece, Ghana, Federal Republic of Germany, German Democratic Republic, Gabon, France, Finland, Fiji, Ethiopia, Egypt, Ecuador, Dominican Republic, Djibouti, Denmark, Democratic People's Republic of Korea, Czechoslovakia, Cyprus, Cuba, Côte d'Ivoire, Colombia, China, Chile, Cape Verde, Canada, Cameroon, Burma, Bulgaria, Brunei Darussalam, Brazil, Benin, Belgium, Barbados, Bangladesh, Bahrain, Bahamas, Austria, Australia, Argentina, Antigua and Barbuda, and Algeria. Contracting Parties include 102 States the aggregate of whose merchant fleet constitute approximately *94%* of the world's merchant fleet's gross tonnage. IMO Status pp. 43-62.

146 Gold p. 49 in reviewing the Collision Convention mentions the 1977 collision in the Tobago Strait between the VLCC AEGEAN CAPTAIN and the VLCC ATLANTIC EMPRESS, resulting in the total loss of one ship, the constructive total loss of the other and an oil spill of close to 300,000 tons.

147 International Convention for Safe Containers, 1972 as amended 1981 and 1983, (Containers Convention), IMO Document p. 15. The Convention entered into force 6 September 1977, the 1981 Amendments to Annex I, 1 December 1981 and the 1983 Amendments to Annexes I and II, 1 January 1984. Ratifying and acceding States include Yugoslavia, Yemen, US, UK, USSR, Ukrainian SSR, Sweden, Spain, South Africa, Saudi Arabia, Rumania, Republic of Korea, Portugal, Poland, Peru, Pakistan, Norway, New Zealand, Netherlands, Luxembourg, Liberia, Japan, Italy, Israel, India, Hungary, Honduras, Guinea, Greece, Federal Republic of Germany, German Democratic Republic, France, Denmark, Czechoslovakia, Cuba, China, Chile, Canada, Byelorussian SSR, Bulgaria, Benin, Belgium, Barbados, Bahamas, Austria, Australia, Argentina and Afghanistan. The Contracting States number 48. IMO Status pp. 253-271.

148 Gold p. 50.

149 21 *International Legal Materials* 1. States Parties include Belgium, Denmark, Finland, France, Federal Republic of Germany Greece, Ireland, Italy, Netherlands, Norway, Portugal, Spain, Sweden and the UK

150 VI *New Directions in the Law of the Sea* 449.

151 Singh, *International Convention of Merchant Shipping* p. 1203, (not in force).

152 160 *United Nations Treaty Series* 223.

153 *International Labour Conference Convention* 145.

154 591 *United Nations Treaty Series* 265.

155 1982 *United Kingdom Treaty Series* 50.

156 IMO Document p. 10.

157 IMO Document p. 50.

158 IMO Document p. 22. Not in force.

159 Article 5 can also have application through its wording for the introduction by ships of Annex I hazardous substances into the Baltic. See Part I Section 4.2.3.1. for discussion.

160 Baltic Sea Environment Proceedings No. 23, Activities of the Commission 1986, Including the Eighth Meeting of the Commission held in Helsinki 24-27 February 1987 (Helsinki Report), pp. 20, 37-40.

161 *Ibid.*

162 *Ibid.*

163 Churchill p. 251.

164 *Ibid.*

165 See for example Article 6 of Barcelona Convention.

166 For the former see generally IMO Status or IMO News for Amendments and Protocols to the main IMO Conventions listed. An example of the latter is the US which developed stricter national rules following the EXXON VALDEZ accident. See United States Oil Pollution Act for Ships signed into law August 18, 1990 obtained from Burlingham, Underwood & Lord, *Report on United States Oil Pollution Laws*, October 1990, One Battery Park Plaza, New York, New York 10004, (OPA).

6. Liability for Damage Arising from Marine Pollution

6.1. Introduction

As seen in Part I Section 2.1.2. Principle 22 of the Stockholm Declaration, generally obliges States to develop international law regarding liability and compensation for transboundary pollution. Unfortunately in the 20 years since that time little progress has been made. One author notes that in several recent international environmental conventions the only reference to State liability is virtually the same statement as in Principle 22.[1] The same can be said regarding *practice*, and probably there is, 'in international practice at least one strong undercurrent towards *less* rather than more explicitness in a State's liability obligations.'[2] In addition to lack of concrete international agreements, see Part II Section 9.3.3. for *problems* in raising liability claims.

6.2. Liability for Pollution Damage from Dumping Activities

As noted the ODC does not contain provisions regarding State responsibility and liability obligations, though the other regional conventions noted in Part I Section 3.2. contain *general* provisions to develop special liability regimes.[3] The LDC has 1 Article, Article X which states,

> 'In accordance with the principles of international law regarding State responsibility for damage to the environment of other States or to any other area of the environment, caused by dumping of wastes and other matter of all kinds, the Contracting Parties undertake to develop procedures for the assessment of liability and the settlement of disputes regarding dumping.'[4]

Article X does not recognize under international law a State responsibility for environmental damage of other States or to other environmental areas, but it 'recognizes that there are *principles* for damage to the environment even beyond the limits of national jurisdiction', and represents the first time such a provision was included in a treaty.[5] Additionally noted however is that Article X is 'extremely primitive' and does not solve the problem.[6] Legal experts associated with the LDC have analyzed the liability schemes of other conventions, including those proposed by the IMO and the Helsinki Convention, however, the majority believed that the need for developing a liability regime was 'inappropriate at this time.'[7]

Another author is somewhat more optimistic regarding Article X and notes that even though this passive view is still held by the majority, by October 1987 there had developed a more positive view of some weight, and LDC Contracting Parties were requested 'to take note' of the different views regarding developing procedures of liability assessment under Article X and 'take action as appropriate.'[8] The author continues calling for positive developments in 'the effective application and interpretation of the LDC provisions,' based upon customary law, the draft Articles of the ILC, LOSC provisions as well as several other conventions and agreements covering liability including those mentioned above.[9]

6.3. Liability for Damage from Land-based and Offshore Installation Pollution

Agreements covering liability for land-based and offshore installation pollution damage are totally lacking world wide with *one* exception, the Nordic Convention, though as noted the regional conventions with the exception of the Paris Convention contain *general* provisions to develop special liability regimes.[10] Liability provisions exist for northwest European waters under an *industrial scheme*, OPOL,[11] and also under a convention covering northwest European waters, the 1976 Liability Convention[12] which however is *not* in force.

6.3.1. The Nordic Convention[13]

As mentioned this Convention contains provisions which allow, under Article 3, any person affected by a nuisance caused by environmentally harmful activities to raise in the appropriate Nordic State both proceedings as to the permissibility of the activities and claims for compensation for damage suffered. Under Article 1 'environmentally harmful activities' mean the discharge from soil or buildings or installations or solid or liquid waste, gas or other substances into water courses, lakes or the sea[14] which may entail environmental nuisance, by water pollution, other effect on water conditions, sand drift, air pollution, noise, vibration, changes in temperature, ionizing radiation, light etc. Thus nuisances from land-based, offshore installation and airborne (from the atmosphere as well) is included. Under Article 4 a State's Supervisory Authority may institute similar proceedings against another Nordic State as long as such a claim may be raised by an authority or other representative for environmental interests. As one author notes, in spite of problems showing the pollution source and a link between the source and the damage suffered, 'it might be desirable to introduce provisions similar to the Nordic Convention in other regions in order to facilitate the bringing of actions by litigants before foreign courts.'[15]

6.3.2. OPOL[16]

OPOL has been in existence since 1975 and has as parties approximately 34 oil companies operating mostly on the *British* continental shelf.[17] The *private* agreement purports to control offshore facilities on the continental shelves and within the jurisdictions of the UK, Denmark, the Federal Republic of Germany, France, Ireland, the Netherlands and Norway.[18]

Any operator who intends to become an operator of an offshore facility within the jurisdiction of the 'designated States' listed may apply to become a Party to OPOL.[19] States may be added to the designated State list by a simple majority of member votes in favour of such amendment.[20] The agreement's provisions roughly parallel the 1976 Liability Convention, much as the industrial TOVALOP and CRISTAL schemes parallel the CLC and the Fund Convention.[21] Generally this agreement is based upon the principle of *strict but limited* liability. If a discharge of oil[22] occurs from an offshore facility[23] operated by an operator who is a Party to OPOL, then the operator must reimburse for any remedial clean up measures taken by public authorities[24] or itself and pay compensation for pollution damage[25] up to an overall maximum of *USD 25,000,000* per incident.[26] This strict liability is subject to exceptions including pollution damage resulting from war, an act of God or from an abandoned well; wholly caused by intentional acts or omissions; by governmental negligence or other wrongful acts or operator compliance with governmental conditions or instructions; and wholly or partially caused by intentional acts or negligence by the person suffering the damage.[27]

Both reimbursement for remedial measures taken and compensation is each limited to USD 12,500,000. Each must be pro rated in the event the limits are exceeded and there is more than 1 claimant. If all of the compensation required is paid from one category without the other USD 12,500,000 being reached, this can be carried over to the other.[28] There is a 1 year Statute of Limitations.[29] If compensation is paid to the claimant under OPOL, then no more actions may be raised by the claimant against the Party or others involved in the incident.[30] Parties to OPOL have to be covered by insurance or show other security to cover the USD 25,000,000 liability. Though under OPOL all member Parties act as guarantees for one another, some consider it questionable whether the injured has a direct action against the other members in the event one fails to pay.[31] Disputes are to be exclusively settled by arbitration, and English law governs.[32]

OPOL's good points from a claimant's view include strict operator liability, and the requirement of company insurance or other financial solvency. From an industrial standpoint its plus points include no joint and several liability; a 1 year period from the date of the polluting incident in which a claim must be made; limited liability; direct loss or damage required thus probably negating any compensation for damage done to fish or birds; no further claims possible after settlement under OPOL; exclusive use of arbitration to settle disputes; no provisions for unlimited liability in case of deliberate operator acts or omissions; no possibility for higher liability or unlimited liability if ascertained reasonable by the State with jurisdiction; English law being the law of the agreement;[33] and in practice very little being paid out as compensation.[34]

6.3.3. 1976 Liability Convention[35]

The 1976 Liability Convention is modelled after the CLC,[36] and its provisions extend the industrial scheme OPOL discussed above. As mentioned however it is not in force. Under the 1976 Liability Convention the operator of an offshore continental shelf installation causing pollution damage[37] is *strictly liable* for damage and remedial measures taken, with the exceptions of damage resulting from war, an act of God, from an abandoned well more than 5 years after it was abandoned, or from an intentional or negligent act done by the person suffering damage.[38] Different from OPOL, operators are *jointly and severally liable* where an installation has more than one operator, where 2 or more installations with different operators are involved in the pollution damage,[39] and where during the course of a polluting incident there is a change of operators.[40] The operator's liability is limited to 40,000,000 SDR's.[41] The operator has *unlimited liability* if it is proved that the pollution damage resulted from his intentional act or omission 'done deliberately with actual knowledge that pollution damage would result.'[42] Additionally, a State may provide for a *higher* limit of liability or *unlimited* liability for operators of installations on its continental shelf which cause pollution damage as long as there is not discrimination based on nationality.[43]

When a State Party is the operator, such State is subject to suit in the Courts having jurisdiction over the area where damage resulted or in the jurisdiction of the Controlling State, and it cannot raise any defenses based upon sovereign immunity.[44] Insurance is compulsory and must provide cover of not less than 35,000,000 SDR's.[45] There is a 1 year Statute of Limitations in which to raise a claim starting from the date of knowledge or the date the claimant reasonably should have known of the damage incurred.[46] In no case shall an action commence more than 4 years after the damaging incident, and if the incident consists of a series of occurrences, the 4 years runs from the date of the last occurrence.[47]

Judgement given by a Court with jurisdiction and not subject to ordinary review must be recognized and enforced in any State Party except where it was obtained by fraud or where the defendant was not given reasonable notice and fair opportunity for a hearing.[48] The State Parties may invite unanimously other States to accede to the Convention which have coastlines on the North Sea, the *Baltic Sea* or the *north Atlantic*, north of 36 degrees north latitude.[49]

As seen, from a claimant's view the 1976 Liability Convention's strong points include, strict operator liability for oil pollution; joint and several liability if more than one operator;[50] provisions for increased or unlimited operator liability by a State with jurisdiction as long as there is no discrimination based on nationality; a Statute of Limitations from the date of knowledge and not the date of incident; compulsory insurance; and judgments given by one State Party recognized and enforced in any State Party. From an industrial view its strong points include limited liability and the possibility of a claim being brought being limited to 1 year.

6.3.4. IMO Mobile[51]

Briefly this *draft* instrument was formed with the goal of regulating 'maritime features to which no special rules apply today' and 'avoiding enactment of widely differing national rules which may cause international confusion and conflict.'[52] Fixed platforms are excluded as are production activities though offshore exploration, exploitation, processing, transport or storage of mineral resources of the seabed or subsoil or ancillary activities are *included*. The IMO Conventions for ships[53] are made applicable to offshore craft where they are in force in the State Parties, though since 4 of the 7 States having registered the largest number of drilling units are *not* maritime convention countries, IMO

Mobile in addition requires these States to apply their national shipping rules to the offshore floating platforms.[54]

Basically legal areas covered by IMO Mobile are the same as those of the IMO Conventions, including collisions, salvage, arrest, limitation of liability, rights in craft, liability for oil pollution, application of national rules, minimum limits of liability for platforms and maritime liens. Though there was voiced interest by some States for an extensive draft Convention covering operational aspects of offshore activities including legal problems concerning oil production, it was felt that these areas 'were not within the natural scope' of IMO work and better left to be dealt with by other international conventions.[55] Limitation of liability is covered by the IMO Mobile but was controversial since the shipping liability conventions were felt by some to be too low for especially the drilling units. Others argued that there is no compelling reason to have different limits, especially since the slow moving platforms have a lesser chance of causing damage than ships. Blowout damage is not covered because 'it is of such a nature and magnitude that it requires special treatment in national law and other conventions.'[56]

6.4. Liability for Damage from Vessel-Source Pollution, Conventions and Agreements

As mentioned previously State liability for damage from vessel-source pollution *is* covered by conventions and agreements and is the only area which presently has anything that resembles a regime. These arrangements are presented below.

6.4.1. CLC and Protocols[57]

The CLC covers damage from *oil* pollution from any escape or discharge from a 'seagoing vessel (or) any seaborne craft' carrying bulk oil in cargo, within the territory including the territorial waters of a Contracting State.[58] 'Oil' means 'any persistent oil such as crude oil, fuel oil, heavy diesel oil, lubricating oil, and whale oil,[59] whether carried on board ship as cargo or in the bunkers of such a ship.' 'Pollution damage' means 'loss or damage caused outside the ship carrying oil by contamination resulting from the escape or discharge of oil from the ship ... and includes the costs of preventive measures and further loss or damage caused by preventive measures.' 'Shipowner' is defined as 'the person registered as owner of the ship.' In relation to the geographical scope for Convention applicability the shipowner need not be a national of a Contracting State nor the ship a flag ship of a contracting State.[60]

The shipowner is *strictly liable* subject to various exceptions, which the shipowner must prove, including damage resulting from war, hostilities, civil war, insurrection, or exceptional, inevitable and irresistible natural phenomena.[61] In the cases there is strict liability, the shipowner's and insurer's liability is limited to 133 SDR's per ton of the ship's total tonnage (GRT) with an upper limit of 14,000,000 SDR's.[62] The liability limits are not available where the pollution damage is 'the result of the actual fault or privity' of the shipowner.[63] The owner of a ship responsible for a polluting incident must deposit, with the Court or other competent authority of any of the Contracting States in which the legal action is brought a fund for the total sum representing the liability limit determined under the CLC.[64] This may be either cash or a guarantee considered adequate by the Court or other authority.[65] The owner of a ship registered with a State which is a Party to the CLC and carrying over 2,000 tons oil cargo must carry insurance or other security covering maximum liability under the Convention.[66] A certificate must be issued by flag State authorities verifying this financial security and must be carried on board ship.[67] State Parties must ensure that ships lacking the necessary certificate are not allowed to trade, and all ships in their ports, wherever registered, must be insured or secured in accordance with these requirements.[68]

Actions must be brought in the Courts of the Contracting State in which the damage occurred, within 3 years from the date of damage, but not later than 6 years after the date of the incident.[69] A Court holding the shipowner's deposit has exclusive competence to distribute the money to those showing valid claims, and if the claims together are greater than the deposit, the Court must distribute the funds in proportion to the damages shown by each claimant.[70] The shipowner's insurer can also claim against the fund for the amount paid by it, previous to the Court's distribution. The shipowner's deposit is the exclusive means to satisfy the oil pollution claims, but a party may pursue other assets in another forum if there are access problems to the Court or the deposit is not really available to the Court.[71] If the shipowner is liable for amounts greater that the CLC's limits, the excess may be satisfied from the Fund Convention noted below. Court judgments where claims are brought must be recognized and enforced in all State Parties.[72] Although one author considers the coverage of the CLC to presently be sufficient, another, though aligned with shipping interests believes the contrary.[73] Both agree however on the *conflict* between the shipping and the oil industries on the burden of carrying the compensation.

As adopted in 1984 the Civil Liability *Protocol* was meant to alleviate some of the CLC deficiencies and consequently increases the limits of liability to 3,000,000 UOA's for ships under 5,000 tons and 420 UOA's per each additional ton for ships above 5,000 tons.[74] The upper limit is increased to

59,700,000 UOA's.[75] Pollution damage in the exclusive economic zone or areas up to 200 nautical miles from baselines is covered as is the cost of preventative measures.[76] Types of damages for which recovery is possible are generally the same as the CLC but modified somewhat, and strict liability remains with the same defenses.[77] More parties are protected from liability including servants or agents of the owner, crew members, pilots, charterer, manager or operator, person salvaging, person taking preventative measures, as well as servants and agents of those listed above.[78] The elimination of the shipowner's limitation and resulting unlimited liability is made more difficult by shifting the burden of proof to the claimant to show the owner or others acted recklessly or intentionally.[79] As before the shipowner must deposit a fund with the State Court where the action is brought, but under the Civil Liability Protocol the shipowner must deposit a fund whether or not a claim has been brought to apply to himself the liability limits.[80] Simpler amendment procedures are offered to increase liability limits by allowing the IMO's Legal Committee to adopt amendments which enter into force 18 months later.[81]

As noted this Protocol as yet to enter into force.[82] Its adoption was controversial, with the Protocol being accepted by a vote of 48 to 0 but with *16* abstentions, those abstaining including large shipping States as Japan, Liberia, Panama, Greece and the USSR[83] One author notes the limits set under the 1984 Liability Protocol and Fund Protocol are considered by U.S authorities to be sufficient to cover possible oil spills in US waters.[84]

6.4.2. Fund Convention and Protocols[85]

The Fund Convention was formed in 1971 to give additional compensation to claims for oil pollution damage greater than the liability limitations under the CLC, and 'to balance the responsibility for oil pollution between shipowners and those with oil cargo interests.'[86] Additionally it is meant to provide compensation in cases coming within the exceptions under the CLC, including cases of no shipowner liability or where there are problems with shipowner finances.[87] Funds come from a levy on imports greater than 150,000 tons per year of oil by oil companies on a pro rate basis in State Parties.[88] The Fund Convention also compensates tanker owners for part of their liability under the CLC.[89] For this to occur however the ship's flag State must be a Party to the Fund Convention, Parties to the Fund Convention must also be Parties to the CLC and indemnification is not available in cases of shipowner intentional misconduct.[90]

The pollutants covered are persistent hydrocarbon mineral oils carried as cargo, or in the bunkers as long as the ship is carrying oil in bulk as cargo.[91]

The damages must occur in the State Parties' territory or territorial seas, although the discharge may occur outside these areas.[92] 'Pollution damage' means the same as under the CLC.[93] Costs of preventative action taken by shipowners may be claimed as pollution damages.[94] Claims must be brought in the State Party where the damage occurred, and it must be proven that one or more ships was responsible.[95] Procedures for bringing claims are similar to those noted above for the CLC.[96] Defenses include war or hostilities, sovereign immunity, insufficient evidence that the damage ensues from the ship's discharge, or the injured party himself caused the damage by is own intentional or negligent acts.[97]

The amount of liability is up to USD 54,000,000, combined with compensation from the CLC. A General Assembly representing all State Parties can decide to increase this to USD 72,000,000.[98] Shipowners are reimbursed for liability under the CLC greater than USD 120 per each ton of ship tonnage or USD 10,000,000, whichever is less, up to and including USD 160 per each ton of ship tonnage or US 16,800,000, whichever is less.[99] Under the 1976 Protocol these were to be expressed in SDR's under the 1976 Protocol, but it has yet to enter into force.[100] One author notes that during the first 6 years of its existence the Fund Convention paid out approximately USD 42,000,000 on *16* claims.[101]

The Fund Protocol from 1984 allows available compensation under the CLC plus Civil Liability Protocol and the Fund Convention to go up to approximately USD 140,000,000, with the Fund triggering into action above approximately USD 62,000,000.[102] Under the Fund Protocol this could be raised to about USD 208,000,000 if the combined quantity of oil imported in 3 of the State Parties total 600,000,000 tons in the previous year.[103] One author notes that it is unlikely this could be met unless the US ratifies.[104]

As with the Civil Liability Protocol, pollution damage in the exclusive economic zone or areas up to 200 nautical miles from base lines is covered as is the cost of preventative measures.[105] As under the Fund Convention under the Fund Protocol each State Party is responsible for making sure all importing oil companies contribute to the fund as required, and failure to do so places the State itself responsible for making up the loss.[106] The same remarks regarding the Civil Liability Protocol concerning breaking up shipowner limitation and simplified amendment procedures apply equally to the Fund Protocol.[107]

As noted this Protocol has yet to enter into force.[108] Its adoption was controversial, with the Protocol being accepted by a vote of 44 to 0 but with *21 abstentions*. Those abstaining included as before large shipping States such as Japan, Liberia, Panama, Greece and the USSR[109]

6.4.3. TOVALOP[110]

TOVALOP was formed in 1969 among tanker owners making up about 50% of the world's tanker fleet to clean up oil spills from participant's ships or to reimburse States for associated clean up costs. The *private* systems are presented here for the sake of background. Sufficient insurance was required not only to cover obligations under TOVALOP but also to reimburse tankers owners for clean up costs. This private system was seen as beneficial since many States initially did not accept the CLC. TOVALOP was amended in 1978 to more closely comply with the by then in force CLC, however if there is CLC coverage, the system does *not* apply.[111]

TOVALOP covers loaded or ballasted tankers operated by owners or charterers who are members in the TOVALOP system. Pollutants covered are persistent hydrocarbon oils. Compensation allowed is to States or individuals for pollution damage caused by discharged oil in the territory or territorial sea of a State where the CLC does not apply. Loss or damage caused by the spillage, preventative measures and subsequent losses arising out of such measures as well as costs in removing a threat without any discharge are also covered. Amount of liability is USD 160 per GRT up to USD 16,800,000 per polluting incident. TOVALOP's members must maintain financial responsibility in accordance with ITOPF standards, and this is generally part of shipowner's P. & I. insurance. Claims must be made to the polluting ship's owner within 1 year, and claims disputes must be submitted to the International Chamber of Commerce within 2 years. Defenses available are the same as under the CLC.

Compared to the CLC, there exists under TOVALOP an absolute right to limit liability, while under the CLC owners whose privity and knowledge contributed to the polluting incident lose this right.[112] Additionally TOVALOP covers ballasted tankers, the CLC only loaded tankers. Bareboat charterers are covered, but not under the CLC. TOVALOP covers in addition to actual damage as under the CLC also threat of damage including precautionary and preventative measures. TOVALOP has a shorter statute of limitations than the CLC, and unsettled TOVALOP claims must go to arbitration rather than to the Courts.[113]

6.4.4. CRISTAL[114]

CRISTAL was formed as a voluntary system for compensating States and persons while waiting for the entry into force of the Fund Convention. Currently it is maintained between USD 3,000,000 and USD 5,000,000 and has become a system to supplement TOVALOP, the Fund Convention and the

CLC.[115] When the Fund Convention came into force in 1978, CRISTAL was changed to be complementary rather than temporary.

It comes into use when the CLC and TOVALOP are not sufficient and does not apply to circumstances covered by the Fund Convention.[116] CRISTAL does not cover compensation and indemnity in those cases of exemption under the CLC as does the Fund Convention.[117] Claims are made to the CRISTAL administration. The pollutants covered are the same as in the Fund Convention. All pollution damage is covered including removal of threat which is not recoverable from the other sources. The agreement is applicable in the territory or territorial sea of any State regardless of where the spill occurred. The ship must be owned or chartered by a participant to TOVALOP, and the oil owned by a participant to CRISTAL. The system is funded by contributions by participants. Generally the liability is up to USD 36,000,000 combined with the other sources but can be increased by the CRISTAL administration to USD 72,000,000.[118] Defenses available include in addition to those in the Fund Convention an intentional act or omission by a third party or the negligence of governments.

After the Civil Liability and Fund Protocols were adopted to raise CLC and Fund Convention limits and extend their scope, CRISTAL participants decided to raise their levels from June 1, 1986 and discontinue the old CRISTAL.[119] The revised CRISTAL however was made contingent upon the new PLATO being put into effect, something which did not occur and, 'the situation is now very confused and unsatisfactory with neither PLATO, CRISTAL nor the Protocols (Civil Liability and Fund Protocols) in effect.'[120]

6.4.5. PLATO[121]

The oil companies through the organization, OCIMF, formed the PLATO/CRISTAL proposals in 1985 due to dissatisfaction with the Civil Liability and the Fund Protocols, and this was to be an interim system while waiting for these Protocols to come into force.[122] It is felt by one author however that while the interim systems TOVALOP and CRISTAL worked well because they were consistent and parallel with the CLC and Fund Convention, PLATO lacked these qualities and thus, 'attempts to force the liability issue on the shipping sector.'[123]

PLATO applies all over the world in contrast to the geographical limitations of the CLC or Fund Convention.[124] Compensation is covered for damage caused by oil pollution, costs of preventative measures and costs of measures to remove threats of pollution higher than the limits placed by CLC or TOVALOP. The limits are divided into pre and post 1990. For pre 1990 the coverage includes USD 10,000,000 for tankers of 5,000 GRT or less. For tankers greater than 5,000 GRT, USD 10,000,000 plus USD 500 per GRT up to a maximum of US 60,000,000. After 1990, the limits are USD 15,000,000 for tankers 5,000 GRT or less. For tankers greater than 5,000 GRT, USD 15,000,000 plus USD 600 per GRT up to a maximum of USD 75,000,000. Linked to the new CRISTAL system, which is the whole idea, the compensation coverage is for pre 1990, USD 50,000,000 for tankers 5,000 GRT or less and USD 50,000,000 plus USD 850 per GRT with a maximum of USD 135,000,000 for tankers greater than 5,000 GRT. After 1990 the coverage is USD 50,000,000 for tankers 5,000 GRT or less. For tankers greater than 5,000 GRT USD 50,000,000 plus USD 1,400 per GRT up to a maximum of USD 200,000,000.

The new PLATO required 50 million GRT enrolled by March 31, 1986 for it to enter into force on 1 June 1986, something as noted which did not occur.[125] The oil companies could possibly have entered their own tonnage, the requirements met and the PLATO/CRISTAL system put into force, however this did not occur.[126]

6.4.6. Liability Conventions Covering Damage from Pollution from Substances Other Than Oil

6.4.6.1 LLMC,[127] and International Convention Relating to the Limitation of the Liability of Owners of Sea-Going Ships (1957)[128]

The LLMC on its entry into force in 1986 between Parties succeeded the 1957 Convention. Claims for *oil* pollution damage, as defined in the CLC and subsequent Protocols, are excluded from this Convention, but claims for pollution damage from substances other than oil are left open. Limitations to coverage for personal claims include ships from 501 to 3,000 tons, 500 SDR's/ton; from 3,001 to 30,000 tons , 333 SDR's/ton; 30,001 to 70,000 tons, 250 SDR's/ton; and ships over 70,000 tons, 167 SDR's/ton. The minimum allowable coverage limitation is 330,000 SDR's. For claims involving property damage, ships 501 to 30,000 tons, 167 SDR's/ton; 30,001 to 70,000 tons, 125 SDR's/ton; and ships over 70,000 tons, 83 SDR's/ton. The minimum allowable coverage limitation is 167,000 SDR's. Claims may be made against shipowners, charterers, managers, operators, salvors and insurers. Limiting liability is prohibited if it is proven that the polluting incident was the result of a intentional and reckless act or omission done with knowledge that loss would result.[129]

The 1957 Convention permits shipowners, charterers, managers, operators, masters and crew to limit their liability for property damage, death and personal injury. The limits are 3,100 gold Francs per ship ton or approximately 207 SDR's per ship ton,[130] of which 2,100 Francs are allotted for personnel claims and 1,000 Francs for property claims. Where a 'limitation fund' is required by the ratifying State, this is generally provided under the ship's P. & I. policy. Limitation is prohibited where the damage was caused by the shipowner's 'actual fault and privity'.[131]

6.4.6.2. HNS Draft Convention[132]

As noted, the shipping liability schemes, the CLC, the Fund Convention, TOVALOP, CRISTAL/PLATO, are limited to damage cause by *oil*, and this is in spite of MARPOL 73/78 regulating pollutants *in addition* to oil. The LLMC and the 1957 Convention are generally considered insufficient to cover liability for these other pollutants due to the liability limits set, and it has been generally felt that a convention is necessary to parallel the CLC and the Fund Convention.[133] A conference was held immediately prior to the meetings on the Civil Liability and Fund Protocols, but no final result was achieved and the HNS Draft Convention was sent back to the IMO Secretariat for further study.[134] One author notes that the failure was due to a 'lack of sufficient negotiating time, new and untried ideas, some basic disagreements among the various industrial components involved in the carriage of hazardous and noxious substances, including a feeling by some that such a convention was premature.'[135] The same author feels, however, that a new liability convention covering these substances will be negotiated within a few years time.[136] In the meantime coverage will have to be met from the usual P.& I. ship coverage, and large claims will often be able to be limited for shipowners by the liability limits set under the LLMC or the 1957 Convention.[137]

6.5. The ILC[138]

One author notes that one positive development in this stagnant area of international law is provided by the ILC.[139] This organization's work will be briefly presented here for background.

The ILC in its meeting of Spring 1988 considered two topics which touch upon this work, International Liability for Injurious Consequences arising out of Acts *not* Prohibited by International Law, and State Responsibility for an International Wrongful Act. The Articles arrived at by the Commission both in draft form and adopted will be briefly presented, and it is hoped will give a general idea of ILC work. Note that developments *continue* with reports given out yearly. However unless these Articles represent customary law or are adopted by States either through conventions or State practice, they are *not* firm international law and as such have little *direct* relevance to transboundary pollution.[140]

6.5.1. International Liability for Injurious Consequences arising out of Acts not Prohibited by International Law

10 draft Articles were developed at the 1988 meeting but with many key terms remaining unsettled.[141] The draft Articles are therefore very much in a state of transition, and the ILC desired State input especially concerning the terms 'risk' and 'harm'. The Articles apply to activities carried out under the jurisdiction of a State or under the effective control of the State, when there exists appreciable risk for transboundary injury.[142] An attempt was made to provide a comprehensive definition of 'dangerous activity' rather than providing a list of such activities.[143] In this connection 'risk' means 'the risk occasioned by the use of things whose physical properties, considered either intrinsically or in relation to the place, environment or way in which they are used, make them highly likely to cause transboundary injury throughout the process;' 'appreciable risk' means 'the risk which may be identified through a simple examination of the activity and the substances involved;' 'activities involving risk' means 'the activities referred to in Article 1;' and 'transboundary injury' means 'the effect which arises as a physical consequence of the activities referred to in Article 1 and which, in spheres where another State exercises jurisdiction under international law, is appreciable detrimental to persons or objects, or to the use of enjoyment of areas, whether or not the States concerned have a common border.'[144] It can be questioned whether the ILC was wholly successful with these confusing and circular definitions.

The State with jurisdiction or control, incurs liability if it knew or had means of knowing that an activity involving risk, was being, or was about to be, carried out.[145] States are free to carry out or permit in their territory any appropriate human activity, however regarding activities involving risk, that freedom must be compatible with other States' rights associated with their sovereignty.[146] Good faith cooperation by States in preventing or minimizing the risk of transboundary injury including notification and information exchange or, if injury has occurred, in minimizing its effects is required, and the duty to co-operate applies to States of origin in relation to affected States and vice versa.[147] States of origin must permit participation by States likely to be affected to allow them to make a risk assessment and play a more effective preventative role.[148] For non-compliance with these cooperative provisions, the ILC urges sanctions be attached.[149] States of origin must take all reasonable preventive measures to prevent or minimize injury that may result from an activity which presumably involves risk and for which no regime has been established.[150] For an injury caused by an activity involving risk there must be reparation for the injury suffered, settled by negotiation between the Parties.[151] Reparation is to prevail where there is no agreed regime between the State of origin and the affected State and is not to be limited to pecuniary damages.[152]

6.5.2. State Responsibility for an International Wrongful Act[153]

In the view of some members of the ILC, transfrontier pollution can come under State responsibility for international *wrongful acts* if the State of origin did not take required precautions when undertaking an activity with a potential for transboundary harm.[154]

Generally the ILC's work in this area is dealt into 3 Parts. Part I, dealing with the origin of international responsibility, was provisionally adopted by the Commission in 1980. Part II, dealing with the content, forms and degrees of international responsibility is presently under consideration by the Commission. Articles 1 - 5 of Part II were *provisionally* adopted by the Commission in 1986. Part III will deal with the settlement of disputes and the implementation of international responsibility. Draft Articles 6 to 16 of Part II and 1 to 5 of Part III and its Annex were sent to the Drafting Committee in 1986.

The international responsibility of a State which in accordance with Part One arises from an internationally wrongful act committed by that State, entails legal consequences set out in Part II.[155] These rules in Part II are subject to other rules of international law relating specifically to the internationally wrongful act in question.[156] In addition international customary law continues to govern the legal consequences of a State's internationally wrongful act not covered by Part II.[157] The legal consequences of an internationally wrongful act of a State set out in Part II are subject, as appropriate, to the provisions and procedures of the Charter of the United Nations relating to the maintenance of international peace and security.[158]

To indicate further the level of complexity at which the ILC operates, quick reference can be made to Articles 5, 6 and 7 the first of which as indicated has been provisionally adopted. Definitions in Article 5 include, 'injured State' which means 'any State a right of which is infringed by the act of another State, if that act constitutes in accordance with Part One an internationally wrongful act of that State.'[159] A State whose action or omission constitutes an internationally wrongful act of a continuing character remains, without prejudice to the responsibility it has already incurred, under the obligation to *cease* such action or omission.[160] The injured State may claim reparation by equivalent (pecuniary compensation) to substitute totally or in part for restitution in kind, provided that such a choice would not result in an unjust advantage to the detriment of the wrongdoer State, or involve a breach of an obligation arising from general international law.[161]

It is hoped from this brief but complicated survey an insight can be obtained into the ILC's proposals and adoptions regarding State liability both for a State's international legal and wrongful acts. As can be observed the legal competence of the Commission seems very high, but the level of agreement low. Over an 8 year period dealing with wrongful acts the Commission has *provisionally* adopted only 5 Articles. If as has been suggested above the work of the ILC is one of the few sources contributing towards a positive development regarding State liability, then the picture is indeed dim. It can only be wondered what the international response will be to the ILC's proposals when completed after many years work?

6.6. Conclusion

The international liability and compensation coverage for all the pollution sources noted previously is *not* positive. Regarding dumping as seen only the global LDC's Article X governs, and the regional conventions contain either non specific provisions or no provisions. Regarding land-based pollution and offshore platform pollution the Paris Convention contains no provisions and the other regional agreements contain only non specific obligations, with the sole exception of the Nordic Convention. Damage from vessel-source pollution *is* covered by several conventions and agreements but only governing oil pollution and with questionable liability limits. Due to this inadequate coverage there have been established by the oil and shipping companies *private* agreements which are characterized by inadequate liability limits or lapse. Although a few conventions have been drafted to attempt to deal with these deficiencies, States have failed to ratify them. Finally, though there is some customary international law developed, this does not seem to be relied upon to a great extent by States.[162] In summary what this means is that liability and compensation for damage from marine pollution from other sources than shipping depend wholly upon *national* legislation covering areas under their respective jurisdictions.[163] For persons suffering pollution damage by States such as Norway this is probably sufficient coverage. For persons suffering pollution damage in other jurisdictions, this is not a fortunate alternative.

Perhaps the fate of the 1976 Liability Convention noted above is most illustrative. Although more than enough marine pollution occurs on a regular basis in the North Sea and surrounding areas to warrant an international agreement covering liability for pollution damage, even such States as the Netherlands, Sweden, Denmark and Norway, normally considered as being environmentally progressive, refuse to ratify this instrument. It appears in this area of international environmental law, concerned with providing *real* channels for economic relief for pollution damage, even catastrophes do not move States to take concrete action.

Notes - Chapter 6

1 Handl pp. 85-86. Also noted is that under the ECE Convention noted above even this general statement does not appear.

2 *Ibid.* emphasis added.

3 Churchill p. 274. OPN 1988 p. 7 notes a liability scheme is also under discussion under the Helsinki Convention. This is covered generally in Article 17 which requires State Parties to *undertake* to develop and accept rules covering responsibility for damage including limits, the determination of liability and possible remedies.

4 Timagenis p. 275, notes in the formation of this provision there was confusion regarding the term, 'responsibility', and submits, it 'should be understood as meaning the general obligation not to cause damage to and/or to protect the marine environment.'

5 Timagenis pp. 274-275, emphasis added.

6 *Ibid.*

7 OPN 1988 p. 7.

8 Kasoulides, George C., 'State Responsibility and Assessment of Liability for Damage resulting from Dumping Operations,' *San Diego Law Review*, Vol, 26, No. 3, (1989) pp. 497-523 at p. 500.

9 *Ibid.* p. 522.

10 Churchill p. 274 and OPN 1988 p. 7.

11 Offshore Pollution Liability Agreement (1976) (OPOL), VI *New Directions in the Law of the Sea* 507.

12 Convention for Civil Liability for Oil Pollution Damage resulting from Exploration for and Exploitation of Seabed Mineral Resources (1976 Liability Convention), VI *New Directions in the Law of the Sea* 535.

13 Convention on the Protection of the Environment between Denmark, Finland, Norway and Sweden, 4 *New Directions in the Law of the Sea* 514 (1975).

14 The use of land, the seabed, buildings or installations in any other way is also included.

15 Churchill p. 281.

16 Offshore Pollution Liability Agreement (1976) (OPOL), VI *New Directions in the Law of the Sea* 507. The following arrangements, OPOL, 1976 Liability Convention and IMO Mobile are presented in order to give background into possible arrangements that may surface for the Barents Sea case study.

17 Very *few* of the companies operating on the Norwegian shelf are members, Hjort, Harald and Gunnar Meyer, *Forurensningsansvar i oljevirksomhet*, Nordisk Institutt for Sjørett (1981) (Hjort), p. 238. Representatives of the British shelf companies include Amoco (UK) Exploration Company, Burmah Oil (North Sea) Limited, BP Petroleum Development Limited, Total Oil Marine Limited, Conoco North Sea Incorporated, Esso Exploration and Production UK Inc., Gulf Oil Production Company, Hamilton Brothers Oil Company (G.B.) Limited, Mobil North Sea Limited, Shell UK Limited, Phillips Petroleum Company, Signal Oil and Gas Company Limited, Siebens Oil and Gas (UK Limited, Texaco North Sea UK Company, North Sea Sun Oil Company Limited, and Cliff Oil Limited.

18 Clause I.3. The Norwegian continental shelf in the Barents Sea would thus already be included within the scope of OPOL for those oil companies listed as Parties. Norwegian domestic law however dominates.

19 Rules, Part II. 2.1.

20 Clause X. This provision would presumably allow the Soviet Union to become a designated State if members were in favour and the Soviets were likewise interested in including their section of the Barents Sea within OPOL.

21 See Section 6.4. Churchill pp. 274-275 probably overrates the similarities between OPOL and the 1976 Liability Convention. The latter has arrangements which appear more 'claimant friendly' than the industrial scheme, OPOL.

22 'Discharge of oil' means under Clause I.11 'any escape or discharge of oil into the sea from a designated offshore facility.'

23 'Offshore facility' is defined under Clause I.11 to include 'any well and any installation or portion thereof of any kind, fixed or mobile, used for the purpose of exploring for, producing, treating, storing or transporting crude oil from the sea bed or its subsoil, and any well used for the purpose of exploring for or recovering gas or natural gas liquids from the seabed or its subsoil during the period that any such well is being drilled (including completion), re completed or worked upon except for normal work-over operations.'

24 A State appears under OPOL to stand strong in being able to collect up to USD 12,500,000 for *remedial measures* taken in containing the oil discharged. This would cover cases of oil in the open ocean even where the oil would do no apparent damage. 'Remedial measures' under OPOL is defined as 'reasonable measures taken by any Party from whose designated offshore facility a discharge of oil occurs and by any public authority to prevent, mitigate or eliminate pollution damage following such discharge of oil or to remove or neutralize the oil involved in such discharge', Clause I.14. Thus if a State having jurisdiction over the area where the discharge occurred determined it reasonable to contain the oil or help it decompose to protect any marine life, it might have a worthwhile claim through OPOL, even though any direct damage could not be shown.

25 'Pollution damage' under Clause I.12 means '*direct loss or damage*, other than loss of or damage to the designated offshore facility involved by contamination which results from a discharge of oil,' emphasis added.

26 'Incident' means under Clause IV.A 'any occurrence or series of occurrences arising out of one event which results in a discharge of oil.' This limit of liability is approximately one-half of that required by the 1976 Liability Convention. It can be questioned whether this amount would be sufficient in the event of a large blowout.

27 Clauses IV.B.1-4 and I.7.B.i.

28 Clause IV.A.1-2.

29 Clause III.

30 Clause VII.

31 Hjort p. 240.

32 Clauses IX and XII.

33 English law is more 'industry friendly' than Norwegian law noted in Part II Section 9.2.4.

34 Churchill Conflicts p. 243 states, '(B)y 1981 only £150 had been paid out under OPOL as far as incidents on the United Kingdom continental shelf are concerned - a reflection of the high safety standards at present prevailing on the British continental shelf.' This amount seems questionably little. As noted in Part I Section 1.3.6. as late as 1986 both the British and the Norwegian sectors were observed to contain more oil slicks than officially reported. Rather than high safety standards being present, it seems rather a lack of reporting, a lack of claimants or a lack of State remedial measures being taken.

35 Convention for Civil Liability for Oil Pollution Damage resulting from Exploration for and Exploitation of Seabed Mineral Resources (adopted 1976) (1976 Liability Convention), VI *New Directions in the Law of the Sea*, 535. As of 1985 6 States have *signed* this Convention, Federal Republic of Germany, Ireland, Netherlands, Norway, Sweden and the United Kingdom. Of the 8 countries attending the Conference adapting the Convention, 2 did not sign it. For the Convention to enter into force, 4 ratifications are required. Of those remaining who did sign, *none* have ratified it in nearly 11 years it has been in existence; so it is doubtful it will come into existence. If in force this Convention under Article 2 would apply in the Barents Sea to all areas and platforms over which Norway has natural resource jurisdiction in accordance with international law.

36 See Section 6.4.1.

37 Pollution damage' is defined under Article 1.6. as 'loss or damage outside the installation caused by contamination resulting from the escape or discharge of oil from the installation, and includes the cost of preventative measures and further loss or damage outside the installation caused by preventative measures.'

38 Article 3.

39 Where all such damage is not reasonably separable.

40 Articles 3.2 and 5.1-2, emphasis added.

41 40,000,000 SDR's are approximately equal to USD 52,000,000. This ceiling is double the amount of OPOL.

42 Article 6.4.

43 Article 15, emphasis added.

44 Articles 11 and 13. Jurisdiction over State Party operators is the same for all operators.

45 Article 8.

46 Article 10.

47 *Ibid.*

48 Article 12.1-2.

49 Article 18, emphasis added. This would seem to allow for an invitation to the Soviets should the other State Parties desire.

50 This allows better coverage for claimants in the event there are questions concerning which operator is responsible. This does not include subcontractors.

51 Draft Convention on Offshore Mobile Craft (IMO Mobile), LEG XXXIV/6(b) 19 December 1977, pp. 2-8.

52 *Ibid.* p. 2.

53 See Section 6.4.

54 IMO Mobile p. 4 lists the US, Panama, Liberia, Venezuela as not being 'maritime convention countries' in 1977. Though some of these countries have ratified some IMO Conventions, still much the same could be said to be valid today. See Section 6.4.

55 IMO Mobile p. 7.

56 *Ibid.* p. 8.

57 International Convention on Civil Liability for Oil Pollution Damage (1969), (CLC), 9 *International Legal Materials* 45. Ratifying, etc. States include Yugoslavia, Yemen, Vanuatu, UK, United Arab Emirates, USSR, Tuvalu, Tunisia, Syrian Arab Republic, Switzerland, Sweden, Sri Lanka, Spain, South Africa, Singapore, Seychelles, Senegal, Republic of Korea, Qatar, Portugal, Poland, Peru, Papua New Guinea, Panama, Oman, Norway, Nigeria, New Zealand, Netherlands, Morocco, Monaco, Maldives, Liberia, Lebanon, Kuwait, Japan, Italy, Indonesia, India, Iceland, Guatemala, Greece, Ghana, Federal Republic of Germany, German Democratic Republic, Gabon, France, Finland, Fiji, Ecuador, Dominican Republic, Denmark, Côte d'Ivoire, China, Chile, Cameroon, Brazil, Benin, Belgium, Bahamas, Australia and Algeria. Protocol to the International Convention on Civil Liability for Oil Pollution Damage 1969, 16 *International Legal Materials* 617 (1976). Ratifying, etc. States include Yemen, UK, United Arab Emirates, USSR, Switzerland, Sweden, Spain, Singapore, Qatar, Portugal, Poland, Peru, Oman, Norway, Netherlands, Maldives, Liberia, Kuwait, Italy, India, Federal Republic of Germany, France, Finland, Denmark, China, Cameroon, Bahamas, and Australia. Protocol of 1984 to Amend the International Convention on Civil Liability for Oil Pollution Damage 1969 (not in force), (Civil Liability Protocol), 15 *J. Mar. L. & Com.* pp. 613-622. Ratifying, etc. States include South Africa, Peru, Federal Republic of Germany, France and Australia. IMO Status pp. 187-204, 205-214, 215-220.

58 Definitions appear in Articles I and II.

59 The 1984 Protocol removes 'whale oil' from the definition.

60 Ganten, Reinhard, 'The International System for Compensation for Oil Pollution Damage, An Assessment Based on the Experiences of the International Oil Pollution Compensation Fund,' *Marius Nr. 62*, Nordisk Institutt for sjørett (Ganten), pp. 7-8 and Gold pp. 114-115 concerning the definitions note that general cargo ships, bulk carriers other than oil carriers and tankers on ballast voyages are excluded from coverage (as are offshore oil rigs); 'persistent oil' is included but vague and undefined; damage caused by explosion or fire is excluded as not being 'contaminating'; and pollution damage occurring in the territory or territorial sea is covered regardless of where the initial spill occurred. Ganten pp. 7-8 continues that it is unclear whether precautionary and pre spill preventive measures are covered, with the prevailing view (in spite of the Scandinavian) that they are *not* covered by the CLC. *Ibid.* pp. 3-4 notes that in respect to the geographical scope of the Convention, post spill preventative measures taken *outside* the Contracting State's territory or territorial sea are also covered. An idea of what this entails can be seen from examples. The TANIO was one pollution incident where preventative measures were taken pumping oil out of the sunken ship outside the French territorial sea, and where neither the P. & I. (protection and indemnity) Club nor the IOPC Fund (International Oil Pollution Compensation Fund) denied responsibility for the owner under the CLC or the Fund Convention. Another incident involved the GINO, which was more problematic since the oil sunk, not polluting the French coast but causing damage to fish in the area of the wreck and consequently causing lost earnings to French fishermen. It was left open whether the shipowner was responsible for costs of such preventative measures.

61 Article III. Additional exceptions include where damage is wholly caused by a third party act or omission done with intent to cause damage, damage partially or wholly caused by an act or omission by the injured party done with intent to cause damage or

negligence, or damage wholly caused by negligent or wrongful acts by any government or other authority responsible for the maintenance of lights or other navigational aids.

62 Article V and Article VII. Under the 1976 Protocol the limits were amended and made expressed in SDR's rather than the Poincare franc as under the original CLC. In February 1988 the exchange rate was 1 SDR to approximately USD 1.3528 which would give the approximate figures of USD 180 per ton of ship tonnage and USD 18,900,000 per incident respectively. See Van Hanswyk, 'The 1984 Protocols to the International Convention on Civil Liability for Oil Pollution Damages and the International Fund for Compensation for Oil Pollution Damages,' *The International Lawyer*, Summer 1988, Vol. 22, No. 2, (Van Hanswyk) p. 322. Eaton, Judith R., 'Oil Spill Liability and Compensation: Time to Clean Up the Law,' *George Washington Journal of International Law and Economics*, Vol. 19, No. 12, 1985 (Eaton), p. 802, notes that the Convention's liability limits are available to any insurer who becomes a defendant in the oil pollution incident, and claims can be made directly against the insurer. Generally the master, crew, servants, agents or other persons cannot be held liable, only the shipowner. Ganten pp. 3-5 points out however that the channelling of liability is incomplete since other questions of liability are not dealt with under the CLC, and in such cases national liability law would apply.

63 Article V. Ganten p. 6 notes that at least up to April 1981 no tanker owner had been held to unlimited liability under the CLC.

64 Article V.

65 *Ibid.*

66 Article VII.

67 *Ibid.*

68 *Ibid.*

69 Articles VIII and IX.

70 Articles V and IX.

71 Article VI.

72 Article X.

73 Ganten p. 21 notes that the CLC's and Fund Convention's upper limits in 1981 were sufficient and were exceeded only by 2 very exceptional major disasters, the AMOCO CADIZ and TANIO, although in the future these limits may be insufficient. Gold pp. 118-119 notes in 1985 that, '(T)he maximum liability limits, originally developed in the early 1970's and revised in 1975, had become quite unrealistic' and '(B)oth shipping and oil interests were in agreement that the upper limits were too low.' Additionally Van Hanswyk p. 324. notes US Congressional hearings have held the CLC and the Fund Convention recovery to be too low.

74 Article VI. The Units of Account (UOA) are SDR's as defined by the International Monetary Fund. The above limits in dollars are approximately USD 4,060,000 and USD 568 respectively, Van Hanswyk p. 324. The definition of ship has been broadened from the CLC to include ships 'constructed or adapted' for carrying oil as bulk cargo, and ships carrying oily ballast, Articles I and II. The following is *background* into the revision of both the CLC (and the Fund Convention) by the Civil Liability and Fund Protocols of 1984, obtained from Gold pp. 118-120 unless otherwise indicated. This has relevance both into problems with such a liability scheme, and also to today's situation since *neither* the Civil Liability Protocol nor the Fund Protocol has yet entered into force. The oil industry wanted the CLC limit, the first level compensation, raised to USD 50,000,000 regardless of ship size, and this be exhausted before the second level, the Fund Convention and CRISTAL were tapped (see Sections 6.4.2. and 6.4.4.). The Oil Companies International Marine Forum (OCIMF) complained that Fund Convention and CRISTAL had been used increasingly for first level pollution clean up compensation. Gold considers the OCIMF lobby as being more united and competitive than

shipping, represented by the P. & I. Clubs and International Association of Independent Tanker Owners (INTERTANKO). The author notes additionally that the oil sector of the shipping industry is very strong minded and ready to develop alternative compensation systems if they felt they could not agree with IMO systems. OCIMF also wanted the Fund Convention's limit raised to USD 75,000,000 per incident, which would provide a combined limit of USD 125,000,000 per incident. The shipping industry wanted the oil industry to carry a larger share of paying compensation, this especially in light of increased oil profits in the 1980's but decreased shipping profits. The shipping industry was against the proposed USD 50,000,000 limit as being too costly for shipowners of small ships. Other proposals by several States for revisions of the CLC and the Fund Convention included coverage of oil from *non tankers* since large bulk carriers and container ships could cause much pollution from bunker oil. This was opposed by both the oil and shipowners as, 'radically changing the basic, well-established principles relating to compensation for damage caused by tankers' and unduly complicating negotiations, *ibid.* Shipping wanted *ballasted tankers* included since pollution damage often comes from discharged dirty ballast and both loaded and ballast voyages are part of the, 'total mode of carrying oil at sea,' *ibid.* The oil companies were against this since dirty ballast is seen as different than cargo damage and thus not covered under the Fund Convention. Other proposals included *non persistent oils* being covered, since these include generally gasoline, kerosene, light diesel and non petroleum oil, where pollution damage can be as much as from the persistent oils. It was also desired that preventative costs and specific conduct where limitation is barred should be covered. See Ganten pp. 21-23 who considers that shipowner's liability should not only be proportional to ship tonnage but also to the aggregate amount of the claims to bring the Fund Convention more into play with the CLC's higher limits. Ganten concludes in fact calling for States *not* to revise the 'generally well functioning system' of the CLC and the Fund Convention, for the questionable advantage of minor amendments which might not even be improvements.

75 Article VI. The amount is approximately USD 80,800,000, Van Hanswyk p. 324. Gold p. 120 notes that ships of 140,000 GRT and over would be at the upper limit. Ganten p. 21 notes that the original CLC limits were set at approximately USD 17,000,000 since this was the limit of the insurance market's capacity. In 1981 however most ships were insured against pollution liability up to USD 300,000,000, an insurance market capacity some say necessitates an increase of shipowner's pollution liability. Ganten also notes a need for a *minimum* limit being set for small ships under the CLC.

76 Article III. Gold p. 121 notes that this new geographical scope, 'may well cause difficulties for States which have problems with the U.N. Convention on the Law of the Sea (LOSC) from which this provision originates,' (parentheses added). Eaton p. 809 notes that the 1984 Protocol definition encompasses sea areas equivalent to 200 mile economic zones for those States which have not proclaimed such.

77 Articles II and IV. Eaton p. 810 notes more specifically that under the Civil Liability Protocol's Article IV damages include not only owner's liability for oil pollution from actual discharges or escapes (CLC Article I) but also 'any pollution damage caused by the ship as a result of the incident.' Additionally noted is that 'incident' has been broadened under Article II to include 'grave and imminent threats' of pollution damage as well as actual discharge, but under Article II recovery for loss of profits or loss due to unreasonable measures taken or to be taken are excluded.

78 Article IV. Protection is unavailable upon a showing of damages from 'a personal act or omission, committed with the intent to cause such damage, or recklessly and with knowledge that such damage would probably result.'

79 *Ibid.* Considering the remoteness of ships on the seas and the problems noted in Part II Section 9.2.3.1. regarding inspection of ships under MARPOL 73/78, this would seem to be an almost impossible burden. Ganten p. 23 notes that 'very serious consideration' should be given to whether the new limits should be unbreakable. Eaton p. 811 notes however the treaty language has been changed where breakage is allowed. Under the CLC Article V the liability limits do not apply if an incident was caused by

the 'actual fault or privity of the owner', which resembles *admiralty* common law. Under the Civil Liability Protocol Article VI the liability limits do not apply under the Article VI 'knowledge' exception noted above, which seems to the author to be a broader exception.

80 Article VI.

81 Article XV. During this time States have the possibility to object. The amendment procedure under CLC Article XV requires a diplomatic conference and State ratification, something which takes much more time.

82 Under Article XIII it is required that 10 States, 6 of which have not less than 1,000,000 units of gross tanker tonnage, have ratified the Civil Liability Protocol, upon which it will enter into force 12 months later.

83 Gold p. 120. The author continues that the Civil Liability Protocol's (and the Fund Protocol's) coming into force much depends upon whether the US ratifies both.

84 See Van Hanswyk, pp. 320, 324-326, 341-342. The Protocols in addition to benefitting the US internally, were adopted by other countries making concessions in Protocol negotiations expecting US ratification, and US ratification would probably cause other States currently waiting for US action, to ratify. *Ibid.* believes that the US is most benefited by ratifying these two Protocols because the total recovery available, 'remains substantially higher than that available under existing federal and state statutes' and the Protocol liability schemes, 'would be a great improvement over the antiquated and piecemeal procedures for recovery under US law.' Also noted is that the US was very active in 1984 Protocol negotiations and 'the effect of its tremendous influence in the proceedings is evident in many of the agreements' provisions.' For an in depth analysis of US law and policy in relation to the 1984 Protocols including *opposing arguments* see Eaton pp. 805-808 and 813-827. The opposing arguments include certain US coastal states claiming that their economic interests in controlling their coastlines, such as protecting beaches, fishing rights and private property, to be threatened by the Protocols. Additional objections include that under the Protocols exceptions exist to strict liability, limits are allowed on damages, and US federal and state law allow broader recovery. The author recommends US ratification of both Protocols however, since they 'channel the liability to the parties conducting the ultrahazardous activity while limiting the amount of damage for which the parties will be liable.' Channelling discourages claims against the State, but limiting liability 'allows socially necessary activities to continue.'

85 International Convention on the Establishment of an International Fund for Oil Pollution Damage (1971) (Fund Convention), 11 *International Legal Materials* 284. Ratifying, etc. States include Yugoslavia, UK, United Arab Emirates, USSR, Tuvalu, Tunisia, Syrian Arab Republic, Sweden, Sri Lanka, Spain, Seychelles, Qatar, Portugal, Poland, Papua New Guinea, Oman, Norway, Nigeria, Netherlands, Monaco, Maldives, Liberia, Kuwait, Japan, Italy, Indonesia, Iceland, Greece, Ghana, Federal Republic of Germany, Gabon, France, Finland, Fiji, Denmark, Côte d'Ivoire, Cameroon, Benin, Bahamas and Algeria. Protocol to the International Convention on the Establishment of an International Fund for Compensation for Oil Pollution Damage 1971, 16 *International Legal Materials* 621 (1976) (not in force). Ratifying, etc. States include UK, Sweden, Spain, Portugal, Poland, Norway, Netherlands, Liberia, Italy, Federal Republic of Germany, France, Finland, Denmark and Bahamas. Protocol of 1984 to Amend the International Convention on the Establishment of an International Fund for Compensation for Oil Pollution Damage (not in force), (Fund Protocol), 15 *J. Mar. L. & Com.* pp. 623-633. Ratifying, etc. States include France and Federal Republic of Germany. IMO Status pp. 233-240, 241-246, 247-252.

86 Eaton p. 803. See Preamble and Articles 2, 4 and 5. Ganten p. 9 notes the case of damages exceeding the limits under the CLC especially by small ships is the most frequent case in which the Fund Convention becomes liable. As was mentioned no tanker owner up to 1981 had been held to *unlimited* liability under the CLC, but 'the theoretical possibility of unlimited liability has always to be investigated very carefully,

by claimants or by the Fund and CRISTAL, if the victims get their additional compensation from either the Fund or CRISTAL,' Ganten p. 6.

87 Article 4.

88 Articles 10, 12, 14 and 22. For a more extensive explanation of this process see Eaton p. 804 and Ganten pp. 12-13.

89 Article 5. Ganten p. 11 notes that the shipowner is relieved of liability under the CLC between 25% and 40%.

90 Articles 5 and 37. Ganten p. 11 notes the Fund Convention also need not compensate if the ship causing the polluting incident did not comply with provisions from IMO conventions including those from OILPOL, SOLAS, the Load Lines Convention or the Collisions Convention. See Part I Chapter 5.

91 Article 1.

92 Article 3.

93 See Section 6.4.1. 'Preventative measures' and 'incident' mean the same as under the CLC. In practice the scope of loss or damage covered by compensation under the Fund is somewhat unsettled. For an interesting article showing Fund practice concerning 19 pollution 'incidents' which have helped to define this scope, see Brown E.D., *The USSR, Eastern Europe and the Development of the Law of the Sea*, Compiled and Edited, with Translations and Annotations by William E. Butler, Professor of Comparative Law in the University of London, Director, Centre for the Study of Socialist Legal Systems, University College, London, (Brown) pp. 275-304. Basically the author concludes with a proposal to the IMO Conference in 1984 to clarify and tighten up the definitions, including making it clear that 'preventative measures' include '(a) measures taken in the geographical area seaward of the outer limit of the territorial sea and (b) preventative or precautionary measures taken prior to an oil spill which does subsequently occur, or prior to a threatened oil spill which does not subsequently materialize.' Additionally proposed is that 'pollution damage' be changed to 'proven economic loss actually sustained as a direct result of contamination,' which 'would ensure that no recovery could be had in respect of speculative claims for economic loss, including that associated with impairment of the environment.' This was due to Fund experience with the Soviet ANTONIO GRAMSCI incident covered by Brown.

94 Article 4. This would include costs of reasonable actions taken by shipowners in minimizing damage by oil pollution.

95 Articles 1, 4 and 7. Pollution from oil rigs and pipelines are excluded. Ganten p. 10 may be overly optimistic when he states, '(I)t is doubtful whether evidence has to be given that the oil escaped from a specific named ship or whether it is sufficient to prove that it came from a laden tanker' or '(I)f it can be proven that oil escaped from a laden tanker, it will normally be possible to prove by means of analyses from which specific ship the oil came.' In practice there have been major problems linking specific oil discharges to specific ships or offshore platforms. See Churchill pp. 264-265, and Part I Sections 1.2.1. and 1.3.6.

96 Articles 6-8. See Ganten pp. 18-20 for the close relationship between the Fund and P. & I. Clubs in settling pollution claims.

97 Article 4.

98 *Ibid.* Eaton p. 805 notes that the higher liability limits are 'a recognition that past incidents have required greater compensation than was available under the CLC and that fluctuating monetary values can have a major effect on liability limits.'

99 Article 5. An shipowner may not obtain indemnification if the damage results from his 'wilful misconduct'. Additionally as long as a shipowner has adequate financial insurance, the Fund establishes itself as guarantor upon request of the owner.

100 Gold p. 115. These figures are somewhat confusing. Van Hanswyk p. 322 writes as though both 1976 Protocols are in force and the SDR's system in use. The figure she presents for the combined compensation between the CLC and the Fund Convention is 45,000,000 SDR's (approximately USD 60,900,000). Churchill pp. 265-266 writes that where the Fund Convention comes into play, the Fund Convention will pay up to 675,000,000 Poincare francs, which was increased in 1979 from the original limit of 450,000,000 francs. Further it is noted that the 1976 Protocol converting francs to SDR's is *not* in force and the Fund Convention will take over in paying shipowner's liability in excess of 1,500 Poincare frances per ton (100 SDR's per ton) or 125,000,000 frances (8,333,000 SDR's), whichever is less.

101 Churchill p. 266.

102 Article 6. Eaton p. 813 notes that the Fund Protocol Article 7 abolishes tanker owner indemnification that transferred part of the liability of the shipowner under the CLC to the Fund, presumably 'due to its potentially large expense.'

103 Articles 4 and 6. Van Hanswyk's figures are as follows. The maximum level of compensation available under the Civil Liability Protocol and the Fund Protocol, 135,000,000 UOA's (approximately USD 182,600,000) per incident; automatically increased to 200,000,000 UOA's (approximately USD 270,500,000) if 3 States parties whose combined annual oil receipts equal or exceed 600 million tons. Churchill p. 267 notes that the maximum limit under the Fund Protocol is raised to 135,000,000 SDR's, which can be increased to 200,000,000 SDR's when certain conditions are met. Eaton p. 812 notes that the Fund Assembly under the Fund Protocol may no longer adjust this limit to a higher amount based on damage experience and changing monetary values.

104 Gold p. 121. See above and also Van Hanswyk pp. 320 and 325 who adds that US ratification would supply approximately 450,000,000 tons which would ensure this increased coverage. In Aftenposten, 9.mai 1989 Uke 19., Nr.207., 130 årg. Administrative Director Tormod Rafgård of INTERTANKO notes that after the EXXON VALDEZ accident it is hoped the US will within 1989 ratify the Fund Protocol (this did not occur), which provides a solid compensation fund in case of oil damages from tankers, since it raised the limit of owner's liability to USD 135,000,000. This in Rafgård's opinion is so large a sum that it would sufficiently cover most large accidents. Another point noted, which partially explains INTERTANKO's interest in the Protocol, is that *cargo* owners are also required to cover a specific amount of the liability in the event of accidents.

105 Article 4. The same comments made concerning the geographical scope of the Civil Liability Convention apply equally here, Gold p. 121.

106 Article 14.

107 Article 33. See also Van Hanswyk pp. 341-342.

108 Under Article 30 the Fund Protocol will enter into force when 8 States, which have a total receipt of at least 600,000,000 tons (measured during the preceding year) of contributing oil, ratify or accede to it.

109 Gold p. 120.

110 Tanker Owners' Voluntary Agreement Concerning Liability for Oil Pollution (1969) (TOVALOP), 8 *International Legal Materials* 497 (1969). The information concerning this agreement is obtained from Gold pp. 116-117 unless otherwise stated. It is administered by the International Tanker Owners Pollution Federation (ITOPF), Staple Hall, Stonehouse Court, 87-90 Houndsditch, London EC3A 7AX, Gold pp. 116-117, ACOPS Yearbook p. 110.

111 Gold p. 116 additionally notes that TOVALOP applies when States, such as the US and Canada, have different legislation than the CLC or where the damaged State has accepted the CLC but the flag State of the polluting ship has not. Ganten p. 15 notes that shipowner double liability is avoided under the CLC and TOVALOP, 'if an incident caused pollution damage to a State Party to the CLC and at the same time to another

State that is not a member of the CLC.' However TOVALOP does not 'prevent the owner from being liable under the CLC in one State and under national law in another State', such complicated cases as this having, 'great practical importance'. Churchill p. 267 notes that TOVALOP (and CRISTAL) still have great benefit for victims of oil pollution damage in approximately *80* coastal States not Parties to the CLC, especially since TOVALOP covers over 90% of the world's oil tankers. Ganten p. 15 believes Parties to TOVALOP to include close to 97% of the world's tanker tonnage.

112 This means that under TOVALOP the owner can limit his liability even where personal fault or privity were the cause of the pollution. Wilful misconduct is however not covered. See Ganten p. 14.

113 Generally Gold describes the CLC and TOVALOP systems as two different types. The CLC is described as an insurance scheme to which flag State ships subscribe usually under their P. & I. schemes. TOVALOP is administered by ITOPF which interprets TOVALOP, handles claims, is directly involved and has much experience in dealing with oil spills.

114 Contract Regarding an Interim Supplement to Tanker Liability for Oil Pollution (1971) (CRISTAL), 10 *International Legal Materials* 137 (1971). This is a compensation agreement including oil companies which ship in excess of 90% of total world crude and fuel oil cargoes. It is administered by the Oil Companies Institute for Marine Pollution Compensation Ltd, in Bermuda and operated on a daily basis by the service company Marine Pollution Compensation Services Ltd. at the same address as ITOPF, Staple Hall, Stonehouse Court, 87-90 Houndsditch, London EC3A 7AX, Gold p. 17 and ACOPS Yearbook, pp. 108, 110.

115 The following information is taken from Gold pp. 117-118 unless indicated otherwise. Gold notes that inflationary pressures made TOVALOP and CLC limits too low and CRISTAL's supplementary character has caused friction between the oil and the shipping industries. Churchill p. 267 considers the supplementary character beneficial since CRISTAL can operate together with the CLC where the Fund is not applicable, providing additional compensation and partially freeing the shipowner from his liability.

116 Churchill p. 267 considers this particular useful since only about 2/3 of the State Parties to the CLC are Parties to the Fund Convention. Ganten p. 16 notes that CRISTAL is meant to be resorted to only after a reasonable exhaustion of compensation possibilities against the shipowner and the oil companies. This means in practice that CRISTAL can make payments only long after the incident, which has caused concern among others CRISTAL members.

117 These include for example war, acts of God, sabotage or wrongful governmental acts.

118 Gold p. 118 adds that CRISTAL also indemnifies tanker owners for liabilities under compensation systems other than the CLC, greater than USD 160 per GRT or USD 16,800,000, whichever is less. CRISTAL's indemnification of tanker owners for the difference existing between TOVALOP or other liability systems, and CLC liability, amounts to liabilities over USD 120 per GRT or USD 10,000,000, whichever is less, up to USD 160 per GRT or USD 16,8000,000, whichever is less. Ganten p. 16 notes that indemnification payable by CRISTAL ranges from 25% to 40% of the shipowner's liability. Churchill p. 267 notes that by the end of 1986 CRISTAL had paid out approximately USD 60,000,000 in compensation.

119 Birnie p. 235. Churchill p. 267 notes the 1986 amendments to TOVALOP and CRISTAL increased the maximum amount of compensation *fourfold*.

120 Birnie p. 235, parentheses added.

121 Pollution Agreement among Tanker Owners (1985) (PLATO). Birnie p. 244 notes that copies of PLATO can be obtained from PLATO Limited, Portland House (6th Floor), Stag Place, London SW1E 5BH.

122 Gold p. 121-122.

123 *Ibid.*

124 *Ibid.* Birnie p. 235 notes a new PLATO arrangement was announced by some tanker owners which raised the previous limits but *not* extending coverage to the exclusive economic zones or environmental damage.

125 Birnie p. 235 notes only 28 million GRT had enrolled by the required date.

126 Gold p. 122. The author notes further that 'PLATO will be resisted by INTERTANKO, as well as the P. & I. Clubs, as it seeks to change the generally accepted trend of a more equitable share in pollution responsibility between ship and cargo.' Further noted is that it is contrary to the Civil Liability and Fund Protocols, it reduces the compensation available and it attempts to shift the liability burden even more on the 'hard-pressed' shipping industry. The author closes with the comment, '(I)t is hoped that the eventual loser will not be the marine environment.' There seems to be a good chance for just this to occur based upon Birnie's observation above that neither PLATO, CRISTAL nor the Civil Liability Protocol nor the Fund Protocol are in effect.

127 Convention on Limitation of Liability for Maritime Claims (1976) (LLMC), IMO Document p. 21. Entry into force 1 December 1986. Ratifying, etc. States include Yemen, UK, Switzerland, Sweden, Spain, Poland, Norway, Liberia, Japan, Federal Republic of Germany, France, Finland, Egypt, Denmark, Benin and the Bahamas. Total number of Contracting States is 16. Both Norway and Sweden note that higher limits of liability are required in their respective national legislation for drilling vessels. IMO Status pp. 305-312. Information concerning this Convention is obtained from Gold pp. 123-124 unless noted otherwise.

128 Gold p. 127, gives a cite of CMI, International Convention on Maritime Law, Antwerp, no date, p. 67. Information concerning this convention is obtained from Gold p. 123 unless otherwise stated.

129 Churchill p. 220 notes that the LLMC increased the limits of shipowner's liability and made clear when the shipowner was not allowed to limit his liability.

130 Gold p. 123 notes that most jurisdictions use the SDR's.

131 Gold p. 123 notes that Courts extended 'actual fault and privity' due to dissatisfaction with the liability limitation or the amounts available, and this has made limiting liability 'increasingly difficult'. Dissatisfaction on both sides led to the LLMC which raised the limitation amounts as well as made conduct barring limitation, intentional act or omission, harder to prove.

132 Draft Convention on Liability and Compensation in Connection with the Carriage of Noxious and Hazardous Substances, (1984), (HNS Draft Convention), 23 *International Legal Materials* 150 (not in force).

133 Birnie pp. 234-235 notes that strict liability of all shipowners is needed established as well as a definition of the geographical and material scope of the damage. The latter is needed since many States consider fishing and exclusive economic zones need be covered and damage to fisheries included.

134 Gold pp. 124-125.

135 *Ibid.*

136 *Ibid.* Birnie pp. 234-235 is not so optimistic noting that the IMO instructions regarding the Draft Convention were, 'to revise the draft into a form that will both provide practical and 'realistic' measures and be 'generally acceptable', which makes it unlike that coverage will be wide or that the negotiations will be speedily concluded.'

137 Gold p. 125 and Churchill p. 268. Birnie p. 234 notes that in the absence of a hazardous and noxious substance liability convention the only remedy to injured persons including fishermen is to bring an action in *national* courts where often strict liability is not available and fault must be proven.

138 International Law Commission, Report of May 9 - July 29, 1988, General Assembly, United Nations, Official Records: Forty-Third Session, Supplement No. 10 (A/43/10) (ILC Report). The Commission consisted of members from Nigeria, Bahrain, Jordan, Iraq, Italy, Argentina, USSR, Canada, Morocco, Egypt, Brazil, Venezuela, Iceland, Jamaica, German Democratic Republic, Ireland, Panama, Cyprus, Sierra Leone, Algeria, United States of America, Kenya, Japan, Poland, India, Madagascar, France, Greece, Mexico, China, Peru, Senegal, Federal Republic of Germany, and Bulgaria. Membership is determined by the U.N. on a geographical basis, though the superpowers have a higher membership rate, conversation with Norwegian Ministry of Foreign Affairs, November 25, 1988.

139 See generally Handl p. 86.

140 Kasoulides pp. 522 notes that the ILC has been working on the codification of the rules of State responsibility for many years, which would seem to imply that at least several of the draft or adopted Articles discussed represent international customary law. See also Henkin pp. 556-578.

141 ILC Report pp. 10-13, 18-44.

142 Article 1. The *scope* of activities covered is controversial, one view favouring limiting the scope to only activities involving appreciable risk, the other favouring encompassment of all activities causing transboundary harm, ILC Report p. 23.

143 *Ibid.* p. 29.

144 Article 2.

145 Article 3.

146 Article 6. This provision was taken from Principle 21 of the Stockholm Declaration, ILC Report p. 36.

147 Article 7.

148 Article 8.

149 ILC Report p. 39.

150 Article 9. 'Prevention' evoked very much discussion, and it was felt that there were three possibilities, prevention might be linked exclusively to reparation; there could be obligations not connected with the eventual harm and its reparation; and the draft could include only norms of prevention. The Special Rapporteur preferred that the draft Articles maintain some linkage to the occurrence of harm, ILC Report p. 41.

151 Article 10.

152 ILC Report pp. 42-43. It is noted however is that if strict liability as a general principle of international law were introduced 'the resistance of a great many Governments was likely to be encountered.' Kasoulides p. 522 notes that the ILC is considering whether State strict liability *should* extend to damage caused by lawful activities, but which are *ultrahazardous* and inherently cause transnational damage. The author applies this to liability from dumping activities, but the same could be applied to offshore drilling and arguably shipping as well as dangerous land-based activities.

153 See generally ILC Report pp. 266-278.

154 ILC Report p. 17.

155 Article 1.

156 Article 2.

157 Article 3.

158 Article 4.

159 Article 5. In particular, 'injured State' includes, 'if the right infringed by the act of a State arises from a bilateral treaty, the other State party to the treaty; if the right infringed by the act of a State arises from a judgement or other binding dispute settlement decision of an international Court or tribunal, the other State or States parties to the dispute and entitled to the benefit of that right; if the right infringed by the act of a State arises from a binding decision of an international organ other than an international Court or tribunal the State or States which in accordance with the constituent instrument of the international organization concerned are entitled to the benefit of that right; if the right infringed by the act of a State arises from a treaty provision for a third State, that third State; if the right infringed by the act of a State arises from a multilateral treaty or from a rule of customary international law, any other State party to the multilateral treaty or bound by the relevant rule of customary international law, if it is established that, the right has been created or is established in its favour, the infringement of the right by the act of a State necessarily affects the enjoyment of the rights or the performance of the obligations of the other States parties to the multilateral treaty or bound by the rule of customary international law, or the right has been created or is established for the protection of human rights and fundamental freedoms; and, if the right infringed by the act of a State arises from a multilateral treaty, any other State party to the multilateral treaty, if it is established that the right has been expressly stipulated in that treaty for the protection of the collective interests of the States parties thereto.'

160 Article 6. Draft Articles 6 and 7 are *new* draft articles proposed to the Commission and are not yet adopted, ILC Report p. 270. Article 6 was added to ensure the preservation of the law since, 'any act of a State not in conformity with an existing rule represents a threat not only to the effectiveness but also the validity, and thus the very existence of the infringed rule, particularly so in the case of an unlawful conduct extending in time,' ILC Report p. 270.

161 Article 7. The internal law of the State which committed the internationally wrongful act may not preclude by itself the injured State's right to restitution in kind. Regarding 'restitution in kind', 2 definitions are recognized, a re-establishment of the situation which existed prior to the wrongful act, *status quo ante*; and the re-establishment of the situation which would exist if the wrongful act had not been committed. A standpoint by the ILC was not taken as to which definition was preferable, however restitution in kind was recognized as the primary form of redress, although pecuniary compensation, reparation by equivalent, was most frequently resorted to. The best form for redress was dependent upon the specific case, and the 'most complete 'satisfaction' of the injured State's interest is the removal of all the injurious consequences of the wrongful act,' *ibid.* pp. 271-272. The injured State has the right to claim from the State which has committed an internationally wrongful act, restitution in kind for any injuries it has suffered, provided such restitution is not materially impossible; would not involve a breach of an obligation arising from a peremptory norm of general international law; and would not be excessively onerous for the wrongdoer State. A wrongdoer State could substitute pecuniary compensation in the case of physical impossibility of restitution in kind. Generally, legal impossibility is not a valid excuse with the exception when restitution in kind is incompatible with a superior international rule such as Charter of the United Nations or a peremptory norm. No impossibility exception is possible based upon internal domestic law, *ibid.* p. 273. Restitution in kind is not to be deemed excessively onerous unless it would represent a burden out of proportion with the injury caused by the wrongful act, or seriously jeopardize the political, economic or social system of the wrongdoer State. The Commission notes however in drafting an Article dealing with 'excessive onerousness', care should be taken not to leave too many loopholes for the wrongdoer State to avoid providing specific reparation, *ibid.* p. 274.

162 See Part I Section 2.1.3.

163 See Part II Section 9.2.4. and Appendix I for a brief outline of Norwegian and Soviet liability legislation governing in the Barents Sea.

7. Joint Contingency Plans

7.1. Introduction

As noted joint contingency plans, plans for dealing with major oil or hazardous chemicals pollution incidents or threats of such, are not covered under many of the international agreements noted above including the Paris Convention, the ODC, the LDC, or MARPOL 73/78 (or the other shipping Conventions)[1] and so have to be supplemented from other sources. The UNEP Conventions and Protocols are an exception. This Section will give a survey of joint contingency plans on a world wide basis, beginning with global, then regional and ending with bilateral. More detail will necessarily be given to the agreements and contingency plans with relevance to the Barents Sea case study. This is not only because of the orientation of this work but also because in contrast to environmental developments in the Arctic and in various parts of the world, no joint contingency plan presently exists covering the Barents Sea, although a general agreement is in the draft state.[2] For a specific idea of the contents of a contingency plan including preparatory and clean up measures, reporting and communications systems and payment procedures, see PART II Section 13.4.3.[3] For problems associated with Arctic contingency plans see PART II Section 9.3.4.

7.2. IMO Contingency[4]

The IMO recently adopted a Convention covering joint and national contingency planning.[5] This Convention is *not* in force and will not be so until 1 year following its ratification by 15 States. The instrument was developed as a direct result of the EXXON VALDEZ accident in Alaska noted in PART I Section 1.1. Under the Convention briefly, all ships must have detailed plans on board regarding *oil* pollution combat and clean up, but does not include pollution incidents involving hazardous and noxious substances. MARPOL 73/78 is to be amended to mandate such plans. States are urged to establish both national and regional oil contingency plans to meet major polluting incidents or threats of such including pre-positioned oil spill equipment, training, communications capabilities and arrangements for coordinating response.[6] International assistance is emphasized including mandatory contributions of equipment, experts and other technical support by member States upon request. Reimbursement procedures are provided for, coordinated research and standardization of equipment included, and financial support for exercises and education in the use of equipment established.

7.3. UNEP Conventions and Protocols[7]

The UNEP Conventions and Protocols have developed as noted cooperative programs for dealing with oil and hazardous chemicals pollution incidents. The Barcelona Convention, Article 9 provides generally that State Parties must cooperate in taking necessary measures for dealing with pollution emergencies whatever the cause, and reduce or eliminate resulting damage. State Parties aware of an incident must notify the Regional Oil Combatting Centre for the Mediterranean at Malta, the IMO and other State Parties liable to be affected.[8] The Barcelona Protocol I[9] is also in force, under which in accordance with Article 3 State Parties must *endeavour* to maintain and promote either individual or joint contingency plans and means for combatting marine pollution by oil and other harmful substances including equipment, ships aircraft and manpower.[10] Under Article 4 monitoring must be developed either individually or jointly to obtain as precise information as possible on polluting incidents.[11] Communications are dealt with in Articles 6, 7, 8, and 9 and include the dissemination of anti pollution information, effectiveness of emergency communications, accident reporting by ships and aircraft, and State reporting of measures taken against polluting incidents. Under Article 10 State Parties may request assistance, including expert advice, nautical facilities and anti pollution equipment and products from other Parties, and State Parties so requested must use their best efforts to render such. Developments under the other UNEP Conventions, especially the Red Sea and Kuwait Conventions, are similar to the Barcelona including the establishment of regional emergency centres. The remaining UNEP Conventions lack the regional centres.[12]

7.4. Bonn Agreement[13]

The Bonn Agreement covers oil (and will cover other dangerous substances under the 1983 Amendment when it comes into force) in the North Sea and English Channel,[14] and has the same Secretariat as the ODC and the Paris Convention. Under this Agreement the ships and aircraft of State Parties are to report to their flag State all large oil slicks and accidents causing or likely to cause pollution which represents a danger for one or more of the State Parties' coasts or related interests, and the flag State must notify other affected State Parties.[15] The North Sea is divided into zones of responsibility, and a State responsible for such a zone in which a slick or an accident occurs must keep the slick under observation and report to other States an appraisal of the situation and the measures taken to combat the oil.[16] A State Party requiring help to combat the oil may request assistance from the other State Parties, and the Parties requested must do their best to render this help.[17] Nothing in the Agreement can be interpreted to contradict any of the States' rights and duties under international law especially as related to preparedness and combat of marine pollution, and nothing regarding the zones of responsibility can be interpreted as a precedent or argument for State sovereignty or jurisdiction.[18] In the absence of agreements regarding costs involved in clean up operations, a Party requesting assistance must pay the assisting Party the costs of the help rendered, and payments in no way affect the requesting State Party's legal rights to claim compensation from a third party.[19] Two or more State Parties may change their common boundaries of their zones of responsibility.[20] The activities in addition to clean up measures carried out under the Bonn Agreement in the past include relations between the Bonn Agreement and the MOU, involving a breakdown of communications and failures to prosecute discharge offenders; contacts with the Barcelona and Helsinki Secretariats; extension of the 1983 Amendment by a Protocol to bring aerial surveillance within the Agreement's; exchange of information between Parties related to aerial surveillance;[21] and possible inclusion of algal blooms under the scope of the Agreement.[22]

7.5. The Copenhagen Agreement[23]

The Copenhagen Agreement appears to have been drafted to deal mainly with *oil* pollution from *ships*. Article 6 requires one State Party to inform another through the competent authority of any offence against anti oil pollution regulations committed within the territorial or adjacent waters by a flag ship of the other State Party. Article 8 requires State Parties to exchange information concerning reception facilities for oily residues from ships, national anti oil pollution regulations and measures, and the proper State authorities to whom information is to be transmitted. Article 7 requires States to assist one another in investigating offenses against anti oil pollution regulations presumed to have been committed within the territorial or adjacent waters of the States, including inspecting the ship's oil record book, official log book, engine room log, and the taking of samples.

The Copenhagen Agreement also is broad enough however to include oil pollution from offshore installations, since nothing is stated to the contrary. Article 1 requires State Parties to inform other State Parties of the sighting of any heavy concentration of oil on the sea which is likely to drift towards the other States' territories. Article 2 requires a State threatened by the presence of substantial oil pollution likely also to cause damage to other State Parties to investigate the oil, the type, approximate quantity, extent, location, direction and speed and communicate these results and information of measures taken to the other States. Under Article 5 a State is required to send out to the other States information concerning heavy concentrations of oil which have obliged it to take anti pollution action, and concerning the measures taken and the results achieved. Article 3 allows a State requiring assistance to save its coast or related interests to call on help from the other State Parties, first to the State also expected to be affected by the oil, and those States called upon must do their utmost to provide help. Each of the States undertakes to set up an emergency service to deal with heavy concentrations of oil at sea under Article 4. The establishment and positioning of oil treatment materials should take place following consultations with the other States, and the other States must be informed of the specifics of the national emergency service. Under Article 8 the State authorities must cooperate directly in the planning and coordination of other measures required of them under the Copenhagen Agreement. The activities in addition to clean up measures carried out in the past under the Copenhagen Agreement include deciding the measures required to combat pollution associated with oil and gas drilling; establishing more efficient sensor and identification systems to detect small discharges (which have increased) and to identify and apprehend responsible ships; and establishing cooperation with other regional agreements including the Bonn Agreement and Helsinki Convention as related to reporting procedures.[24]

7.6. NORBRITPLAN,[25] DENGER,[26] MANCHEPLAN,[27] Mexico-US Agreement[28]

NORBRITPLAN, the bilateral contingency plan between Norway and the UK for use in the North Sea will not be examined extensively here, since an adapted version can be viewed in PART II Section 13.4.3. In contrast to the Agreements above, NORBRITPLAN is a *comprehensive* contingency plan with very *specific* measures.[29] Briefly the main points covered entail the geographic scope, types of incidents, action required, principles of co-operation, emergency organizations, higher level decisions, danger zones, communications, pollution reports, country guidelines for offshore operators, control of airspace, definitions and abbreviations.

Under the Helsinki Convention Annex VI[30] there was by 1987 established a joint contingency plan, DENGER between Denmark and the Federal Republic of Germany covering the German Bight of the North Sea and the Western Baltic Sea. Similar plans were however under preparation between Denmark and the German Democratic Republic, the Germany Democratic Republic and Sweden, as well as between the German Democratic Republic, Poland and the USSR, and the USSR and Finland.[31] Basically DENGER resembles the Bonn Convention both in thoroughness and in being restricted to only *oil* pollution.[32] As under the Bonn Convention zones of responsibility are established, in this case, the 'exterior zone' and the 'quick response zone'.[33] Principles of reporting, coordination and responsibility are addressed in Article 4. The responsibility for initiating joint actions in the North Sea and in the 'quick response' zones of the Baltic, normally rests with the State in whose zone the incident occurs, while as in the rest of the Baltic normally responsibility rests with the State who feels a threat. General principles of combatting oil, including observation surveillance, measures and actions, are addressed under Article 5, financial matters under Article 6 and yearly exercises under Article 7.

MANCHEPLAN, the Anglo-French Joint Maritime Contingency Plan for the English Channel, part of the Celtic Sea and Dover Strait to the western limits of the Bonn Agreement, resembles most NORBRITPLAN in levels of specificity. Briefly MANCHEPLAN encompasses in 3 Parts and 20 Annexes geographic region, principles of coordination, governmental intervention, relations with the press;[34] maritime search and rescue and pollution clean up and environment protection, entailing definitions, national organization, coordination and cooperation, preparatory measures and operational procedures;[35] diagrams of communications, channels and decision taking levels, search and rescue and pollution;[36] search and rescue resources, France and the UK;[37] pollution clearance resources and facilities, France and the UK;[38] emergency specialized resources, France and the UK;[39] governmental intervention;[40] joint action in foreseeable disasters, search and rescue and pollution;[41] communications plan (COMPLAN);[42] relations with the news media;[43] forms of messages, search and rescue and pollution;[44] lists of pollutants (glossary);[45] control of airspace;[46] geographic grid for facility of operations (Manchegrid);[47] and measures to be taken in the event of problems in a traffic separation scheme following an accident.[48]

The Mexico - US Agreement resembles most the Barcelona Protocol I both in terms of specificity and in covering discharges of hazardous substances.[49]

The Agreement is defined to cover each Party's sea including the shorelines within the respective maritime boundaries and within 200 nautical miles measured from the baselines.[50] State Parties commit themselves to develop national monitoring systems and contingency plans,[51] and to cooperate to avoid and combat adverse effects by pollution on the marine environment and undertake to exchange information.[52] Joint responses are to be applied only upon agreement by the Parties,[53] rights and obligations under other treaties and positions with regard to the Law of the Sea are not affected,[54] and technical Annexes may be added to the Agreement as necessary.[55] The Annexes cover such topics as on scene coordinators, and functions and responsibilities of such;[56] joint response teams and their functions and responsibilities;[57] joint response centres;[58] operational phases including discovery, notification and alarm, incident evaluation, consultations and agreement on joint response, containment and measures against the spread of the pollutant, cleanup and recovery;[59] response and communication including system of rapid notification, initiation of join response, situation reports and termination;[60] and finally the names of national coordination and auxiliary agencies.[61]

7.7. The Norwegian - Soviet Agreement[62]

The Norwegian - Soviet Agreement is a bilateral agreement wherein Norway and the Soviet Union make general proclamations of cooperation in environmental protection, including the Arctic marine environment, but take on few concrete obligations to promote it.[63]

Both countries consider that the Arctic environment has a specially sensitive character, and consequently there exists a need to protect and preserve it.[64] Both must mutually cooperate towards solving important environmental problems and attaining an ecological balance.[65] The cooperation involves research into damaging environmental effects and the realization of measures to counteract them, and is to encompass marine environmental pollution, transboundary air pollution, environmental surveillance and the exchange of information regarding industrial wastes.[66] Both Parties are to cooperate to introduce new technology in pollution prevention, and other cooperative measures may as well be instituted.[67] Concrete measures are called for to carry out further agreements between the Parties, and are to include scientific experiments, bilateral and international conferences, meetings and symposiums, and the exchange of specialists, scientific information, documentation and research results and administrative experience.[68] The establishment of contacts and cooperation between the relevant organizations is to be encouraged by the

Parties, and questions of a commercial, financial or legal nature are to be negotiated by the relevant organizations, which can enter into contracts and agreements covering specified areas.[69]

One of the most important measures appearing is the establishment of a joint Norwegian - Soviet Commission to initiate the measures mentioned as well as prepare regulations and make recommendations.[70] The Commission met in August 1988, again in April 1989 and will continue regularly.[71]

7.8. Arctic Joint Contingency Plans[72]

In the Arctic, Canada, Denmark/Greenland, the US and the USSR have all entered joint contingency plans with the exception as noted of Norway and the USSR in the Barents Sea. A brief mention will be made of these plans as well as the Norwegian - Russian Draft Agreement for the sake of background.

7.8.1. The US - USSR Agreement for the Bering and Chukchi Seas[73]

The US - USSR Agreement covering the Bering and Chukchi Seas was entered into May 11, 1989 and an idea of this Agreement can be obtained from PART II Section 13.4.2., where it is used as a model for the *proposed* AGREEMENT between Norway and the Soviet Union in the Barents Sea. The Soviet - US regime is two level, and as can be seen the initial Agreement more resembles a bilateral Bonn Agreement than the specific contingency plan NORBRITPLAN. *Attached* to the Agreement however is a contingency plan though less comprehensive than NORBRITPLAN. The provisions covered include objectives to jointly develop preparedness measures and systems for discovering and reporting pollution (both oil and hazardous chemicals), for preventing the spread of such pollution, and for allocating resources for response to a pollution incident. Other provisions include State responsibility, principles and procedures, dispersants to be used, mechanisms for invoking the plan, joint response teams, the on scene commander, response actions, discovery and alarm, evaluation, containment, countermeasures, cleanup and disposal, documentation and cost recovery, joint response centres, a rapid alert system, warning messages, situation report requirements, revocation of joint response, post incident reports, competent authorities, definitions and amendments.

Probably the most interesting aspect of the US - USSR Agreement is its *disregard* for jurisdictional problems associated with a marine boundary dispute in its area of application, which as noted in PART II Section 8.3. has been in contrast to the Norwegian position relating maritime boundaries to contingency plans. This question is moot now since as will be seen in PART II Section 9.2.5. the US and the USSR recently, Spring 1990, settled their marine boundaries dispute.

7.8.2. The US - Canadian Agreement for the Beaufort Sea[74]

The joint US - Canadian Agreement is a fairly extensive contingency plan, more along the lines of NORBRITPLAN with hazardous chemicals added covering in addition to the Beaufort Sea, the Great Lakes, the Atlantic Coast, The Pacific Coast, and the Dixon Entrance. Each area is addressed specifically in a separate Annex to the main Agreement; the Beaufort Sea plan being presented in Annex Four. Points in the US - Canadian Agreement encompass purpose, objectives, scope, abbreviations and definitions; joint policy and responsibility including funding, invocation of the Plan, and responsibilities for Canadian and US Federal Agencies; planning and response elements including joint response centres and teams, on-scene commanders and deputies; response operations including discovery and alarm, evaluation and Plan invocation, containment and countermeasures, clean-up and disposal; reports and communications including rapid alert system, situation report requirements, revocation of incident and post incident reports; public information including on-scene news office and interim public information arrangements; administration; and amendments. Annex Four, consists of provisions including responsibility areas, designated joint response centres and on-scene commanders. As with the US - USSR Agreement jurisdictional problems associated with a marine boundary dispute in the Plan's direct area of application are more or less ignored, each Party being required to take anti pollution measures consistent with the Plan in marine areas believed to be under its jurisdiction. As will be seen in PART II Sections 9.2.2. and 9.3.4. the Beaufort Sea boundary conflict resembles the Barents Sea conflict with State assertions of sector lines and equidistant lines as marine boundaries in areas geologically promising for oil discoveries.

7.8.3. The Canadian - Danish Agreement and Contingency Plans for Nares Strait, Baffin Bay and Davis Strait[75]

The Canadian - Danish Agreement is a general environmental cooperative agreement which has as objectives to jointly protect the marine environment from pollution, including oil and hazardous chemicals, of the Nares Strait, Baffin Bay and Davis Strait areas. Generally it undertakes to establish cooperation in scientific research programs regarding marine environmental quality, in developing methods to measure marine pollution and in identifying and monitoring ship traffic. Attached to the Canadian - Danish Agreement are Annexes A and B dealing respectively with contingency plans for offshore installation pollution and shipping pollution incidents. These plans each resemble roughly the Bonn Agreement as to level of specificity. The status of these plans is somewhat unclear however, with one author noting that as of January 1987 there had not been any formal research initiatives implementing this Agreement, presumably also including contingency planning.[76] The plans do not encompass Arctic areas under jurisdictional dispute as those Agreements noted above, since the Canada - Denmark/Greenland marine boundary dispute, north of 82 degrees north latitude, lies to the north of the Nares Strait, Baffin Bay and Davis Strait areas.

7.8.4. The Norwegian - Russian Draft Agreement[77]

As noted the Norwegian National Television announced October 28, 1991 that a joint contingency agreement between Norway and the Soviet Union covering the Barents Sea was presently under development, and this despite the condition asserted by the Norwegian Ministry of the Environment that such a plan was dependent upon a definite maritime boundary being negotiated in the Barents Sea.[78] This draft is similar to the US - USSR Agreement but lacking the PLAN part. Additions have as well been made from NORBRITPLAN and the Finnish - Soviet Draft. Presently the draft is being examined by the Norwegian Ministry of the Environment and will soon be sent to the correspon-ding Russian ministry for evaluation.[79] The original time frame was for this Agreement to come into force sometime within the first half of 1992, though there have been delays. The breakthrough came in May 1991 with President Gorbatsjov's visit to Oslo to receive the Nobel Peace Prize, whereby a statement was made, probably involving the maritime boundary conflict, whereby the Norwegian government no longer felt threatened with this kind of agreement as before. Though no PLAN part exists in the draft, in practice most operational functions are in place including cleansing equipment and notifica-tion systems. The exact 'areas of concern' are not known, nor how the jurisdictional issue is dealt with.[80]

7.9. Conclusion

As can be seen contingency plans compared to most of the other areas of international environmental law are moderately developed. Generally the largest deficiency other than not having any joint plan at all is that hazardous chemicals are not covered. This is changing however with many of the newer plans encompassing these substances. Hopefully this development will continue with emphasis being placed on improving technology, allowing oil and chemical gathering in seas over 1.5 meters, on researching methods of preventing pollution, and on improved organization, allowing the equipment and manpower to come into action with relative ease.[81]

With general international environmental law having been presented in PART I, that applicable specifically to the Barents Sea will be addressed in PART II including the problems inherent with these existing regimes, and proposals for betterment.

Notes - Chapter 7

1 Churchill p. 263 notes that on a limited level MARPOL 73/78 under Article 8 and Protocol 1 requires ships involved in a pollution incident to report the details to the authorities of a State Party (though which one is unclear), which must notify the flag State and any States liable to be affected.

2 This is as seen in PART II Section 8.3. due to the absence of a definite maritime boundary. The Norwegian National Television however announced October 31, 1991 that a joint contingency agreement without a definite plan between Norway and the Soviet Union covering the Barents Sea is presently under development and is expected in force by summer 1992. The Norwegian Department of the Environment in correspondence August 13, 1992 reported that an Agreement is finalized and is presently awaiting signature. See Section 7.8.4.

3 PLAN is proposed for the Barents Sea.

4 International Convention on Oil Pollution Preparedness, Response and Cooperation (IMO Contingency). Can be obtained from the IMO.

5 IMO News No. 3, 1990, p. 1., and Aftenposten 3. desember 1990 Uke 49., Nr.554., 131.årg., s. 10.

6 IMO News, 1:1991 pp. 1-3 notes that the conference which drafted IMO Contingency also adopted a resolution calling for the Secretary-General of the IMO to initiate early implementation of Article 12 which includes an outline of IMO tasks and responsibilities under the agreement. This was especially relevant during the recent oil pollution disaster during the Gulf war. Under this the IMO established a Coordination Centre which coordinated offers of assistance from Australia, Belgium, Canada, Denmark, Finland, France, Germany, Greece, Japan, Netherlands, New Zealand, Norway, Poland, Spain, Sweden, the UK, the US and the USSR, with requests for help from Bahrain, Iran, Qatar and Saudi Arabia. The IMO also assisted in establishing an international fund for clean up of the oil slick to which Japan, Luxembourg, the Netherlands, Switzerland, the UK and the US contributed.

7 See PART I Section 3.2.3.

8 Churchill p. 263 notes that during the first 10 years of the Centre's existence it was notified of over 100 alerts and accident reports.

9 Protocol concerning Cooperation in Combatting Pollution of the Mediterranean Sea by Oil and Other Harmful Substances in Cases of Emergency (1976), 15 *International Legal Materials* 306, Ratifying, etc. States include Yugoslavia, Turkey, Tunisia, Syria, Spain, Morocco, Monaco, Malta, Libya, Lebanon, Italy, Israel, Greece, France, EEC, Egypt and Cyprus. IMO Status pp. 323-324.

10 Emphasis added.

11 Such incidents are defined under Article 1 to include grave and imminent danger to the marine environment the coast or related interests of one or more of the Parties due to the presence of massive quantities of oil or other harmful substances resulting from accidental discharges or accumulation of small discharges which are polluting or threatening to pollute the sea.

12 Churchill p. 263.

13 Agreement of 9 June 1969 for Cooperation in Dealing with Pollution of the North Sea by Oil, II *New Directions in the Law of the Sea* 632. Ratifying, etc. States include Denmark, France, Federal Republic of Germany, Netherlands, Norway, Sweden, UK

14 Articles 1 and 2. Churchill p. 263 notes this agreement is 'more modest' than the UNEP Conventions.

15 Articles 1 and 5. General objectives of the Bonn Agreement incorporated in the proposed Arctic Counter Pollution Protocol can be viewed in PART II Section 13.1.1.

16 Article 6 and Annex.

17 Article 7.

18 Article 8.

19 Articles 9, 10 and 11. The reimbursement provisions were added by the 1983 Amendment.

20 Article 17.

21 The observation was made that use of the surveillance results for *prosecution* purposes was *not* within the competence of the Bonn Agreement and that this matter would have to be taken up with the MOU.

22 Agreement for Cooperation in Dealing with Pollution of the North Sea by Oil, Twelfth Meeting of the Contracting Parties, Copenhagen, 13-16 September 1988, BA12/14/1-E, pp. 1-16, obtained from SFT, November 1988. This information is given as background to give an idea of parallel activities under other contingency plans. Further activities include a comparison of *reimbursement* for clean up actions and those arrangements in other bilateral or multilateral agreements; thorough revision of the Bonn Agreement's counter pollution manual; staff, alarm equipment and operations exercises; parallel activities being pursued in MANCHEPLAN, the EEC, the DENGER Plan, the IMO, the Helsinki Convention and the Copenhagen Agreement (see Sections 7.5. and 7.6.); dispersants and publication of a list of authorized chemicals; guidelines for sampling and analysing oil spills; discussion of pollution incidents including the English PIPER ALPHA accident and the lifting from the North Sea the ANNA BROERE carrying 1,000 tons of dangerous chemicals; and finally implications of LOSC for the Bonn Agreement.

23 Agreement between Denmark, Finland, Norway and Sweden Concerning Cooperation in Measures to Deal with Pollution of the Sea by Oil (1971) (Copenhagen Agreement), 822 *United Nations Treaty Series*, 324.

24 Rapport Från Det Adertonde Nordiska Plenarmøtet Inom Københamnsavtalet I Vasa Den 29.8.-1.9.1988 Angående Samarbete I Fråga Om Åtgärder Mot Oljeførorening Av Havet, Utkast 25.10.1988, sider 1-16, obtained from SFT, November 1988. This information is given as background. Other activities include technical developments including new and more effective collection booms; the acquisition of new oil pollution equipment; research programs, results and aerial surveillance to examine oil, industrial and agricultural discharges, algae, seals and the weekly position of the fishing fleet; developments within the Helsinki Convention including aerial surveillance, sensor systems, research and development to combat spillages of oil and other harmful chemicals under cold weather and ice conditions, and the establishment of guidelines to combat accidental spillages from offshore platforms; developments with the Bonn Agreement including work conducted with dispersion chemicals and the probability of coordinated aerial surveillance programs over the North Sea; developments within the EEC, DENGER, and within the draft bilateral Finnish - Soviet agreement covering marine pollution; common training programs among Scandinavian lands, the Soviet Union and Poland; possible zones of State responsibility, discharge statistics; measures to improve regional cooperation; and finally possible expansion of the Copenhagen Agreement to include combat of harmful chemicals.

25 Obtained from the Norwegian Ministry of Foreign Affairs.

26 Obtained from the Danish Ministry of the Environment.

27 Obtained from the British Department of Transport.

28 Mexico - United States: Agreement of Cooperation regarding Pollution of the Marine Environment by Discharges of Hydrocarbons and other Hazardous Substances (Mexico - US Agreement) (1980), 20 *International Legal Materials* 696.

29 Its scope is limited solely to oil pollution however.

30 See PART I Section 3.2.2.

31 Helsinki Secretariat correspondence August 1987),

32 Very close ties between both the Bonn Agreement and the Helsinki Convention are in fact noted in Article 1 where it is stated that since the contingency plan creates no additional obligations, there is no present need for a legal instrument to formalize such cooperation.

33 Article 3. Article 1 defines the DENGER region in the North Sea to be the inner area of the Danish and German responsibility zones under the Bonn Agreement with the line of demarcation coinciding with the Bonn Agreement boundary.

34 Part 1.

35 Parts 2 and 3.

36 Annexes 2 and 3.

37 Annexes 4 and 5.

38 Annexes 6 and 7.

39 Annexes 8 and 9.

40 Annex 10.

41 Annexes 11 and 12.

42 Annex 13.

43 Annex 14.

44 Annexes 15 and 16.

45 Annex 17.

46 Annex 18.

47 Annex 19.

48 Annex 20.

49 Article 1.

50 Article VII.

51 Article III.

52 Article IV.

53 Article VIII.

54 Article IX. It is unclear if LOSC is meant in this Article.

55 Article X.

56 Annex I.

57 Annex II.

58 Annex III.

59 Annex IV.

60 Annex V.

61 Annex VI.

62 Overenskomst mellom Kongeriket Norges Regjering og Unionen av Sovjetiske Sosialistiske Republikkers Regjering om Samarbeid På Miljøvernområdet, januar 1988. Obtained from the Norwegian Ministry of the Environment.

63 See PART I Section 4.2.5. however for more specific provisions regarding air pollution.

64 Preamble.

65 Articles I and II.

66 Article II. It is under this Article a joint contingency plan covering the Barents Sea can be formed.

67 Articles II and III.

68 Article III. Under Article VII the technical and scientific information exchanged by the Parties must be held confidential if so requested. It may be used freely in experiments unless limitations are placed on its use.

69 Articles VI and V.

70 Article VI.

71 Conversation with the Chief Consultant Per Antonsen, Norwegian Ministry of the Environment, September 26, 1988. In answer to a question regarding whether any concrete measures were established during the joint Commission's meeting concerning marine pollution in the Barents Sea, Mr. Antonsen responded that contacts are still in the process of being established. An expert group has begun meeting over environmental oil field technology, conversation with SFT Chief Engineer Jens Koefoed, August 1, 1989. Generally according to Joint Protocol 1 and Protokoll fra det Andre Møte i den Blandete Norsk-Sovjetiske Kommisjon for Samarbeid på Miljøvernområdet 10.-14. april 1989 (Joint Protocol 2), the main weight of Commission work seems placed on measures to study and combat transboundary *air pollution*, including an examination of pollution discharges from industries, cities and electric power station especially in the Norwegian - Soviet border area. An examination of pollution in watercourses is also included. Under Joint Protocol 2, pp. 3-5 it is noted that an expert

group has been established to work on establishing practical measures of surveillance, and prevention and limitation of oil pollution from oil rig and shipping activities in the Barents Sea. Further an expert group is to be established to work on ecological aspects, including sea birds and biological surveillance of other components of marine biota.

72 See PART II Section 9.2. Figure 1 for the specific Arctic geographic areas discussed.

73 Agreement between the Government of the United States of America and the Government of the Union of Soviet Socialist Republics concerning Cooperation in Combatting Pollution in the Bering and Chukchi Seas in Emergency Situations, signed May 11, 1989; and United States of America and Union of Soviet Socialist Republics Joint Contingency Plan against Pollution in the Bering and Chukchi Seas (US - USSR Agreement), obtained from the US State Department, Bureau of Oceans and International Environmental and Scientific Affairs, August 1989 and January 3, 1990 respectively.

74 Canada - United States Marine Pollution Contingency Plan for Spills of Oil and Other Noxious Substances (US - Canadian Agreement), 29 *United States Treaties* 2569, Contingency Plan and Annex Four (Beaufort Sea) obtained from the Canadian Ministry of External Affairs, December 1, 1989.

75 Agreement between the Government of Canada and the Government of the Kingdom of Denmark for Cooperation Relating to the Marine Environment, 1986 (Canadian - Danish Agreement), 23 *International Legal Materials* 269.

76 Lamson Arctic pp. 7-8.

77 The following information is obtained from SFT October 28, 1991.

78 See PART II Section 8.3.

79 The corresponding Soviet Ministry was recently dissolved.

80 As noted the Norwegian Department of the Environment reports in a correspondence 13 August, 1992 that the Agreement is in the final stages and is currently awaiting signature. The Agreement is reportedly fashioned upon the US - USSR Agreement and ignores the jurisdictional issue. The PLAN part is fashioned after the NORBRITPLAN.

81 See Aftenposten 28 mars 1989, Uke 13. Nr.139. 130 årg. s. 3 for an account of poor operational organization of contingency routines and equipment as related to the EXXON VALDEZ accident in Alaska. Further noted is that almost ideal weather conditions existed for at least 24 hours following the accident which but for the poor organization would have allowed pollution prevention to take place.

Part II - The Barents Sea - A Case Study

8. The Barents Sea - Oil, Shipping, Politics, Fisheries and Pollution in the Arctic

8.1. Introduction

As noted a case study was to be made of the Barents Sea including a review of the international marine environmental agreements presented in Part I governing this area, any deficiencies in this legal coverage and proposals forwarded to cover these deficiencies. It was shortly discovered however that any proposals should be *broadened* to cover the entire Arctic region. This was chiefly because environmentally there were important benefits gained through *regional* measures, which dealt more effectively with natural processes in the Arctic by which pollution from one incident could affect *all* the Arctic States.[1] In addition the legal deficiencies found concerning the Barents Sea were generally also the same as for the Arctic, and it was a short step to enlarge the Barents Sea proposals to encompass the entire Arctic. As will be presented in Section 8.3. it was also discovered that any bilateral anti pollution agreement covering the Barents Sea was always *overshadowed* by Norwegian scepticism to bilateral agreements with the Soviets due to the Soviet military presence. A solution to Norway's scepticism could be a regional Arctic agreement including Norway, Norway's Arctic NATO allies and the Soviet Union.[2]

Consequently what follows is first in this Chapter a brief survey of oil and shipping activities, political considerations, fisheries and ecological considerations in the Barents Sea. Specific statistics of pollution will be presented when available. Secondly a brief summary is presented in Chapter 9 of the marine environmental conventions from Part I governing in the Barents Sea and the Arctic. The conventions involved are generally the same for both areas. Then follows a discussion of the inadequacies of this conventional coverage. Finally in Chapter 10 follows a discussion of the issues favouring an Arctic regional agreement, both legal and non legal, and why it is felt a comprehensive Arctic marine environmental agreement is superior to a series of bilateral agreements, including bilaterals governing the Barents Sea. From this it is hoped a framework is established for supporting the proposed Arctic Convention and subordinate Protocols, presented and discussed in Part II Chapters 11 - 17.

8.2. Oil and Shipping Activities in the Barents Sea

The Barents Sea, outlined in Figure 8.1 and defined for the purpose of this project as the area described by the coordinates indicated in Figure 8.2, has for many years been the site of extensive fishing activities both by Norway and the Soviet Union. Many settlements in northern Norway are in large part based upon the fishing conducted in this sea.[3] Recently however the focus of economic interest has shifted to include new activities. The continental shelf of the Barents Sea has proven to be geologically promising for oil.[4] Additionally both the Soviet Union and Norway have become interested in opening a northern Arctic sea route, the Northeast Passage, to shipping connecting Europe to Asia and North America.[5] To provide background into the extent of each countries' oil exploration and shipping activities in the Barents Sea characterized by ice, hard weather and winter polar darkness and some of the reasons behind these new activities, each country will be examined separately.

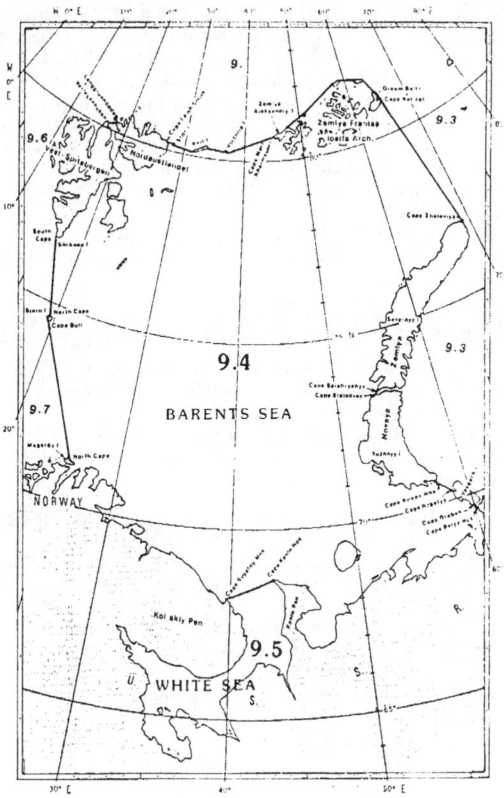

Figure 8.1.[6] The Barents Sea and White Sea.

BARENTS SEA

The Barents Sea, situated off the northeastern coast of Norway and the northwestern coast of USSR, is bounded by Vest-Spitsbergen and Nordaustlandet on the Northwest, by Zemlya Frantsa Iosifa Archipelago on the Northeast, and by Novaya Zemlya on the East. Its limits are the following :

On the North :

A line joining Verlegenhuken (80°04′N - 16°16′E), the northern extremity of Vest-Spitsbergen, eastward to Langgrunnodden (80°08′N - 17°46′E), the western extremity of Nordaustlandet;
thence from Langgrunnodden eastward, through Nordaustlandet, to Cape Leigh Smith (80°10′N - 26°50′E), on the northwestern coast thereof;
thence a line joining Cape Leigh Smith eastward, through Stor Island (80°08′N - 27°50′E), Kvit Island (80°10′N - 32°40′E), and Victoria Island (80°10′N - 36°45′E), to Cape Meri Kharmswort (80°36′N - 14°55′E), the western extremity of Zemlya Aleksandry Island, the western island of the Zemlya Frantsa Iosifa Archipelago;
and thence from Cape Meri Kharmswort eastward, along the northern coasts of this archipelago, to Cape Kol'zat (81°00′N - 65°20′E), on the eastern coast of Green Bell Island.

On the East :

A line joining Cape Kol'zat southward to Cape Zhelaniya (75°57′N - 68°35′E), on the northern coast of Severnyy Island (Novaya Zemlya);
thence from Cape Zhelaniya southwestward, along the western coast of this island, to Cape Serebryanayy (73°21′N - 54°05′E), the southwestern extremity thereof;
thence a line joining Cape Serebreyanyh southwestward to Cape Stolbovoy (73°17′N -53°53′E), on the northwestern coast of Yuzhnyy Island (Novaya Zemlya);
thence from Cape Stolbovoy southward, along the western coast of this island, to Cape Kusov Nos (70°28′N - 57°07′E), the southern extremity thereof;
thence a line joining Cape Kusov Nos southeastward to Cape Rogatyy (70°15′N - 58°26′E), the western extremity of Vaygach Island);
thence from Cape Rogatyy southeastward to Cape Greben' (69°40′N - 59°59′E), the southern extremity thereof;
and thence a line joining Cape Greben' southeastward to Cape Belyy Nos (69°36′N - 50°12′E), on the northwestern coast of USSR (*the common limit with the Kara Sea, sec. 9.3*).

On the South :

From Cape Belyy Nos westward, along the northwestern coast of USSR, to Cape Kanin Nos (68°39′N - 43°18′E), the northwestern extremity of Kanin Peninsula;
thence a line joining Cape Kanin Nos southwestward to Cape Svyatoy Nos (68°08′N -39°46′E), on the northeastern coast of Kol'skiy Peninsula (*the common limit with the Kara Sea, sec. 9.5*);
and thence from Cape Svyatoy Nos northwestward, along the coasts of USSR and Norway, to North Cape (71°10′15″N - 25°47′E), on the northern coast of Mageröy Island, off the northern coast of Norway.

On the West :

A line joining North Cape, in Norway, northwestward to Cape Bull (74°20′40″N - 19°05′E), the southern extremity of Björn Island;
thence from Cape Bull northward, along the eastern coast of this island, to North Cape (74°31′15″N - 19°06′30″E), the northern extremity thereof;
thence a line joining North Cape, the northern extremity of Björn Island, northwestward to South Cape (76°28′N - 16°37′), the southern extremity of Sörkapp Island, off the southern coast of Vest-Spitsbergen (*the common limit with the Norwegian Sea, see 9.7*);
and thence from South Cape northward, along the eastern coast of Vest-Spitsbergen, to Verlegenhuken (80°04′N - 16°16′E), the northern extremity of this island.

Figure 8.2. Sub-divisions of the Arctic Ocean.[7]

8.2.1. The Union of Soviet Socialist Republics[8]

The Soviet Union has since the early 1980's become more and more dependent upon the export of oil. Since this time oil has given 60% or more of the Soviet income (not counting gold or weapons). During the past 15 years the Soviet Union's import from the West increased especially in grain, machinery and technology, and the trade is expected to continue since portions of the Soviet society have become dependent upon the West.[9] Hence the Soviet government will continue to export oil in order to try to gain the 'hard currency' needed to pay for Western imports that are necessary for the economic plans. A fall in the world's oil market price however does not result in a cutback in oil *exploration* in the Soviet Union as is the case elsewhere. A lower market price means increased oil export since the Soviets have few other export alternatives.[10] At the same time reports indicate that Soviet oil *production* continues falling.[11] For the first half of 1990 there occurred a 4% reduction of production from the same period last year to 291 million tons, equivalent to 11.64 million barrels/day on the average (which is however still by far the world's highest production of oil).[12]

From this it is possible to see some of the reasons for the Soviet interest in oil and gas in the Barents Sea. According to geological estimates a potential of 2200 million tons of oil lie in the European part of the USSR, which includes the Barents Sea. Developments in adjacent on shore areas in the USSR also indicate the possibility of oil and gas finds in the Barents Sea.[13]

Until recently little was known of the actual Soviet exploration and of the results obtained. Now however there is taking place Soviet cooperation with Western countries in geological data gathering and processing, and conceptional development study.[14] Additionally it is recently reported that the American oil company Texaco signed a joint oil exploration and production agreement with 'Arkhangelskgeologija' concerning the Archangels area and the Soviet Barents Sea continental shelf in which discovery, extraction and refining of oil is to be covered.[15] The Norwegian Geophysical Company of Norway (GECO) owned by Schlumberger has an agreement with the Soviet company Laboratory of Regional Geodynamics (LARGE) whereby LARGE has shot seismic and GECO has processed the data and is assisting LARGE in marketing the processed data. Additionally the Norwegian Trondheim-based Continental Shelf Research Institute IKU and the Murmansk-based NIImoregeophysica and P/O Soyusmorgeo, (Ministry of Gas and Oil) have worked together since 1988 to market geophysical date acquired by both organizations in the Barents Sea.

There are several areas known in the Soviet Barents Sea where wells have been drilled. These areas are shown in Figures 8.3 and 8.4 below and include estimates of the number of wells appearing in each:

Murmansk High	2-3
Goose Bank	1-2
Severo-Kildinskaya	1
Pechora River	1-2[16]

See Figure 8.5, 'Soviet Mobile Offshore Drilling Rigs' for FNI's estimate of future Soviet drilling activity in the Barents Sea.

Recently and somewhat unexpectedly on August 21, 1987 a Soviet oil tanker was reported carrying oil from the Soviet island Kolgujev in the Barents Sea to the port of Kandalaksja on the Kola peninsula, marking the first commercial oil activity to take place in the Barents Sea.[17] Additionally reported was that according to the Soviet party newspaper, Pravda, there were 35 wells drilled in the Kolgujev area and of these 15 presently were oil producing.[18] An underwater pipeline, 4 km. long was used to carry the oil from Kolgujev to a buoy where the loading was conducted. An even more recent report noted the discovery of the world's largest underwater gas field in the Soviet Barents Sea along with 2 oil fields containing together 100 million tons.[19] See Figure 8.4. *Shtockmanovskoye* gas field is reported to contain 3-4 trillion cubic meters of gas, in which there may be significant volumes of oil present in thin underlying zones. Western companies interested in possible joint development with the Soviet Ministry of Oil and Gas include Norsk Hydro of Norway, DuPont Services of the Netherlands (which owns Conoco), and the Finnish companies Wartsila, Imatran Volma and Neste. A pipeline to Western Europe is planned. The *Murmanskoye* (Murmansk High) field contains 300 billion cubic meters gas and is under consideration for development as is a possible large oil discovery on *Admiralteyskaya* off the northern part of Novaya Zemlya. Another oil discovery is located west of the southern tip of Novaya Zemlya close to the Kolgujev field noted above. Also east of Kolgujev island a number of smaller but commercially promising fields may be developed.[20]

In spite of these finds it is acknowledged that in the Barents Sea the Soviets experienced severe technological difficulties involved with exploratory activities, and results obtained are poor when compared to Western standards. The poor results are attributed to inexperience in operating the sophisticated equipment mostly purchased from Finland, and to organizational problems with supplies and services, resulting in long delays.

Based upon the Soviet need for oil and the large amounts of mobile offshore drilling rigs destined for the Barents Sea noted on Figure 8.5, coupled with technological inexperience and poor organization, Soviet authorities were thought to be faced with three difficult choices concerning Barents Sea development. They could continue with previous methods and take a chance on enormous waste, especially in cases of large discoveries. They could postpone exploration until after the year 2000 in order to build up expertise both technologically and organizationally.[21] They could decide to allow *foreign* participation in oil exploration and production.[22]

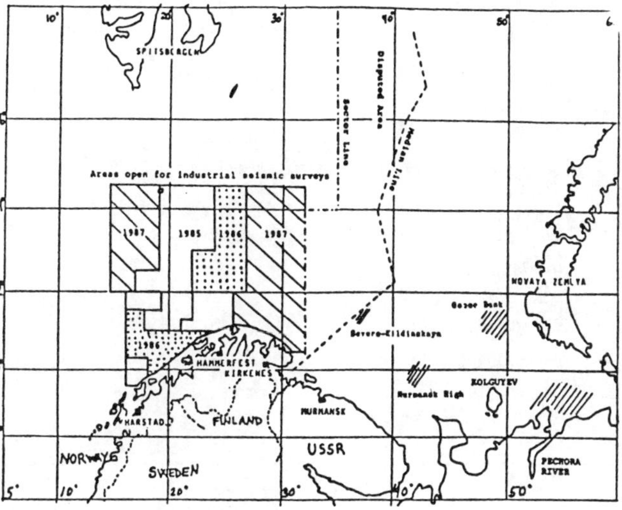

Figure 8.3.[23] Exploration Areas in the Barents Sea.

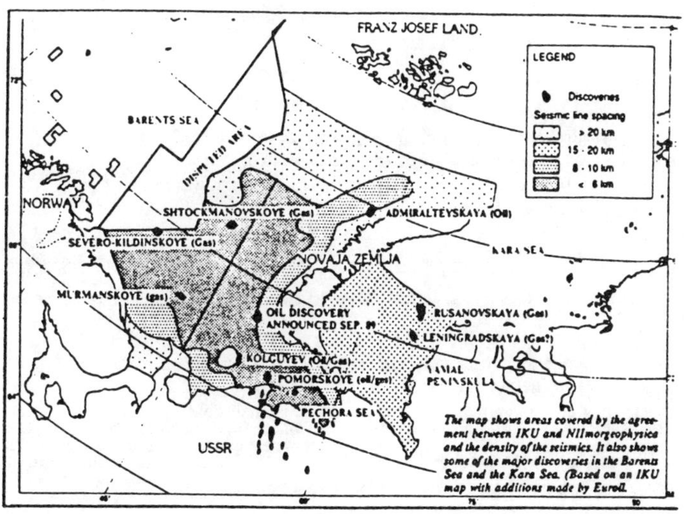

Figure 8.4.[24] Major Discoveries in the Barents Sea.

Area	Until summer 1986	Configuration summer 1986	1986-1990 Expected future supplies	1990
Barents Sea[1]	1982: *Valentin Shashin* ds 1982: *Viktor Muravlenko* ds 1985: *Kolskaya* ju 1985: *Shelf-4* ss	2 drillships 1 jack-up 1 semi-sub (4)	86-7: Shelf ss 87-8: 2 jus 87-90: 2-5 additional rigs	9-12
Sakhalin[2]	1983: *Mikhail Mirchink* ds [1,3] 1980: *Okha* ju[4] 1984: *Ekhabi* ju[3] 1985: *Sakhalinskaya* ju[1]	1 drillship 3 jack-ups (4)	86: Shelf ss[5] 87-90: 1-2 additional rigs	6-7
Caspian Sea[6]	1981: *Kaspmorneft'* ss[7] 1981: *Shelf-1* ss[7] 1983: *Shelf-2* ss 1986: *Shelf-3* ss[8] 1974: *Apsheron, Azerbaydzhan and Khazar* ju's[7] 1975: *Baku* ju[7] 1978: *60 let Oktyabrya* ju[7] 1981: *60 let Azerbaydzhana* ju[7] 1985: *40 let Pobedy* ju[9]	4 semi-subs 7 jack-ups (11)	86-90: 5-10 additional rigs[10]	16-21
Black and Azov Seas	1981: *Sivash* ju[7,11]	1 jack-up[12] (1)	86-90: 0-3 additional rigs (1-2 Romanian)	6-9
Baltic Sea[13,7]		1 jack-up	86-90: 0-2 additional rigs	1-3
TOTAL		21 units (13 jack-ups, 5 semi-subs, 3 drillings)	Soviet domestic prod. Romanian jus[14] Brazilian jus[15]	35-52

cont...

[1] See description in this chapter

[2] A.N. Trutnev: 'Sakhalinskoye Uskoreniye', *Gazovaya promyshlennost,* 5, 1986, p.7.

[3] The rigs have also been operating off Vietnam. *Petroleum News,* December 1985.

[4] *Oil&Gas Journal (OGJ),* 12 October 1981, p.170. The Okhan platform was built in Japan.

[5] *Sotsialisticheskaya industriya,* 14 September, 1986.

[6] OGJ, 19 May 1986.

[7] OGJ, 12 October 1981, p.16. The Dutch Khazar is the prototype for the following Kaspii series of Soviet built jack-ups. The *60 let Azerbaydzhana* is an improved version which can drill 6.500 metres in 70 metres of water.

[8] *Economicheskaya Gazeta,* 14, 1986, p.1; according to several sources the Shelf-3 was completed long ago, but this recent article proves the opposite.

[9] *Soviet Geography,* April 1985, p.293.

[10] *OGJ,* 19 My 1986; Azerbadzhany local officials ask to receive about 20-5 more rigs.

[11] *Sotsialisticheskaya industriya,* 14 May 1983.

[12] There is uncertainty concerning the characteristics of other installations operating in the Black Sea. The *Golitsyno* series consisting of 5 rigs has been referred to both as fixed platforms and as semi-subs; see for example *Gazovaya Promyshlennost,* 12, 1985. In any case, the waters in these areas are very shallow and the rigs probably not very sophisticated; thus we exclude them from this list of 'real' mobile drilling rigs.

[13] *International Petroleum Encyclopedia,* vol 18, 1985, p.232.

[14] *Petroleum Economist* (PE), April 1984, p.146.

[15] *Norges handels og sjøfartstidende,* 7 March 1985. However, recent reports indicate that this project is cancelled.

Foreign rigs:		
	Dutch	(1) *Khazar*
	Japanese	(1) *Okha*
	Singapore	(1) *Ekhabi*
	Finnish	(6) *Kaspmorneft, V.Shaskin, V.Muravlenko, M.Mirchink, Kolskaya, Sakhalinskaya*

ju=Jack-up. A Floating drilling platform with legs. The platform floats while being moved. But the legs can be lowered and the deck jacked up to allow the platform to stand on the ocean floor while drilling.
ss=Semi-submersible. A drilling platform consisting of a body under water for stabilization and at deck with drilling rig. The deck is attached to the body by pillars minimizing the diameter at the water-line, thus reducing the movements of the vessel.
ds=Drillship. A ship with a drilling rig, particularly for drilling in deep waters. To stabilize the ship it has to be equipped with a dynamic positioning system.

Figure 8.5.[25] Soviet Mobile Offshore Drilling Rigs.

As recently as 1987 it was believed that the first choice was being followed, however the new joint Soviet - Western developments noted above indicate rather the third choice, at least in the initial stages of development. It must be noted as a total picture in spite of the Soviet 'glasnost' and cooperative interest, a part of the previous information noted covering the Soviet Barents Sea, number of wells, geologic results, production figures, developmental plans, governmental strategies, are Western estimates, and the picture is clearing but is not yet complete. Especially when political developments in the Soviet Union are so unstable it seems difficult to state definitely the direction of Soviet oil activities in the Barents Sea, especially as regards Western participation.[26]

Regarding *shipping*, most of that noted above regarding the Soviet need for 'hard currency' can be applied to shipping activities and the proposed opening of the Northeast Passage.[27] See Figure 8.6. President Gorbatsjov's Murmansk speech of 1 October 1987 declared an interest in conditionally opening the Northeast Passage to allow shipping a connection from Europe to Asia and North America via a passage north of the Soviet Union.[28] It is difficult to say much specific regarding this possibility as research is currently taking place, through a joint Norwegian - Soviet pilot project was due complete in Spring 1991.[29] Basically the attractiveness of such a passage is that the distance between the Atlantic and the Pacific Oceans can be cut to close to half. The problem however as can be imagined is *ice*. The Soviet Union has a total of 93 ice-breakers which can be used in different capacities to hold the passage open.

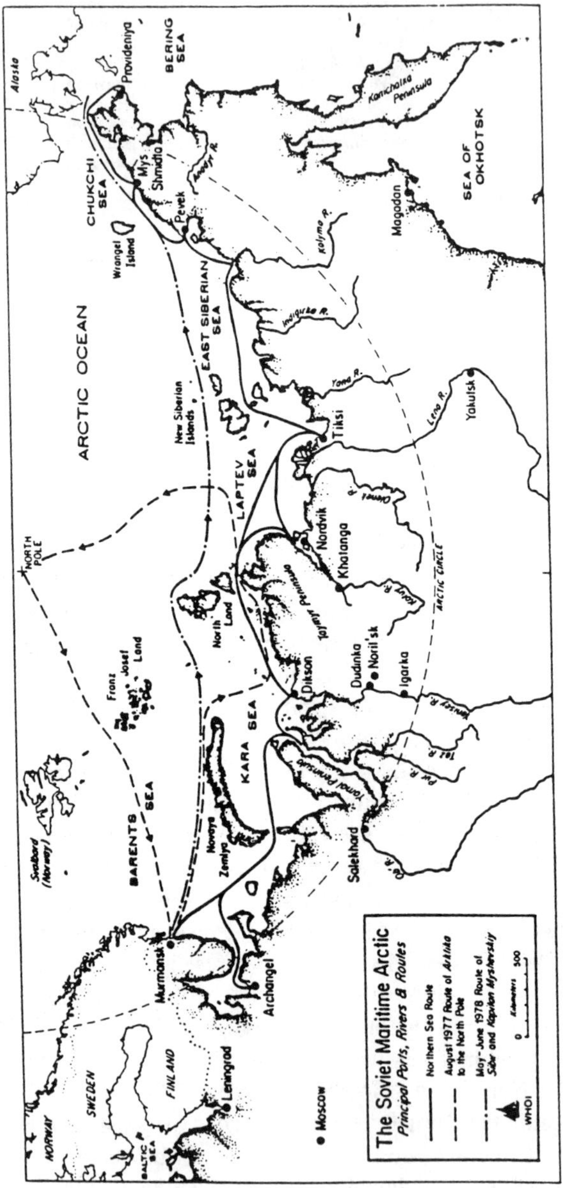

Figure 8.6.[30] The Soviet Maritime Arctic.

See Figure 8.7. Currently the entire route is held open approximately 4 months per year, and the difficulty to hold the passage open increases towards the east. It is claimed that the western part from Murmansk to the Yenisey River up to Dudinka is held open 11 to 12 months.[31] The Laptev and East Siberian Seas are held open for 5 to 6 months.[32] The SA-15 and the nuclear LASH indicated on Figure 8.7 escorted by nuclear ice breakers can increase these periods of open passage. Most of the shipping activities currently takes place in West Siberia from the city of Dudinka on the Yenisey River, exporting metals from smelters west to Murmansk. Yearly transport is estimated at 250 ships, 600 trips and 6 to 8 million tons, though these statistics are somewhat unclear. One Soviet author notes the annual *increase* in ship traffic through the Northeast Passage is approximately 10% with 5 million tons being freighted in 1985.[33] Other statistics note that many of the ships used make 3 round trips, however as many as 53% are damaged by ice.[34] Ice conditions were especially difficult in the East Siberian and Chukchi Seas in 1988 with one icebreaker, the ARKTIKA, taking 29 days from Murmansk to Pevek, even with the escort of two other major icebreakers, the YERMAK and the ADMIRAL MAKAROV.

In terms of commercial profitability, it is estimated that it costs the Soviets approximately USD 20,000 per day to hold the passage open, and it is probable that the Soviets will use their own fleet to ship wares, or if not, to demand high transit fees. Regardless, for the Barents Sea this means that ships from all corners of the world will travel through the Norwegian sector to take wares both to and from Murmansk; and eastward from Murmansk in the Soviet sector there can be expected heavier ship traffic if this project comes into being as anticipated.

16 -	Polar
36 -	Subarctic
	20 - Merchant Marine
	5 - Soviet Navy
	5 - KGB Maritime Border Troops
	2 - Polar Research
7 -	Salvage
13 -	River
	6 - Shallow-draft
	7 - Extreme Shallow-draft
3 -	Large Harbour
1 -	Conventional LASH
1 -	Nuclear LASH
16 -	SA-15 Icebreaking Cargo
93	**Total**

Figure 8.7. Soviet Icebreaker Fleet.[35]

8.2.2. Norway[36]

In 1986 the Organization for Oil Producing and Exporting Countries (OPEC) changed their oil marketing strategy from that of defending world oil prices to increasing their share of the market, and this resulted in a fall of the world's price for oil per barrel. In response the Norwegian government stipulated that if OPEC would agree upon certain conditions that would stabilize the price at a reasonable level, Norway would limit its production to contribute to this price stability. Norwegian production limitation resulted in large repercussions for the Norwegian economy. The income to the State, measured in percentages obtained from the petroleum sector in the form of taxes and license fees, dropped from *19%* in 1985 to *7%* in 1986 (51 billion Norwegian crowns in 1985 to 18 billion Norwegian crowns in 1986). Norway cannot expect to derive *1/2* of the income per year as obtained during the period 1981-1985, before 1989-1995. However, even with these cutbacks and the price for oil lying around US USD 18.00/barrel, the profitability of the petroleum sector is greater than that of most other Norwegian industries.[37]

From this brief background it is understandable that there is a Norwegian interest in the Barents Sea in spite of the considerable costs involved in discovering and producing oil in the Arctic. Although Norway's economic problem is different in form from the Soviet problem, it has parallels. The economy is to a large degree dependent upon the export of oil.

For Norway the Barents Sea is the last unexplored area in which a large oil find can be made, and the government steadily grants certain sectors opened for oil exploration. Because of Norway's economic situation in relation to the world's oil market, combined with the speculative nature of oil exploration, both the Norwegian authorities and the oil companies are dependent upon a large discovery being made for exploration to continue in the Barents Sea. A new system was therefore initiated whereby certain so-called 'strategic blocks' were opened in previously unopened yet geographically promising areas, and the advantage gained by the oil sector was one of time.[38] See Figures 8.8 and 8.9. The usual procedure for opening an area for exploration, involving environmental impact studies, was not followed allowing a quick determination to be made of a promising structure's suitability. At the same time normal procedures for opening general areas in the Barents Sea, complete with the environmental studies and hearings, were pursued in parallel. A vote by the Parliament on the opening of the areas 'Barentshavet Syd,' 'Troms II,' 'Troms III' and the southern section of 'Finnmark Vest' by New Years 1988-1989 was hoped for.[39] The government was prepared to allot even more strategic blocks in 1988 in other promising but unopened areas.

Briefly, in the intervening years current to 1990 the exploration continued in the Barents Sea on a scale less than hoped for by the Norwegian oil interests and with negligible results, in spite of discoveries noted above in the Soviet Barents Sea.[40] Figure 8.10 shows the total number of wells drilled 1965 to 1986 in Norwegian waters, including those in the Barents Sea (Nord Norge), and it can be seen that 30 wells were drilled in the north.[41] Drilling plans for the Barents Sea included drilling 5-7 exploratory wells each of the seasons 1987-88 and 1988-89 on a year around basis, and if a discovery were made then increasing the number of projected exploratory wells. This was not realized.[42]

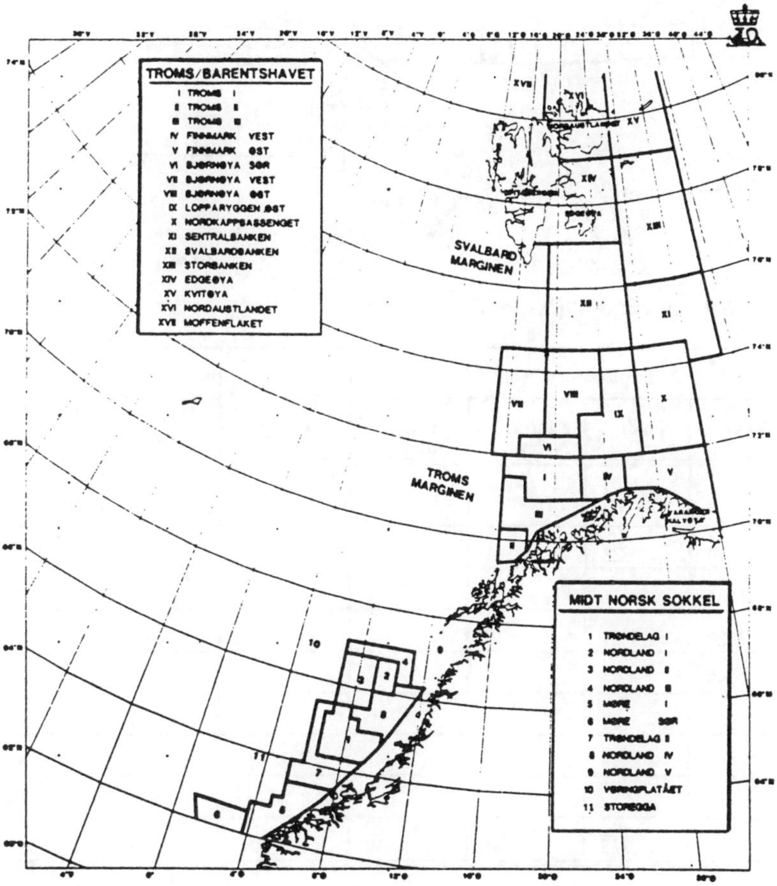

Figure 8.8. Licensing Areas in Norwegian Offshore Oil Activities.

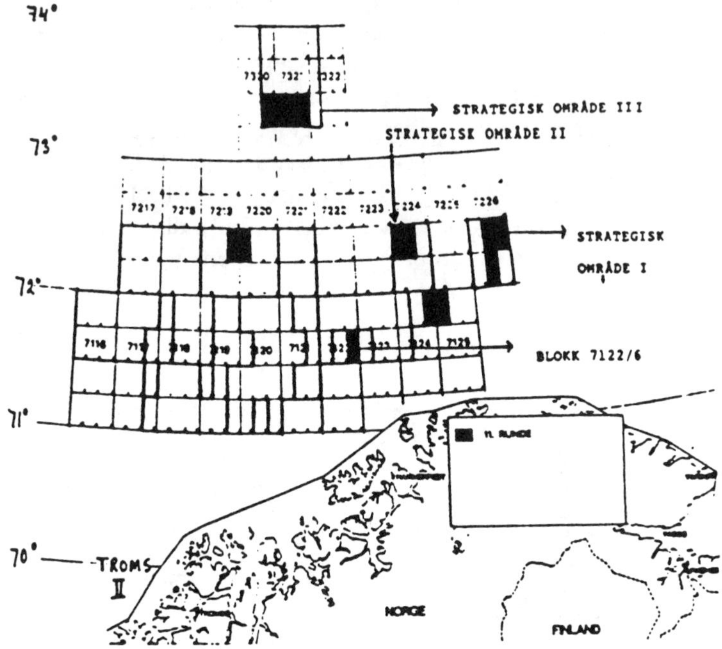

Figure 8.9. Allocated Blocks for Drilling.[43]

Exploratory Well						
Year	**North Sea**	**Sentral Norway**	**North Norway**	**Total**	**Appraisal Well Total**	**Total Drilling Activity**
1965-79	174	0	0	Average 11,6/year		
1980	23	1	2	26	10	36
1981	21	2	3	26	13	39
1982	27	5	4	36	13	49
1983	21	7	5	33	7	40
1984	22	6	7	35	12	47
1985	13	10	7	30	20	50
1986	14	10	2	26	10	36
	315	**41**	**30**	**386**	**85**	**297**

Figure 8.10.[44] Exploratory and Appraisal Wells 1965 - 1986 including Those in the Barents Sea.

Figure 8.11 shows the allotted blocks on the Norwegian continental shelf, including 'Tromsflaket' in the Barents Sea. The recent 13th round of bids offers new blocks for drilling directed mainly towards exploration in the Barents Sea with 25 blocks available, including new areas opened in 'Finnmark Øst'.[45] Though most of the oil companies are still optimistic about oil finds in the Barents Sea especially with the prospect of so many new areas, however, 2 companies, Arco and Unocal have 'left the scene altogether', and it is expected more will follow unless definite finds are made.[46] The Norwegian Minister of Oil and Energy is also optimistic noting that, 'the likelihood of commercial discoveries (should be increased) by continuing exploration within known areas as well as testing new areas in the region.'[47]

8.3. Political Considerations[48]

Political considerations in the Barents Sea, though of utmost importance to the management of pollution problems, will not be dealt with extensively here. This element, further complicated by the enormous social, economic and political changes presently taking place in the Soviet Union, will be left to the experts noted who are better equipped to analyze military and political strategies. The goal here will be to give a brief survey of various political factors involving the Barents Sea and their relation to environmental regimes.

The Soviet Union's most important naval fleet, the Northern Fleet, operates from 12 bases on the Kola Peninsula, which necessarily makes political considerations in the Barents Sea very important.[49] See Figure 8.12. The placement of the Soviet bases is necessitated by the fact that geographically this is the only area where Soviet battleships can reach open sea without sailing through straits controlled by opponents.[50] Additionally from the northern bases the Soviet Union could deter adversaries such as the US, Western Europe or China. As late as 1987 the Northern Fleet consisted of 220 surface ships and 203 submarines, with 17% of the Soviet surface ships and 57% of Soviet submarines based here. It is believed that this fleet has as objectives, to contribute to maintaining the balance of power between the superpowers; in war destroy or neutralize Western aircraft carriers before they can attack Soviet targets; cut off supply lines between the US and Western Europe; offer backup for land operations and for the army; defend the Kola bases; protect transit routes within the Barents Sea so that the SSBN submarines have safe access to Arctic stations including under the ice; and support Soviet *foreign policy*.[51]

To achieve these goals the surface fleet must be able to move in and out of the Barents Sea without hinderance, but during winter *ice* limits transit to approximately 330 kilometres from the ice edge to Kola. Any narrowing down of this operational area and channelling of naval transit to more easily identifiable areas by the placement of oil rigs for example is a problem for the Soviets, though the ice free area on the Norwegian side is wider and the use of sub sea wells could lessen this problem.[52]

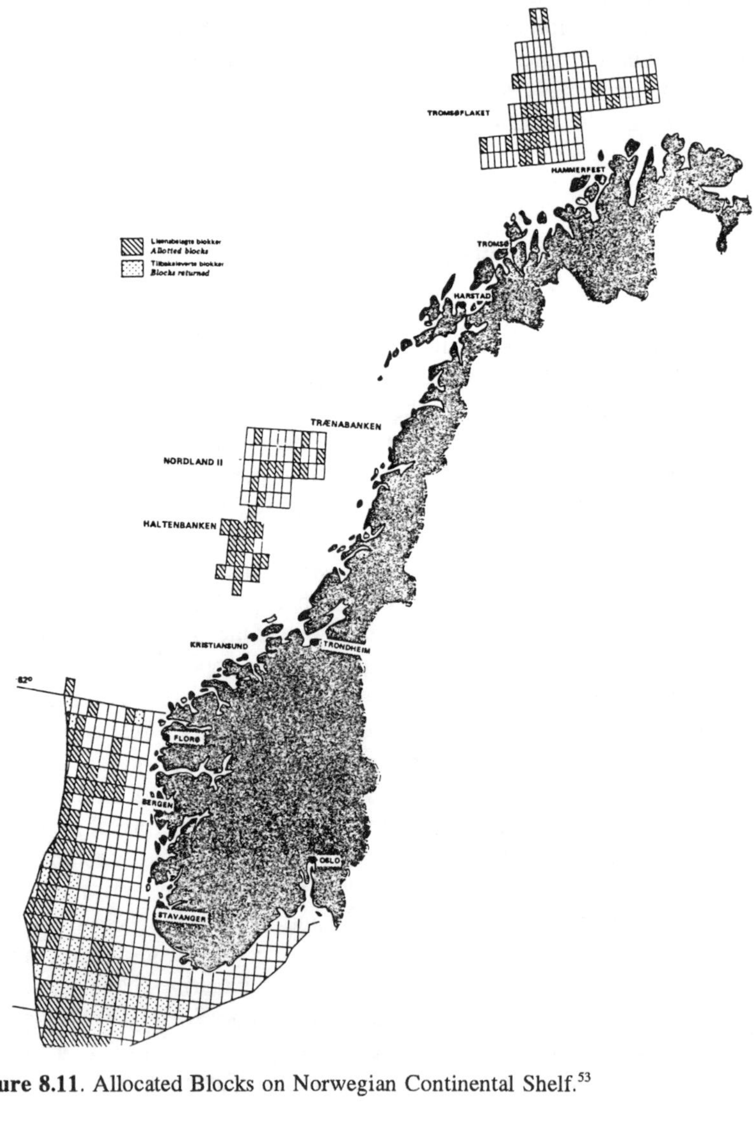

Figure 8.11. Allocated Blocks on Norwegian Continental Shelf.[53]

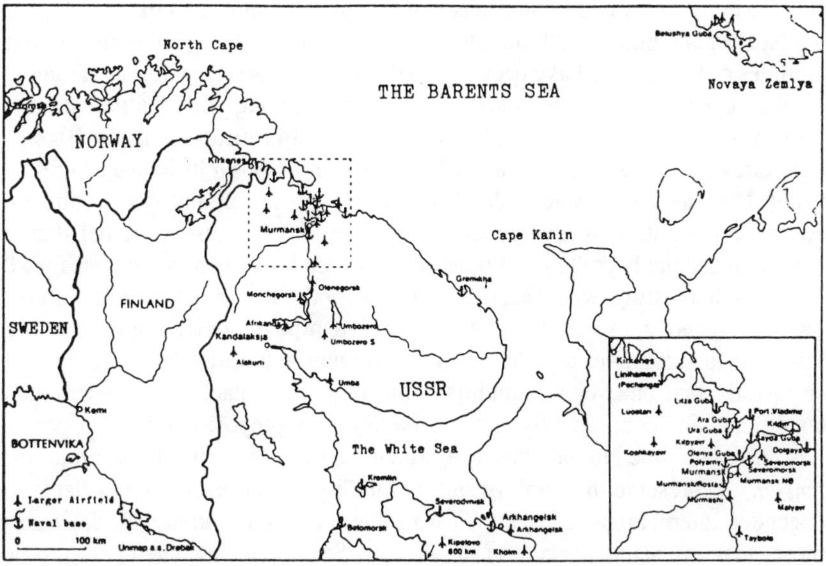

Figure 8.12.[54] Soviet Military Bases.

The ice presents a different situation for Soviet submarines since they can go under, though there are limitations due to the depth of icebergs and pack ice limiting maneuverability. The Arctic station areas for the submarines include the Barents Sea, the East Greenland Sea and the areas north of the Greenland, Iceland and UK gap, as well as the Arctic Ocean in the marginal ice zone and the centre with an ice thickness of under 3 meters. Safe transit routes, which are vital for the Northern Fleet to protect to maintain its strategic deterrent effect, due to bottom topography run between Bjørn Island and Northern Norway, and between Franz Josef's Land and Novaya Zemlya.[55] Any economic activities which thus interfere and channel the submarines, as with surface vessels could cause the Soviet military problems in carrying out the objectives noted above.[56]

Though as seen above up to now there has been relatively limited oil and gas exploration and production activities as well as shipping in the transit areas noted, any increase of activities could bring interference problems to a head. The Soviets cannot directly control and plan shipping and oil activities on the Norwegian side. Geological structures are fixed and wells will be drilled wherever there are finds. However the Soviets could attempt to influence developments through a bilateral Soviet - Norwegian exploration, production and shipping lane agreement. Along these lines several authors note that until very recently the Soviet Union has continuously attempted to exert *influence* on Norwegian policy and interests in the North.[57] On Spitzbergen Soviet attitudes and behaviour have been directed towards influencing the development of Norwegian politics, new legislation and the carrying out of administrative measures, and similar actions taken regarding control measures in the Barents Sea 'Grey Zone' fishing agreement.[58] The Soviet proposal in January 1988 by Prime Minister Rysjkov for joint Norwegian - Soviet oil and gas production and processing in a 'special economic zone' in the Barents Sea rather than a definite maritime boundary, negotiations of which have been in progress for 20 years, is interesting when seen in light of both the Soviet military transit interests noted above and the history of attempted Soviet domination. In response to these overtures however Norwegian authorities have maintained steadfastly that Norwegian administration must take place under Norwegian conditions, and consequently rejected the Rysjkov proposition.[59]

At the same time Norway has as a primary goal in the North to promote *low tension*, 'to preserve the stability and tranquillity, and to avoid moves liable to engender international conflict.'[60] This has included making official statements that any activity taken on the Barents Sea continental shelf, 'was to be guided by civil, peaceful and ordinary economic principles under firm Norwegian management and control';[61] and officially acknowledging that the Soviet presence on Svalbard besides for economic reasons, also is motivated by, 'the strategic location of the archipelago and a wish to observe that developments do not run counter to Soviet interests.'[62]

What this means very simply in the Barents Sea is that in practice Norway performs a balancing act, attempting to defend its interests in a definite and firm, yet appeasing manner. This policy includes preventing conflicts through Norwegian - Soviet *cooperation*, in specific areas which are 'non threatening' to Norwegian interests.[63] This has resulted in several bilateral Norwegian - Soviet agreements covering the Barents Sea, in spite of the attempted domination noted above, in areas such as search and rescue, Arctic research, atomic accident warnings, *general environmental* concerns and the prevention of maritime incidents, but does *not* include as yet a joint contingency plan.[64] Although one thought might be that the contingency plan is not included because clean up operations may involve the use of military equipment, this is countered by the fact that several of the agreements noted also might encompass use of military equipment, at least theoretically. It is thus difficult to say exactly what type of bilateral agreement threatens Norwegian interests. Elements of agreements however which appear important to Norwegian interests include that the provisions be legally binding, precise, balanced, mutual and verifiable.[65]

One recent development in Soviet policy which counters much noted above, at least in terms of environmental concerns, is that the Soviet Union appears to have 'decoupled' security from a 'military strategy core.'[66] This means that the Soviets may be disconnecting military from non military issues and seem to be attempting to identify issues where cooperation may result in more progress.[67] The West apparently started decoupling security Arctic concerns from other interests several years prior to the Soviets, and all the Arctic rim States appear to be moving towards a situation, 'in which military-strategic conflict may come to exist side by side with non-military cooperation.'[68]

In spite of this development Norwegian scepticism regarding certain forms of bilateral cooperation with the Soviet Union however remains. Although it is difficult to predict, based upon Norway's history of not entering a specific Barents Sea contingency plan without the establishment of a definite maritime boundary as well as the statement made by the Ministry of the Environment it can probably be said that definite and effective *bilateral* Soviet - Norwegian environmental agreements encompassing the Barents Sea will not be entered without the fulfilment of the same condition.[69] Before this occurs if Norway is interested in achieving environmental protection for the Barents Sea, it will be through cooperative agreements including other western States and NATO members along with the Soviet Union. Norway has invited the Soviets to become members of regional international agreements covering land-based, sea-bed and dumping pollution sources in the Barents Sea, and there has been mentioned a possibility for the establishment of a cooperative surveillance agreement among all the Arctic rim States, Denmark/Greenland, Canada, Norway, the US and the Soviet Union.[70] Perhaps this is the direction environmental protection on all levels in the Barents Sea will proceed.

8.4. Fisheries and Other Interests[71]

This Section deals exclusively with *Norwegian* fisheries data. No attempt has been made to collect data on the Soviet fishing activities except on a general level. One author notes generally that the average fish taken in the Barents Sea, Svalbard area and northern areas of the Norwegian Sea are approximately 2.2

million tons per year over a 20 year period with a top reached of 4 million tons in 1977.[72] This catch in the last half of the 1980's however has fallen drastically to approximately 1/3 of the 1968-1988 average due mainly to a collapse of the capelin stock in 1985-1986. The cod stock on the average has been relatively stable, but now it is also in a less than desirable condition,[73] and most of the other commercial stocks are maximally exploited. This drastic decrease of fish in the Barents Sea has dramatically weakened the economic support structure for Northern Norwegian communities.[74] The Norwegians and the Soviets have dissimilar interest in the Barents Sea fish stocks, with the Soviets preferring capelin and a special cod type (kolmule), and the Norwegians preferring the Norwegian - Arctic cod, although recently the latter has risen in interest for the Soviets due to possible sales to Western markets. The most important fisheries conflict between the Soviets and the Norwegians involve questions related to the type of authorized equipment and forms for fishing control. This includes the size of the mesh used in trawler nets,[75] and the extent of coastal fishing and trawler free zones along the Norwegian coast.

Partially covering the Barents Sea is the area known as 'Barentshavet Syd' shown in Figure 8.13. The main fish types in this area in which the Norwegians have been interested comprise bottom fish which include the Norwegian - Arctic cod, the Norwegian - Arctic haddock, the pollack, the ling, the blue ling, the brosme, and the flounder; pelagics which include the capelin and the herring; and shrimp and scallops. The Norwegians have as well conducted whaling operations in this area, concentrating on the minke whale. A brief account of each of these groups will be given including a geographic description of where in the 'Barentshavet Syd' area they are mainly found, as well as some idea of amounts in tons taken in later years. It should be noted that the statistics given are *approximate*. The idea behind this presentation is to give a general idea of which geographic marine areas conflicts could be expected to arise should they be proposed for shipping lanes or for oil exploration and production.

Generally, the southern and eastern sections of the 'Barentshavet Syd' area together with areas southwest and southeast of Bjørn Island are the most important for fisheries. See Figure 8.13 (and Section 8.2. Figure 8.1.). The northern coast of Norway is shown for reference. No attempt will be made to differentiate between the different methods of fishing such as trawling and line fishing. For the bottom fish industry the two major types caught are the Norwegian - Arctic cod (torsk) and the Norwegian - Arctic haddock (hyse), though there exist many bottom fish. See Figures 8.14, 8.15 and 8.16 below.

It was expected in 1987 that these cod and haddock stocks were on the upswing and mid 1970 figures would again be realized, however as noted this has not occurred. Specifically the major areas in the 'Barentshavet Syd' where bottom fish are caught from Figure 8.16 include areas 03 - Øst-Finnmark; 12 - Nordkappbanken; 13 - Thor Iversens Bank (within the Barentshavet Syd area). It is also expected that area 20, southwest and southeast from Bjørn Island can become an important area if an increase of fish stocks occurs. For the shrimp fishing industry the 'Barentshavet Syd' area on the average contributed 25% of the total Norwegian shrimp catch for the years 1980-1983. The main geographic areas where most of the catch occurred, from Figure 8.16, 03 - Øst-Finnmark; 12 - Nordkappbanken (only moderate); 13 - Thor Iversens Bank (most important); 15 - Sentralbanken (increasing importance); 20 - Bjørn Island (southeast).[76] For the capelin fishing industry the 'Barentshavet Syd' was a very important area in which 35% to 40% of the total Norwegian winter catch took place. The summer catch took place in the northern Barents Sea, though outside of the 'Barentshavet Syd'. From Figure 8.16 the main areas of catch are 03 - Øst-Finnmark; 12 - Nordkappbanken; and 13 - Thor Iversens Bank. Additionally in area 20 - Bjørn Island, there has been conducted winter capelin fishing in the southeast section and summer capelin fishing in the northern sections.[77] As noted in later years the capelin was so drastically reduced that fishing was prohibited in 1987, and it is expected that this will continue for some years to enable the stock to replenish itself.[78] There is presently little herring catch conducted in 'Barentshavet Syd' since the stock was almost depleted in the late 1960's and the legal size limit per fish was set to 25 cm. The possible catch of herring in the Øst-Finnmark area along the coast and in the fjords possibly would not be in the areas where oil activities are conducted, however, young herring are susceptible to oil pollution damage and seismic activities, which could result in future conflicts between the fisheries and the oil activities.

Norway has conducted whaling operations in the 'Barentshavet Syd', mostly for the minke whale. The areas where most of the catch has taken place, from Figure 8.16, was in areas 03 - Øst Finnmark; 12 - Nordkappbanken; and 20 - Bjørn Island. There has been little whaling activity in areas 13 - Thor Iversens Bank; and no activity in 15 - Sentralbanken. In the later years whaling activities have been sharply limited, and for the period 1988-1990 the Norwegian government decided that only whaling conducted for research will be allowed.[79] The goal is to study the stock with the goal of continuing whaling later if numbers and conditions allow.[80] Norway has also conducted large scale sealing up to the late 1970's taking yearly over 100,000 animals. The EEC stopped all import of skins from young seals in 1983 and presently is working to stop import of all seal skins which will essentially close the market for such products.[81]

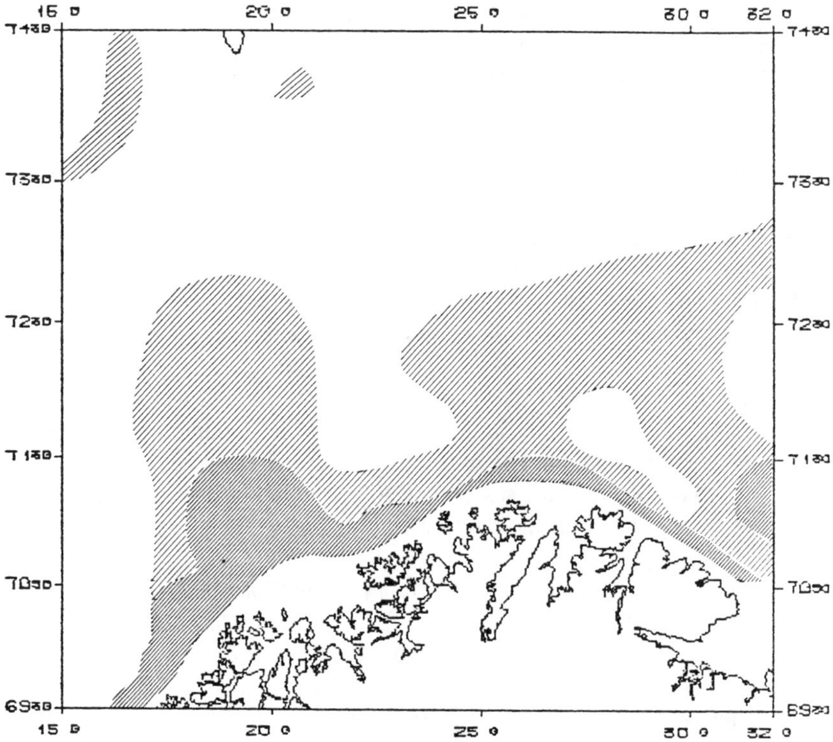

Figure 8.13.[82] Map Indicating the Most Important Fishing Banks in 'Barentshavet Syd' Area.

Year	I Area 'Barentshavet Syd'	II All areas north of 62°N	I in % of II
1977	239	650	36
1979	133	591	22
1981	100	606	17
1983	67	489	14

Figure 8.14. Total Catch Statistics.
Figures are given in 1000 tons for the bottom fish catch in 'Barentshavet Syd' area, and for all areas north of 62 degrees N. Bottom fish consist primarily of Norwegian - Arctic cod and Norwegian - Arctic haddock.

	Barentshavet Syd				North for 62°N			
	1977	**1979**	**1981**	**1983**	**1977**	**1979**	**1981**	**1983**
Cod	78	60	85	76	66	55	54	56
Haddock	11	26	26	11	7	12	10	5
Pollack	9	12	7	11	22	24	27	31
Other	2	2	2	2	5	9	9	8
	100	100	100	100	100	100	100	100

Figure 8.15.[83] Specific Catch Statistics.
Figures are given in percentages of bottom fish catch for 'Barentshavet Syd' area and for all areas north of 62 degrees N.

An industry with a possible future in the 'Barentshavet Syd' is that related to scallops. Little is presently known of the extent of the colonies, however those discovered in the vicinity of Bjørn Island may prove to be commercially feasible for development.

The coast of Øst Finnmark has revealed colonies though of little commercial value.

Besides the industries described above there exist as seen in Part I Section 1.2. other systems in the Barents Sea which could be affected and damaged by marine pollution including primary production (phytoplankton), zooplankton, birds, mammals and fish, benthic habitats and the land-freshwater-marine interrelation as well as ice. The list is by no means complete and could well include energy flow patterns and climatic changes as well.[84] The direct negative effects of oil pollution upon these is considered important since it is known that all these systems have a relation to each other, man included. The definition of 'ecosystem' given by one author perhaps conveys this idea best, '(A) system formed by the interaction of a community of organisms with their (sic) environment.'[85] Thus a change of a community of one or more of marine organisms due to a polluting vessel discharge or a platform blowout may have consequences for segments of the entire system. The Arctic ecosystems are often described as being 'fragile'[86] though at least one scientist considers them on the contrary to be at least as tough as others.[87]

It is hoped by this brief overview that an idea of the main types of fishing conducted in the Barents Sea can be obtained as well as an idea of some of the other ecosystems present. It is these industries and ecosystems which might come into conflict with pollution from any oil, shipping, dumping and land-based activities, and the biological and economical damage of which would give a real basis for any international anti pollution conventions and agreements governing the Barents Sea as well for as any proposals made for bettering legal coverage.

8.5. Oil and Other Pollutants in the Barents Sea[88]

As seen in Part I Section 1.2. generally there is expected little difference between the behaviour of oil in a temperate sea and the Barents Sea, especially in the proposed ice free oil leases in the southern Barents Sea concerning the processes evaporation, emulsification, sedimentation and adsorption. Dissolution in the Barents Sea would be less; photo-oxidation would be increased during the summer and decreased in autumn, winter and spring; auto oxidation would be minor; and microbial degradation slow. As noted the most difference would be oil being stranded on low energy beaches and in ice zones in the north. Currents patterns capable of carrying any oil spills are indicated in Figure 8.17.

With respect to fisheries mentioned in Section 8.4. which may be affected, briefly the 'Barentshavet Syd' area shown in Figure 8.13 can be compared with the areas IV, V, VII, VIII, IX, X, in Figures 8.9, 8.10 and 8.11 in Section 8.2.2., delineating oil exploration sections in the Norwegian Barents Sea, and seen to roughly correspond. Thus the areas which are expected to be leased for oil exploration are also areas containing the Norwegian - Arctic cod, the Norwegian - Arctic haddock, the pollack, the ling, the blue ling, the brosme, and the flounder; pelagics which include the capelin and the herring; and shrimp and scallops. Shipping traffic to and from the *Northeast Passage* discussed in Section 8.2.1. could also be expected to traverse the important

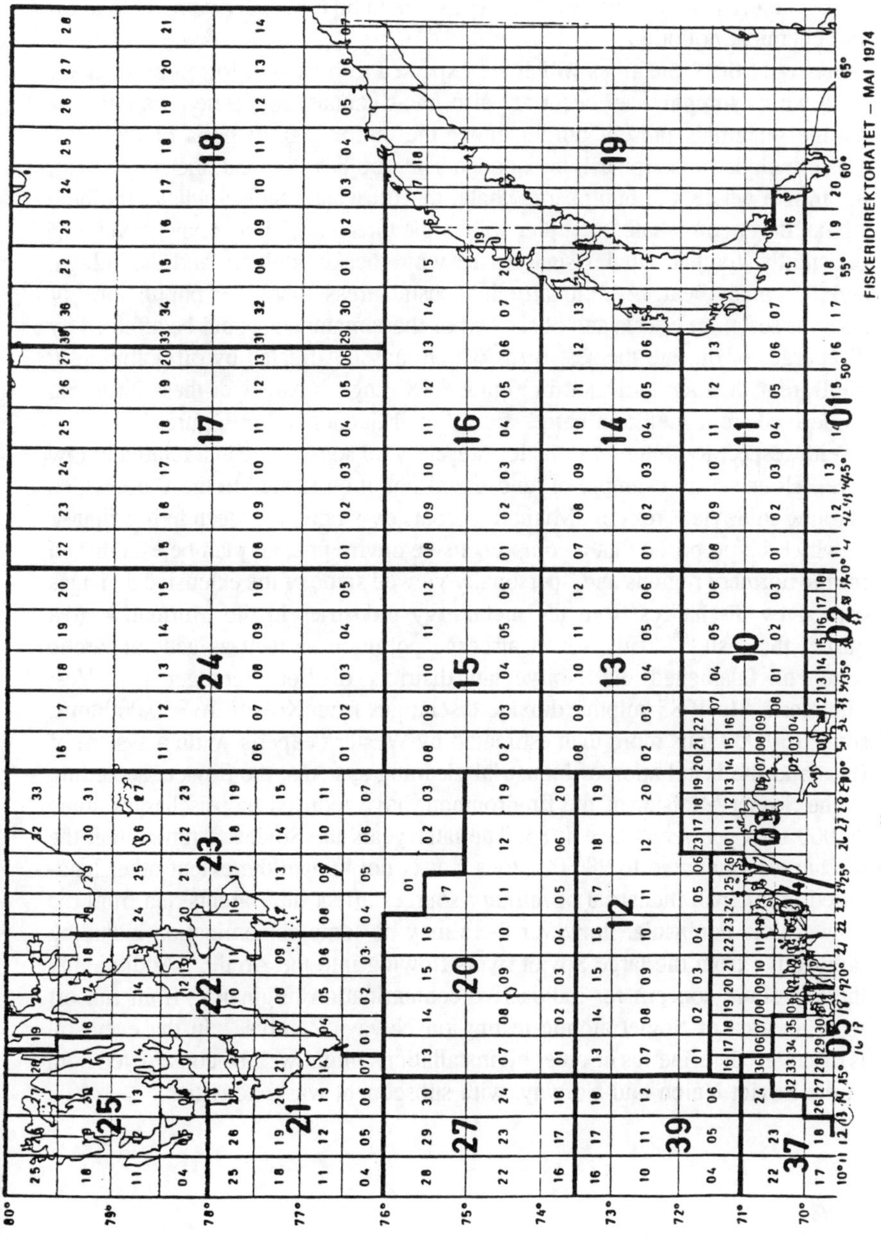

Figure 8.16.[89] Statistical Areas and Locations in the Barents Sea.

fishing areas of 03 - Øst Finnmark, 12 Nordkappbanken and 13 Thor Iversens Bank. Young fish or larvae in these areas would be especially vulnerable to oil and chemical pollution.

Generally the same areas which are expected to be leased for oil exploration or in which shipping can occur are also areas containing different varieties of marine mammals present during either the entire year or parts of the year. These include the harp seal, the hooded seal, the bearded seal, the ringed seal, the minke whale, the humpback whale, the fin whale, the sei whale, the blue whale, the sperm whale, the killer whale, the Greenland right whale, several of the smaller toothed whales such as the white-beaked dolphin and the beluga, and the polar bear. Additionally in coastal areas there are populations of harbour seals, grey seals and otters. All of these mammals could be affected by discharges of oil into the sea. Birds which may be affected by oil pollution in the Barents Sea are indicated in Figure 8.18. English names of these birds can be derived from the Latin names from Part I Section 1.2.6. Figure 8.11.

With respect to pollution statistics presently all sources are at a relatively low level. However an example of land-based pollution related to the Barents Sea is found in Soviet President Mikhail Gorbatsjov's famous speech in Murmansk in which he proposed that a comprehensive environmental plan be established for the northern regions and, 'personally viewed some of the extensive damages caused by discharges from the metallurgy industries in the Murmansk area during the visit.'[90] This Soviet airborne pollution as well as general 'arctic haze' has damaged the Norwegian districts of Sør Varanger and Vest Finnmark.[91] In 1985 sulphur dioxide discharges from Kola were 350,000 tons, some 150,000 tons more than estimated by Western experts. With a system of fines and the installation of industrial cleaning systems, the Soviets according to the Soviet Minister of the Environment have reduced the discharges some 70,000 tons per year, since 1985. The later years the Soviets estimate that the discharges amounted to 280,000 tons.[92] It is not known the extent other land-based municipal, industrial or military sources affect the Barents Sea from the Soviet Kola Peninsula, however there may be some contamination including radioactive from the large Soviet rivers flowing into the Arctic.[93] Additionally there is some concern for radioactive contamination originating from run off from an area of Soviet atomic testing on Novaya Zemlya.[94] In the event oil is discovered, refineries or other oil installations might well be constructed both in the Soviet Union and Norway, with subsequent waste activities.

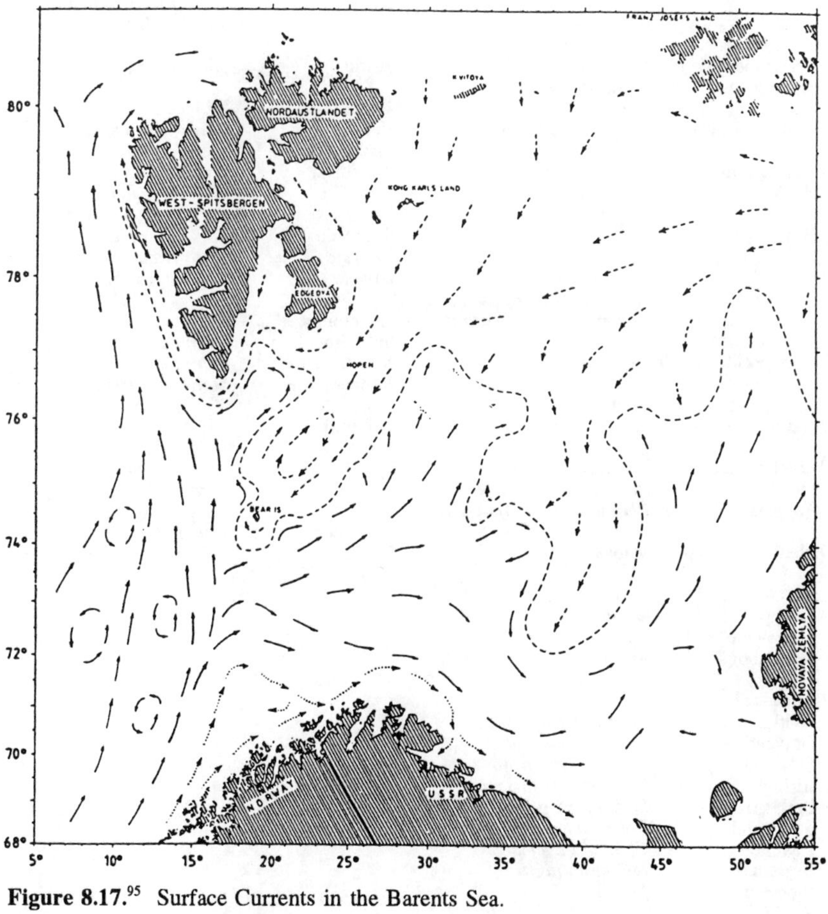

Figure 8.17.[95] Surface Currents in the Barents Sea.

Orden: Lommer - Gavrilformes

Lomfamilien:	Gavidso
Smålom	_Gavia stellata_*
Storlom	_Gavia arctica_*
Islom	_Gavia immer_*
Gulnebblom	_Gavia adamsii_*

Orden: Lappedykkere - Podicipedformes

Lappedykkerfamilien	Podicipedidae
Toppdykker	_Podiceps cristatus_*
Gråstrupedykker	_Podiceps griseigena_*
Horndykker	_Podiceps auritus_*

Orden: Stormfugler - Procellarilformes

Stormfuglfamilien	Procellariidae
Havhest	_Fulmarus glacialis_
Grålire	_Puffinus griserus_*
Havlire	_Puffinus puffinus_*
Stormsvalefamilien	Hydrobatidae
Havsvale	_Hydrobates pelagicus_
Stormsvale	_Oceanodrama leucorhoa_

Orden: Pelikanfugler - Pelecaniformes

Sulefamilien	Suldiae
Havsule	_Sula bassana_
Skarvefamilien	Phalcrocoraeidae
Storskarv	_Phalacrororax carbo_
Toppskarv	_Phalacrororax aristotelis_

Orden: Andefugler - Amseriformes

Andefamilien	Anatidas
Knoppsvane	_Cygnus olor_
Sangsvane	_Cygnus cygnus_
Kortnebbgås	_Anser brachyrhynchus_ A
Grågås	_Anser anser_
Kanadagås	_Branta canadensis_
Hvitkinngås	_Branta leucopsis_ A
Gravand	_Tadorna tadorna_
Stokkand	_Anas platyrhynchos_
Ærfugl	_Somateria mollissima_
Praktærfugl	_Somateria spectabilis_
Stellerand	_Polysticta stelleri_* A
Havelle	_Clangula hyemalis_*
Svartand	_Melanitta nigra_*
Sjøorre	_Melanitta fusca_*
Kvinand	_Bucephala clangula_*
Siland	_Mergus serrator_
Laksand	_Mergus merganser_*

Orden: Vademåke og vadefugler - Charadrilformes

Snipefamilien	Scolopacidae
Svømmesnipe	_Phalaropus lobatus_*
Polarsvømmesnipe	_Phalaropus fulicarius_* A
Jofamilien	Stercorariidae
Polarjo	_Stercorarius pomarinus_* A
Tyvjo	_Stercorarius parasiticus_
Fjelljo	_Stercorarius longicaudus_* A
Storjo	_Stercorarius skua_
Måkefamilien	Laridae
Hettemåke	_Larus ridibundus_
Fiskemåke	_Larus canus_
Sildemåke	_Larus fuscus_
Gråmåke	_Larus argentatus_
Polarmåke	_Larus glaucoides_ A
Svartbak	_Larus marinus_
Krykkje	_Rissa tridactyla_
Ismåke	_Pagophilia eburnea_ A
Ternefamilien	Sternidae
Splitterne	_Sterna sandvicensis_
Makrellterne	_Sterna hirundo_
Rødnebbterne	_Sterna paradisaea_
Alkefamilien	Alcidae
Lomvi	_Uria aalge_
Polarlomvi	_Uria lomvia_
Alke	_Alca torda_
Teist	_Cepphus grylle_
Alkekonge	_Alle alle_
Lunde	_Fratercula arctica_

Figure 8.18.[96] Marine Birds. Birds Found in the Barents Sea Area Are Indicated with 'A'.

With respect to vessel-source pollution the speculated opening of the Northeast Passage could result in substantial ship traffic, depending upon the requirements and fees set by the Soviets, sailing directly through the Barents Sea with ensuing marine pollution.[97] In addition with the advent of oil production in the Barents Sea it can be expected that one of the alternatives for freighting oil in this region is by tanker.[98] With respect to the probability of oil spills from tankers it is estimated that for the geographic area from approximately midway between Harstad and Hammerfest (Tromsø) to the Soviet border (see Section 8.2., Figure 8.3 the probability per year for a tanker related *accidental* discharge is *.19*. The probability is .01 within 5 to 50 nautical miles from land; .13 less than 5 nautical miles from land and .05 in the harbour.[99] From these figures it can be seen that the possibility for an accidental discharge from a vessel is approximately 1 in every 5 years. The probability for accidental oil discharges due to ships may *possibly decrease* somewhat during the period 1985-2000 due to the effect of MARPOL 73/78 and an increased control over ships.[100] Although the statistics do not cover the possibility of *illegal* discharges, these will be affected at least as much as the accidental discharges and will similarly possibly decrease.[101] Neither the probability for oil discharges from ships other than oil tankers, nor the quantity of the discharges themselves is addressed however, and accidental tanker oil discharges has been few with only 3 resulting in discharges over 2000 cubic meters in Norwegian waters.[102]

With respect to dumping in the Barents Sea the dumping of obsolete military weapons and equipment from the large Soviet military bases on the Kola Peninsula as well as industrial waste could be a problem, the extent of which though is unclear.[103] In addition a foreseeable problem will probably be that of disposal of offshore oil field scrap, similar to what has occurred in the North Sea.[104]

The types of relatively small discharges from offshore installations noted in Part I Sections 1.3.4. and 1.3.6. are presently occurring in the Barents Sea both from Soviet and Norwegian units. In addition, one might only conjecture what impact such a blowout as IXTOC I might have on the bays, estuaries and fjords bordering the Barents Sea.[105] As indicated, there is always the chance of a blowout which can over a short period of time discharge many tons of oil into the sea. According to figures developed by the Norwegian Department of Oil and Energy for oil activities in the Troms II area, the probability for an uncontrolled oil blowout is 1 per 1,800 discovery and relief wells.[106] Internationally only about 7% of blowouts documented have consisted of oil, the rest being primarily gas; therefore the probability for a blowout, *either* of oil or gas is 1 per 126 wells.[107] Additionally even the oil blowouts have been small with less than 5% of the foreign blowouts being in the order of 240,000 cubic meters or more.[108] In spite of the uncertainty involved making such estimations, it is thus expected that with approximately 50 exploratory and relief wells being drilled per year *1* oil blowout can be expected in *30-40* years.[109] For the first 450 exploratory wells drilled on the Norwegian continental shelf drilled up to 1983, there were 3 blowouts, all gas, which is in accordance with statistics registered on a world basis.[110]

Another probability appraisal for oil platform blowouts for different fields on the Norwegian continental shelf 1985 to 2000 is given in Figure 8.19. These figures include the probability for uncontrolled blowouts occurring and being equal to .0018 to .0036 per drilled well for exploratory, relief and production wells, and during production .0003 per well per year.[111] Additionally historically only 1 of 10 blowouts consists of oil and of this, only 1/3 result in a large quantity.[112] These estimates roughly compare with those given above. Of importance in the different fields is how much oil and gas they are expected to contain. The figures for the areas from Troms northward are based upon results obtained from Tromsø shelf where much gas is expected to lie. Another variable mentioned which adds uncertainty to these figures is how much the hard climate and the deep ocean drilling conditions are expected to effect the chances for oil spills.[113]

		1984		2000	
		>2000t	**<5000t**	**>2000t**	**<5000t**
1	Ula - Ekofisk - Valhalla	0.003	0.008	0.004	0.009
2	Statfjord - Murchinson - Oseberg - 34/4 - 34/7 - Gullfaks	0.003	0.008	0.005	0.012
3	Trøndelag - Haltenbanken - Vestfjorden	0.001	0.003	0.002	0.004
4	Troms II	-	-	0.0003	0.0007
5	Troms I - Bjørnøya	0.0006	0.001	0.001	0.001
6	Finnmark	-	-	0.0005	0.0001
7	Barentshavet	-	-	0.0007	0.002
		0.009	0.021	0.014	0.032

Figure 8.19.[114] Probability for Large (greater than 20,000 tons) and Small (less than 5,000 tons) Oil Pollution per Year in Fields on the Norwegian Continental Shelf in 1984 up to the Year 2000. There was no activity in the field Troms II or the fields offshore from Finnmark and in the Barents Sea in 1984.

As indicated above in Section 8.2.1. the Soviet oil operations in the Barents Sea have been characterized by human mistakes and inexperience. If as seen in Part I Section 1.3.4.1. approximately 92% of blowouts studied had an element of human error involved, one can only wonder what this means in terms of blowout possibilities on the Soviet side, combined with the hard weather conditions. On the Norwegian side this could as well be a factor.

8.6. Conclusion

In summary in spite of incomplete, uncertain and varying statistics,[115] it seems clear that if the Barents Sea in not already being polluted it is certain to be exposed to oil and hazardous chemical pollution from especially marine shipping, if the Soviets open up a route through the Arctic to the Pacific, and to a lesser extent from offshore oil operations both on the Norwegian and Soviet sides if activities are trapped up due to larger commercially feasible discoveries being made. Pollution from land-based sources and dumping seem also to be hovering on the horizon, with the former likely to increase if large finds are made, and refineries, factories and larger population concentrations are established in the surrounding areas. The dumping of atomic wastes and military equipment by the Soviet bases on the Kola Peninsula seems clouded in secrecy and it is anybody's guess whether it has been carried on extensively in the past, and is likely to continue. Dumping has a chance of increasing in the offshore sector on the Soviet side should commercial production be initiated.

Several of the international regimes viewed in Part I regulate marine pollution in the Barents Sea, somewhat more predominantly on the Norwegian side. These as well as international customary law are presented in Chapter 9.

Notes - Chapter 8

1 The idea of Arctic regional environmental protection is not new; it has been proposed several times that Canada, Denmark/Greenland, Norway, the United States and the Soviet Union enter into a regional environmental protection agreement covering the Arctic. See St. meld.nr. 26 (1972) for the Norwegian offer to hold an international conference on the protection of the Arctic environment, and Harders Marine p. 297 concerning the United States' offer in 1970 and the Canadian proposal in 1972. None of these conferences were convened.

2 Although Norwegian fears have noticeably lessened due to the recent developments in the Soviet Union, there remains a certain wariness of the Russian State to which Norway shares a border. Therefore the arguments presented still apply, albeit to a lesser degree and directed more against Russia now than the Soviet Union. Speech by Norwegian Military Chief of Staff, Admiral Torolf Rein, 'Norsk forsvarspolitikk fram mot år 2000', Atlanterhavskomiteens Tromsø seminar, 24 October 1991.

3 Hoel, Alf Håkon, 'Ressursforvaltningen i nordområdene,' *Norsk utenrikspolitisk årbok*, 1989 (Hoel), s. 35. See Part I Section 2.2. Note that the citations given in the work do not contain any included footnotes unless otherwise noted.

4 Euroil Vol. 1, Issue 5, August 1990 (Euroil) pp. 30-32 and 53-54.

5 Østreng, W. 'The Changing Mood of the Kremlin', *International Challenges*, (Østreng Mood) pp. 10-14 at p. 12.

6 Obtained from '*Limits of Oceans and Seas*, International Hydrographic Organization', Special Publication 23, Draft 4th edition 1986 (Hydrographic), p. 200, which notes, '(T)he limits described in this publication have been drawn up solely for hydrographic purposes. It must not be construed as having any legal or political connotations whatsoever.'

7 Obtained from Hydrographic p. 201.

8 The following information is obtained from *Soviet Oil and Security Interests in the Barents Sea*, Bergesen, Moe and Østreng, Fridtjof Nansen Institute (Østreng), pp. 6-9, 12, 25, 34-45. 38-39, 49-51 and 54 unless noted otherwise.

9 Due to the drastic political developments in the Soviet Union since 1990 in which the Soviets are attempting a restructuring of their society to a market economy, their dependency on the West might be considered even more pronounced than indicated here.

10 Euroil p. 32 notes however that some Soviet economists consider it acceptable to lower the oil production because too many barrels produced are sold for prices which do not even cover the expenses. Government figures indicate an increase in *cost* of oil production of 160% over the past 10 years. Future strategies include prioritizing profitable production which would keep up a high level of hard currency earning export, while domestically improving energy efficiency and conservation including changing to gas and other fuels. It is noted however there will be continued increased exploration in frontier areas as well as exploration and enhanced recovery operations in existing oil fields.

11 *Ibid.* President Mikhail Gorbatsjov backed action to remove the bottleneck in the oil and gas sectors, and Soviet oil production in the first 6 months of 1987 increased 2% from 309 tons with respect to the same period the previous year.

12 Euroil p. 32. Gas production on the other hand increased 3% in the first 6 months of 1990.

13 Bergsager, Egil, 'Hydrocarbon Resources and Energy Politics in Arctic Areas Offshore Norway and USSR.' Updated version of paper given at Offshore Northern Seas. Stavanger 20-24/8 1986 (Bergsager).

14 Euroil p. 30. Cooperation in marine equipment supply is also taking place.

15 Sovjet Nytt 28. september 1990 nr.24, s. 2. Euroil p. 31 notes that *no* production sharing agreements have been arranged so far for the Barents Sea, the Kara Sea or other areas of the Soviet Arctic.

16 This list was current in 1987.

17 Aftenposten, 21. august 1987 Uke 34. Nr.382. 128.årg. s. 1, 2 and 8. The Norwegian Minister of Oil and Energy admitted that he had no previous knowledge of this commercial development.

18 *Ibid.* For the years 1986-1990 production from the Kolgujev field was expected to reach 1/2 million tons per year, making this field 'economically interesting'.

19 See Euroil pp. 30-31 (current to August 1990).

20 Euroil p. 32 notes that regarding other Soviet Arctic Seas the northern part of the Kara Sea may be perhaps one of the world's most interesting offshore areas and that the Rusanovskaya discovery off the Yamal peninsula may contain as much gas as the Shtockmanovskoye.

21 This alternative could conflict with export plans mentioned earlier.

22 This would conflict with the Soviet military interests in the area, see Section 8.3. and additionally would involve increased economic dependency on the West.

23 Obtained from Fig. 2.1., Østreng p. 34.

24 Obtained from Euroil p. 30.

25 Obtained from Table 2.1. Østreng pp. 38-39.

26 See for example Aftenposten 21. desember 1990, Uke 51., Nr.586., 131.årg., s. 1, 2, 9 in which it is noted that the Soviet Foreign Minister Sjevardnadse stepped down and warned of the return to dictatorship.

27 The information concerning the Northeast Passage is obtained from a FNI lecture given by US Coast Guard officer Lawson Brigham (Brigham), October 30 1989 at Fridtjof Nansen Institute, Fridtjof Nansens vei 17, P.O. Box 326, N-1324 Lysaker, Norway (FNI), and from conversations with FNI Research Fellow, Olav Stokke, unless noted otherwise.

28 Aftenposten 2 oktober 1987, Uke 40., Nr. 454., 128 årg., s.8.

29 Aftenposten 18. desember 1990, Uke 51., Nr.580., 131.årg., s. 12, and Nordlys 18.oktober 1990, Nr.241, 89.årgang., s. 12. It is noted that FNI together with the Soviet Ministry for the Shipping Trade is directing the pilot project and will have a chief role in the main project as well. The pilot project encompasses 3 studies including environmental impact, economics and political considerations. Norway is interested in the opening of the Northeast Passage among other reasons because of the possibility for industrial and service functions being situated in Northern Norway as well as the large possibilities for Norwegian shipping. See also Nordlys 25 oktober, 1989, 'Nord-Norge mot År 2000, Drøm blir virkelighet?', s. 5-7 and Nordlys 10 oktober 1990, Nr.241., 89 årgang., s.12.

30 Obtained from Figure 1.1. Principal ports, rivers, and routes for the Soviet Maritime Arctic, *The Soviet Maritime Arctic* (1991), p. 3.

31 Armstrong T., 'The Northern Sea Route,' *Polar Research*, Vol. 25, (157) (Armstrong), p. 247. It is additionally noted that Igarka, the timber port farther up the Yenisey, has a much shorter season of 4 to 5 months.

32 Armstrong, T., 'The Northern Sea Route,' *Polar Research*, Vol. 25 (157) (Armstrong) p. 248 notes that sea shipping in and out of Tiksi lasted in 1988 for 4 months from 25 June to 25 October. See Figure 6. In the eastern part statistics for a 10 year average indicate open periods for Pevek - 125 days, for Mys Shmidta - 100 days, and for Provideniya 165 - days. Pevek handled 200 ships in 1988. FNI notes technically it is possible to hold the passage open for longer periods, even the entire year, but is not economically profitable. See Nordlys 18 oktober 1990, Nr.241., 89.årgang, s. 12.

33 Roginko p. 1.

34 Armstrong p. 248. The author however notes that this figure may represent only ships sailing the central or eastern Arctic areas.

35 Obtained through FNI from Lawson Brigham, US Coast Guard Officer, 3 October 1989.

36 The following information is obtained from St.meld.nr. 40, Åpening av Barentshavet Syd for letevirksomhet, (1988-89) (St.meld.nr. 40) s. 5-9, 29, 32-34 and 42 unless noted otherwise.

37 Due to the crisis in the Gulf War, the price of oil on the world market Autumn 1990 fluctuated drastically to as much as USD 40.00/barrel. The price in mid December 1990 was approximately USD 26.75/barrel, International Herald Tribune, No. 33,530 December 14, 1990, p. 14. Thus Norwegian oil profits were greater than anticipated.

38 See St.meld.nr. 79, s. 18-19.

39 See Figure 9.

40 Euroil pp. 52-54.

41 In the summer, 1987 two further wells were drilled, one receiving much attention in the press, see Aftenposten 15. juni 1987, Uke 25, nr.266, 128 årg., and Nordlys, 22.juni 1987, nr.140, 86. årgang. The Ministry of Oil and Energy in a conversation of July 27 1987, noted that although the sandstone structure contained oil, there was a lack of the necessary pressure for commercial production.

42 Tabell 10 from the Norwegian Bureau of Statistics, *Oil and Gas Activity*, 3rd Quarter 1990, Statistics and Analysis, (Oil and Gas) p. 35 shows for the years 1987, 1988, 1989 and 1990 up through September respectively 36, 29, 28 and 24 wells drilled on the Norwegian continental shelf. Gas finds have been uncovered but interest for present development is somewhat low due to other gas fields lying further to the south. The main Norwegian interest in exploratory drilling in the Barents Sea remains oil. A recent find however by the Norwegian company Statoil has again raised expectations for oil activities in the North though exact figures have not yet been released. See Aftenposten 5. desember 1991 Uke 49. Nr.562., 132 årg. s. 4.

43 Figure 8 obtained from Map no. 1. St.meld.nr. 40. p. 8, (1988-89). 'Barentshavet' translates into 'Barents Sea', and 'Midt Norsk Sokkel' into 'Sentral Norwegian Shelf'. Figure 9 obtained from Map no. 2. St.meld.nr. 40, (1986-87). 'Strategisk Område' translates into 'Strategic Area' and 'Norge' into 'Norway'.

44 Obtained from Tabell 6.1. St.meld.nr.40(1988-89), s. 42. Out of the discovery wells drilled in new geological structures, 318, 129 showed either oil or gas in varying amounts. This gives a discovery rate of 40% which is considered high internationally, and the Norwegian authorities want to continue with this rate while trying to hold the expenses at a reasonable level.

45 Euroil pp. 52-54. See Figure 9. Nordlys 9.november 1990 Nr.260., 89.årgang, notes that only 4-5 new licensees will be dealt out to the oil industry on the continental shelf off of Finnmark due to fishery and environmental concerns.

46 Euroil pp. 52-54.

47 *Ibid.* p. 44. parentheses added.

48 The following information is obtained from Østreng pp. 57-87 unless noted otherwise. See also Holst, J. 'Petroleum i norsk sikkerhetspolitikk', *Internasjonal Politikk*, nr. 2, 1986 (Holst) and Karstad, Arne, 'Security Aspects of Norwegian Oil and Energy Policy,' Address by the Norwegian State Secretary for Defense, to the Conference on Norway, OPEC and the Middle East at SAS Royal Hotel, Bergen, 8 October 1987 (Karstad).

49 The Northern Fleet has its bases from the Pechenga Fjord near the Norwegian border eastward, with 10 of the 12 close to the seaward approach to Murmansk.

50 Though the information may be somewhat dated, with present East - West adversative positions being more diffuse, it is not completely irrelevant. See Brundtland, A., 'Sovjetisk utfordring og vestlige svar - Norsk sikkerhetspolitikk på nye vilkår?' *Internasjonal Politikk*, nr. 3, 1989, s. 135 where it is noted that it can be assumed that unprovoked Soviet use of force directed solely against Norway can presently be ruled out, although it is another question whether military confrontations between the superpowers and alliance systems could be moved from central Europe to northern Europe, from land to the sea.

51 'SSBN' means nuclear strategic ballistic missile submarine. Another type, 'SSN' means nuclear attack submarine. Note that the next to last objective is especially important to counter the US forward based naval strategy whereby fighting the Soviets in their own waters in the initial phases of a war was a primary goal. Note that the size of one of the SSBN's, the *Typhoon*, has a height of between *25 to 30 meters*, keel to tower. This enormous size, besides limiting areas of accessibility to ocean areas with depths of approximately 68 meters, also conjures up images of possible collisions in the Barents Sea with oil platforms, ships or fishing boats and equipment with subsequent serious consequences. Kibsgaard, 'Militære operasjoner og mulige interessekonflikter i de nordlige havområdene,' *Internasjonal Politikk*, Nr. 2-3, 1988, (Kibsgaard) s. 37, adds to the military objectives noted, specifically support amphibian operations and test new weapon systems and operations.

52 See Section 8.2. Figure 1. In winter from Bjørn Island to Northern Norway is approximately 430 kilometres wide. The West cannot make much military use of oil rigs. This is due to their unsuitability as listening posts because of the noise and as missile launching pads because of explosion danger. Additional drawbacks include lack of security for the crew in the event of war, and little informational value compared to that obtained by other methods. Kibsgaard s. 41 notes however that oil platforms *can* be used for military activities through the mounting of radar and warning equipment to pick up electronic transmissions, something which can be carried out even in war. Also noted is that shipping, fishing and civil air transport would stop in war and surveillance activities from such vessels would also stop necessitating other measures.

53 Obtained from Figure 1 the Norwegian Bureau of Statistics, *Oil and Gas Activity*, 3rd Quarter 1990, Statistics and Analysis, p. 20.

54 Obtained from Figure 3.5, Østreng p. 62. See also Kibsgaard s. 36-37.

55 *Ibid.*

56 Brigham notes that holding the Northeast Passage open for longer periods of the year can have possible considerations for the moving of Soviet warships and for search and rescue of submarine personnel. FNI notes that Soviet suspicions in allowing international ships in its northern waters seem to have been put away for good. Nordlys 18 oktober 1990, Nr.241., 89.årgang, s. 12.

57 See Frydenlund, K., 'Lille land - hva nå? (Frydenlund), s. 102-103 and Tresselt, P., 'Norsk-sovjetiske forhandlinger om avgrensning av kontinentalsokler og økonomiske soner', *Internasjonal Politikk*, nr. 2-3, 1988 (Tresselt), s. 89-90.

58 Tresselt s. 90. The 'Grey Zone' agreement is a Norwegian - Soviet compromise covering fishing regulations that took into consideration important banks in boundary areas. Hereafter 'Spitzbergen' will be denoted by the Norwegian name, 'Svalbard' since it lies under Norwegian administrative control.

59 *Ibid.*

60 Østreng W., 'Delimitation arrangements in the Arctic Seas, Cases of precedence or serving of strategic/economic interests?' *Marine Policy*, April 1986, Vol. 10, Nr. 2, (Østreng Delimitation) p. 133 at 152.

61 *Ibid.* This was to waylay any possible Soviet fears that future shelf activity was for military purposes.

62 *Ibid.* citing *UD-informasjon*, No. 25, 13 July 1977, published by the Norwegian Ministry of Foreign Affairs.

63 Frydenlund s. 98.

64 See Part I Sections 4.2.5., 7.7. and 7.8.4 for the bilateral environmental agreement and draft contingency agreement. Chief Consultant Per Antonsen of the Norwegian Ministry of the Environment stated in an interview September 26, 1988 that as long as there was an unclear maritime boundary between Norway and the Soviet Union in the Barents Sea, there would be no bilateral marine environmental protection agreement for

the area. Recently Soviet vice Minster of Foreign Affairs Kovaljov forwarded a proposal for an encompassing bilateral agreement including expanded economic, environmental and energy measures, something in which Norway is interested. The maritime boundary question was not addressed although both governments acknowledge that negotiations are progressing. Aftenposten 12.desember 1990. Uke 50. Nr.579. 131.årg. s. 7. See also Aftenposten 3.november 1990, Uke 44. Nr.504., 131.årg., s. 12 and Nordlys 13 november 1990, 89 årgang, s. 14 which note positive progress in boundary negotiations.

65 St.meld. nr. 101 (1981-1982) pp. 8 and 17. These elements are applied to arms limitations agreements but may perhaps be extrapolated to other international agreements as well including anti pollution provisions.

66 Østreng Mood p. 12.

67 *Ibid.* pp. 11-12 bases this statement upon President Gorbatsjov's Murmansk speech of 1 October 1987, in which a desire was indicated to distinguish between military and non military Arctic issues, directed towards cooperation in Arctic regional environmental issues, in the exploitation of natural resources, in the conditional opening of the Northeast Passage, and in the establishment of a northern unified energy program including invitations to Norway and Canada as well as others to participate in and negotiate oil and gas joint ventures on the northern Soviet continental shelf.

68 *Ibid.* p. 14.

69 As seen the environmental bilateral Norwegian - Soviet agreement above is very *general*, with most activities directed towards air pollution emanating from the Kola Peninsula and damaging the Finnmark district in Northern Norway (an area with a definite boundary).

70 St.meld.nr. 39 (1974-1975) p. 48.

71 The following information concerning Norwegian fisheries in the Barents Sea is obtained from Fangstdata Barentshavet Syd Dokument Nr. 04-01-01, Olje- og Energidepartmentet Konsekvenssutredninger for Barentshavet Syd, Georg Oppedal, Fiskeridirerktoratet, April, 1987 (Oppedal) pp. 1-9, 16-18, 54, 84 and 93 unless noted otherwise.

72 Hoel s. 35. Another author notes that 4% of the world's annual catch was obtained from the Barents Sea, which was 50% of the total Norwegian and 12% of the total Soviet catch respectively, Østreng Delimitation pp. 133-134.

73 Hoel s. 35 notes the 1990 quotas are the lowest ever.

74 *Ibid.* s. 36 notes that causes of the fisheries crisis in the Barents Sea are related to administrative problems of the Norwegian - Soviet Fisheries Commission, biological conditions of the capelin, heavy fishing from trawlers for the past 30 - 40 years, natural mortality, illegal fishing and the illegal dumping of small fish.

75 *Ibid.* The Soviets want small size mesh since they have mostly young, smaller fish in their areas, while the Norwegians prefer larger mesh to allow more rational fisheries management through the capture of larger fish.

76 See also Oppedal, Table R1, Tabellvedlegg Reketrålfiske, showing the main geographic areas, the total catch of shrimp (for 1983) and the percentage of total catch per area.

77 For an idea of the total catch of capelin made in 1984 per area in tons, see Oppedal, Table L3 Tabellvedlegg Loddefiske.

78 Capelin fishing in the Barents Sea was to be opened again to a limited extent in January 1991. Nordlys 3 desember 1990, Nr.280, 89.årgang. s. 5.

79 Hoel s. 40-41 notes that in 1989 17 animals were killed for research purposes. The political dynamics of Norwegian whaling as related to a threatened US boycott of Norwegian fish products in US markets, estimated to affect 10% of total Norwegian fish exports, are additionally noted. Most important in the author's opinion however is the whales' (and seals') annual consumption of commercial fish estimated to run several times the yearly Norwegian catch.

80 *Ibid.*

81 *Ibid.*

82 Obtained from Figur 0, Oppedal s. 9. Areas are designated which are important for cod, haddock, shrimp, and capelin caught in the winter season. The lesser fisheries will be covered below.

83 Both Figures 14 and 15 were obtained from Charts Nos. I and II Oppedal s. 16.

84 See generally Engelhardt pp. 1-31. See also *Our Common Future, The World Commission on Environment and Development*, Oxford University Press (1987) (Our Future) p. 155 where it is noted that environmental considerations (including conservation of species) is justified aesthetically, ethically, culturally, and scientifically, but for those who want it justified exclusively economically, 'the economic values inherent in the genetic material of species are alone enough to justify species preservation.'

85 Obtained from Engelhardt p. 1. from *Random House Dictionary*.

86 See for example Hakapää pp. 220-221. The Barents Sea actually is classified by Engelhardt p. 1 as 'sub Arctic' as opposed to an Arctic zone proper since it is a region of mixed water types. For the purposes of this work however, the word 'Arctic' is meant to include the Barents Sea.

87 Engelhardt p. 27. The author gives as reasons that individual species numbers are large and the same ecosystem extends over very large geographic areas allowing damage in one area to be repaired by immigration from adjacent areas. Although the author is referring to 'Arctic' in its narrowest sense, it can probably be assumed that the same idea may be applied to the Barents Sea, due to the large geographic areas and the large number of individual numbers within the species. It is noted further, '(T)his of course is not to say that there is no danger of pollution in the Arctic; far from it. It means simply that the dangers of pollution are of the same sort and magnitude as elsewhere. Ecosystems are resilient. Extinction of species has been very frequent in the past, without the help of mankind. If species are lost, the ecosystem heals over and a new balance is achieved. The loser, today, is mankind.'

88 Although as noted most of the available studies are restricted to 'oil', which will be followed here, similar effects could be expected for other hazardous chemicals.

89 Obtained from Figur 9, Oppedal s. 93.

90 Aftenposten 2. oktober 1987, Uke 40., Nr.454., 128 årg., s. 8. See Part II Section 10.1.2.1. for more information concerning land based pollution related to the Arctic generally.

91 Roginko pp. 16-17. See also Aftenposten, 2 oktober 1987, Uke 40, Nr.454, 128 årg. s. 8. Aftenposten, 8 mars, 1988, Uke 10, Nr.113, 129 årg., s. 49, and Aftenposten, 9 mars, 1988, Uke 10 Nr.114, 129 årg., s. 14. To date the airborne pollution has been heavy metals and sulphur dioxide transmitted as dust through the atmosphere.

92 Nordlys, onsdag 18 mai 1988, nr. 113, 87 årgang.

93 See Part II Section 10.2.1. and Harders Marine p. 286.

94 See Aftenposten 5. oktober, Uke 42. Nr.470., 131.årg., s.4, and Aftenposten 22 oktober, Uke 43., Nr.482., 131.årg., s. 12. Nordlys 13. november 1991 Nr.263, 09.årgang, s. 19 notes that a storage system is planned for Nova Zemlya for atomic waste produced by the over 200 atomic reactors on the Kola Peninsula. The storage system will take atomic waste from the Northern Fleet, the ice breakers and atomic power stations.

95 Obtained from Fig. 2. Proceedings of the Sixth Conference of the Comité Arctique International, 13-15 May 1985, Ed. Rey, L. and V. Alexander (1989), p. 330.

96 Obtained from Tab. 7.1. Anker-Nilssen p. 34.

97 See Part II Section 10.1.2.1. for more information concerning vessel-source pollution related to the Arctic generally.

98 Norwegians plan on using a buoy loading system to tankers (as the Soviets at Kolgujev Island) if Barents Sea oil fields are small. Undersea pipelines to land will be built if the fields discovered are large. Conversation with Consultant D. Nygaard, Department of Oil and Energy, July 27, 1987. Consequences involved with the tanker alternative can be seen by the EXXON VALDEZ accident ran aground in the Prince William Sound. See Aftenposten 28.mars 1989. Uke 13. Nr.139. 130.årg. s. 3. Although Nygaard indicated use of *super* tankers is not expected, interesting points in the newspaper article allow for speculation concerning possible causes and consequences of such tanker accidents in the Barents Sea. See Part I Section 1.3.2.

99 Veritec Report No. 84-3372, juli 1984, 'Sannsynligheten for oljesøl ved tankskips-uhell og ved kontinentalsokkelvirksomhet (blowouts) i norsk farvann i perioden 1985-2000', (Veritec) pp. 19-21. For the same geographic area the estimated accidental discharge per year in 1984 was *.17*, and in 1979 was *.21*.

100 *Ibid.* p. 33. See Part I Section 5.2.3. for MARPOL 73/78.

101 Veritec p. 33.

102 *Ibid.* p. 21.

103 There is alleged dumping of atomic wastes off Novaya Zemlya and of military weapons such as mustard and nerve gas east from Murmansk, Aftenposten 24 desember 1990, Uke 52., Nr.510., 131.årg., s. 10.

104 See Churchill Conflict pp. 204-205. In the future the removal of offshore installations will become important due to the concentrated naval ship traffic as well as the expected commercial ship traffic in the Barents Sea.

105 See Part II Section 10.1.2.1. for more information concerning offshore activities as related to the Arctic generally.

106 St.meld. nr. 79, p. 11. See Section 8.2., Figures 8, 9, and 10. The figure could be lower due to the future use of subsea wells in the Barents Sea. However as noted in Part I Section 1.3.4. completed subsea wells contribute to a higher pollution risk by a factor of 10 to 20 times.

107 *Ibid.* Usually gas blowouts do little environmental damage, although there is a danger to personnel and equipment due to ignition and explosion, see Part I Section 1.3.4.1.

108 *Ibid.*

109 *Ibid.* As seen in Section 8.2. Figure 10 the number of exploratory and relief wells drilled in the Barents Sea has been well under the 50 well figure used, averaging approximately 5 per year since 1980.

110 St.meld. nr. 79, p. 11.

111 Veritec p. 23.

112 *Ibid.*

113 *Ibid.*

114 Obtained from Tab. 4.1. Anker-Nilssen p. 21 based upon data from Veritec.

115 Nordlys 12. september 1991 Nr.210, 90.årgang, s. 26 notes that Soviet and Norwegian researchers have formed an agreement to establish pollution levels in the Barents Sea emanating from industry, rivers, atomic tests, military dumping and the Gulf Stream.

9. Scope of International Marine Environmental Law Governing the Barents Sea and the Wider Arctic Region

Though it can generally be deduced from Part I Chapters 2 - 7 which international customary and general principles of law and conventions and agreements govern in the Barents Sea and wider Arctic, it was considered beneficial to provide a brief yet specific outline here. From this it hopefully can be easier to keep this law in perspective while discussing the legal deficiencies in Section 9.3.

9.1. Customary International Law and General Principles

The 1982 LOSC[1] provisions apply to the Barents Sea and the Arctic to the extent they represent international customary law which as seen in Part I Section 2.1.4. is controversial. The progressive provisions are discussed below in Part II Chapters 10 - 17 as related to each pollution source, contingency plans and State liability, and provide a framework and support if not a strictly legal foundation to the formation of an Arctic regional agreement with subordinate Protocols. This support would as well be valid to the same extent for any bilateral agreements entered into by the Soviet Union and Norway in the Barents Sea.

As seen although it is generally acknowledged that a State has responsibility under treaty, custom or other for an act or omission which results in a breach of a legal obligation, still States seem reluctant to either make or allow claims for pollution damage based upon solely custom.[2] Thus it seems at this point *unlikely* that international customary law as related to liability will play much of a role in either the Barents Sea or the Arctic. In addition the Soviets do not utilize the ICJ and the US did but has recently withdrawn; so attempting to bring a liability dispute under this Court would probably be fruitless.[3] The Soviets did acknowledge liability and paid compensation, arguably limited to conventional coverage, for damage done by their Satellite to Canadian territory, as well as by a Soviet submarine to Norwegian fishing equipment.[4] Norway on the other hand seems *reluctant* to demand compensation for damages from the Soviets with the exception of damaged fishing gear. To the author's knowledge no Norwegian claim has ever been raised for damages suffered from radioactive particles associated with the Chernobyl accident or for damages in Northern Norway from sulphur dioxide and heavy metals from the Soviet industrial cities on the Kola Peninsula noted in Part II Section 8.5. The US demanded and Canada payed compensation for pollution damage in the *Trail Smelter* arbitration. Denmark/Greenland's history of compensation demands or payments for environmental damage is not known. Thus in the Barents Sea and the Arctic it would be anybody's guess, lacking a direct liability *convention*, whether compensation would be demanded and paid for damage caused by transboundary pollution.

The various States' legislation, Court decisions and declarations possibly affecting international customary environmental law should be analyzed, but as noted this is far beyond the scope of this book.[5] However since Arctic States' national law has relevance to the Barents Sea case study presented, very briefly the Arctic States' legislation and declarations possibly affecting international customary law are as follows. See also Section 9.2.3.1. for Arctic State implementation of Part I shipping conventions.

Soviet State practice is considered by some as being an important *source* of customary law.[6] Soviet legislation covering pollution from vessels and from dumping follows *closely* the LOSC (though not totally), and can fit into the definitions given above required for the formation of a norm in customary law.[7] Also considered a key factor is the Soviet status as a super power. In practice, 'great powers will have more influence than others on the formation of customary rule.'[8] In addition the Soviet Union also is a leading maritime State which, 'is in a position to disrupt the desired stability of expectations by military, economic or political means.'[9]

The arguments made relating State super power and maritime power status to the formation of customary international law could as well be made for the US The U.S., though more environmentally conservative than probably both Canada and the Soviet Union, for its part has acknowledged in a Presidential Proclamation that a State's jurisdiction and responsibility over the natural resources, environmental control and conservation in the exclusive economic zone are almost *identical* to LOSC provisions.[10] It was indicated in the Presidential Proclamation that except for LOSC provisions dealing with deep sea mining, the LOSC, 'generally confirms existing maritime law and practice' and the US 'will recognize those rights and interests' consistent with the LOSC.[11] Noted further is that the US as well as other States argue that 'the Convention does not so much create positive law in the non-sea bed areas as simply incorporate existing law that will continue to be applicable to all States, not because of the treaty, but because of the customary law underlying the treaty.'[12] Further '(T)here exists an international law of the sea totally independent of the Law of the Sea Convention. And, indeed, the first evidence for the existence of such a body of law lies in the fact that today virtually all countries are applying rules of international laws reflected in the Treaty text even though the Treaty is not in effect.'[13] Finally, as noted previously, even if LOSC provisions are not automatically accepted as customary international law, they are recognized as the best available evidence of customary law.[14]

As regards Canada, Denmark/Greenland and Norway not much analysis has been conducted regarding their effect on the formation of international customary law. It seems however that they generally follow other States including the superpowers as regards their relationship to the LOSC.[15]

Thus, it would be anybody's guess concerning the coverage in the Barents Sea and the Arctic offered by customary law and general principles of law for all the legal areas associated with transboundary pollution control, dumping, land-based, offshore installation, vessel-source and contingency plans. Due to the vagueness and complexity of the principals involved, effective coverage will have to be achieved through conventions.[16]

9.2. Central Conventions with Application to the Barents Sea and the Arctic

9.2.1. Dumping

The LDC[17] covers deliberate marine dumping of waste substances, is a global agreement and is ratified by both Norway and the Soviet Union. It is applicable in the Barents Sea. but does not include dumping related to offshore oil activities. It as well covers the entire Arctic region, and Canada, Denmark and the US are Parties.[18]

The ODC[19] is a regional agreement which regulates the marine dumping of harmful substances. Its geographical jurisdiction includes the Barents Sea, as well as sections of the Greenland Sea, the Arctic Ocean and the Norwegian Sea, comprising in total approximately 25% of the Arctic area up to the North Pole. The ODC vaguely includes dumping activities related to offshore oil activities. Norway and Denmark have ratified this Convention, and the Soviet Union is expected to ratify.

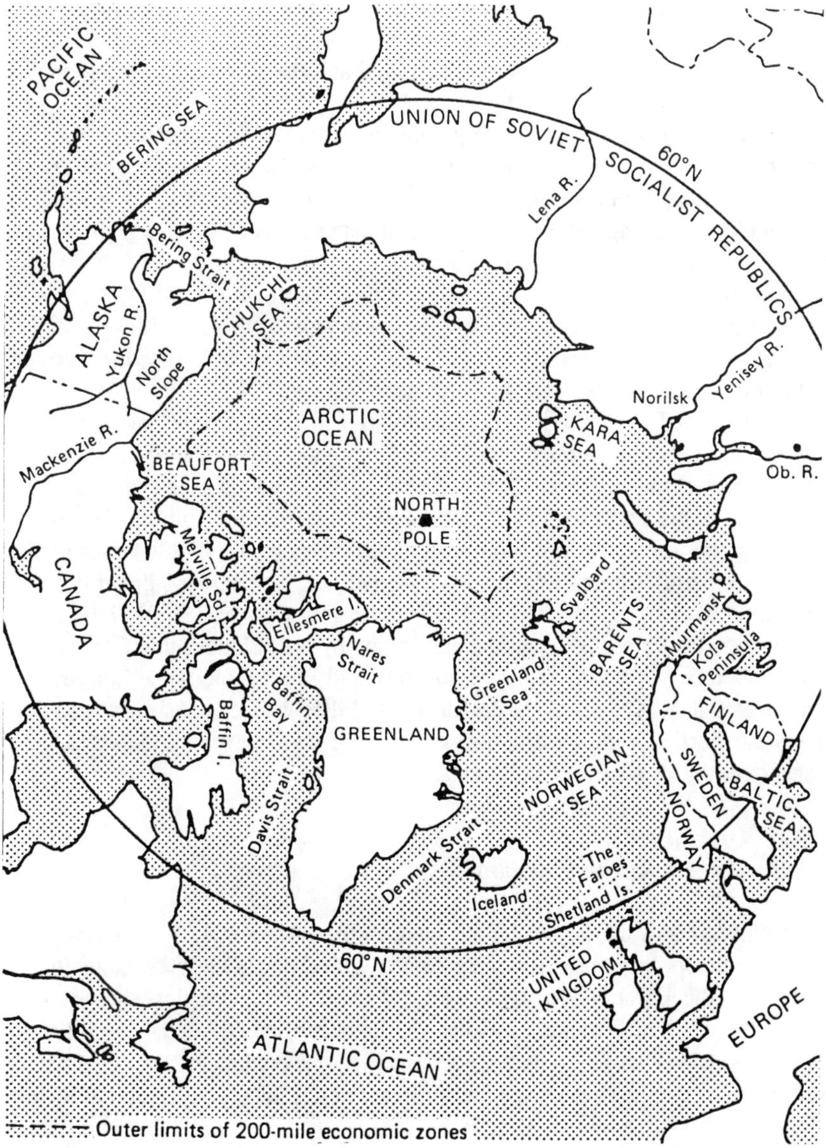

Figure 9.1. The Arctic Ocean.[20]

9.2.2. Land-based, Offshore Installation and Airborne Pollution

The 1958 Continental Shelf Convention[21] is a global convention ratified by Norway and the Soviet Union and is applicable in the Barents Sea.[22] It is as well ratified by Canada, Denmark/Greenland and the US and applicable in the Arctic.

The Paris Convention[23] is a regional agreement which provides for the prohibition or regulation of land-based marine pollution and airborne pollution.[24] Its geographical application is practically the same as the Oslo Convention including the Barents Sea, sections of the Greenland Sea, the Arctic Ocean and the Norwegian Sea, comprising approximately 25% of the Arctic area up to the Pole. Norway and Denmark have ratified, and the Soviet Union is expected to ratify. Marine pollution arising from offshore drilling installations as well as refineries, reception facilities and other land-based sources of pollution is covered.[25]

The 1979 ECE[26] is a regional convention with State Parties including the Eastern and Western European countries, the Soviet Union, the US and Canada. All the Arctic rim States are Parties. The Conventional coverage comprises all areas within the national jurisdiction of the State Parties, thus in the Arctic including the territorial seas and possibly the exclusive economic zones of the Arctic rim States. Some 21 States have pledged to reduce their sulphur dioxide emissions by 30% from 1980 to 1993 under this Convention. Various other States have also pledged to reduce emissions by 50%. Under this Convention the Cooperative Programme for Monitoring and Evaluation of the Long-range Transmission of Air Pollutants in Europe (EMEP) was established to provide the State Parties with information concerning concentrations, deposits, and transboundary flows of air pollutants.

9.2.3. Vessel-source Pollution

Both Norway and the Soviet Union have ratified MARPOL 73/78[27] covering vessel-source pollution with various Annexes, and this is the main Convention governing the Barents Sea. It as well covers the Arctic region since it is a global convention. Denmark and the US have ratified MARPOL 73/78 with various Annexes, while Canada has *not* ratified.[28] Denmark has however made a reservation that its accession is until further notice subject to reservation with regard to the obligations of Greenland.

OILPOL and its Protocol[29] together with the 1958 Geneva Conventions[30] generally require State Parties to prevent oil pollution of the high seas resulting from navigation as well as from oil activities on the continental shelf. The

States may also regulate pollution in the contiguous zone bordering territorial waters, but no specific pollution control is established. All of these Conventions are global and apply to both the Barents Sea and the Arctic. MARPOL 73/78 has superseded OILPOL for all Arctic States but Canada and Denmark/Greenland (since Denmark/Greenland acceded subject to reservation). All the Arctic States are bound by the 1958 Geneva Conventions either through ratification or through these Convention's status as international customary law.

The Intervention Convention and its Amendment[31] allow coastal States the right to take 'reasonably necessary steps' on the high seas to combat pollution or pollution threats resulting from marine incidents. This is a global Convention and applies to both the Barents Sea and the Arctic. All the Arctic States are Parties to the Convention but Canada, and it would be expected that these countries might put these provisions to use to protect among other interests their fisheries. All the Arctic rim States but Canada are also Parties to the Protocol.

SOLAS[32] dealing with safety provisions including fire extinguishing, radar and steering systems for ships is a global Convention with application to both the Barents Sea and the Arctic. All Arctic rim States have ratified this Convention, and all of the Arctic rim States have ratified the Protocol but Canada.

INMARSAT[33] contributes to distress and safety of life at sea communications and radio determination capabilities. These systems have relevance and coverage of both the Barents Sea and the Arctic. INMARSAT is ratified by all the Arctic rim States.

STCWS[34] deals with basic standards of crew training and ship operation including in ice conditions. It is global and applies to the Barents Sea and the Arctic. All the Arctic rim States are Parties with the exception of the U.S.

The Load Lines Convention[35] dealing with the overloading of ships, is a global Convention and with application in both the Barents Sea and the Arctic. Parties include all the Arctic rim States.

The Collisions Convention[36] dealing with ship separation zones, is a global Convention with application to both the Barents Sea and the Arctic. Parties include all the Arctic rim States. The relevance to the Arctic is due to the naval fleets, fishing boats, commercial vessels in the projected Northeast and Northwest Passages and offshore production resulting in congested seas in some areas.

The Containers Convention[37] dealing with the safe handling and shipping of containers, is a global Convention with application to both the Barents Sea and the Arctic. Parties include all the Arctic rim States.

See Part I Section 5.4.4. for several IMO and ILO Conventions with indirect environmental relevance to the Barents Sea and the Arctic. These deal with such concerns as minimum crewing standards, wages, hours of work, accommodations, continuity of employment and dangerous goods.

9.2.3.1. Brief Summary of Each Arctic Rim State's Ratification of Main Shipping Conventions and Implementation into National Legislation

Since vessel-source pollution agreements are complicated a brief and easily accessible description of each Arctic rim State's ratification of anti pollution shipping Conventions and the relevant national legislation will be presented.[38] Most of the Arctic States are in addition large shipping powers, and therefore their legislation will govern a substantial tonnage of the world's oil tanker fleet and freighters through their flag ship jurisdiction.[39]

Norway has a territorial sea of 4 miles with the corresponding jurisdictional rights and as well claims a 200 mile exclusive economic zone.[40] Norway has ratified both OILPOL and MARPOL 73/78, with all Annexes of the latter but Annex IV. MARPOL 73/78 was substantially incorporated into Norwegian law through the Pollution Control Act of 13 March 1981.[41] Norway has ratified the CLC, the Fund Convention, the Intervention Convention and Protocol, the STCWS and the Port State Control Memorandum 1982. Pollution damage liability is listed as 'almost strict,' and those potentially liable include the vessel, master and owner-operator but not the cargo.[42] Polluting substances include oil and other substances. Sanctions include fines, with imprisonment of up to 3 months for serious violations. Higher prison terms are possible in aggravated circumstances and those involving danger to human life or health. Companies may be additionally fined on a strict liability basis. A certificate evidencing financial responsibility is required.

The Soviet Union claims a 12 mile territorial sea as well as a 200 mile exclusive economic zone.[43] It has ratified both OILPOL and MARPOL 73/78, including all the Annexes of the latter, the CLC, the Intervention Convention, the STCWS, but *not* the Fund Convention nor the Port State Control Memorandum 1982. Liability for pollution damage is strict, with the vessel, master and owner-operator being held potentially liable. Polluting substances include oil and other substances, and a certificate evidencing financial responsibility is required. Sanctions for violations are emphasized as being *very* high, up to SUR 25,000 and imprisonment up to 2 years for discharging any pollution substance into Soviet waters.[44] Regulations are strictly enforced.[45] In several cases Soviet authorities have made very high claims for resource damage and for expenses in *restoring* waters to their clean condition.[46]

Canada claims a 12 mile territorial sea and a 200 mile fisheries zone. Arctic water are a special zone. Canada has ratified OILPOL and the STCWS. Canada has *not* ratified MARPOL 73/78, the CLC, the Fund Convention the Port State Control Memorandum 1982 nor the Intervention Convention. Under Canadian federal law[47] liability is 'almost strict', and the vessel's master or owner-operator are liable for pollution damage. Pollution substances include oil and other substances. Sanctions include severe fines though not imprisonment, and evidence of financial responsibility is through contributions to a national fund. Enforcement is strict. Legislation listed is divided depending upon latitude. South of Latitude 60 degrees N, both the Canada Shipping Act. Pt. XX and Regulations, and Fisheries Act and Regulations apply. *North* of 60 degrees N the Arctic Waters Pollution Prevention Act and Regulations apply.[48] The Regulations include the Arctic Shipping Pollution Prevention Regulations, the NORDREG Canada-Arctic Traffic System Rules. Canadian Arctic Waters are defined north of 60 degrees North to be out to 100 miles from the coast, except between Canada and Greenland where the boundary is the median line. The Arctic waters are divided into 16 zones according to ice hazard to shipping, classified by the Shipping Safety Control Order. Discharging *all* polluting substances is prohibited in these waters, and discharges observed, accidental or deliberate must be reported to NORDREG through the Canadian Coast Guard. An 'Arctic Pollution Prevention Certificate' is required as is evidence of financial responsibility, a 'Declaration'[49] and a 'Special Policy Endorsement' to the Certificate of Entry from the insurer to the owner. Due to the complexity of the law, the severity of the fines and the strict enforcement, one author notes it is, 'advisable that owners seek P. & I. club advice well in advance to trading in this area.'[50]

The US has recently expanded its territorial sea from 3 nautical miles to 12 and recognizes a 200 mile exclusive economic zone.[51] It has ratified both OILPOL and MARPOL 73/78, the Intervention Convention, but *not* the CLC, the Fund Convention, the STCWS or the Port State Control Memorandum 1982. Federal legislation listed includes in addition to the Clean Water Act and the Deepwater Port Act noted above, the Act to Prevent Pollution from Ships, the Trans-Alaska Pipeline Authorization Act, the Port and Tanker Safety Act, and the Refuse Act, the Marine Protection, Research and Sanctuaries Act.[52] There is 'almost strict' liability, and the vessel's master or owner-operator are liable for pollution damage. Polluting substances covered are oil and other substances. Fines and not imprisonment are allowable sanctions. Generally standards set are at least equal to MARPOL 73/78. Administration of the federal law is by the US Coast Guard (USCG) and the Environmental Protection Agency (EPA). The USCG has the authority to inspect ships, among other things for compliance of safety standards and ensure anti pollution equipment is on board, and check oil record books and make arrests. A certificate of financial responsibility, showing insurance, surety bond or self insurance, is required of ships using US waters or ports over 300 grt. This amounts to USD 150 per grt or USD 250,000 for ships with oil or hazardous substances for cargo, which ever amount is greater. Even more is required for ships coming in under the applicable provisions of the Deepwater Port Act or the Trans-Alaska Pipeline Authorization Act. Alaskan State legislation includes the Trans-Alaskan Pipeline Authorization Act and the Alaska Water Pollution Control Act. Liability is unlimited and fines of up to USD 100,000,000 and imprisonment up to 12 months are possible sanctions. Generally Alaska's pollution jurisdiction, type of legislation, notification requirements and evidence of financial responsibility are different or additional to Federal requirements.

Denmark/Greenland claims a 3 mile territorial sea, and a 200 mile fisheries zones.[53] Denmark has ratified OILPOL, MARPOL 73/78 (with reservation), the CLC, the Fund Convention, the Intervention Convention, the STCWS and the Port State Control Memorandum 1982. Liability is limited, and the vessel's master or owner-operator are liable. Polluting substances include oil and other substances. Fines and imprisonment up to 12 months are possible sanctions. A certificate is necessary to show financial responsibility. Relevant legislation listed includes Acts 130 (1980), 134 (1978), 324 (1975) and 290 (1972). There is 'almost strict' liability, and the vessel's master or owner-operator are liable for pollution damage. Polluting substances covered are oil and other substances. Fines and not imprisonment are allowable sanctions. Generally standards set are at least equal to MARPOL 73/78.

9.2.4. Liability and Compensation Regimes

Regarding dumping the only coverage afforded both the Barents Sea and the Arctic is Article X of the LDC which merely requires the Parties to undertake to develop rules and procedures in accordance with the principles of international law for the assessment of State liability for environmental damage.[54]

Regarding land-based pollution the only Convention dealing with liability for such is the Nordic Convention, and the only Arctic rim State a Party to the Nordic Convention is Norway. Thus this Convention has no effective application to the Barents Sea or the Arctic as such.

There is no liability agreement between Norway and the USSR covering polluting incidents from oil rigs causing damage in the Barents Sea. Likewise for the Arctic there is no liability agreement covering polluting incidents from oil rigs causing damage. There perhaps can be found private oil company arrangements in use on the US and Canadian Arctic continental shelves similar to OPOL[55] on the British North Sea shelf, although this in unconfirmed.[56]

With regard to shipping the following international global liability conventions are in force and are applicable in both the Barents Sea and the Arctic: the CLC,[57] where Norway, the Soviet Union and Denmark are Parties, yet neither the US nor Canada are; and the Fund Convention[58] where Norway, the Soviet Union and Denmark are Parties, yet neither the US nor Canada are. For instances where the CLC or the Fund Convention do not apply, the private regimes established by tanker owners or oil companies to limit their liability, TOVALOP, CRISTAL and PLATO apply.[59] There exist additionally the LLMC and the 1957 *International Convention Relating to the Limitation of the Liability of Owners of Sea-going Ships* covering pollution damage caused by substances *other* than oil or radioactive matter. Norway and Denmark are the only 2 Arctic States members of the LLMC, and there are only an additional 14 Contracting States. The IMO and many States are interested in establishing the HNS Draft Convention despite widespread disagreement and the existence of these 2 Conventions.[60]

In summary what this means in the Arctic is that liability and compensation for damage from marine pollution from *other* sources than shipping depend wholly upon *national* legislation covering areas under their respective jurisdictions. In the *Barents Sea* briefly the Norwegian provisions seem to be a good set of rules preferable both to OPOL and the 1976 Liability Convention type systems due foremost to *unlimited strict liability*, and a possibility for claims for loss and damage to fishing grounds.[61] The channelization of responsibility to the licensee seems to be an advantage as long as the licensee is solvent. This is probably no problem due to the economic size of the oil companies, but from a claimant's point of view joint and several liability covering operators, licensees and subcontractors would be better, since full claims could be brought against companies more easily reached and most solvent. The requirements of compulsory insurance and it not being necessary for the incident doing damage to be located on the Norwegian continental shelf are additional positive features. Also the Norwegian legislation has been in use in the North Sea and can therefore more easily be put in use in the Barents Sea. Regarding Soviet provisions an idea can be obtained of the relevant legislation from Appendix I Sections F.2., J.3., Supplement to J.4. dealing with compensation.[62] Unfortunately due to the vagueness of the Decrees not enough can be said regarding Soviet marine liability legislation to make a judgment concerning the adequacy of the coverage of the Barents Sea. For the rest of the *Arctic* not enough is known to make a judgment concerning adequacy of coverage of liability and compensation for damage from marine pollution from sources other than shipping by Danish, Canadian and U.S./Alaskan national legislation.

9.2.5. Joint Contingency Plans

Presently there is a draft joint contingency agreement for dealing with pollution incidents in the Barents Sea under examination by the Norwegian Ministry of the Environment. Norway was not expected to enter a contingency plan prior to the establishment of a definite marine boundary line between Norway and the Soviet Union in the Barents Sea. Meetings are also taking place between the Soviet Union and Norway under the bilateral, Norwegian - Soviet Agreement wherein marine pollution is under discussion.[63] The Copenhagen Agreement[64] applies in the Barents Sea but is limited to Scandinavian cooperation. Under it Norway could request assistance from the other Scandinavian countries to take anti oil pollution action in its territorial sea if feasible and necessary. The Bonn Agreement[65] and NORBRITPLAN,[66] instruments applying in the North Sea, do not apply to the Barents Sea but could serve as models in forming the future joint Norwegian-Soviet contingency *plan*. IMO Contingency,[67] urging States to establish both national and regional oil contingency plans, will not have application to the Barents Sea and the Arctic until it is comes into force

In the Arctic in the Beaufort Sea a contingency plan has been entered into between the US and Canada.[68] An agreement and contingency plan has been entered into between the US and the USSR in the Bering and Chukchi Seas,[69] and a joint contingency plan has been established between Canada and Denmark/Greenland in the Baffin Bay/Davis Strait area which allows the other country access to its own waters for pollution prevention operations.[70] All of these contingency plans have been entered into in spite of marine boundary conflicts, some of them similar to the Norwegian - Soviet conflict in the Barents Sea.[71]

9.2.6. Bilateral Arctic Agreements

In addition to the Conventions and Agreements mentioned above there exist several cooperative bilateral agreements touching upon environmental issues between the Arctic rim States.[72]

The United States and the Soviet Union entered into the 1972 US - USSR Agreement[73] and the 1973 US - USSR Ocean.[74] These cover among other things co-operation in studying the Arctic ecological systems and in exchanging information on marine pollution prevention, including ship design, traffic control, shore facilities and offshore oil drilling safeguards. Also included is a joint study of legal measures for environmental protection in the Bering Sea.[75] Canada and the USSR in 1984 entered the Canadian - Soviet Technical

Agreement[76] based upon an earlier agreement of 1971. Activities include developing strategies, regulations and standards to preserve quality of the northern environment and developing improved methods of monitoring environmental pollution in Arctic oil and gas fields.[77] Additionally, the Soviet Union has recently approached Canada with a proposal for a general environmental agreement.[78]

The agreements between the USSR and Norway include the aforementioned Norwegian - Soviet Agreement, and the 1988 Norwegian - Soviet Technology Agreement.[79] Other Norwegian - Soviet bilaterals include the 1988 Rescue at Sea Agreement,[80] the 1988 USSR - Norway Nuclear Notification Agreement,[81] and the USSR - Norway Agreement on Credit[82] These are all mentioned here due to their relevance to Norwegian policies in the North mentioned in Part II Section 8.3.

The Canadian - Danish Agreement in addition to providing for the joint contingency plan mentioned makes commitments to protect the Baffin Bay/Davis Strait region from marine pollution, to undertake joint research programs, develop pollution measure methods, identify and monitor ship traffic and protect the economic and social welfare of inhabitants from oil pollution risks.[83] The exchange of scientific information, co-operation on vessel information and vessel traffic management, compensation and access to the other country's waters for pollution operations are also covered.[84]

Finally, between the US and Canada there exist several bilaterals including the 1988 US - Canada Agreement on Arctic Cooperation (US Canadian Arctic) which covers cooperation in navigation by icebreakers in the Arctic, in understanding and avoiding disturbing the Arctic environment; the 1985 Addendum to the 1970 Memorandum of Understanding, between the US Department of Transportation and Transport Canada (US - Canadian Transport), providing for cooperation in developing Arctic ship designs, regulations, ice navigation procedures and safety of Arctic ship operations; and the 1982 Memorandum of Understanding between the US Coast Guard and the Canadian Department of Environment (US - Canadian Understanding), covering cooperation in pollution control research, including the behaviour of oil spilled in the Arctic and the development of pollution response equipment.[85]

9.2.7. Other Environmental Conventions with Application to the Barents Sea and the Wider Arctic Region

Finally a regional agreement which all 5 Arctic nations adopted on the initiative of the Soviet Union is the 1973 Polar Bear Agreement.[86] This prohibits the taking of polar bears, except by local people and for scientific and conservation purposes and is considered the major *precedent* for Arctic regional environmental cooperation.[87]

9.3. *Inadequacy* of Present Legal Coverage in both the Barents Sea and the Wider Arctic Region

This Section is probably the most important of the entire work and should be represented by at least a Chapter. It would be desirable to be able to present a *comprehensive* critique both of a legal and practical nature of international environmental coverage. Unfortunately probably due to the infancy of marine environmental protection, little *critical* information exists. Consequently, that what is presented, in addition to obvious geographical and definitional deficiencies by which large geographic as well as legal areas are not even encompassed by agreements, are general observations made by legal experts as well as the author. Where more specific information is available, it will be presented. This has only noticeable application for shipping, where an in depth summary of legal and practical problems are presented, dealing chiefly with MARPOL 73/78. The depth of this presentation perhaps reflects the *lack* of interest and experience accorded control of the other pollution sources. This is an area needing much additional 'in the field' research.

9.3.1. Offshore Platforms, Dumping, Land-based and Airborne Pollution[88]

In the *Barents Sea* until just recently the problem of coverage under the Paris and Oslo Conventions of offshore oil platforms, land-based pollution and dumping was *acute*, because only the Norwegian side was encompassed.[89] There was no indication that either the member States nor the Soviet Union were interested in the Soviets becoming members of these Conventions, and consequently anti pollution agreements covering these sources were required. If and when the Soviets accede to the Paris Convention and the ODC, only a liability agreement and an in force contingency plan will be lacking in terms of pollution coverage in the Barents Sea.[90]

Although the coverage offered by the Paris Convention and the ODC is certainly better than nothing and the experience gained is very useful,[91] it is still felt that the Barents Sea could be better served by completely new agreements covering these pollution sources, chiefly because of political conflicts. Briefly an idea of these problems can be obtained from a 1987 Norwegian report, and although this report concerns only the Paris Convention specifically, generally the results noted can possibly be extrapolated to other international pollution regimes as well including the ODC and the LDC.[92]

For an outsider it is difficult to obtain an idea as to the effectiveness of the Paris Convention in practice. The Paris Commission meetings are closed and only official representatives are allowed to be present.[93] The Commission's files also are closed, and reports of meetings do not refer to the individual Parties involved in discussions, following the so-called 'principle of anonymity.'[94] In addition the Par.Comm. reports covers activities dated almost 6 to 18 months previously and are made public even later, indicating a definite time lag in available information.[95] The Paris Commission is characterized by arriving at few binding Resolutions. The ability for a country to put forth its preferences is dependent upon the delegation's ability to present the case and present documentation as well as the administrative effort and resources the country chooses to allocate to the Commission. Another point is the EEC's apparent and frequent *blockage* of Commission Resolutions, and the consequent limitation placed on the Commission's ability to adopt Resolutions. A special point is the ranking of the member countries in terms of level of activity in the Paris Commission related to interest in stronger environmental regulation. On top both in terms of activity and interest are Sweden and the Netherlands. The Federal Republic of Germany, Denmark and Norway have a preference for stronger regulations, but are more passive.[96] The UK is very active, but prefers weaker controls. France and Belgium are passive and favour weaker controls. A proposal for making the Commission more effective includes increasing the national administrative resources, together with sending delegates from a higher administrative level.[97] Additionally due to the problems experienced with the EEC, a suggestion has been made that environmental considerations be given a more prominent place in EEC deliberations, which could possibly allow more binding Resolutions to be adopted in the Paris Commission.

In addition to this one author notes that in recent years much environmental experience has been gained from offshore installation blowouts and other polluting incidents yet there lacks a sharing both by States and corporations of 'their experience with these problems with a view to assessing the 'state of art' of this sector of marine pollution control.'[98] Further, since many States involved in offshore activities have not yet adopted any regional agreements, 'coastal States with related interests would benefit from concluding a treaty providing, *inter alia*, for updating and permanent adaption to technological progress.'[99]

As seen States are reluctant to be bound in international agreements dealing with land-based pollution because of a variety of problems. These include this source being the most 'national' source of pollution, a lack of uniform approach of regulation, a large variety of different pollutants which differ from country to country, and overall monitoring difficulties.[100] In addition one author notes that though there now lacks a body of knowledge and experience with respect to land-based pollution, this will gradually develop over time, and this needs be set down in global and regional agreements.[101] As related to refineries, which is a site of environmental risks for a coastal State, the same author notes that appropriate environmental standards related to design, construction and operation do not exist at an international level, and this is needed.[102]

For dumping the problems in practice which can be expected for the Barents Sea are illustrated in observations concerning the North Sea.[103] Both the LDC and ODC prohibit without a special permit the dumping of containers, scrap metal and other bulky wastes liable to seriously obstruct fishing or navigation. However, as seen, the disposal of oil field scrap, including empty drums, steel piping, drifting marker buoys, weights, ropes, cables and hawsers related to the exploration and exploitation of seabed mineral resources, would not by definition be covered by the LDC, Article III (c). Similarly under the ODC definition, Article 19 (1)(a) and (2), it could be maintained that the disposal of such scrap is a, 'discharge incidental to or derived from the normal operation', and therefore not constitute dumping.[104] In addition one author notes that the LDC 'seems to have a number of grave deficiencies.'[105] In addition to the sovereign immunity exception noted, several 'grey list' substances should be 'black listed', there is not provision for uniform prescription and enforcement of standards, and the text is weak and ambiguous.[106] Further noted is that there lacks a comprehensive and systematic monitoring of the effects of dumping globally, in spite of the fact some of 'the most dangerous substances on this planet' continue being dumped in the seas.[107]

In the United Kingdom the legislation implementing the LDC and the ODC generally prohibits the dumping of waste from British ships and structures without a license. This legislation is however, 'less relevant than appears at first sight', and has not been very effective in preventing dumping in the British offshore oil sector, due to weak provisions and lack of enforcement.[108] It does not apply to the construction and maintenance of pipelines or installations, although consents given for the laying of pipelines and for the erection of installations usually contain conditions relating to the discharge of debris. The dumping of debris from supply vessels is also a problem, since such dumping can be argued under the Act to be incidental to or derived from the normal operation of the ship, the same as under the ODC. Even when dumping is covered and hence illegal, it largely has gone unpunished because of the 'difficulties of detecting and proving the commission of an offense and because of the slack enforcement of the Act generally.'[109] Unfortunately the way is open for the Soviet Union to follow in the Barents Sea the U.K.'s model of implementation and enforcement of the LDC and the ODC when it is ratified. The actual practices of the Soviet Union regarding dumping in the Barents and Kara Seas are not known, and as noted in Part II Section 8.5. there are serious allegations of Soviet dumping of radioactive materials, war gasses and obsolete military equipment. Unfortunately as noted in Part I Section 3.2.1.5. neither Convention (nor as well as the 1958 High Seas Convention) will prove of much assistance if the dumping is carried out by the military, since under the sovereign immunity exception military and other government activities are exempt.[110] Norway's general legislation prohibiting ocean dumping also was ineffective, when in the late 1970's underwater examinations showed large amounts of debris on the seabed.[111] The Norwegian Government consequently required the relevant licensees to clean up and passed the Royal Decree of October, 26, 1979 which dealt with the dumping of offshore oil field scrap. Debris is now required to be stored in containers and transported to land, record books must be kept and reception facilities must be provided on shore. Licensees can be required to remove any scrap found around installations or on supply vessel routes. In addition the Petroleum Directorate's regulations of September 23, 1981 provide for the cleaning up around abandoned wells and the removal of all structures to a depth of 5 meters below the seabed. Because of the measures taken, 'the problem of debris on the Norwegian continental shelf is a declining one.'[112]

As noted the Oslo Commission of the ODC in 1977 adopted a Recommendation, for cleaning the sea bed of oil field scrap and considered the Recommendation was being respected since the number of compensation claims by fishermen decreased.[113] Thus there may be *partial* observance of ODC regulations as related to offshore oil field dumping. Concerning other types of dumping not much information is available, and in fact only the ODC appears to have information available through annual reports of the Oslo Commission.[114] These show that dumped industrial waste and sewage sludge have declined, dredged spoils remains the same and trace metals occurring in these wastes have also declined.[115] Both a British Parliamentary Committee in 1986 and the Norwegian Ministry of the Environment in 1989 however consider pollution damage resulting from present dumping practices to be minimal in the ODC area of jurisdiction.[116]

The Barents Sea would thus be better served by agreements which do not bring with them the European environmental political problems caused by environmentally conservative countries and organizations which have little or no relation to the Arctic.[117] New agreements also would allow better terms to be negotiated with better environmental coverage to alleviate the problems found to exist under the ODC, the LDC and the Paris Convention, and could as well apply better to Arctic conditions. For platform pollution provisions could include measures regarding safe design and construction and the adequate training of personnel and be directed solely towards platform pollution and not 'tacked on' as part of another agreement as was done under the Paris Convention. The ODC's problem with a vague definition regarding coverage of offshore oilfield activities as well as LDC's omission of offshore oil field activities completely would be avoided by specifically strengthening the definition to include such activities.[118] State obligations for all the agreements could be upgraded to be more mandatory. Phrases could be used such as 'States shall', rather than 'States shall undertake', 'agree to', 'shall endeavour' and 'shall take account of.'[119] Black lists and grey lists could be made more stringent, whether covering land-based, offshore installation or dumping pollution. Possibly this could include deleting 'grey list' substances (allowed upon special permit) altogether and listing all substances as 'black list' substances (totally forbidden).[120] Additionally the sovereign immunity exception allowing all military dumping (and possibly State oil platform discharges as well) something which needs tightening up in a new agreement, though it would be difficult. Also the exceptions allowing the dumping of 'black list' substances by special permit, by incineration, if trace contaminant, and if rendered harmless by the sea could be closed.[121] Other measures needed include having open and reliable information channels regarding the operation of the agreements in practice.[122]

For the wider *Arctic* region, regarding offshore platform pollution and land-based pollution, the problem is more *severe* since as seen above there are no international conventions or agreements in effect, with the exception of the Paris Convention, and until the Soviets ratify this Convention, the only area of the Arctic covered is the Norwegian Barents Sea, parts of the Greenland Sea and the Arctic Ocean to the north of these. Even when the Soviets ratify, approximately 75% of the Arctic remains uncontrolled, and where controlled, the same arguments against the Convention's use as above apply. Regional Arctic agreements covering offshore platform and land-based pollution are clearly needed including better environmental coverage and better application to Arctic conditions. For dumping the LDC is in effect as is the ODC in its limited area, but the same arguments against their use as above apply. A better agreement which clearly covers dumping associated with offshore oil and gas activities and harmful 'grey list' substances has to be negotiated. As well the dumping of wastes might be prohibited on Arctic islands and the European continent as well, if these wastes reach Arctic waters.[123]

As noted airborne pollution in the *Barents Sea* is covered by the Paris Convention Protocol, the ECE Convention on Long-range Transboundary Air Pollution and the bilateral Norwegian - Soviet Agreement. Part of the *Arctic* is as noted before covered by the Paris Convention and hence Protocol when it comes into force. The Arctic areas under jurisdiction of the Arctic rim States are covered by the ECE Convention. While there is progress internationally regarding this source of pollution, problems in both the Barents Sea and the wider Arctic region have included vagueness in the agreements with few controlling concrete measures or regulations and specific emission standards. Where there have been specific emission reductions agreed to, the largest polluters, the U.K., Poland and the US have not consented.[124] Additionally 80 monitoring stations have been established in some 22 ECE countries but have monitored mainly latitudinal and not longitudinal upper air movements; and it also appears that the monitoring needs to be done along the entire perimeter of the Arctic Ocean.[125] This progress could be continued by provision being made in an Arctic agreement for strict compliance by the Arctic States. Not much more is proposed here (under the Arctic Land-based Protocol) both because of the work being accomplished under these Conventions and the fact that many more States than just the Arctic States are involved.

9.3.2. Vessel-source Pollution

Many of the same arguments as above hold in criticism of MARPOL 73/78. It is a global convention and a more regional agreement would probably better serve the Barents Sea and the Arctic, since the States involved would be fewer. In addition the possibility exists for a better agreement being negotiated with better coverage to alleviate the problems existing under MARPOL 73/78. This includes briefly problems of State enforcement practices and of compliance by member States in providing reception facilities for vessels carrying their discharges from port to port.[126]

It is an issue however whether there are more advantages to be gained by forming a new agreement. A global regime through the IMO is in place with correspondingly much more regulation than noted for the other pollution sources. A special area can be declared under MARPOL 73/78 for the Arctic under Annex I (oil), Annex II (noxious liquid substances) and Annex V (garbage) with more stringent rules, including complete discharge prohibition of various substances.[127] Thirdly some experts consider the MARPOL 73/78 rules sufficient for dealing with vessel-source pollution, and what is needed is compliance by the Member Parties.[128] Fourthly, the IMO, the governing body of MARPOL 73/78, is such an established body with much experience that with compliance by the Arctic rim States which are as well shipping nations,[129] this regime with the Arctic as a special area is arguably sufficient to protect the Arctic. Since this pollution source area is the most heavily regulated, it is *doubtful* that shipping nations would want to go outside the established system. Other arguments against a new agreement include MARPOL 73/78 having definite standards followed by shipping nations, including those which have not even signed LOSC (the US and the Federal Republic of Germany), and Article 234 of LOSC governing ice-covered areas would thus be precluded against these States, whereas MARPOL 73/78 regulations would not.[130]

Arguments for *forming* a new agreement, in addition to those mentioned initially, include the following: A new agreement could establish uniform requirements for design and equipment, cargo limits and crewing standards for ships navigating in ice-covered areas within exclusive economic zones; establish uniform and stringent norms and standards for pollutant discharges from ships; coordinate pilotage services, weather and ice forecasts in the Arctic;[131] provide for the determination of the extent of liability for pollution caused by ships of either country within the exclusive economic zone of the other; provide for cooperation and assistance in combatting pollution emergencies and liquidation of their consequences; provide for cooperation in training and proficiency test the Arctic ship personnel; provide for joint use of satellite systems of navigation in the Arctic (TRANSIT, NAVISTAR), communications (INMARSAT), search and rescue (COSPAS - SARSAT); provide for research cooperation in the design, construction and testing of the Arctic ships; and provide for joint research or exchange of data on the effects of pollution on Arctic marine ecosystems.[132] Additionally while MARPOL 73/78 discharge standards for 'special areas' are more stringent than normal MARPOL 73/78 standards, they are *less* stringent than Canadian regulations for shipping in the Arctic.[133] Also MARPOL 73/78 standards for special areas rely on the implementation of existing agreements, and jurisdiction of the coastal state in its territorial sea and of the flag state on the high seas remain unchanged under MARPOL 73/78.[134] Along these lines the specific problems of MARPOL 73/78 in the Arctic can be stated simply in terms of alteration costs, enforcement, verification, military and flag of convenience ships, reception facilities and surveillance. Briefly these entail the following.

9.3.2.1. MARPOL 73/78[135]

In the Arctic exist a large number of *immune military* vessels including the Soviet fleet from Kola, US and Canadian polar submarines as well as other NATO[136] vessels. Military vessels are exempt from the application of MARPOL 73/78 through Article 3 (3) as well as having a general sovereign immunity under customary international law. Denmark, Norway, the Soviet Union and the US by their ratification of MARPOL 73/78 however, 'shall ensure by the adoption of appropriate measures not impairing the operations or operational capabilities of such ships (military) owned or operated by it, that such ships act in a manner consistent, so far as is reasonable and practicable, with the present convention.'[137] It is not known the extent which military ships discharge oil, chemicals, containers and other scrap, sewage or garbage into the Arctic, but it is a factor which should be addressed, due to the numbers of these ships operating in this area.[138]

Under MARPOL 73/78 Annex I, Regulation 12 each Party undertakes to ensure adequate *reception* facilities for residues and oil mixtures from tankers and other ships. This provision has often not been complied with by State Parties, and the lack of adequate and reasonably priced port reception facilities has been one of the main complaints from the shipping industry.[139] It is not known specifically the state of development of Arctic reception facilities, but generally Norway has provided such and the Soviet Union is in the process of doing so.[140]

The cost of *converting* existing tankers or building new tankers with segregated ballast tanks (SBT's), dedicated clean ballast tanks (CBT's), crude oil washing (COW) and inert gas systems (IGS's) and double sides and double bottoms for use in the Arctic is also a problem which States and shipping firms are going to have to come to terms with.[141] Since all Arctic countries but Canada have ratified MARPOL 73/78 Annexes I and II, they are required to ensure that their oil and chemical tankers and other ships comply with convention provisions covering ship structure, equipment and operations, including SBT's, COW's and IGS's. This perhaps is being controlled, but experience indicates otherwise.[142] Proper control by the Arctic States of their flag vessels would benefit the Arctic including the Barents Sea.[143]

Surveillance of illegal discharges in the Arctic will be a difficulty due to the large ocean spaces involved, the hard climate and the presence of polar darkness during parts of the year. This results in problems both for putting contingency plans into operation and in gathering sufficient circumstantial evidence which can be used in inspecting, arresting and bringing proceedings against suspected violators. An alternative used now in the Norwegian North Sea is aircraft surveillance operated by SFT.[144] Since there are presently used civilian satellite systems to monitor Soviet military activities on Kola Peninsula, these and similar systems might also be used to monitor pollution activities in the Barents Sea and the wider Arctic Region.[145]

Flag of convenience ships represent an environmental problem on a world wide basis and can become so for the Arctic. A flag of convenience State is characterized by the following. The country of registry allows ownership and/or control of its merchant vessels by non citizens; access and transfer to the registry is easy; taxes and fees are low; manning by non-nationals is permitted;

the country of registry is small with no national requirement for the registered shipping; national income is substantially increased by registry fees; and the country of registry does not have the power nor the administrative machinery to impose any effective governmental or international regulations.[146]

7 States have been classified as flag of convenience States, Liberia, Panama, Singapore, Cyprus, Somalia, Bermuda and the Bahamas.[147] The two main flag of convenience States in the past have been Liberia and Panama.[148] Cyprus and the Bahamas are presently both open registers, and these four represent approximately *25%* of the world tonnage. Not all of this tonnage represents oil tankers, but oil tankers and bulk/oil carriers comprise 69.7% of the Liberian fleet, 31.8% of the Panamanian fleet and 26.3% of the Cyprus fleet.[149] Beneficial owners of open register fleets have home countries including the US (29.7%),[150] Hong Kong (19.7%), Greece (13.1%), Japan (10.7%), Germany (3.3%), Norway (2.9%), as well as others (20.6%). Of these fleets 36% (Liberian), 58% (Panamanian), 88% (Cyprus) are 10 years old or older.[151] In terms of expected losses of gross tonnage, Panama and Cyprus are listed as *very unsafe*,[152] and Liberia, *relatively unsafe*.[153]

As can be deduced, by lax governmental supervision the ship owner avoids more stringent labour, tax, structural and equipment regulations, and consequently the *cost* of transporting oil and other cargo is reduced. With beneficial owners in the U.S., Hong Kong, Japan, Germany and Norway among others it seems conceivable that these ships could attempt to use the Soviet Northeast Passage if allowed to.[154] What all this means in terms of vessel-source pollution in the Arctic is obvious, and these ships will present a substantial environmental problem. Strict supervision is needed regarding both illegal discharges and the design, construction equipment and manning standards in order to keep these ships out of the Arctic as much as possible.[155]

All State *enforcement* practices under MARPOL 73/78 both of illegal discharges and of construction, design, equipment and manning standards are wanting. Under Article 4 the content of coastal jurisdiction and the geographical extent of this was left to LOSC to decide.[156] The coastal States can potentially inspect foreign ships in their ports, but their competence in out of port waters is vague under Article 4 (2).[157] Flag States under Article 4 while required to supervise the construction, design equipment and manning standards in flag ships, have *no* requirement for inspecting their ships for discharge violations.

Under Article 5 the central and most effective enforcement feature of MARPOL 73/78 is the certificate of a ship's compliance with both technical and operational standards. However, the port State competence to question the certificate is limited as it has to prove 'clear grounds' for inspecting the ship under Article 5(2).[158] Thus an *upgrading* of *port* State inspection of Certificates is required as well as upgrading of detainment procedures by the port States of ships showing substantial non compliance with MARPOL 73/78 standards. As noted previously flag States gain questionable incentive to comply with the MARPOL 73/78 annual survey requirement.[159] An apparent strong stance of US port State authorities however in boarding ships and controlling both certificates and oil record books gave positive results with an 88% rate of compliance.[160]

The ship classification societies, which do the inspections for the States and issue the certificates, sometimes include shipowners whose interests lie in not controlling their ships properly since it costs them additional money.[161] Consequently the quality and reliability surrounding the inspection and the issuance of Certificates can vary, and an emphasis upon State supervision of ship classification societies is needed.

Monitoring devices for discharge verification are not yet developed and available which are reliable and reasonably priced. Easily alterable logbooks are therefore relied upon, and compliance is dependent upon the ship operators being conscientious. Violations are impossible to verify.[162] It seems difficult to imagine that this technology is not yet developed, considering the sophistication of the oil field technology in use in the North Sea.[163] Perhaps the required technology could be developed in Norway as environmental technology and sold on the world market at reasonable prices.[164]

In spite of the oil record books of MARPOL 73/78 Annex I and the cargo record books of Annex II having system flaws in that they are open to fraudulent entries, certified copies are admissible in judicial proceedings as evidence of the facts stated in the entry. The log book system should be coupled to a *inaccessible* oil discharge monitoring and control system, of which certified copies of both be admissible in judicial proceedings.

Under MARPOL 73/78 Article 6(5) a port State may, upon *request* concerning an alleged violation which has occurred anywhere supported by 'sufficient evidence', inspect a ship within its jurisdiction and can issue a report to the flag State and the State Party requesting it. Port authorities do not have authority however to initiate an inspection themselves of the oil or cargo record books,[165] and a port State may *not* bring legal action for violations occurring outside its jurisdiction, which may be the major flaw of MARPOL 73/78.[166] Physical inspections are left mostly to the flag States, which are subject to their conscientiousness. Meeting the 'sufficient evidence' requirement is difficult. The claimant carries the burden of proof and cannot give prima facie evidence and thus shift the burden to the defendant vessel to rebut the claim, and this strongly limits port authority inspections.[167] Even if there exists 'sufficient evidence', port authorities still have the discretion not to inspect the oil and cargo record books, and MARPOL 73/78 does not really give any incentives to port authorities to make inspections.[168] In sum this system not too reliable.[169]

9.3.2.2. Other Conventions[170]

Not all the shipping conventions appear in this Section due to the lack of practical information. Although more information could undoubtedly be obtained from the IMO concerning these conventions, more *critical* information is needed in order to obtain a good picture of their functioning, and this seems lacking. Considering the problems with State enforcement practices under MARPOL 73/78 and expected under LOSC[171] it can only be surmised what kind of practice exists under the other shipping conventions. This would be a fruitful area for further research.

Under the Intervention Convention and Protocol, problems include both the criticism that coastal States are allowed too much discretion in taking preventative actions against marine casualties on the high seas, and that coastal States are too limited in their rights to take such action.[172] Canada as noted is lacking as a Party, but based on Canada's history of Arctic environmental protection, it probably considers itself as falling under the latter point.

The SOLAS Conventions, amendments, Protocol and Resolutions were difficult to obtain an overall view of, and this is unfortunate for such an important subject as safety of life at sea. The IMO in its publication, IMO Document, lists 7 different pamphlets covering SOLAS.[173] The practical effect of ships following these agreements would be interesting.

The same might be said of the satellite agreements INMARSAT, the satellite navigation systems TRANSIT and NAVISTAR, the satellite search and rescue system COSPAS - SARSAT, the universal coverage communications systems LORAN-C - TSJAJKA linkage and OMEGA, the position fixing system (VLF)OMEGA) and the ice and ocean conditions surveillance and monitoring Landsat and NOAA National Environmental Satellite Services. It was difficult to obtain a clear picture of the satellite systems with potential use in the Arctic. In the IMO publication no mention is made of TRANSIT, NAVISTAR, COSPAS - SARSAT, while reference is made to INMARSAT, the International Convention on Maritime Search and Rescue (1979) and 1975-76 International Conference on the Establishment of an International Maritime Satellite System.[174] This lack of ready and information could cause problems for Arctic shipping.[175] Additionally specific problems with the various systems include INMARSAT besides being limited to coverage is also limited in terms of access and costs, (ULF) OMEGA is not yet fully operational, NAVISTAR is operated by the US government and accessibility may be restricted, and getting navigational fixes from TRANSIT can take undue time.[176]

SCTWS, though setting out standards of crew training and ship operation in ice and making recommendations, lacks measures for *testing* the level of skill or proficiency.[177]

The problems involved with the International Convention for the Unification of Certain Rules of Law Relating to Assistance and Salvage at Sea include that pollution risks in salvaging a tanker may be higher than loosing the ship. Additionally salvors have become more reluctant to follow Convention provisions due to unfavourable arbitral awards, considering the costs and equipment risks involved. Some salvors have even been found negligent in causing pollution damage from wrecks.[178]

All in all it can probably be said that *environmentally* a new Arctic agreement covering the points mentioned would be most beneficial. One author notes that because of recent developments in law and technology there is a need to update shipping laws since they were designed when the only interests protected were crews, ships and cargoes.[179] Today coastal State environmental interests need be taken into account.[180] In terms of State *compliance* however, relying chiefly on MARPOL 73/78 provisions and the IMO Conventions would be the most feasible. A new shipping agreement will be recommended for environmental considerations, but it should be remembered that it is subject to strong influence by the already existing IMO regime, and the possibility of its being ratified could be questioned.

9.3.3. Liability Regimes[181]

The complete lack of effective international liability regimes for transnational pollution damage in the Barents Sea and the Arctic region is readily apparent.[182] Though in existence, international liability regimes covering pollution damage from shipping are outdated and inadequate. The limits of compensation of the CLC and the Fund Convention do not take into account inflation nor damage occurring in the exclusive economic zones or fishing zones or environmental damage as such.[183] The 1984 Protocols to the CLC and the Fund Convention take into account these elements but are not ratified by any Arctic rim State and are not in force.[184] The voluntary schemes set up by tanker owners and cargo owners, TOVALOP, CRISTAL and PLATO are unsatisfactory and the situation confused with neither PLATO nor CRISTAL in effect.[185] Additional problems include that it is often difficult to identify the vessel causing the pollution, it may be difficult to establish a causal link between this pollution and the damage suffered, and there are difficulties proving fault on the part of the shipowner, something which most legal systems require.[186] Other problems include difficulties bringing an action before the courts of a person's own State where the shipowner is a foreign national because these courts may be reluctant to assume jurisdiction, enforcing a judgment, compensation awarded to the person damaged exceeding the financial resources of the shipowner, and the shipping defendants' limitation of liability probably not covering damages from a large catastrophe.[187]

Damage from pollution from land-based pollution is note covered over the entire Arctic.[188]

Pollution damage from offshore installations is not covered over the entire Arctic, with a possible exception of private oil company arrangements similar to OPOL on the Canadian and US continental shelves.[189] Even if there were an Arctic liability regime in force general problems of *proving* damages as well can be envisioned similar to those encountered with discharges from ships noted above. Similar problems include in the case of oil and chemical slicks remaining at sea it is difficult to imagine who possible claimants could be, unless specific equipment such as boats, fishing gear, other equipment, is damaged. It seems scientifically proven that very young fish are damaged by oil; so if a discharge drifts into an area of young fish, affected fishermen might have claims for damage if allowed under the national legislation.[190] In other cases science cannot show definite damaging effects of oil upon adult fish in open ocean outside of the laboratory, and therefore if a discharge covered a productive fishing area, the fishermen would stand less of a chance of collecting damages since insufficient damages could be shown.[191] If fish caught were tainted by oil or chemicals, these damages might appear more direct and perhaps damages could be obtained. Damage to animal and plant life both on land and in the sea could be difficult to show if no economic loss could be shown.[192] Problems might also occur of proving which installation the pollution came from. These type problems appear lesser the larger the discharge involved, but for smaller discharges, it might be difficult tracing the oil or chemicals back to the origin due to current and weather conditions.[193] In the absence of a common responsibility of all licensees in the general area, similar to that established for the dumping of scrap,[194] a problem will exist of locating the defendant.

Liability for pollution damage from dumping as noted is limited to Article X of the LDC. Since dumping associated with offshore oil activities is excluded under the LDC, so would liability of such dumping be excluded. As well the State obligation of 'undertaking to develop' is so loose that it is doubtful it will be of much help.[195] The ODC has no liability provision and apparently no attempts have been made to develop rules and procedures.[196] As above there could be envisioned causation and ship identity problems. Additionally regarding the high seas questions arise as to who will be the claimant, how is the damage assessed, and what will be the competent Court? How will victims be compensated where no person liable for the damage has been traced? Due to these questions and to special insurance problems involved, 'a need exists for coordination of national and international approaches to the problems of responsibility, liability and compensation for pollution damages.'[197]

As noted State practice indicates that States do not usually use customary law to forward a claim for environmental damages. Thus except for the inadequate shipping liability conventions, possible private owner liability schemes and unused general provisions of international customary law, there are *no* liability regimes in the Arctic covering damage for marine pollution. In conclusion the development of principles and procedures of responsibility for transnational pollution damage in the Arctic is a serious international problem, which the Arctic States should try to resolve.[198]

9.3.4. Bilateral Norwegian - Soviet Contingency Plan; Arctic Contingency Plans

As was seen previously a joint contingency plan for pollution clean up operations is *lacking* in the Barents Sea though there is a draft under consideration. Although under the Norway - Soviet Agreement the possibility for establishing one is present, this has yet to be done due to the lack of a definite boundary line on the continental shelf and in the exclusive economic zone.[199] On a wider scale encompassing the entire Arctic Region, as noted bilateral contingency plans have in fact been one of the most common specific environmental agreements to appear. Thus with the exception of the Barents Sea the Arctic picture concerning contingency plans is somewhat optimistic, though what is lacking is an overall Arctic coordination and harmonization of these plans.[200] The US - Canadian Agreement is in force from 1974 covering the Beaufort Sea,[201] the Canadian - Danish Agreement from 1983 covering the areas between Canada and Greenland,[202] and the US - USSR Agreement covering the Bering Sea and the Chukchi Sea from 1989.[203] Between the neighbouring Arctic littoral States it is only Norway and the Soviet Union which *do not* have a contingency plan worked out for combating pollution incidents.

Interestingly enough there are also *boundary disputes* between the State Parties in the agreements noted above.[204] Canada and Denmark/Greenland have not yet settled their Arctic boundary north of 82 degrees 13 minutes N latitude where this meets with the 60th meridian of longitude.[205] The agreement apparently is not troubled by this, covering areas to the south and being based upon a definite boundary; however it is noted that perhaps one of the reasons why the boundary has not been established further north is that the

States may be, 'waiting for more precise information on the geological nature of the Lomonosov Ridge.'[206]

The US - USSR Agreement also is based upon a definite maritime boundary between the two countries following approximately the 169th meridian of longitude, but ending at 72 degrees, 00 minutes N.[207] The continental shelf extends much further north than this point under the Chukchi Sea however, and 'there remains the method for delimiting the shelf north of that point.'[208] The US - USSR Agreement says nothing about this,[209] and since the Chukchi Sea is specifically mentioned in the agreement, it seems curious that this boundary question is not even referred to. The permanent ice covered area also occurs just north of Wrangle Island, but with the rapid development of offshore oil technology it seems possible to expect drilling even in these areas in the future.[210]

Finally, the dispute in the Beaufort Sea between Canada and the US is perhaps the most interesting, since the maritime boundary in question appears to be in the area of actual oil drilling and not in the Arctic pack ice.[211] The Canadians have been using the 141st meridian of longitude as its western boundary for jurisdiction covering among other things the issuance of oil and gas exploration permits and control over marine pollution, something which the US has 'acquiesced in' since 1965.[212] Because of the geography involved, the equidistance method favours the Americans. Drilling for oil and gas has been intensified in recent years, and consequently the arrival at a definite continental shelf boundary line should be expected.[213] In terms of a contingency plan the area appears 'hotter' than even the Barents Sea, due to the apparent higher scale drilling activity in a disputed area. Additionally if the Americans acquiesce in Canada's use of the 141st meridian of longitude, this seems contrary to their position in the conflict with the Soviets noted above where they have definitely *not* supported 'sector claims' in the polar regions.[214]

As noted Norway *has* entered several bilateral agreements with the Soviet Union,[215] including the general environmental agreement noted. In spite of these agreements Norway at the same time seems to proceed with much caution and wants no part of a condominium arrangement with the Soviets on Svalbard nor cooperative oil agreements with the Soviet Union in the Barents Sea.[216] It is therefore difficult to predict just how far Norway will go in forming bilaterals with the Soviets.[217] It seems the more specific, cooperative and involved the agreements might tend to be, the less interest the Norwegians have.

Because of the indefiniteness of the Norwegian interest in bilateral agreements a regional Arctic agreement with a sub protocol covering a contingency plan is probably preferably for the Barents Sea. Norway could enter a comprehensive and specific agreement without being fearful of too much Soviet influence over Norway in the Barents Sea.[218] As noted in Part II Chapter 13 however one of the proposals made is that Norway enter a bilateral contingency plan with the Soviet Union, independent of the boundary conflict and in conjunction with a regional Arctic agreement. This is based upon the idea that the Norwegian fears of Soviet domination are overrated, since all the other Arctic contingency plans contain a weaker State party involved in a marine boundary conflict.

9.3.5. Summary

As seen above there are *problems* of varying degrees concerning international conventional coverage of almost all of the sources of pollution as well as contingency plans and liability regimes, both in the Barents Sea and in the wider Arctic. It is therefore obvious that improvement is needed. In proposing improvements, however, the alternatives seem to be either a series of bilateral agreements between the Arctic rim States or a regional Arctic agreement. It has been shown that at least for Norway a regional agreement is probably preferred, mostly because of security reasons as related to the Soviet Union. Are there other reasons however indicating an Arctic regional agreement is the optimum solution? These and other questions will be addressed below.

Notes - Chapter 9

1 United Nations Convention on the Law of the Sea (LOSC), 21 *International Legal Materials* 1261. Ratifying States include Fiji, Zambia, Mexico, Jamaica, Council for Namibia, Ghana, Bahamas, Belize, Egypt, Côte d'Ivoire, Philippines, Gambia, Cuba, Senegal, Sudan, Saint Lucia, Togo, Tunisia, Bahrain, Iceland, Mali, Iraq, Guinea, Tanzania, Cameroon, Indonesia, Trinidad and Tobago, Kuwait, Yugoslavia, Nigeria, Guinea-Bissau, Paraguay, Democratic Yemen, Cape Verde, Sao Tome and Principe, Cyprus, Brazil, Antigua and Barbuda, Kenya, Zaire, Somalia, and Oman. OPN December 1989 p. 1. The Treaty will inter into force 1 year after the deposit of the 60th ratification.

2 See Part I Section 2.1.3.

3 The Soviet Union and the US do not accept the Court's jurisdiction under the Optional clause 36(2) and (3) of the ICJ Statute though Denmark, Norway and Canada do. See Fleischer s. 158 and Akehurst, M. *A Modern Introduction to International Law*, 6th Ed. (1989) (Akehurst) p. 249.

4 Norwegian National Broadcasting System, Television Report, September 1990.

5 See Churchill pp. 21-22 for an expansive list of references indicating State practice including national Yearbooks of several States and a study conducted by the University of Virginia under M.H. Nordquist giving the legislative history and a commentary of every LOSC article.

6 Franckx pp. 156-159.

7 *Ibid.* The Soviet Union was the first country to legislate national rules covering vessel-source pollution and dumping in its exclusive economic zone following signature of the LOSC.

8 *Ibid.* pp. 158-159. See also Butler, W.E., 'Anglo-Soviet Links and the Law of the Sea', *The Law of the Sea and International Shipping*, Anglo-Soviet Post-UNCLOS Perspectives, Ed. W.E. Butler, Oceana Publications, Inc. (1985), p. v., who notes, 'Soviet interpretations and applications of the 1982 UNCLOS (LOSC) will have an impact far beyond the States mostly associated with her; within the Convention framework, Soviet State practice in all its forms will be the benchmark for dozens of other countries who have ratified or will ratify the Convention,' parentheses added. See Appendix I for an outline of Soviet Edict and Decrees dealing with the marine environment including the exclusive economic zone found in Butler, Wm. *The USSR, Eastern Europe and the Development of the Law of the Sea*, Compiled and Edited, with Translations and Annotations by William E. Butler, Professor of Comparative Law in the University of London, Director, Centre for the Study of Socialist Legal Systems, University College, London (Butler), Sections F.2., F.3., F.1.a., G.1.b., G.3., G.4., J.2., J.3., J.4., J.5., J.6., and J.7.

9 Franckx pp. 158-159. A subsidiary point is also made that there is a better chance for different States legislating similar rules and hence developing customary law if those States have a common legal 'point of reference', which the LOSC provides.

10 Belsky p. 748, citing Malone, 'Who Needs the Sea Treaty'?, 45 *Foreign Aff.* Spr. 1984 (Malone) at 44, 59-60, noting Malone is the Assistant Secretary of State for Oceans International Environmental and Scientific Affairs and Special Representative of the President to the Law of the Sea Conference. The Presidential Proclamation appears at Proclamation No. 5030, 48 Fed.Reg. 10, 605 (1983) codified at 3 C.F.R. Section 5030, 22 *International Legal Materials* 465 (1983). Further Admiral William L. Schachte, Jr. in a speech before the Conference of the Law of the Sea Institute, Malmö Sweden, August 1991 noted that, '(T)he United States believes that the 1982 LOS Convention reflects customary international law for protecting and preserving the marine environment.' Reproduced in *Council on Ocean Law, Special Report*, August 1991 p. 3. In spite of the presence of the disclaimer that the views expressed by the author were solely the author's and not necessary those of the US Navy nor of the US Department of Defense, Admiral Schachte noted that the speech was cleared by US government officials and accurately reflected US government policy, conversation August 8, 1991. Admiral Schachte was a member of the US Delegation to the final sessions of the U.N. Conference on the Law of the Sea.

11 Belsky p. 753.

12 *Ibid.*

13 *Ibid.*

14 *Ibid.*

15 The Norwegian Ministry of the Environment has for example recommended to the Ministry of Foreign Affairs that Norway ratify the LOSC, conversation Project Manager Atle Fretheim Norwegian Ministry of the Environment, August 2, 1989. This has not occurred, at least one of the reasons presumably being because no other industrial State with the exception of Island has ratified.

16 See Part I Sections 2.1.5. and 2.2.5.

17 11 *International Legal Materials* 1294. See Part I Section 3.2.

18 IMO Status pp. 353-364.

19 11 *International Legal Materials* 262. See Part I Section 3.2.

20 This figure was first published in *Marine Policy*, Vol. 11, No. 4, October 1987, p. 289. Østreng Delimitation p. 144 notes that under the Soviet sector line claim some 43% of the Arctic Ocean would be under Soviet jurisdiction.

21 499 *United Nations Treaty Series* 311. See Part I Section 4.2.1.

22 As noted, Article 6 of this Convention is the basis of both the Norwegian and the Soviet claims to different boundaries in the Barents Sea. The Parties have agreed to also establish a marine boundary for the economic zone uniform with the continental shelf boundary. The Norwegians claim a line based upon the equidistance principle, calculated form the northernmost points on the continent and the eastern most points of the Svalbard archipelago, and which would give the Soviets 60% of the Barents Sea continental shelf and the Norwegians 40%, Tresselt pp. 79-80. The Soviets claim the sector line, due to special circumstances, including coastal shape, coastal length, geological conditions, size of local population, ice conditions, economic interests including fisheries and shipping, special environmental considerations, special conditions involving the Spitzbergen Treaty and special Soviet strategic considerations, *ibid.* The importance of this unresolved conflict in terms of oil reserves, disregarding fisheries, military and political considerations, is reflected in a comments made by Director General Evgene Reschetnijak, head of exploration and production of oil and gas in the northern Soviet areas, who expects that 25% of the Soviet Barents Seas oil reserves lie in the disputed area, Nordlys november 3, 1987.

23 13 *International Legal Materials* 353. See Part I Section 4.2.2.

24 Airborne pollution is included under the classification of land-based pollution since as seen in Part I Section 4.1. this has been the usual practice. As noted a Protocol to the Paris Convention has been adopted covering airborne pollution from land-based sources but is not yet in force. Airborne pollution is monitored however by most member countries.

25 See Part I Section 4.2.2.

26 18 *International Legal Materials* 1442. See Part I Section 4.2.5.

27 MARPOL 73, 12 *International Legal Materials* 1315; MARPOL 78, 17 *International Legal Materials* 546. The Soviet Union has ratified *all* 5 Annexes, while Norway has ratified all Annexes but Annex IV. IMO Status pp. 63-98. The Annexes I-V deal respectively with the prevention of pollution from oil, hazardous chemicals, hazardous chemicals in containers, sewage and garbage. See Part I Section 5.2.3.

28 Denmark has ratified all of the Annexes. The US has ratified all the Annexes with exception of IV. IMO Status pp. 63-98 and OPN June 1991 p. 6.

29 9 *International Legal Materials* 1. See Part I Section 5.2.2.

30 1958 Convention on the High Seas, I *New Directions in the Law of the Sea* 257; 1958 Convention on the Territorial Sea and the Contiguous Zone, I *New Directions in the Law of the Sea* 1; and 1958 Convention on the Continental Shelf, I *New Directions in the Law of the Sea* 257. See Part I Sections 4.2.1. and 5.2.1.

31 9 *International Legal Materials* 25 and 16 *International Legal Materials* 1103 respectively. See Part I Section 5.3.1.

32 14 *International Legal Materials* 959, 17 *International Legal Materials* 579 and *Gassetta Ufficiale della Repubblica Italiana* Suppl. 260. See Part I Section 5.3.2.

33 15 *International Legal Materials* 1051 and 15 *International Legal Materials* 233. See Part I Section 5.3.3. for discussion of INMARSAT and of other communication and navigational regimes under development.

34 IMO Document p. 49. See Part I Section 5.3.4.

35 IMO Document p. 33. See Part I Section 5.4.2.

36 IMO Document p. 46. See Part I Section 5.4.3.

37 IMO Document p. 15. See Part I Section 5.4.4.

38 The material presented here is taken from Gold pp. 81-82, 87, 98, 101, 104, 107 and 108 unless otherwise indicated. Some inconsistencies were found between Gold's information and that found in Churchill, IMO Status and other sources. In addition there are various unclarities presented by Gold, for example p. 107 showing that Norwegian sanctions include only fines, and on pp. 86 and 107 that prison terms are possible. 'Almost' strict liability is not defined. Where possible, changes were made on the basis of information found in the other sources. It is beyond the scope of this project to present a comprehensive survey of the national legislation dealing with vessel-source pollution, and as noted in the Preface more research is needed.

39 For a summary of other shipping nation's legislation which would affect flag State jurisdiction of respective oil tankers and freighters expected in the Arctic, see Gold pp. 79-89 and 91-111 and Churchill pp. 343-359.

40 Gold writes that Norway claims only a 200 mile fisheries zone, while Fleischer p. 98 writes that a 200 mile exclusive economic zone with corresponding jurisdiction over measures against pollution was established through Norwegian law of 17 december 1976. Ulfstein, Geir, *Økonomiske Soner - Hva Nå?*, Universitetsforlaget, (Ulfstein) (1982) p. 53 substantiates this.

41 Further legislation cited includes Act of 9 June 1903, Chapter 11 as amended in 1979; Maritime Code of 1893 as amended in 1974; Oil Pollution Regulations of 26 June 1980.

42 Bull, Hans Jacob, 'Petroleumsrett, Knophs Oversikt over Norges Rett', 9.Utgave, 1987, Universitetesforlaget (Bull) pp. 682-693; and Bull, Hans Jacob, 'Erstatningsansvar i petroleumsvirksomheten,' *Tidsskrift for Rettsvitenskap*, 1-2/1986 (Bull Erstatning) pp. 84-118 note that there is strict liability for the shipowner for all pollution from the ship under Sjøl. kap. 12.

43 Gold notes that a 200 mile fisheries zone only is claimed. See Appendix I for an outline of Soviet Edicts and Decrees. Gold lists national legislation including Decree No. 5590/1974 amended 1980; Decisions No. 118/1974; No. 247/1978 and No. 1226/1980; and Decree of 18 March 1981.

44 ABC Bank in OSLO reports the exchange rate as of January 1 1989 to be *10.90* rubles to the Norwegian crown. (Approximately 6.5 crowns equal USD 1.00). According to Nordlys 14. desember 1988, 87 årgang s. 17, the international exchange rate is approximately 1 ruble to 1 Norwegian crown. This is explained in the ruble not yet being convertible on the international market and therefore being reduced in the conversion. Consequently the fine would be approximately 25,000 crowns or $3846. In line with the ABC bank figure the ruble's value is 10 times as high within the Soviet Union itself, with the *effective* fine within the Soviet Union being approximately 250,000 crowns or $38,460.

45 From Butler Sections F.2., F.3. and J.4. it appears the Soviet Union is one of the few States clearly implementing pollution control in its 200 mile economic zone. See Appendix I.

46 See Butler Section J.7. See Appendix I.

47 Only Canadian federal law will be addressed here. Gold notes that navigation is federally regulated under the Canadian Constitution and hence provincial law is not at issue as in the U.S.

48 Only Canadian Arctic Waters legislation will be briefly examined here.

49 This gives information about the ship including the name of the owner and the name of the ship, specifics of the trip, amount of polluting substances on board, name of cargo owner, name of approved insure and amount of liability coverage.

50 Gold p. 82. As can be seen Canada is one of the few States clearly implementing pollution control similar to a 200 mile economic zone in a 100 mile Arctic Waters zone.

51 Churchill p. 351. Note that only US *federal* law will be discussed here and not state law with the exception of Alaska.

52 Note that on August 18 1990 President Bush signed into law the very comprehensive OPA which briefly imposes strict liability on vessel owners, operators and bareboat charters for unlawful oil spills with possible unlimited liability. The OPA does not preempt stricter State liability laws. See Burlingham, Underwood & Lord, Report on United States Oil Pollution laws, October 1990 for a compendium of current federal and state statutes and a comparison of coverages under the shipping liability conventions and agreements (see Part I Section 6.4.); an analysis of OPA provisions; and a timetable of dates by which tankers operating in US waters must be constructed or retrofitted with double hulls. Aftenposten 1 november 1990 Uke 44., Nr.499., 131.årg., s. 17, notes that international insurance companies do not want to insure oil tankers which use US harbours because of the new regulations with unlimited liability.

53 Churchill p. 344. This can perhaps be different in Greenland itself, which is the point of the project with respect to Denmark. More research is necessary to determine whether there is special anti pollution legislation which applies to Greenland and if so the specific provisions.

54 See Part I Section 6.2.

55 VI *New Directions in the Law of the Sea* 507. See Part I Section 6.3.2.

56 Dr. R. Engelhardt, Director, Circumpolar and Scientific Affairs Directorate, Canadian Department of Indian and Northern Affairs reports in a conversation May 22, 1990 that liability arrangements are under development by several oil companies operating in the Beaufort Sea.

57 9 *International Legal Materials* 45. See Part I Section 6.4.1. for a discussion. Also appearing is the Protocol of 1984 to Amend the International Convention on Civil Liability for Oil Pollution Damage 1969, which would be applicable in the Arctic if it were in force. It raises the limits of liability and hence the amount of compensation available and extends the geographical area of application into the exclusive economic zones and fishing zones. *None* of the Arctic States have ratified this Protocol.

58 11 *International Legal Materials* 284. See Part I Section 6.4.2. Also appearing is the Protocol of 1984 to Amend the International Convention on the Establishment of an International Fund for Compensation for Oil Pollution Damage, which raises the liability limits and extends the geographical area of application into the exclusive economic zones and fishing zones. It would be applicable in the Arctic if it were in force, yet *none* of the Arctic States have ratified this Protocol.

59 8 *International Legal Materials* 497, 10 *International Legal Materials* 137, and Plato Limited, Portland House, 6th Floor, Stag Place, London SWIE 5BH, respectively. See Part I Sections 6.4.3, 6.4.4. and 6.4.5. respectively.

60 See Part I Section 6.4.6. and Churchill p. 268 who notes that until such time as a new Draft Convention on Liability and Compensation in Connection with the Carriage of Noxious and Hazardous Substances by Sea (HNS Draft Convention) can be agreed upon by States and ratified, a shipowner may often be able to limit his liability under the LLMC and the 1957 Conventions.

61 A brief overview of the Norwegian liability laws regulating damage from marine oil and dumping activities including dumping is as follows. See generally Bull pp. 682-693; Bull Erstatning pp. 84-118; Churchill Conflict pp. 205-211; and Ørebech, Peter, *Om allemannsrettigheter*, Osmundsson Forlag as., (Ørebech), Kap. 3-5. Petrl. Sections 1.2. and 1.4. give the possibility for Norwegian law to be utilized in areas outside of the continental shelf in accord with international law or agreements with foreign States, excluding however Svalbard's internal waters and territorial seas. The licensees are responsible to third parties for damages done by subcontractors in job related oil activities to the same degree as the tortfeasor, but with the same defenses available, Petrl. Section 60. In offshore operations normally there is a group of licensees with a central participant, the 'operator', acting on behalf of the group. The liability is first

'channelized' to the operator and thereafter to the licensees in case of operator insolvency or reluctance to accept liability. Thus assignees, equipment suppliers, and employees cannot be sued. All the licensees seem to be given a more central place in the Norwegian liability scheme than that of the 1976 Liability Convention (see Part I Section 6.3.3.) with its liability channelization against the operator with little mention made of other licensees. The licensees have *unlimited* strict liability for damage or loss cause by oil pollution from production, drilling, maintenance, hotel and loading platforms, pipelines and wellheads with exceptions including acts of God, war and other *force majeure* events, Petrl. kap. V. Both the claimant's contributory negligence and intentional actions are as well taken into account under Erstl. Section 5.1. Remedial measure taken are as well covered, as are damages cause by clean up chemicals. These rules come into force when damages occur on the Norwegian mainland, in the territorial waters or on the continental shelf, or affect Norwegian ships, fishing equipment or installations on adjacent ocean areas. There is no condition that the equipment causing the damage be placed on the Norwegian continental shelf, pollution from installations on other countries' continental shelves doing injury under Norwegian jurisdiction would as well be covered. There are in addition *proposed* special liability rules for oil companies for damages caused to fishing interests, NOU 1986:6. These consist partly of compensation rules for expropriation of a fishing ground by the placement of oil rigs, loading platforms, pipelines etc., and partly of rules covering damages caused by pollution and scrap. Prior to this general principles of Norwegian tort law would allow fishermen to receive compensation for damage to equipment from debris, but there would be problems in identifying the tortfeasor or in economically bringing an action. Additionally in a negligence action there are problems of proving the damage was due to a breach of the duty of care or that a statutory duty was not observed. In 1976 the Norwegian Government established a fund to deal with claims for compensation for damage caused by dumped offshore scrap. This fund covers all claims by Norwegian fishermen whether or not the tortfeasor can be identified. If the tortfeasor can be found, then the fund can bring an action demanding reimbursement. The fund also covers loss of fishing time caused by seismic surveys and gives compensation for scrap brought ashore or a position marked on a map. There is a 25% or Norwegian crown 500 deductible, whichever is smaller, for every claim. Under the proposed new scheme mentioned added to the Petroleum Law a licensee would have strict liability for all damage caused to fishermen by scrap. In the event the origin was unknown, all licensees would be jointly liable. Claims would be handled through a special board with appeals possible through the regular court system. Insurance is required by Norwegian authorities to cover oil activities and possible liability incurred, Petrl. Section 43. Professor Bull notes in a conversation December 1987 regarding the 'reasonableness' of the limits set in OPOL, (see Part I Section 6.3.2.) the 1976 Liability Convention and the unlimited liability under Norwegian law, that the *limited* liability was related to the insurance coverage obtained by the oil companies. The Norwegian authorities felt that oil company *assets* plus insurance could as well be taken into account and seen as a source of recovery to meet the Norwegian unlimited liability. Guarantees by other companies could as well be demanded or extra insurance required of the licensees. Floating platforms or similar installations have the same liability as ships regarding collision and (ship)owner liability, Sjøl. kap. 8, kap. 10. Liability is limited in these cases but the amount is higher than for ships, 20,000,000 SDR's (approximately USD 27,500,000) plus 12,000,000 SDR's (approximately $16,700,000) if there are personal injuries.

62 See these same Sections in Butler for more information.

63 See Part I Sections 7.7. and 7.8.

64 See Part I Section 7.5.

65 See Part I Section 7.4. Though the Bonn Agreement applies to the North Sea south of 61 degrees north latitude, Norway has apparently forwarded a type of invitation for Soviet association along with the invitation noted to join the ODC and the Paris Convention, Joint Protocol 1, p. 3. The Soviet response indicated that all 3 of these Agreements do not cover all the regional needs and hence a concrete cooperative Arctic program at the regional level is to be preferred, *ibid.*

66 See Part I Section 7.6.

67 See Part I Section 7.2.

68 US - Canadian Agreement, 29 *United States Treaties* 2569. Contingency Plan, Annex Four (Beaufort Sea), obtained from Canadian Ministry of External Affairs, December 1, 1989.

69 US - USSR Agreement, Obtained from the US State Department.

70 Canadian - Danish Agreement, 23 *International Legal Materials* 269. See also Lamson Arctic p. 8 and Pharand pp. 752-753.

71 The US and the USSR recently settled their maritime boundary dispute June 1, 1990 in the North Pacific, Bering Sea and Chukchi Sea up to approximately 72 degrees N. For details of this interesting settlement with special areas of jurisdiction in each other's exclusive economic zones, see OPN June 1990 pp. 3-4.

72 See Osherenko pp. 2-3 and Roginko pp. 6-7.

73 Agreement on Co-operation in the Field of Environmental Protection (US - USSR Cooperative), 11 *International Legal Materials* 761, 1408.

74 USSR - US Agreement on Cooperation in Studies of the World Oceans (US - USSR Ocean). See Osherenko pp. 2-3.

75 *Ibid.* and Harders, J. Enno, 'The Arctic Ocean: Environmental Protection and Circumpolar Cooperation - An Assessment and a Proposal for a Legal Regime', Seminar Proceedings, Munich, Federal Republic of Germany, *Energy Law* 1986 (Harders Seminar) p. 521.

76 USSR - Canada Protocol on the Development of a Programme of Scientific and Technical Cooperation in the Arctic and the North, (Canadian - Soviet Technical Agreement). Obtained from the Canadian Department of External Affairs.

77 Other activities include recommending ways to protect the environment from major oil spills; determining and predicting effects of these spills; studying adverse effects of developing Arctic gas hydrate beds; softening the environmental impact of oil and gas pipeline construction; studying the diffusion of petroleum hydrocarbons and other pollutants in water; studying oil oxidizing microorganisms in Arctic seas; and modelling degradation of petroleum hydrocarbons in northern waters.

78 Conversation Canadian representative of Ministry of External Affairs, Associate Professor Geir Ulfstein, University of Oslo and author December 8, 1987.

79 Avtale mellom Norges Allmennvitenskapelige Forskningsråd og USSR's Statskomite for Vitenskap og Teknikk om Teknisk Vitenskapelig Samarbeid om Utforskningen av Arktis og Nordområdene, januar 1988, (Norwegian - Soviet Technology Agreement. Obtained from the Norwegian Ministry of Foreign Affairs. For the former see Part I Sections 4.2.5., 7.7 and 7.8.4.

80 Overenskomst mellom Kongeriket Norges Regjering og Regjeringen i Unionen av Sovjetiske Sosialistiske Republikker om Samarbeid ved Ettersøking av Savnede og Redning av Nødstedte Mennesker i Barentshavet. Obtained from the Norwegian Ministry of Foreign Affairs.

81 Avtale mellom Regjeringen i Kongeriket Norge og Regjeringen i Unionen av Sovjetiske Sosialistiske Republikker om Tidlig Varsling av Atomulykker og om Utveksling av Informasjon om Atomanlegg. Obtained from the Norwegian Ministry of Foreign Affairs.

82 Not available from the Norwegian Ministry of Foreign Affairs.

83 Roginko p. 22 and Harders Seminar p. 522.

84 *Ibid.*

85 *Ibid.*

86 Agreement on Conservation of Polar Bears, 17 *International Legal Materials* 13.

87 Harders Seminar p. 523.

88 See Part I Sections 3.2. and 4.2. for discussion of provisions from Paris Convention, ODC, LDC and other agreements referred to in this Section.

89 As noted the LDC would encompass the Barents Sea but excludes dumping related to offshore oil field activities.

90 The Norwegian Department of the Environment was not willing to venture a guess as to when accession might be expected, but emphasized, 'soon'. Conversation with Chief Consultant Per Antonsen Norwegian Department of the Environment, November 1989.

91 As seen only Article 3(c)(iii) of the Paris Convention deals specifically with pollution from offshore oil platforms, but the programs and procedures have been developed under this structure making the regime probably one of the most advanced in the world. See generally Oslo-Paris Comm. and Par.Comm.8, Par.Comm.9 and Par.Comm.10. The broad experience gained under the Paris Convention in the North Sea since its inception in 1974 covering monitoring for example should be studied and implemented and improved upon to better relate to the Arctic. See however Part I Section 1.3.6. for surveillance and reporting problems of platform discharges however.

92 Sætevik. The following information is obtained from Sætevik pp. 2, 115, 126, 131, 171, 178-185, 201 and Vedlegg 5 unless noted otherwise.

93 Conversation with Project Leader Atle Fretheim, Norwegian Ministry of the Environment, August 11, 1988. Mr. Fretheim agreed somewhat with Sætevik's conclusions, though not with all data and arguments presented. He argued that environmental gains are not totally measured by the number of Resolutions adopted, actual gains also being achieved by the exchange of information, for example from Scandinavia to the UK This over time it was felt does affect actual national performance. Additionally argued is that in the Paris Commission Denmark is noted for being 'difficult' and achieves little through this strategy.

94 Project Manager Fretheim and the Paris Commission's Secretary, Peter Hayward, report that the Norwegian Ministry of the Environment however has an open policy regarding providing information covering Paris Commission meetings in which they participate.

95 See generally Par.Comm. 9 and 10.

96 Sætevik notes that Norway's passivity is at least partially intentional, based upon a note found in the Ministry of the Environment's files.

97 Sætevik notes that the tendency seems to be that States send delegates from 'lower' levels with less competence, which seems to indicate that the Commission and hence marine pollution is given lower priority. Par.Comm.9 p. 2 notes however that both the Paris and the Oslo Commissions have agreed to work towards holding a meeting at the Ministerial level in 1991.

98 Johnston pp. 412-413.

99 *Ibid.* This is to be based upon LOSC Articles 76 and 208. See Part II Section 12.1.2. The author also specifically proposes that UNEP play a central role 'promoting the conclusion of a treaty providing for environmental standards relating to offshore exploration and exploitation of mineral resources.'

100 Churchill pp. 277-278.

101 Johnston p. 412. See Part I Section 4.2.3.2. The author notes much of the relevant knowledge and experience will come from the UNEP Regional Seas Programme and recommends that States agree on global and regional rules, standards and recommended practices and procedures consistent with LOSC Article 207(4), and devote attention to problems of prohibiting or controlling global and regional levels of 'toxic, harmful or noxious substances, especially persistent substances accorded priority status in Article 207 (5), with UNEP playing a central role. See Part II Section 16.1.2. for LOSC provisions.

102 Johnston p. 413. The author suggests that when coastal States negotiate a refinery contract with foreign oil companies, they should have access to a model contract with standard environmental safeguard clauses, possibly drafted by the U.N. centre on Transnational Corporation.

103 See generally Churchill Conflict pp. 204-205.

104 Though platform removal is a distant problem its solution in the Barents Sea will largely depend upon the solutions arrived at in the North Sea. Similar to the North Sea States both Norway and the Soviet Union have ratified only the 1958 Continental Shelf Convention and not the LOSC and will be likely to follow the North Sea practice.

105 Johnston p. 413.

106 *Ibid.* The author concludes that, 'In the absence of reliable scientific data, it is not possible to measure the environmental shortcomings of the London Convention, but the best evidence available suggests that the minimum global standards created by it are *seriously inadequate*,' emphasis added.

107 Johnston p. 413. The author recommends various organizations including UNEP, IMO, WHO, IAEA and others to come up with effective monitoring programs and assist States in developing regional agreements consistent with LOSC Article 207 with standards similar to the Helsinki Convention. Johnston p. 212 notes that there lacks on a general level lack of monitoring marine pollution from all sources. This is needed to follow the consequences of discharging noxious wastes into the seas, to identify the current sate of pollution especially in highly vulnerable areas, and to control the efficiency of existing regulatory systems. To be effective the monitoring has to be internationally coordinated for the sake of comparison, and this is only possible, 'if methodologies and procedures are harmonize in regard to standards and measurements (e.g. inter calibration of instruments).' See Part I Section 3.2.2. See Part II Section 15.1.2. for LOSC provisions.

108 Churchill Conflict pp. 204-205.

109 *Ibid.* Additionally phrases such as 'good oil field practice' and prohibitions to 'interfere unjustifiably with navigation or fishing' have proven too vague to restrain the dumping.

110 The qualification to the exception 'as long as such ships and aircraft are operated in a manner consistent with the Conventions' provisions' probably would not be much of a hinder to such dumping practices. Johnston p. 413 notes that many dangerous toxic substances and *most* radioactive are carried in government ships. Churchill p. 270 notes that Parties to the LDC in 1983 adopted a resolution for a 2 year moratorium on the dumping of radioactive waste pending a scientific study into the environmental damage caused. The study was not conclusive a another resolution was adopted in 1985 for a continuation of the moratorium pending further studies, and though these resolution are *not* legally binding they appear to be followed. There is in addition controversy between Parties to the LDC regarding sub-sea-bed disposal of radioactive wastes, which some States considerable suitable. *Ibid.* notes that no agreements have been reached on this issue, but that State Parties have agreed no such disposal should take place until it is proven to be 'technically feasible and environmentally acceptable.'

111 Churchill p. 270.

112 *Ibid.*

113 See Part I Section 3.2.1.

114 Churchill p. 272.

115 *Ibid.*

116 *Ibid.* citing House of Lords Select Committee on the European Communities, *Dumping of Waste at Sea*, HL Paper (1985-86) 219, p. 21. Conversation with Project Manager Atle Fretheim, Norwegian Ministry of the Environment, August 2, 1989. See however Aftenposten, 17. januar 1990, Uke 3. Nr.27, 131 Årg. s. 52 and Nordlys 18 january 1990 Nr.15, 89 Årgang, s. 15 regarding a Greenpeace action in the North Sea against the authorized UK dumping of wastes from a power station and the planned dumping of 50,000 tons wastes from the pharmaceutical industry. The Norwegian, Swedish and Danish Ministries of the Environment have protested against these authorized dumpings, but the British government maintains there do not exist sufficient possibilities for disposing of the waste on land.

117 Several 'environmentally slow' countries were also noted in a conversation with Chief Engineer Jens Koefoed, August 1, 1989, SFT regarding the workings of the Paris Convention in the North Sea. Relations with European countries regarding airborne pollution on the other hand is highly relevant to the Arctic. See Part I Section 4.2.5. and Part II Section 10.2.1.

118 See Part II Section 3.2.1.

119 The LDC has a somewhat stronger use of language and is considered by Hakapää p. 98 to be a worthwhile environmental convention with the exception of omitting offshore oil field activities.

120 Norwegian authorities are still in the process of discovering the scope of toxic effects on fish from drilling platforms in the North Sea, and knowledge of the effects in Arctic waters is still at an elementary stage if that. See Part I Sections 1.2.7. and 1.3.6.

121 See Part I Section 3.2.1.5. and 3.2.1.6.

122 See Part I Sections 3.2.1. and 4.2.2. for monitoring requirements under the ODC and the Paris Convention. See Roginko pp. 26-27 and Churchill p. 272 who equally note problems obtaining information, as well as Sætevik noted above.

123 See Roginko p. 26.

124 Stokke pp. 13-14. In the Arctic the relevant pollutants from this source include sulphur dioxide, nitrogen oxides and heavy metals which can adversely effect both vegetation and animal life such as reindeer. Specific agreements have been reached wherein 21 States pledge to reduce their sulphur dioxide emissions by 30% (1980 quantities) by 1993 and others by even 50%.

125 See Roginko p. 17 where it is stated that a such organization along the entire perimeter of the Arctic Ocean with the exchange of data, 'would greatly contribute to the solution of the problem of monitoring transboundary transmission of pollutants in the Arctic atmosphere' since it would allow for introducing a unified system which regulated pollutants discharges in the atmosphere and hence reduced transboundary flows. Roginko thinks this system additionally could be used for monitoring and exchanging information under EMEP on Arctic radiological concentrations.

126 See below.

127 Regulation 10(1), Annex 1; Regulation 1(7) Annex II; Regulation 5(1) Annex V; Regulation 1(9) Annex I. See Part I Section 5.2.3.1. The special rules include the total prohibition of discharge of oil from oil tankers and other ships 400 grt and over, which must be retained on board and discharged at reception facilities.

128 Birnie p. 234.

129 As noted above Canada has not ratified MARPOL 73/78.

130 See Part I Section 5.2.3., Part II Section 11.2. and Roginko p. 14. Roginko notes however that LOSC needs ratification since the Arctic clause Article 234 allows Arctic rim States to exercise more control over foreign shipping which threaten the marine environment, while MARPOL 73/78 does not grant coastal States any special control over compliance with foreign shipping of its provisions.

131 Johnston p. 414 believes further that States and organizations such as the IMO gradually establish effective systems of vessel traffic management under the authority allowed States in their exclusive economic zones under LOSC Articles 234, 211 (5) and (6). See Part II Section 14.1.2. for LOSC provisions.

132 Roginko pp. 11-14.

133 *Ibid.*

134 Harders Seminar pp. 511-512. Johnston p. 415 takes it further and demands that UNEP assist coastal States in making and maintaining an inventory of sites which because of vulnerability may warrant the establishment of 'special areas' in the *exclusive economic zone* under LOSC Article 211(6) to preserve them from vessel-source pollution. He calls for the IMO to advise governments on legislative techniques for establishing 'special areas' for developing guidelines for preserving marine ecosystems under coastal State jurisdiction. See Part II Section 14.1.2. for LOSC provisions.

135 See Part I Section 5.2.3. for discussion of MARPOL 73/78 provisions referred to in this Section.

136 See Henkin pp. 995-96 where it is generally noted that the North Atlantic Treaty was signed in Washington on April 4, 1949, T.I.A.S. 1964. Original members included Belgium, Canada, Denmark, France, Iceland, Italy, Luxembourg, The Netherlands, Norway, Portugal, The United Kingdom, and the United States. Greece and Turkey acceded in 1951, the Federal Republic of Germany in 1955, and France withdrew its military contingents in 1966 but remains a member. It is not known whether other WARSAW PACT vessels than Soviet are located in the Arctic, especially after the political developments of 1990 whereby this organization's influence was much diminished. For completeness sake however, see Henkin pp. 995-996 where it is noted that the 20 year Treaty of Friendship, Cooperation and Mutual Assistance was signed at Warsaw on May 14, 1955 with the parties including Albania, Bulgaria, Czechoslovakia, the German Democratic Republic, Hungary, Poland, Rumania and the U.S.S.R.

137 Article 3 (3).

138 See Sovjet-Nytt, 21 juli 1989, Nr.20, s. 6 where it is noted that enormous amounts of oil products are generally released by military vessels in the world's oceans. Johnston p. 416 calls for States and international organizations to develop rules establishing a duty for 'owners' of wrecks to notify the existence, location and contents of these wrecks to an organization such as the IMO as well as a duty to recover the wrecks whenever feasible. This could have application to the Arctic, especially the Barents Sea and the narrow straits and shallow passages of both the Canadian and the Soviet Arctic areas.

139 See Gold pp. 60, 159-161. The Norwegian Ministry of the Environment however notes that very few complaints come to Norway (or other countries) from the shipping industry regarding lack of reception facilities. Conversation Project Manager Atle Fretheim August 2, 1989.

140 Conversation Chief Engineer Jens Koefoed, SFT, August 1, 1989. However Norwegian *fishing boats* operating in the Barents Sea rather than delivering their used motor oil to facilities where fees are required often dump this oil into the sea. Conversation fisherman, Tromsø, September 1988 and Nordlys 27 november 1989, Nr. 275 88.årgang, s. 7. If Norwegian disposal prices are too high, the same activities could be expected of the shipping industry. Johnston p. 415 notes additionally a lack of network of facilities for ships carrying oil or hazardous substances to call and be

provided emergency assistance without threatening populated or environmentally sensitive areas and calls upon States to establish such a network globally, which might have application for the Arctic.

141 See Curtis p. 86. See also Aftenposten 9. mars 1990, Uke 10., Nr.115., 131.årg., s. 12. where it is noted that upgrading a tanker to meet environmental considerations costs an additional 15 million dollars from the original 33 million dollars. Johnston p. 414 goes even further and demands that States and international organizations *accelerate* the development of *new* standards related to ship design, construction, manning and equipment especially as directed towards the coverage of hazardous substances. This is based upon LOSC Article 211 and the innocent passage Articles 19, 21-25 which require that coastal State standards be in compliance with international standards including MARPOL , SOLAS, ILO Convention No. 147 of 1976 concerning Minimum Crewing Standards, and STCWS, provisions of which the author wants upgraded. See Part II Section 14.1.2. for the LOSC provisions.

142 Wang pp. 334-335 notes that the traditional maritime States have their share of substandard ships in addition to the flags of convenience States. This is in spite of MARPOL 73/78 (and LOSC) emphasis on flag State enforcement, requiring Parties to survey their flag ships annually or on an unscheduled basis.

143 *Ibid.* notes that one measure included in former IMO Conventions that failed is having every State report to the IMO on their implementation of international law, including flag State enforcement. This has not been complied with by the member States, and it is noted that the legal means of monitoring flag State enforcement are limited, thus requiring political and economic means also be considered to ensure motivation. These means could include exerting political pressure by international organizations such as the IMO; establishing bilateral and regional enforcement systems among States against each other's ships; increasing domestic political and public opinion for giving oil pollution a higher priority among States; oil company screening of tankers before they charter; and giving foreign aid to developing countries to help them economically and incentive wise to bring their tanker fleets within international standards.

144 Conversation Chief Engineer Jens Koefoed, Norwegian State Pollution Control Authorities, August 1, 1989.

145 See Nordlys 17 mars 1990, Nr.65, 89.årgang, s. 33. See also Aftenposten 2. mars 1990, Uke 9., Nr.103., 131.årg., s. 6. where it is noted Norway plans an ocean surveillance centre in Bergen which will use satellites, buoys and ships for gathering information covering marine pollution, flooding, algae blooms, and other developments. This is coupled to a super centralized computer which collects the data and gives notification of discharges and flow prediction analysis.

146 See Bergstrand, S. and R. Doganis, 'The Impact of Flags of Convenience (Open Registries)', *The Law of the Sea and International Shipping: Anglo-Soviet Post-UNCLOS Perspectives*, Ed. W.E. Butler, Centre for Socialist Legal Systems, University College, London (1985) (Bergstrand) p. 414 and Dempsey p. 63.

147 Bergstrand p. 425.

148 The statistics that follow appear in Bergstrand pp. 418-422. For a short time Singapore was an open registry but was closed in 1982. Dempsey p. 40 adds Honduras to the list of flag States characterized by ships of poor condition, inadequately trained crew and frequent collisions and notes that flag of convenience ships lay back many disastrous oil spills including the TORREY CANYON, the ARGO MERCHANT, and the AMOCO CADIZ.

149 Wang p. 331 notes that flag of convenience States have the most tonnage and the largest numbers of tankers. He also notes that Liberia covers nearly 90% of the convenience tanker fleet.

150 *Ibid.* p. 332 notes that Liberian and Panamanian fleets are so overwhelmingly owned and controlled by the US that it can be said that their maritime policy-making is chiefly made by the major American tanker owners or in Washington D.C.

151 Bergstrand pp. 418-422. The world average is 22%. Nordlys 12 mai 1989, Nr.107, 88.årg., s. 24. recently reported that the Norwegian fleet's average age has increased from approximately 6 years to 15-16 years old, which can result in an decreased security at sea. The Norwegian Directory of Shipping has consequently drastically increased their control of technical and safety equipment aboard ships.

152 *Ibid.* pp. 418-422. Table 8. Greece and South Korea are also included.

153 *Ibid.* India is also included.

154 Nordlys 4.april 1990, Nr.80, 89.årgang, s. 22, indicates that the present Soviet intention is to use its own ships to freight Western tonnage through the Northeast Passage, and this system is expected to continue for a time.

155 In Bergstrand pp. 424-432 measures proposed against these ships include States jointly implementing legislation to effectively prosecute by port State authority polluting discharges in their exclusive economic zones and *high seas*. There is at least an argument that a State could bring legal proceedings for illegal discharges in the exclusive economic zones under LOSC provisions. See Part II Section 14.1.2. Bringing such proceedings for illegal discharges on the high seas is something new, and this provision is contrary to MARPOL 73/78 and LOSC provisions giving weight to flag State jurisdiction. This provision would probably be the cause of Arctic State non ratification of the Arctic Vessel-Source Protocol and therefore these measures could be included only if the Arctic States were receptive. Other provisions include a requirement that a certain percentage of the crew be nationals and there be a degree of beneficial ownership.

156 Wang pp. 317-319. Johnston p. 416 calls for the coastal States with responsibility for environmental protection of *ice-covered* areas of their exclusive economic zone to cooperate and exercise their authority with respect to appropriate aspects of vessel construction and design, discharge controls and routing of traffic to ensure protection of these Arctic areas. The coastal State authority is based upon LOSC Article 234, see Part 14.1.2.

157 Wang pp. 317-319.

158 *Ibid.*

159 See *ibid.* pp. 320, 334-335, where it is noted Conventions have tried to apply sanctions by appealing to the flag State's comparative self interests, wherein the flag State conducts its annual surveys responsibly to avoid the port authority's ad hoc spot survey which might indicate the issuance of false certificates and thus cause 'loss of face' internationally. The effects of this can be questioned however.

160 See *ibid.* pp. 317-319. Curtis p. 707 notes that in the US for a 7 month period the US Coast Guard boarded 2045 ships and *88%* of the hips complied with certificate and oil record book regulations. This would indicate that a strong stance by Port State authorities is bringing positive results regarding compliance with MARPOL 73/78 standards, indicating the same is also possible in other jurisdictions.

161 *Ibid.* See also Aftenposten 28. mars 1990, Uke 13., Nr.147., 131.årg. s. 15. where the problem as regards Norway is mentioned.

162 See Curtis p. 695 and Wang p. 315.

163 See for example *Noroil*, April 1989.

164 See Aftenposten 20 mars 1990, Uke 12, Nr.134, 131.årg., s. 10. where it is noted that the development of environmental technology is currently in vogue.

165 A request may be made by the port State however.

166 See Wang pp. 315-316.

167 *Ibid.*

168 *Ibid.*

169 Problems regarding Annex II following the 1978 Protocol include a continuation of the problems noted above, specifically impractical procedures, unavailable technology, administrative problems with checking compliance, constraints on trade, inadequate chemical reception facilities and excessive costs for efficient tank stripping systems, this in spite of the fact that an accident involving some of the chemicals appearing in Annex II's 'Category A' high hazard list might be more disastrous than an oil spill, Curtis p. 704.

170 See Part II Sections 5.3. and 5.4. for discussion of provisions from other IMO Conventions referred to in this Section.

171 See Section 9.3.2. and Part II Section 14.1.2.3.

172 Gold p. 38.

173 IMO Document p. 87.

174 IMO Document p. 86.

175 Perhaps on a practical level it may not be so confusing. People who use these systems in the Arctic seem to have little trouble differentiating them and knowing their limits. Conversation with ship's electrician, Tromsø, March 23, 1990.

176 Lamson pp. 4-5. Lamson Arctic pp. 70-71 notes that Arctic navigation and communications are characterized by special problems some of which could be remedied by better research and funding. These problems include a need for hydrographic surveying and charting, installation of additional aids, beacons and buoys, and the implementation of universal coverage communications systems such as LORAN-C and OMEGA. Special problems caused by Arctic atmospheric and environmental conditions include magnetic wandering, disturbance and error; ionospheric disturbances including 'flare blackouts'; polar cap absorption and storms; and bad visibility caused by local fog, storm icing or abnormal refraction.

177 Lamson p. 6.

178 Gold pp. 48-49.

179 Johnston p. 417.

180 *Ibid.* The author notes in particular that collision should cover movable structures as well as ships and salvage rules besides being concerned exclusively with the hull and cargo should also take into account a pollution risk from ships in distress.

181 See Part I Sections 6.2. - 6.4. for discussion of liability provisions referred to in this Section.

182 National legislation is of course in force in the areas of State jurisdiction and in the Norwegian case consists of good coverage, though it could be better as regards vessel-source pollution. See Section 9.2.4. and Bull Erstatning pp. 682-693.

183 Birnie p. 235.

184 Johnston pp. 415-416 calls for the IMO to form 'an international convention relating to damage caused by marine pollution providing for rules concerning responsibility and liability, uniform determination of costs to be reimbursed for remedial measures, assessment of damages cause to natural resources, especially to those without market value and concerning the establishment of expeditions standard procedures for rapid compensation.

185 *Ibid.*

186 Churchill pp. 264-265.

187 *Ibid.* and Birnie p. 235.

188 Johnston p. 237 notes that one of the main difficulties related to land-based pollution is that it is basically a collective or generic phenomena, and virtually impossible to identify the person or category of persons responsible. The author also notes that therefore the control of land-based pollution is therefore more dependent on tax incentive or tax deterrent systems.

189 See Section 9.2.4. As noted there is in existence the 1976 Liability Convention covering *oil rig* activities in the North Sea. This would not apply to the Arctic both because of its geographical area of application (only the Norwegian Barents Sea) and because it is *not* in force. It is mentioned here because the ratification problems associated with this convention can be expected to be the same for the Arctic. States signing the Convention include the Federal Republic of Germany, Ireland, the Netherlands, Norway, Sweden and the United Kingdom. *None* have ratified. The Netherlands, Norway, Sweden and possibly the Federal Republic of Germany have 'good' environmental reputations, yet refuse to become Parties. Though it is difficult to say exactly why, a comparison with Norwegian and British law, and the 1976 Liability Convention might partially provide an answer. Norway's legislation covering offshore liability and compensation is characterized by unlimited, strict liability of the operator. See *ibid.* Under the U.K.'s legislation the doctrine of strict liability of the operator is probably not recognized and possible causes of action are limited to negligence, trespass, breach of statutory duty and nuisance. See generally Bull Erstatning and Hjort s. 210-226. There are differences as well in damages recoverable, and overall the Norwegian legislation can be said to be more 'claimant friendly' and the U.K.'s more 'oil industry friendly'. In short the 1976 Liability Convention might be too conservative for Norway to ratify and too progressive for the UK to ratify. 'Claimant friendly' States however have the option under the 1976 Liability Convention to raise the liability limits even to unlimited; so this is not the entire answer. It might be simply that all States in spite of their rhetoric are reluctant to agree to too binding liability and compensation regimes to which oil companies upon whom States are dependent for State revenue would be subject.

190 This assumes that a definite amount of damage could be assessed as well as the argument countered that the fish are owned by no one and hence no one can collect damages.

191 See Part I Section 1.2.3.

192 Bull Erstatning p. 110.

193 See Part I Section 1.2.1.

194 See Section 9.2.4. and Churchill Conflicts pp. 204-205 for a description of a fund established to cover compensation demanded for damage caused by offshore scrap.

195 Kasoulides pp. 497-523 on the contrary puts much faith in the LDC's Article X, customary law and the work of the International Law Commission on international liability for ultrahazardous activities and codification of rules of State responsibility. Specifically is mentioned the development of liability and compensation rules and procedures under the LDC as an *actual* process as well as Resolution LDC 2(9) in which State Parties agreed not to resume dumping of low-level radioactive material before the adoption of an 'adequate and comprehensive' regime.

196 *Ibid.* p. 513.

197 Timagenis p. 64. The author directs most of this discussion to vessel-source pollution, but the problems noted seem equally relevant to dumping.

198 See Roginko p. 15.

199 See Part I Sections 7.1. and 7.8.4. Johnston p. 411 notes that under LOSC Article 199 each coastal State has an obligation to develop its own national contingency plan, though preferably it should be done together with its neighbours so that a basis for *joint* continency planning is provided as required by Article 199. See Part II Section 13.2.2. for LOSC provisions.

200 Additional problems of existing contingency plans are of a more technical nature wherein it is doubtful that with today's technology much clean up can be done in sea areas with ice, or if waves are over 2.5 meters high. See Westergaard pp. 250-252. Additionally there may be operational and organizational problems at the national level indicated by inadequate initial responses especially in Alaska in the EXXON VALDEZ accident but also in Norway in the Sognfjord accident. See Aftenposten 31.januar 1990, Uke 5., Nr.51, 131 årg., s. 12, and Aftenposten 26.oktober 1989, Uke 43., Nr.494., 130., respectively. See also Stokke p. 10.

201 See Part I Section 7.8.2.

202 See Part I Section 7.8.3.

203 See Part I Section 7.8.6. Norway and Denmark/Greenland are not included here since there is not yet talk of oil exploration in the area between Jan Mayan and Greenland.

204 Pharand pp. 752-757. There is also a maritime boundary dispute between Denmark/Greenland and Norway in the area between Jan Mayan and Greenland. See Østreng Delimitation p. 137.

205 This point is just north of the permanent ice covered zone according to *Gyldendals Store Verdensatlas*, Gyldendal Norsk Forlag (Atlas) s. 12.

206 Pharand p. 752. The respective areas of responsibility are defined in Article 1 as areas of Nares Strait, Baffin Bay and Davis Strait lying to either side of the dividing line established by the 1973 Continental Shelf Delimitation Agreement.

207 *Ibid.* at 756. It appears that both the continental shelf boundary and the maritime boundary are the 1867 Convention line up to 72 degrees north latitude, which is just north of Wrangel Island. Another author however notes that the boundary in the Bering Sea is *not* definite due to the different application of cartographic techniques, causing an area of approximate 15,000 square nautical miles to be in dispute, Antinoir, Camille M., 'The Bering Sea_ A Maritime Delimitation Dispute between the United States and the Soviet Union,' *Ocean Development and International Law*, Vol, 18, No. 1, (Antinori) 1987 pp. 1-11. The Soviets claim the rhumb line method is the one intended by the 1867 treaty negotiators while the US claims the great circle method was intended. Since the area in dispute is considered one of the most promising boundary areas of the US outer continental shelf for hydrocarbon exploration and is of 'great economic, environmental and strategic importance', the dispute has some similarity to the Norwegian - Soviet dispute. Note however that the US - USSR Agreement does not address this issue. As noted the US and the USSR just settled their maritime boundary dispute.

208 Pharand p. 757.

209 See Part II Section 13.2.3. for discussion of Article II.

210 *Gyldendals Store Verdensatlas*, Gyldendal Norsk Forlag, (Atlas) s. 12. The US Department of State, Bureau of Oceans and International Environmental and Scientific Affairs, was contacted concerning these questions where it was substantiated in a correspondence of January 3, 1990 that the boundary question does not affect the contingency plan.

211 Atlas s. 12.

212 Pharand pp. 754.

213 *Ibid.* at 753.

214 *Ibid.* at 756. It also seems possible that US acquiescence could affect the Norwegian - Soviet dispute, as it gives weight to the sector theory propounded by the Soviets, see 9.2.2. and Tresselt s. 75-93. The US State Department maintains however that the US has never deviated from its equidistant line principle in the Beaufort Sea. Correspondence from US State Department, Bureau of Oceans and International Environmental and Scientific Affairs, January 3, 1990.

215 See Section 9.2.6.

216 See Part II Section 8.3.

217 Conversation with Professor Geir Lundestad, University of Tromsø, November 1987, where it was stated there was no unclassified literature indicating Norwegian policy regarding bilateral agreements with the Soviets.

218 See generally Østreng, pp. 57-85, and Young, Oran R., 'The Age of the Arctic,' *Foreign Policy*, 1985-86, No. 61, (Young) pp. 160-179.

10. Proposal: A Comprehensive Arctic Regional Antipollution Agreement

10.1. Introduction

In order to consider other arguments giving support to an Arctic regional environmental agreement, a broader picture than just the legal is needed. In the Arctic including the Barents Sea there are various conflicts present including economic, environmental, political-strategic and legal.[1] Briefly the scenario is that there are expected to lie hugh oil reserves in the offshore areas, which if developed with new technology could bring not only large economic gains but also the corresponding environmental damage.[2] On top of this the Arctic States have different political systems, 2 of which are the world's superpowers standing at opposite ends of the political spectrum though the Soviet Union is changing. Both superpowers regularly patrol nuclear submarines under the polar ice, and Canada may to do the same. Additionally, as noted there are boundary line disputes involving every one of the five littoral States.[3]

These conflicts, especially the political-strategic one, at first sight seem to preclude the idea that any kind of a regional environmental agreement might succeed. Logically it would seem simpler for the Arctic States to enter bilateral arrangements, if seen in light of these differences. It seems less complicated to keep the players as few as possible when negotiating. To counter this idea, however, it must be noted that there are regional pollution control agreements which contain countries with strong and bitter political-strategic differences, more so than in the Arctic.[4] As noted Libya and Israel, and Greece and Turkey are UNEP Convention members in the Mediterranean. Iran and Iraq are convention members in the Persian Gulf. Saudi Arabia and South Yemen are members of the Red Sea agreement, and the United States and Cuba members of the Wider Caribbean agreement. The Helsinki Convention includes 3 former Warsaw Pact States, including the U.S.S.R., and 4 Western States. 2 of the Western States are NATO members (Denmark and West Germany) and 2 are nonaligned (Finland and Sweden). Though alliances are presently changing these arrangements are noteworthy both because of the political frictions, and because these States meet *regularly* to solve environmental problems. With models such as these, especially in view of the fact 3 of the Arctic rim States, Denmark, the Soviet Union and the U.S. are themselves parties to 'frictional' agreements, the political-strategic argument against regional conventions loses some of its force.[5]

Environmentally a regional agreement may be preferable, since the entire region would be managed and protected *uniformly* and more efficiently, which would not be the case with a series of bilateral agreements. In addition there are economic reasons pointing towards a regional agreement being optional. Although the world is 'greener' than before and more interested in preserving the environment, still countries are very interested in the *costs* and will proceed into regional agreements more readily when the benefits gained outweigh the costs.[6]

10.2. Discussion

Though some of the processes briefly described in this Section have already been presented in Part I Chapter 1, this is considered necessary in outlining the arguments in support an Arctic Regional Agreement.

10.2.1. Arctic Vulnerability

Several authors, chiefly Canadian and Soviet,[7] mention the fact that Arctic *current* patterns may cause any pollution spills or accidents from offshore oil activities or Arctic shipping to affect the waters of *all* Arctic nations, and this may cause the Arctic States to eventually enter a legal regime which coordinates Arctic environmental protection.[8] See Figure 10.1. As can be seen from Figure 10.1, there are several drift patterns of pack ice and water in the Arctic. It can be seen that the Beaufort Gyre, 'a vortex of pack ice which moves in slow, clockwise circuits over Alaskan and Canadian waters,' in cycles of 7 to 10 years,[9] eventually merges with the major circulation pattern, the Transpolar Drift Stream, which carries ice from the east Siberian Sea over the Pole along the east Greenland coast and into the Atlantic Ocean. Simplified surface current patterns over Norway and the Soviet Union can be viewed in the Barents Sea from Part II Section 8.5., Figure 8.17.

Although as seen in Part I Sections 1.2.3. through 1.2.7. not much is known regarding the behaviour of oil and other pollutants in the Arctic, it is known that pollutants are broken down slowly due to low temperatures, and because of this they are slowly released from ice and snow into the water over long distances and covering vast areas through the sea drift patterns.[10] Since most drifting ice from the Arctic drifts into the Northern Atlantic, this area may as well be affected by the pollution, and consequently Arctic States may be affected by pollution discharged, 'a thousand miles away from its coasts' with immense areas of the marine environment disrupted, much more so than would occur in lower latitudes.[11]

The Arctic climate is harsh and characterized by low water temperatures, large areas which are ice covered most of the year, and sharp contrasts of solar radiation during the polar day and the polar night periods. This harsh climate necessary affects the life present, the 'tough' life forms found in the Arctic seas often exist on the verge of the possible and consequently even minor pollution is 'extremely dangerous' for the Arctic.[12]

Besides the toxic effects pollutants have on the Arctic, effects may also occur physically and chemically to the ice and snow cover. Oil films on the ocean surface slow down evaporation of the water by 50% and ice growth by 1.3 to 1.9 times, which results in increases of *melting* of sea ice.[13] Reflection of solar radiation on ice and snow is as well reduced, resulting in even more melting, and it has been found that ice and snow melt 5 to 10 times faster in polluted areas than in non polluted ones.[14] On a large scale if much pollution were to occur, this could hypothetically cause faster melting and a shift to the north of the ice edge, which could result in climatic changes through changes of the heat and moisture exchange between the ocean and the atmosphere, and the ocean and the land, and also result in changes in the intensity and direction of cyclonic winds.[15] As noted oil
pollution also reduces the amount of diluted *oxygen* in Arctic waters, which is particularly critical since Arctic waters are characteristically rich in oxygen, and the majority of the commercial fish such as salmon are highly sensitive to such changes.[16]

Thus Soviet researchers, if not Canadian to a lesser extent, consider the Arctic environment fragile, easily put out of balance and capable of being easily damaged. 'One major oil tanker accident alone or a blowout of an oil platform may drastically change the environmental situation in the entire Arctic Ocean.'[17] Although as seen this view is not without controversy, the *precautionary principle* should be of application here.

This principle means basically that, in case of doubt as to the effects on the marine environment, preventive or remedial action *should* be taken. In other words any decisions made should err on the safe side.[18] In this case this would mean to follow the Soviet Arctic research stressing the vulnerability of the Arctic and implement stringent measures as soon as a detrimental effect to the Arctic marine environment becomes *plausible*.[19]

Against this background, it must be noted that currently human activities are on the upswing in the Arctic with all the polluting consequences. In addition to those Barents Sea activities noted in Part I Chapter 8 commercial shipping in several Arctic areas is on the increase.[20] Currently the U.S. is moving oil through a pipeline from the North Slope to Valdez, Alaska and then tankers down to the west coast of the continental US The U.S. is also considering plans on moving oil from the Bering Sea and the Chukchi Seas down to the Pacific markets and moving Beaufort Sea oil and gas to eastern markets through the Canadian *Northwest* Passage.[21] Arctic shipping between Europe and the Pacific through *both* the Northwest and the Northeast Passages (as noted) is as well being planned for the future.[22] Also offshore oil exploration is being conducted or is in the planning stages in the Barents Sea (Norway, Soviet Union) (as noted), the Beaufort Sea (Canada, U.S.), the Bering and the Chukchi Seas (U.S., Soviet Union).[23] Military activities up until recently in the Arctic have also been increasing and posing threats to the marine environment through nuclear accidents and dumping.[24]

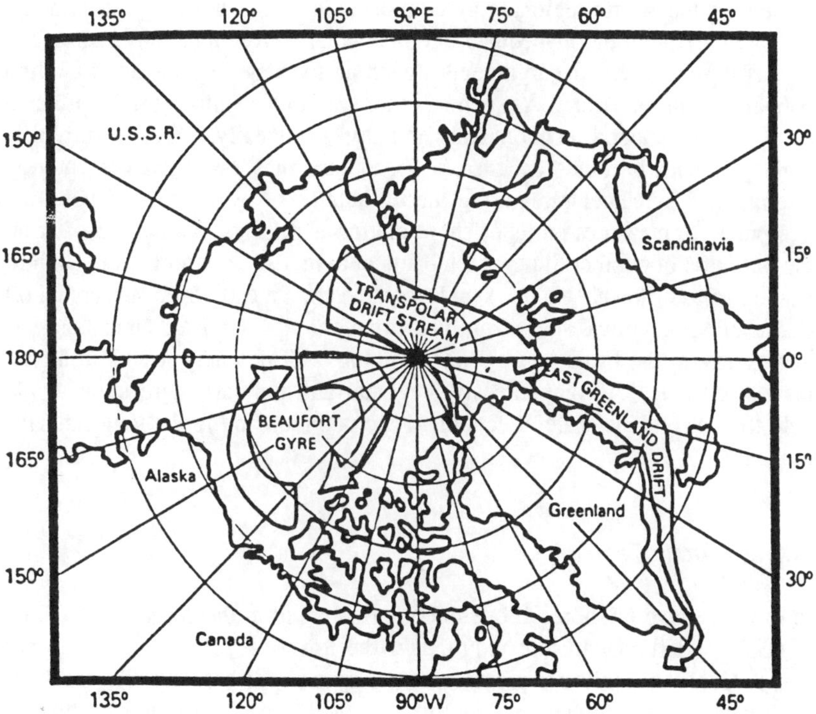

Figure 10.1. Arctic Drift Patterns.[25]

Land-based pollution in the Arctic mostly has its source from the large Soviet rivers, the Yenisey, the Lena and the Ob, the Yenisey being one of the most polluted rivers in the Soviet Union due mainly to the mining and metal processing.[26] Presently there is little pollution added from Canadian, U.S. and Norwegian rivers.[27] Airborne pollution in the Arctic is imported, mostly from industrial centres in Eurasia and America, which forms Arctic haze in the winter.[28] The Arctic haze can be comparable to smog in large cities, can be as much as 6 km deep and is 20 to 40 times higher in concentration in the winter than in the summer due to the sharp drop in the temperature which causes larger deposits. Sulphur dioxide, sulphates, nitrogen oxides and heavy metals are the chief pollutants, and reports of effects on vegetation and reindeer have been noted.[29] Local effects can be worse, such as those noted in Part II Section 8.5. on the Soviet Kola Peninsula around the industrial cities of Murmansk and Nikel.[30]

From the above activities it can be seen that there are already pollution problems in the Arctic, and they will probably become worse without effective regulation.[31]

Viewing the Arctic regional vulnerability combined with the precautionary principle and with the inadequate environmental conventional coverage noted in Part II Section 9.3. the arguments are strong for State regional anti pollution cooperation in the Arctic. As soon as more than two countries can be affected by pollution incidents, economically and environmentally, a regional multilateral agreement becomes of interest. A country may 'save' in not entering international agreements requiring the installation of expensive anti pollution equipment or measures on its offshore platforms or ships, but may 'pay' from the pollution of another State which drifts onto its fish spawning grounds, bird nesting areas or towns. No one stands to gain economically from substantial oil or hazardous chemical spills from platforms or ships. Oil or chemicals lost is money lost to the owners or operators and as well probably results in higher insurance rates.[32] A comprehensive regional anti pollution agreement would help to control wide scale effects of Arctic pollution and in the long run save money.[33]

10.2.2. Arctic Technology

In order to have offshore oil operations or shipping in the Arctic, technology is needed additional to that required in the lower latitudes, and this is not cheap. One need only see pictures of the special enclosed oil rigs used for exploration in the Barents or Beaufort Sea, or the special Arctic ice breakers used by the Soviet Union, Canada and the U.S. to get an idea of what is required. Even more extensive technology will be required as offshore oil and gas operations move into areas with pack ice, and the Northeast Passage and the Northwest Passage are held open for longer periods of the year.[34] To acquire this technology, engineering expertise related to Arctic conditions is required, and at this point of time this is *limited.*[35] If cooperative efforts are made among the Arctic countries to gain the needed expertise through research and development, all involved would save substantial time and money, since duplicate efforts would be reduced.[36] The countries involved actually have a good starting point in this regard, and several have as noted entered into bilateral Arctic technological agreements.[37] Specific regional cooperative projects envisioned include, 'Arctic ice breaking capacity, navigation and communications systems, ship traffic management and emergency response capabilities.'[38] This latter area includes developing *environmental* technology as related to pollution incidents from oil platforms and ships and subsequent clean up operations. To the extent that Arctic environmental technology could be regionally researched and developed to prevent and clean up pollution spills, which may damage fish spawning grounds and decimate animal populations, all stand to gain.

10.2.3. Living Resources Migration Patterns

A third reason for Arctic State regional cooperation involves the transboundary migration patterns of living resources. Perhaps of most economic interest is that mentioned previously involving fish. Although no effort will be made here to illustrate the different migration patterns of Arctic fish, one need merely consider the accounts mentioned briefly in Part II Section 8.4. of Norwegian and Soviet efforts to effectively manage the Norwegian Arctic cod with its different movements in the Barents Sea to obtain an idea of the extent of this transboundary problem. Migration of different species similarly takes place in the Chukchi and the Beaufort Seas, and can cover the waters of 2 or 3 Arctic States. Of environmental significance are the birds, whales, seals, walruses and polar bears in the Arctic, many of which have migration patterns which cross the boundary lines of two or more States.[39] In addition many of these animals as noted have special roosting, spawning or habitat areas, the destruction of which by oil or hazardous chemicals would mean the destruction of a large amount of that population. Several bird and animal species are already identified as requiring special protective status, and special areas have been designated to ensure their survival.[40]

In terms of offshore oil activities or Arctic shipping it is obvious that their placement in relation to protected animal areas is crucial. An example is given in Canada in which 38 locations across the North were indicated where oil activities should be *prohibited* completely, 17 where activities were prohibited unless approved by an environmental assessment board, 8 where exploration could proceed only following assessments of the specific site and *only 1* where oil exploration could proceed unopposed.[41] Norway has as well planned areas where oil exploration is limited either completely or allowed only during specific times of the year so that fish spawning or bird populations will not be threatened, and the U.S. (Alaska) appear to be in the process of doing the same.[42] It is not known specifically what Denmark/Greenland have done in this respect, but the point is that because of the transboundary nature of both pollution and many Arctic animals, the most optimum solution for the protection of the animals would be regional cooperation. It helps little if Canada or Norway for example have progressive policies regarding the placement of oil rigs or Arctic shipping lanes, if other countries do little or nothing. There will always be species of fish, birds or mammals which would suffer from one Arctic State's less than satisfactory policies. Regional Arctic cooperation would ensure an exchange of knowledge of animals habitats and patterns and a more equal application of protective measures such as restrictive areas or times for drilling activities or shipping.

10.2.4. Native Population

A fourth consideration in favour of a regional anti pollution cooperation involves *native* human populations, dependent upon birds, fish, seals and whales which may be adversely affected by offshore oil developments and Arctic shipping. As related to the Inuits in Canada and Alaska primary food sources include ringed and bearded seals, bowhead and beluga whales, walruses, caribou, migratory birds and freshwater and ocean fish.[43] Changes in the availability of these species because of pollution spills, besides causing food problems also may threaten Inuit survival culturally, 'because hunting, fishing, and trapping are core activities which embrace important norms and values such as sharing, bonding, identification, pride, nutrition, and adventure.'[44] Presumably these experiences can be extrapolated to Greenland and Soviet Siberia as well and to some extent to Northern Norway with folk groups who are dependent upon renewable marine resources.

Another negative affect which should be considered is that when jobs are created, such as in the offshore oil industry, native populations in the area can become dependent upon this source of income, and traditional skills lost. When the oil dries up, this work force either has to move to another area or return to the traditional norms, the latter, which experience has proven, is difficult.[45]

More than bilateral State cooperation is needed in the Arctic to deal with both the transboundary migrational nature of animals hunted or fished by the native populations and the Arctic circulation patterns. A look at Figure 10.1 and Figure 8.17 from Part II Section 8.5. shows that oil or chemical spills north of Norway and the Soviet Union could cause problems not only to animals, fish and peoples north of these States but also to the same east of Greenland. A spill in the Beaufort Sea, besides affecting animals, fish and peoples north of Canada and Alaska could also later cause problems for the same north of the Soviet Union and Norway and east and west of Greenland. Regional cooperation in the placement of the oil rigs and Arctic shipping routes, as well as the timing of the explorative drilling, and cooperative contingency plans for dealing with any spills could alleviate some of these problems for native populations.[46]

10.2.5. Security

In addition to the Norwegian security problem mentioned in Part II Section 8.3. favouring a regional anti pollution agreement the same argument might as well apply to varying degrees to the other Arctic countries. Perhaps the U.S. would not have problems entering bilateral environmental agreements with the Soviets, since both presumably respect one another, politically and militarily. Canada since it is a smaller country however *might* have some reservations entering specific bilaterals with the Soviets.[47] Denmark/Greenland might have the same reservations as Norway. Denmark/Greenland and Norway might also have some reservations entering bilaterals with the U.S. because of fears of undue political influence in the North.

10.2.6. Other Considerations

In addition to the above arguments the regional agreement has other advantages as well, mostly legal and quasi legal considerations. Briefly the 1982 LOSC, the 'Constitution of the Law of the Sea', must rely on regional *supplementation* in order to be implemented. The LOSC's provisions as will be seen though progressive are so general that specific provisions are needed for successful pollution control and prevention. The differences of types of pollution and degrees vary according to the geographic area and require differences in control. Regional organizations are more familiar with the problems involved and distribute equipment to those areas where accidents may occur. Additionally a regional approach would involve only those countries directly involved with a given type or area of marine pollution.[48] Though these arguments could be forwarded in favour of bilateral cooperation, as will be seen regional cooperation under LOSC provisions seems more specifically addressed and preferred.

From this background a brief summary will be given of the *legal foundation* in international law upon which a regional Arctic environmental agreement could be based.

10.3. Legal Foundation for Arctic Regional Agreement

10.3.1. Rules of International Law

Generally authoritative writings, judicial decisions, general principles of law, and the practice of States and conventions provide guidance to the determination of international environmental law which would give legal support to the formation of a regional agreement in the Arctic.[49] Additionally the statements and declarations of international organizations and international conferences, 'soft law', play an important role, though they are of no direct legal status.[50] State *practice* and *conventions* are chiefly addressed here. General principles of law, authoritative writings, judicial decisions and 'soft law' will be mentioned, though to a lesser degree. All of these sources give a legal foundation of varying degrees to the formation of a regional Arctic environmental agreement.

10.3.2. General Principles

As seen in Part I Sections 2.1.3. and 2.2. general principles of international law, such as 'good neighbourliness', 'one must so use his own as not to do injury to another' and the prohibition against 'abuse of rights' along with the *Corfu Channel Case* and *Trail Smelter Arbitration* impose State responsibility, although inadequately, where transboundary pollution results in substantial damage.[51] Additionally there are procedural obligations, including the duty of a State to notify, exchange information, consult and negotiate regarding transboundary pollution.[52]

10.3.3. Customary Law and Norms - Pollution Conventions, LOSC

The conventions and agreements mentioned in Part I lend support to Arctic cooperation among the Arctic rim States as legal precedent. Additionally as noted in Part I Section 2.1. a series of State practices, exemplified by these treaties and agreements, at a certain point in time and supported by the writings of legal scholars and the acknowledgment of States, passes from national actions into international customary law, which would be binding on all States.

Sections of the LOSC reflect State practice and thus presumably represent international customary law covering marine pollution and apply to the Arctic Ocean binding on all States.[53] As noted the problem is the *extent* to which LOSC provisions represent international law, with one author claiming *all* Articles from 192 to 237, another only Article 192 and 194(2), and still another lying somewhere in between.[54] The sections which can *at least* be said to have passed into customary law include the principles, 'States have the obligation to protect and preserve the marine environment,' and States are 'not to cause damage by pollution to other States and their environment, and [are obliged to assure] that pollution ... does not spread beyond the areas [under their jurisdiction],'[55] though there might easily be others. The LOSC pollution Articles are presented specifically in Part II Section 11.2. as well as in each individual Chapter 12 - 17 in the discussion of the Arctic Convention and sub Protocols, in the event that they represent international customary law.

In the opinion of one author apart from *codifying* international customary law, the LOSC also creates *norms* for the formation of regional agreements which could be applied to the Arctic.[56] Generally, even though there are 42 *ratifications* out of a necessary 60 for the Convention to enter into force, as of December 1989,[57] all of the Arctic rim States with the exception of the United States have *signed* this instrument and thus have indicated their willingness, 'to refrain from acts which would defeat the object and purpose of the treaty.'[58] The U.S. has however as noted indicated in a Presidential Proclamation that American maritime policy in the economic zone will be generally in line with the Convention.[59] It therefore appears that in addition to any *obligations* required under customary law in forming an Arctic regional agreement the Arctic States have the opportunity to follow *recommendations* made by LOSC for forming Arctic environmental regional agreements in at least 2 major areas.[60]

The first deals with *co-operation* in protection of the marine environment. Co-operation in marine protection is represented by Articles 207 and 208, which declare that marine pollution policies for land-based sources and sea-bed activities are to be harmonized at the *regional* level as well as on the global level. Under Articles 210 and 211, States are to establish global and regional rules for the control of ocean dumping and international rules for vessel-source pollution, and are to promote routing systems to minimize the threat of accidents. In Article 212 States are asked to 'endeavour' to control atmospheric pollution of the marine environment. Article 200 requires States to co-operate in scientific research concerning marine pollution. Finally, Article 201 requires

States to co-operate in establishing the scientific criteria for the formulation of rules and recommended practices to control marine environmental pollution.

The second area, the most compelling rule for Arctic regional co-operation, centres around Articles 122 and 123 concerning *semi-enclosed seas.* Article 122 defines a 'enclosed or semi-enclosed sea' as a gulf, basis or sea surrounded by two or more States and connected to another sea or the ocean by a narrow outlet or consisting entirely or primarily of the territorial seas and exclusive economic zones of two or more coastal States. Article 123 declares that States bordering an enclosed or semi-enclosed sea *should cooperate* with each other in the exercise of their rights and in the performance of their duties under LOSC. To this end they *shall endeavour,* directly or through an appropriate regional organization (a) to coordinate the management, conservation, exploration and exploitation of the living marine resources and (b) to coordinate the implementation of their rights and duties with respect to the protection and preservation of the marine environment.[61] Thus if the Arctic could be considered a semi-enclosed sea, the Arctic rim States under Articles 122 and 123 have close to a *mandatory* duty to cooperate and attempt to establish joint agreements to protect and preserve the Arctic environment.

In determining whether the Arctic qualifies as a 'semi-enclosed' sea, geographic and treaty practice are considered. Though the Arctic has been described by some as geographically semi-enclosed, the treaty practice of 60 some States in the Baltic Sea, the Mediterranean, the Red Sea, the Persian Gulf and the Wider Caribbean, indicates that a semi-enclosed sea is to be bordered by land by about 90%[62] Thus geographically the Arctic Ocean is necessarily excluded. Article 122 also requires 'a narrow outlet' to another sea or ocean, and thus the Arctic Ocean is excluded due to the open Greenland Sea, the Norwegian Sea and the Bering Sea outlets. Additionally the informal understanding among the delegation of the Law of the Sea Conference was to exclude the Arctic from the semi-enclosed rules.[63] So it would appear that the more stringent LOSC regulations covering 'semi-enclosed' sea could not be applied to the Arctic. The whole matter however needs more research before anything definite can be decided, something which is beyond the scope of this project.[64]

Apart from this question however, the Arctic does qualify as a *regional* sea; coming under the Articles mentioned above dealing with regional measures. Also of application to the Arctic is Article 234 governing *ice-covered area* areas. Generally this Article gives the coastal State the right to regulate the entry, passage, conditions of the operation of the ship, the design, construction, manning and equipment of foreign ships in ice-covered areas of a State's exclusive economic zone.

As can be seen, these LOSC provisions generally advise the formation of *regional* anti pollution agreements with inherent State obligations and which if accepted by the Arctic States have application for the Arctic.

10.3.4. Soft Law

In addition to general principles and customary international law and recommendations supporting regional cooperation applicable to the Arctic, three instruments *may* serve as 'quasi-legal guide posts'[65] which persuade States to co-operate in marine management. They also may contribute to the formulation of rules of customary law as the expression of the *opinio juris* of States.[66] These soft-law rules include the Stockholm Declaration, the 1979 UN Resolution on Co-operation in the Field of the Environment, and the 1980 World Conservation Strategy.[67]

Under the 1979 UN Resolution on Co-operation in the Field of the Environment States are urged to exchange information and consult regularly concerning shared resources and to co-operate by means of joint contingency plans. It requires States to consider the transboundary environmental effects of domestic developments, to provide foreign residents equal access to administrative and judicial proceedings and to join in environmental studies. It also asks for speedy, effective and binding procedures for dispute settlement.

As noted under the 1972 Stockholm Declaration the protection of and co-operation for internationally significant eco-systems is requested of all States. Co-operation to protect living resources, for the regulation of transboundary industrial activities, for the management of shared water resources, and the development of compensation schemes for transboundary pollution is also sought.

The World Conservation Strategy of 1980 prepared by the International Union for Conservation of Nature (IUCN) urges littoral States to facilitate environmental protection through bilateral or regional agreements, due to the sensitivity of the Arctic and the conservation problems involved. Also under this Strategy the Arctic is declared 'a priority sea'.

10.4. Conclusion and Recommendation

It is hoped by this discussion that a general idea can be obtained of the advantages inherent in an Arctic regional anti pollution agreement and the legal support for such an agreement, involving Canada, Denmark/Greenland, Norway, the Soviet Union and the United States. Although the support afforded by international law is perhaps weaker than might be expected, it must be pointed out that relying upon vague and general provisions in conventions and customary law, as well as recommendations and requests for legal support and precedence, is by no means uncommon in international law.[68] Also as noted the LOSC Articles will be presented in more detail in Part II Chapters 11 - 17.

A proposal, in consideration of a regional agreement, is to follow the comprehensive UNEP model.[69] This is because as seen it consists of an umbrella convention which sets forth a *general* obligation to co-operate in marine pollution, followed with Protocols which set forth *particular* measures for controlling pollution from specific sources. The advantage of using a general Arctic Convention with Protocols, includes the argument that the Convention can more easily be ratified by States since it does not contain specific commitments, yet it allows the States at the same time to take a stance in favour of environmental protection. The Protocols *do* contain specific points regarding marine pollution and can be ratified and attached to the central Convention by the various Arctic States at a later date depending upon the political priorities set.[70] From an environmentalist's point of view, the entire document, Convention and Protocols covering *all* sources of pollution, should of course be adopted at the same time. This however, in international law, seems to be seldom the case and not politically feasible. Environmental progress seems to be made by gradually inducing States to enter general arrangements, and then enter consequently more specific agreements as time progresses and conditions allow.[71]

Some experts note, '(A)lthough it is highly unthinkable that the independent Arctic rim States will ever require UNEP's assistance, the approach to adopt an umbrella convention, followed by protocols covering the various pollution sources, appears to be of value for an Arctic accord.'[72] This would seem especially true since the United States and Canada respectively developed their bilateral environmental agreements with the Soviet Union in this manner.[73]

What follows are descriptions and discussions of the proposed Arctic Convention with 5 Protocols. These are consistent with what is considered from the above to be the optimum environmental legal regime for the Arctic.

Notes - Chapter 10

1 See Harders Seminar pp. 494-501, 527-529, 532. The conflicts noted in Part II Chapter 8 related to the Barents Sea can possibly be extrapolated to the entire Arctic.

2 Harders Seminar pp. 498-499.

3 See Pharand pp. 749-750, 752-756.

4 Harders Seminar pp. 528-529, 532.

5 Roginko p. 7 notes in addition that it would be hard to overestimate the contribution Arctic regional environmental cooperation would have to develop and strengthen confidence in the region, 'to the cause of creating a genuine zone of peace and international cooperation in the European North.'

6 Norway is a prime example. The country is considered a leading environmental land, represented by Prime Minister Gro Harlem Brundtland, Chairman of the study, *Our Common Future*, The World Commission on Environment and Development, 1987, (Brundtland Report), yet at the same time containing very strong oil and shipping interests upon which a substantial part of the economic is based. See Part II Section 8.2.2.

7 The Soviet Union is the country with the largest amount of research conducted devoted to the Arctic. See Norges Offentlige Utredninger (NOU) 1989:9 p. 57 which notes additionally that a reasonable assumption can be made that approximately 1/2 of all Arctic research presently conducted is carried out by the Soviet Union.

8 VanderZwaag, David and Cynthia Lamson, 'Ocean Development and Management in the Arctic: Issues in American and Canadian Relations,' Conference paper, *Transit Management in the Northwest Passage: Problems and Prospects*, Dalhousie Ocean Studies Programme, Halifax Canada, 1984 (VanderZwaag) p. 67 and Harders Seminar p. 500. See also Roginko p. 5, where it is noted, '(T)he general pattern of water and ice circulation in the Arctic Ocean is such that the path of pollutants from a majority of most possible pollution sources (continental shelf and coasts of Alaska, Canada, the Barents Sea, etc.) will run practically across the entire Arctic Basin.'

9 Lamson Arctic pp. 51-52.

10 Roginko pp. 4-5 notes that ice and snow cover absorb many pollutants readily. Lamson Arctic pp. 51-54 notes subsurface water movements may as well be significant and play a role in transporting pollutants over the Arctic.

11 Roginko pp. 4-5. See however Lamson Arctic pp. 53-54, where it is stated that Arctic science is not very well developed with respect to understanding and predicting water circulation and ice movements, and while it is known that surface waters movements are caused by prevailing winds, other factors such as tides, fresh water inflow, atmospheric pressure gradients, bottom topography, lateral boundaries, Coriolis and centrifugal forces and degree of ice cover also play roles which presently are not understood.

12 Roginko p. 2. The author notes that the specific biological features of Arctic marine life include shorter food chains, poor species composition, dependence on ties with other oceans and clear fluctuating patterns of organic matter production. As noted previously however, Oil Sea p. 488 declares, '(T)he potential impact of a major oil spill on an Arctic ecosystem can presently not be estimated with confidence.' This is due to 'large gaps (present) in the data base on Arctic biology, there existing only slim understanding of biological events during the brief summer open-water season and virtually no understanding of winter events,' parentheses added. In spite of this fragmented knowledge, it is felt that, 'the Arctic marine environment is not nearly as fragile and pristine as it was once popularly thought to be' and 'northern marine biota no more fragile than their more temperate counterparts.'

13 Roginko p. 3.

14 *Ibid.*

15 *Ibid.* Additionally more snow could fall on land areas to the south due to the increased evaporation from the ice free area. It is noted that it is far more likely that oil spills in the Arctic will occur *infrequently*, and their effect on the physical processes in the Arctic is presently re evaluated from lasting decades to lasting from periods of a *few days to 1 to 2 years*. In spite of this, '(I)t should be stressed, however, that the impact of oil spills on fauna and flora of the Arctic seas will be negative and possibly fatal in any case.'

16 *Ibid.* p. 4.

17 Roginko p. 2.

18 Hey, Ellen, 'The Precautionary Approach' Implications of the revision of the Oslo and Paris Conventions,' *Marine Policy*, (Hey), July 1991, p. 245. The author argues that implications of the precautionary approach for international marine environmental policy and law could include (for a revision of the ODC and Paris Convention) a reversal of the burden of proof, placing the burden of proof with those who urge that action is not required.

19 *Ibid.*

20 Roginko p. 1.

21 *Ibid.*

22 *Ibid.* See Part II Section 8.2.1.

23 See Stokke pp. 8-9.

24 Churchill p. 243 and Hakapää pp. 52-54. See Nordlys, 3. februar 1990, Nr. 29, 89 årgang, s. 13, regarding possible leakage of radiation from the Soviet Mike submarine which sunk in the Barents Sea April 7, 1989. Mentioned as well is a Soviet desire for an international institute to inspect all sunken atomic submarines, including 6 in the Atlantic, 2 U.S. and 4 Soviet, comprising a total of *50* atomic weapons and *9* reactors on the Atlantic sea bed. Roginko p. 1. notes that worldwide, hundreds of accidents have occurred either from ships or submarines carrying nuclear weapons or operating with nuclear reactors. Aftenposten 28.november 1991 Uke 48. Nor. 550. 132.årg notes that though the Soviets do not admit to the dumping of radioactive waste, a special group of Russian and Norwegian experts was formed to investigate the possibility of dumping *11,000 tons* of radioactive waste into the Kara Sea. One Russian official maintains that the dumping of atomic waste in the Kara Sea occurred as late as 1990, Nordlys 13.november 1991, Nr.262. 99.årgang, s. 19.

25 Obtained from Lamson Arctic p. 52.

26 Harders Marine p. 286. It is noted that *51%* of the Soviet Union's land territory drains into the Arctic Ocean. Aftenposten 22 november 1991, Uke 47. Nr.540 and Nordlys 22.november 1991, Nr.271, 90.årgang s. 48 note that one researcher indicates that the Arctic Ocean contains *4 times* as much radioactivity as it should based upon normal radioactive sources. This concerns especially Cesium 137, and one theory maintains that the Soviets dumped radioactive waste into the Russian rivers which was carried out to the Arctic seas. Roginko pp. 1-2 believes otherwise noting that Arctic waters continue to be the *least* polluted in the world. Soviet research studies indicate that concentrations of pollutants such as detergents and pesticides in the Soviet Arctic seas, including the mouths of the Siberian rivers, are insignificant and have no upward trend. Chlorinated hydrocarbons, transported from neighbouring seas and the Atlantic Ocean, have been found only in marine organisms in the Barents Sea at levels much lower than international levels. Concentrations of petroleum hydrocarbons in the Arctic seas offshore the coasts of the USSR generally do not exceed the maximum allowable .05 mg/l. Oil pollution occurs locally and mainly along oil transportation routes, mouths of large rivers or in areas with intensive shipping.

27 Stokke p. 8.

28 Roginko pp. 16-17 and Stokke pp. 12-13.

29 Stokke pp. 12-13.

30 *Ibid.* considers the environmental threat from this source is not acute today since most of it is absorbed into the oceans.

31 A graphic description of these activities is given by the comprehensive figure appearing in *Northern Perspectives,* Vol 18, No. 3, September - October 1990, Canadian Arctic Resources Committee, pp. 6-7.

32 Roginko pp. 16-17. Noted additionally is that '(I)t is well known that prevention of environmental disasters costs the society much less than mitigation of their effects. And such effects may be very grave for the Arctic environment.'

33 *Ibid.* p. 7 notes additionally that Arctic State environmental cooperation 'would demonstrate to the world community at large the sincerity of national commitments to preserve the Arctic ecosystem affecting climatic and other environmental conditions on a global scale.'

34 See Nordlys, 25 october, 1989, Nord-Norge Mot År 2000, 'Suksess eller mareritt', 'Drøm blir virkelighet', s. 5-7, 36-38; Nordlys, 24 august 1989, Nr.194, 88 årgang, 'Sovjetisk gigantfunn i Barentshavet'; Nordlys, 16, februar 1990, Nr.40, 89 årgang, s. 64.

35 Lamson Arctic pp. 67-68.

36 See Roginko p. 7 where it is noted that since, 'existing facilities to simulate Arctic conditions and test (engineering) designs are few in number, the expedience of resource pooling on an international basis is obvious,' (parenthesis added).

37 See Part II Section 9.2.6.

38 Lamson Arctic p. 68. Roginko p. 7 notes that cooperation among the Arctic States could make it possible to 'substantially enhance' Arctic navigation safety and reduce vessel-source pollution through the national pooling of resources and coordination in ice and weather forecasting, search and rescue operations, the employment of satellite remote sensing systems, and responding to marine environmental emergencies.

39 See Nordlys 4.april 1990, Nr.80, 89.årgang, s. 56 where it is noted polar bears wander over large areas from Greenland via Svalbard to Franz Josef Land and Novaya Zemlya.

40 Lamson Arctic pp. 57-61 notes that Canada and the U.S. have identified 25 sites along the Canada - Alaskan Beaufort coast and northeast Chukchi Sea that require special status. Some have already achieved some degree of protection while studies are being conducted concerning others. Norway as well has established special areas for birds, fish and whales, see Backer, Inge Lorange, *Naturvern og naturinngrep, Forvaltningsrettslige styringsmidler*, Universitetsforlaget (Backer), s. 111 for a review of Norwegian wildlife reservations on Svalbard. The Soviet Union has legislation establishing the same, see Butler, Section J.4., Edict on Intensifying Nature Protection in Areas of the Far North and Marine Areas Adjacent to the Northern Coast of the U.S.S.R.

41 Lamson Arctic p. 60.

42 See St.meld.40 pp. 19-43 and Lamson Arctic p. 61 respectively. The latter also notes that ice edges and polynyas at a minimum require seasonal protection from offshore oil activities or shipping due to their significance to whales, seals, bears and birds.

43 Lamson Arctic pp. 62-63.

44 *Ibid.* All folk groups who are dependent upon renewable marine resources will be affected in a similar manner, and all the Arctic rim States have such groups.

45 The withdrawal of the Hudson Bay Company and the economic and health affects, including alcoholism, on native populations in Canada is a good example.

46 An additional interesting point made by Lamson Arctic pp. 66-67 involves the need for regional Arctic research to be *in touch* with the native communities rather than serving just industrial and national interests. Because the native populations are more in tune with the environments of their area, they can more effectively contribute to Arctic research, including monitoring offshore development, whale and fish populations, wildlife harvesting and utilization studies, and this knowledge could be distributed throughout the Arctic. It is noted however that although, 'circumpolar native peoples have raised questions about the need for transboundary environmental research, the governments of Canada, the U.S. and Denmark appear to be steadfast in maintaining cooperative research at arm's length.' This argument is also relevant for native populations in Northern Norway and the Soviet Union.

47 The Canadian Ambassador to Norway, Mr. Graham Mitchell noted in a conversation August 11, 1990 that there was little fear in Canada of Soviet domination in the Arctic. However another Canadian official notes the Canadian - Soviet Technical Agreement although apparently very comprehensive, is also very *superfluous*. Conversation Dr. R. Engelhardt, Canadian Department of Indian and Northern Affairs, May 22, 1990. Though this of course could be accounted for by many reasons, it may also indicate a reluctance on the part of Canada to enter *specific* bilateral agreements with the Soviets, similar to the Norwegians.

48 Harders Marine p. 291.

49 Article 38(1), Statute of the International Court of Justice, Møse, Erik, 'Folkerettslige tekster,' Universitetsforlaget A/S (ICJ Statute) 1986 p. 74 at pp. 77-78. This applies to the determination of all international law.

50 See Henkin pp. 91-114.

51 Churchill p. 245 and Harders Seminar pp. 506-507. Churchill also adds to these Article 2 of the 1958 Geneva High Seas Convention to state a principle that States cannot allow their citizens to discharge substances out into the seas which can cause damage to the citizens of other States.

52 Harders Seminar pp. 506-507.

53 See Part I Section 2.1.4.

54 *Ibid.*

55 Articles 192 and 194(2) respectively.

56 Harders Seminar p. 519.

57 OPN December 1989, p. 1. See Part II Section 9.1.

58 Article 18(a), Vienna Convention.

59 See Part I Section 2.1.4.

60 Harders Seminar pp. 519-521 notes 4 major areas. In addition to the 2 mentioned above also noted are Articles 63 and 65, urging States to cooperate in the *management* of fish and marine mammals and other transboundary living resources; and Articles 200, 243-257, 275-277, and 143, which oblige States to cooperate through bilateral or multilateral agreements to create favourable conditions for marine scientific *research*, to establish regional research centres and to integrate the work of scientists studying ocean phenomena. Additionally in regard to the straits in the Arctic Northeast and the Northwest Passages, Articles 41-44 recommend that user States and coastal States cooperate in the establishment and maintenance of navigation and safety aids or other improvements for international shipping.

61 Emphasis added. Additional phrases of Article 123 call for States (c) to coordinate their scientific research policies and undertake where appropriate joint programs of scientific research in the area, and (d) to invite, as appropriate, other interested States

or international organizations to cooperate with them in furtherance of the provisions of this Article.

62 Harders Seminar p. 524-525.

63 *Ibid.*

64 Concerning this point *ibid.* notes '(L)ittle analysis has been done yet.'

65 Harders Seminar p. 517 and Roginko p. 8.

66 See Henkin pp. 91-114.

67 See Part I Section 2.1.2. and generally VanderZwaag.

68 See generally Henkin pp. 10-113.

69 See Part I Section 3.2.3.

70 See Roginko pp. 10 and 23 and Harders Seminar pp. 513-515. Roginko as well suggests the model worked out in the Baltic Sea, the Helsinki Convention, and refers to President Gorbatsjov's Murmansk speech declaring, 'the experience of joint measures on protection of the Baltic marine environment, which are being implemented by the commission of seven coastal states, is worth extending over the whole ocean and sea areas of the planet's northern zone.'

71 Roginko p. 18 notes additionally that efforts towards establishing an environmental regime in the Arctic should proceed cautiously since, '(I)f implemented too quickly and without sufficient national preparation, such proposals could be counter-productive.'

72 Roginko p. 11 and Harders Marine p. 298. Professor David Caron, Boalt Law School, University of California, Berkeley, in conversation of July 24, 1990, notes that the U.S. is dissatisfied with the UNEP Caribbean agreement and therefore can be expected to proceed cautiously in any regional Arctic agreement.

73 Harders Marine p. 298. The formation of an Arctic regional agreement may already in fact be under way. Roginko pp. 9 and 23 notes that in June 1989 scientific organizations from 8 Arctic and Nordic countries signed an agreement to create a non governmental, International Arctic Science Committee (IASC), a possible 'precursor to the establishment of international environmental regimes in the Arctic.' In addition Finland proposed and held a conference in 1989 among the 8 Arctic and Nordic States to conclude an agreement on joint action to protect the Arctic environment, and a follow up meeting was held in Yellowknife Canada in Spring 1990. Professor Caron (above) notes however that the U.S. did not send lawyers as members of its delegation to Yellowknife, and the U.S. interest might therefore be somewhat questioned. Norwegian interest might also be questioned as evidenced by extreme reserve at the Yellowknife meeting, at least as seen though Canadian eyes, conversation Stephen Hazell, Director, Canadian Arctic Resources Centre, August 10, 1990. OPN July 1991 p. 5 notes that a general agreement calling for cooperation in protecting and monitoring Arctic wildlife and Arctic environment was recently signed by Canada, Denmark, Finland, Iceland, Norway, Sweden, the Soviet Union and the US Development of programs covering control of oil, radiation, sewage, acid rain, heavy metals, noise pollution and contingency plans for oil spills and other accidents is planned.

11. Discussion of Arctic Convention[1]

11.1. Introduction

In keeping with the suggestion noted in Part II Section 10.4. advocating a regional Arctic Convention with attached Protocols similar to those developed by UNEP, several proposals will be presented.

A general Arctic Convention will be presented but *not* in its entirety. An idea of the basic characteristics however will be offered and reference made to the Barcelona Convention, the instrument upon which the Arctic Convention is patterned.[2]

A Protocol dealing with offshore oil installations, 'Protocol for the Protection of the Arctic Region against Pollution Resulting from Exploration and Exploitation of the Continental Shelf and the Sea-Bed and Its Subsoil' (Arctic Oil Rig Protocol), will be presented in its *entirety* since it covers a central theme of this project.[3] A bilateral agreement dealing with a joint contingency plan proposal will also be presented in its *entirety* since that is what is presently lacking in the Barents Sea. Both of these proposals appear respectively in Part II Chapters 12 and 13 with corresponding discussions.

Additionally discussions of Protocols covering vessel-source pollution, dumping, land-based pollution, and civil liability will be presented in Part II Chapters 14, 15, 16 and 17 respectively, though full documents will *not* be presented. An outline of the proposed arrangement appears below.

	Arctic Convention[4]	
Protocol[5] Offshore Oil Instal- lations	*Protocol*[6] Counter- pollution (Bilateral agreements)	*Protocol*[7] Vessel- source
Protocol[8] Dumping	*Protocol*[9] Land-based	*Protocol*[10] Civil Liability

11.2. Legal Basis for Arctic Convention under LOSC[11]

Arguments and the general applicable law supporting an Arctic Regional Seas agreement was noted in Part II Sections 10.2. and 10.3. A more in depth presentation of the regional LOSC provisions supporting the formation of the Arctic Convention is as follows.

As noted, under Article 192, '(S)tates have the duty to protect and preserve the marine environment.' Under Article 194 States are required to take individually or *jointly* as appropriate measures consistent with Convention provisions to prevent, reduce and control marine pollution from any source, including measures to minimize to the fullest possible extent: the release of toxic, harmful or noxious substances from land-based sources, the atmosphere or by dumping; pollution from vessels, particularly measures for preventing accidents and dealing with emergencies, ensuring safety operations, preventing discharges, and regulating the design, construction, equipment, operation and manning of vessels; pollution from installations and devices used in exploration or exploitation of sea bed and subsoil natural resources, particularly measures for preventing accidents and dealing with emergencies, ensuring safety operations, and regulating the design, construction, equipment, operation and manning of such installations or devices. States shall endeavour to *harmonize* their policies in this connection, and the measures taken must include those necessary 'to protect and preserve rare or fragile ecosystems as well as the habitat of depleted, threatened or endangered species and other forms of marine life.'[12] Under Article 197 States *shall cooperate* on a global or as appropriate on a regional basis, directly or through international organizations in formulating and elaborating international rules, standards, recommended practices and procedures for protecting and preserving the marine environment. These must be consistent with LOSC provisions and take into account characteristic regional features.

Article 199 requires (shall) States to *jointly develop and promote* contingency plans to respond to marine pollution incidents.[13] Articles 207, 208, 210, 211, 212 deal with the adoption by States of national anti pollution legislation and the establishment of global and *regional* rules, standards and recommended practices and procedures to prevent respectively land-based pollution, pollution from national sea bed activities, dumping, pollution from vessels and atmospheric pollution. The standard of compliance required of the States in forming regional anti pollution rules differ based upon the source of pollution.

Article 207 (4) (land-based pollution) declares States, '*shall endeavour* to establish global and regional rules, standards and recommended practices and procedures.' Under Article 207 (1) States are required to adopt national legislation to prevent, reduce and control marine pollution from land-based sources but shall merely *take into account* internationally agreed rules, standards and recommended practices and procedures.[14]

Article 208 (5) (pollution from national sea bed activities) declares States, '*shall establish* global and regional rules, standards and recommended practices and procedures.' Under Article 208 (1), (2) and (3) States are required to adopt national legislation and take other measures to prevent, reduce and control marine pollution from sea bed activities, and these laws, regulations and measures shall be *no less effective* than the international rules, standards and recommended practices and procedures.[15]

Article 210 (4) (dumping) declares States, '*shall endeavour* to establish global and regional rules, standards and recommended practices and procedures to prevent, reduce and control' pollution from dumping. Under Article 210 (1), (2) and (6) States are required to adopt national legislation and other measures to prevent, reduce and control marine pollution from dumping, and these laws, regulations and measures shall be *no less effective* than the global rules and standards.

Article 211 (1) (vessel-source pollution) declares States, '*shall establish* international rules and standards to prevent, reduce and control' marine pollution from vessels. Under 211 (2) States are required to adopt national legislation to prevent, reduce and control marine pollution from their flag vessels, and these laws and regulations shall at least *have the same effect* as that of generally accepted international rules and standards.

Article 212 (3) (airborne and atmospheric pollution) declares States, '*shall endeavour* to establish global and regional rules, standards and recommended practices and procedures to prevent, reduce and control' marine pollution. Under 212 (1) and (2) States are required to adopt national legislation to prevent, reduce and control marine pollution from or through the atmosphere *taking into account* internationally agreed rules, standards and recommended practices.

Articles 213, 214, 216, 217, 218, 219, 220 and 222 generally deal with the enforcement by States of their laws and regulations adopted in accordance with the above Articles and the enforcement as well of applicable international rules and standards.

Under Article 235 (1), (2) and (3) States, '*shall be liable* in accordance with international law,' 'are responsible for the fulfilment of their international obligations concerning marine protection and preservation' and '*shall ensure* that recourse is available' for liability proceedings for marine pollution damage caused by natural or juridical persons under State jurisdiction. States '*shall cooperate*' in implementing existing international law and further developing international responsibility and liability law.

Under Article 234 as briefly mentioned coastal States are given the right to legislate non discriminatory anti pollution regulations in *ice-covered areas*. Under this Article the State may legislate regulations covering ship design, construction, manning or equipment in excess of accepted international rules and standards as long as they are non discriminatory and have due regard to navigation.

Under Article 200 joint studies and research programs must be promoted among States for the purpose of exchanging information and data regarding marine pollution. In addition under Article 204 States must endeavour to observe, measure, evaluate and analyze the risks or effects of marine pollution and keep under surveillance the effects of any licensed activities to determine whether they are likely to cause marine pollution. Under Article 205 reports showing the results obtained of such studies are required published or provided by States to competent international organizations for distribution.

From the above to the extent that these provisions can be said to be international customary law and recommendations for norms it can be seen that regional cooperation is generally called for among States to protect and preserve the marine environment and would give a legal basis for forming a regional agreement in the Arctic.[16] Specifics are lacking however, and through the use of such legal phrases as 'shall endeavour' and 'take into account', the call is a weak one.[17]

11.3. Arctic Protocols as Related to LOSC Provisions

If thought of as a continuum of State obligation, ranging from most mandatory to least mandatory, for establishing regional rules and standards and related national legislation under the LOSC provisions, a classification can be arranged as follows: offshore platform pollution, contingency plans, vessel-source, dumping, land-based pollution and finally liability regimes.

The LOSC provisions covering *offshore* platform pollution declares States 'shall establish' global and regional rules, and national legislation shall be 'no less effective' than the international provisions; covering *contingency* plans, States 'shall jointly develop and promote'; *vessel-source* pollution, States 'shall establish' international rules and standards, and national legislation shall 'at least have the same effect as generally accepted international provisions'; *dumping*, States 'shall endeavour to establish', and national legislation shall be 'no less effective'; *land-based* pollution, States 'shall endeavour to establish' international provisions, and while establishing national legislation 'take into account' the international provisions; *atmospheric* pollution, States 'shall endeavour to establish' international provisions and while establishing national legislation 'take into account' the international provisions; civil *liability*, States 'shall cooperate in implementing and developing existing international law', while at the same time 'shall ensure' recourse is available in its courts.[18] This ranking is reflected in the chart covering the Arctic Protocols in Section 11.1. above, beginning with the Arctic Oil Rig Protocol and ending with the Arctic Liability Protocol.

Concerning the LOSC provisions with most mandatory regulation, covering offshore platforms and contingency plans, it has been seen that in the Arctic control of the former is generally wanting with the exception of the Paris Convention, while the latter is in the developing stage. The 2 Arctic Protocols presented first and in their entirety are those dealing those with pollution from seabed activities and contingency plans.[19] Concerning vessel-source pollution the establishment of global or regional provisions are mandatory under LOSC, and MARPOL 73/78 is already in force and encompasses the Arctic. As seen previously a Protocol for the Arctic regarding ship pollution is proposed next, envisioning continued reliance upon MARPOL 73/78 but with various additions including LOSC enforcement provisions to attempt to repair deficiencies.[20] From these in the spectrum of State obligation, comes dumping. The establishment of dumping provisions is not mandatory under LOSC, yet still the LDC is already in force and encompasses the Arctic. The ODC is also in force and encompasses part of the Arctic. As seen the LDC excludes dumping related to offshore oil activities, thus an Arctic Dumping Protocol is proposed with this important addition as well as other improvements.[21] It is not known to what extent the Arctic States would be interested in ratifying such a Protocol, but the environmental need is there and it was considered simpler than attempting to amend the LDC to include dumping from offshore oil activities. Continuing on the spectrum to lesser State mandatory obligations comes land-based pollution, which traditionally States have been reluctant to regulate. Correspondingly LOSC provisions require States only to endeavour to establish such global and regional rules, and presently it is only the Paris Convention in force partially covering the Arctic. Because of this reluctance it is not known to what extent the Arctic States would ratify the proposed Arctic Land-based Protocol upgrading the Paris Convention and covering the entire Arctic.[22] As noted below 3 of the 5 Arctic rim States are or will be Parties to the Paris Convention; so it seems possible for at least these 3 to ratify such an agreement. Finally in the spectrum comes State liability for pollution damage, a legal area where traditionally States show very little interest. Though the LOSC provisions require State cooperation, the goal of the cooperation is so nebulous in reality that States would have to do very little to implement and develop existing international liability law and at the same time not obligate themselves. Correspondingly there is no liability agreement among States in existence covering marine pollution damage in the Arctic and it seems doubtful that the Arctic rim States would ratify such an agreement. Yet the Arctic Liability Protocol is presented here for the sake of completeness.[23]

From this discussion perhaps the UNEP model explained previously is better understood under which States are allowed to enter a general environmental convention and then when the time is politically right, to enter into more specific and environmentally desirable Protocols. Under this system the Arctic rim States could possibly at an early state form Arctic contingency plans and enter the Arctic Oil Rig Protocol covering pollution from offshore oil platforms while at the same time waiting to enter the agreements to which they felt there were problems being bound such as the Arctic Land-based Protocol and the Arctic Liability Protocol.

11.4. Discussion of Proposed Arctic Convention

11.4.1. Introduction

In using the Barcelona Convention as a pattern for forming the Arctic Convention it must be noted that this is mainly due to the form this Convention has in relation to its Protocols and the *experience*, both legal and environmentally, UNEP has acquired in establishing such instruments. Since it is the 'umbrella' Convention which gathers the States together in environmental cooperation, it was thought beneficial to briefly reproduce the main points in informal Articles to give an idea of the general cooperative measures among the State Parties. The 24 Articles presented generally parallel those in the Barcelona Convention, and discussion appears simultaneously with the Articles. Additions and deletions relevant to the Arctic are made which are noted, however changes and comments are not made to the extent done with the Protocols, since the generality of the Barcelona Convention is desired preserved. For further development of an Arctic Convention, inputs from national legislation would be desirable as suggested in the Foreword. Additionally inputs from the Helsinki Convention might be desirable, both because it is also a comprehensive Convention covering all pollution sources, and because it contains 2 of the Arctic rim States as Parties, the Soviet Union and Denmark, which might give valuable experience in East - West environmental cooperation.

11.4.2. Articles

The *Preamble* gives general aspirations of the Arctic rim States. These include that the Contracting Parties, conscious of the economic and cultural value of the marine environment of the Arctic Region; fully aware of their responsibility to preserve this common heritage for the benefit of present and future generations; recognizing the threat posed by pollution to the marine environment, its ecological equilibrium, resources and legitimate uses; mindful of the special hydrographic and ecological characteristic of the Arctic Region and its particular vulnerability to pollution; appreciating a need for recognition and tolerance of the differences in legal political, socio-economic systems between the Arctic; recognizing the 1973 Polar Bear Convention and the various bilateral environmental and technical agreements entered into by the Arctic rim States;[24] noting that existing international environmental conventions do not cover sufficiently all aspects and sources of marine pollution nor meet the special requirements of the Arctic Region; realizing fully the need for close cooperation among the Arctic rim States and international organizations concerned in a coordinated and comprehensive regional approach for the protection of the marine environment in the Arctic Region; have agreed as follows.

Article 1 covers the geographic coverage of the Arctic Convention. This should be defined to include all ocean and sea-bed areas north of 60 degrees North, with the exception of the Norwegian Sea between Greenland and Norway.[25] In this area the delineation should be set farther north, possibly at the Arctic circle, 66 degrees, 30 minutes, or even at 70 degrees North, due to the more moderate climatic conditions provided by the Gulf Stream. Within this circle the Arctic Region would thus include the territorial seas and the 200 mile exclusive economic zones of the Arctic littoral States, the high seas beyond the exclusive economic zones, and the respective Arctic rim States' continental shelves.

Article 2 covers definitions of 'Pollution', 'Organization' and 'Arctic rim States'. The definition of pollution suggested is, 'Pollution' of the marine environment means the introduction by man, directly or indirectly, of substances or energy into the marine environment, including estuaries, which results or is likely to result in such deleterious effects as harm to living resources and marine life, and marine ecosystems, hazards to human health, hindrance to marine activities, including fishing and other legitimate uses of the sea, impairment of quality for use of sea water and reduction of amenities.'[26] 'Organization' means the body designated as responsible for carrying out Secretariat functions pursuant to Article 13 of this Convention. Realistically, the Organization probably would not take on all the duties required of it under the different Protocols, and sub organizations specializing in the different pollution sources could be established upon the ratification and coming into force of each Protocol. It is considered important to have a comprehensive Organization covering the administration of all the programs and measures dealing with all the sources of pollution for among other things the sake of efficiency, uniformity and the exchange of information.[27] The definition of 'Arctic rim States' in this Convention and its sub Protocols is meant exclusively to refer to the countries of Canada, Denmark/Greenland, Norway, the United States and the Soviet Union. This term is used synonymously in this Convention and its sub Protocols with the terms 'Parties', 'Contracting Parties', 'Arctic States', 'Controlling States' and 'States', and all are meant to refer to these five countries.

Article 3 provides that the Contracting Parties may enter into bilateral or multilateral agreements, including regional or sub regional agreements for the protection of the marine environment of the Arctic Region against pollution as long as they are consistent with the Arctic Convention and in accordance with international law.

Article 4 provides generally that the Parties must individually or jointly take all appropriate measures in accordance with the provisions of the Arctic Convention and those Protocols in force to which they are Party to prevent, abate and combat pollution of the Arctic Region and to protect and enhance the marine environment in that Area. Additionally the Parties must cooperate in the formation and adoption of Protocols, prescribing agreed measures, procedures and standards for the implementation of this Convention. Further the Parties pledge themselves to promote within competent international bodies measures concerning the protection of the marine environment in the Arctic Region from all types and sources of pollution.

Article 5 covers pollution caused by dumping and provides that the Parties take all appropriate measures to prevent and abate pollution of the Arctic Region caused by dumping from ships and aircraft.

Article 6 covers vessel-source pollution and provides that the Parties take all measures in accordance with international law to prevent, abate and combat pollution of the Arctic Region caused by discharges from ships and ensure the effective implementation in that Area of the rules which are generally internationally recognized relating to the control of this pollution.

Article 7 deals with offshore platform pollution and provides that the Parties take all appropriate measures to prevent, abate and combat pollution of the Arctic Region resulting from exploration and exploitation of the continental shelf-land, the sea bed and its subsoil.

Article 8 deals with land-based pollution, including airborne and atmospheric, and provides that the Parties take all appropriate measures to prevent, abate and combat pollution of the Arctic Region caused by discharges from rivers, coastal establishments or outfalls, or pollution emanating from any other land-based sources within their territories.

Article 9 covers cooperative action in pollution emergencies and provides that the Parties cooperate in taking the necessary measures for dealing with pollution emergencies in the Arctic Region, whatever the causes of such emergencies, and reducing or eliminating resulting damage. Additionally, any Party which becomes aware of any pollution emergency in the Arctic Region, without delay must notify the Organization and, either through the Organization or directly, any Party likely to be affected by such emergency.

Article 10 covers monitoring and provides that the Parties shall endeavour[28] to establish a multilateral pollution monitoring system for the Arctic Region. For this purpose the Parties must designate the competent authorities responsible for pollution monitoring within their jurisdictional areas and participate in international arrangements for monitoring in areas beyond national jurisdiction. The Parties undertake to cooperate in the formulation, adoption and implementation of such Annexes to this Convention as may be required to prescribe common procedures and standards for pollution monitoring.

Article 11 deals with scientific and technological cooperation and provides that the Parties as far as possible cooperate directly or through competent regional or other international organizations in the fields of science and technology, and exchange data as well as other scientific information for the purpose of this Convention.[29] The Parties undertake as far as possible to develop and coordinate their national research programs relating to all types of marine pollution in the Arctic Region and to cooperate in the establishment and implementation of regional and other international research programs for the purposes of this Convention. Also the Parties undertake to cooperate in the provision of technical and other possible assistance in fields relating to marine pollution.

Article 12 covers liability and compensation and provides that the Parties undertake to cooperate as soon as possible in the formulation and adoption of appropriate procedures for the determination of liability and compensation for damage resulting from the pollution of the marine environment deriving from violations of the provisions of this Convention and applicable Protocols.

Article 13 deals with duties of the Organization and provides that the Organization is designated by the Parties responsible for carrying out the following Secretariat functions: convene and prepare meetings of the Parties provided for in Articles 14, 15 and 16; transmit to the Parties notifications, reports and other information received in accordance with Articles 3, 9 and 19; consider inquiries by, and information from the Parties, and consult with them on questions relating to this Convention and the Protocols and Annexes;

perform the functions assigned to it by the Protocols; perform such other functions as may be assigned to it by the Contracting Parties; ensure the necessary coordination with other competent international bodies, enter into such administrative arrangements as may be required for the effective discharge of Secretariat functions.

Article 14 deals with meetings of the Parties and covers the frequency and the functions involved. These latter include briefly to review the implementation of the Convention and the Protocols; review the inventories carried on by the State Parties on the state of marine pollution and its effects in the Arctic; consider reports submitted by the Parties under Article 19; adopt, review and amend as required the Annexes to this Convention and to the Protocols in accordance with Article 17 procedures; make recommendations regarding the adoption of any additional Protocols or amendments to this Convention or the Protocols in accordance with Articles 15 and 16; establish working groups to consider matters related to this Convention and the Protocols and Annexes; consider and undertake any additional action required for the achievement of the purposes of this Convention and the Protocols.

Article 15 deals with the adoption of additional Protocols and provides that the Parties at a diplomatic conference may adopt additional Protocols to this Convention in accordance with Article 4. Such a conference for the purpose of adopting additional Protocols must be convened by the Organization at the request of 3/5 of the Parties.[30] Pending the entry into force of this Convention the Organization may, after consulting with the signatories to this Convention, convene a diplomatic conference for the purpose of adopting additional Protocols.[31]

Article 16 deals with the amendment of the Arctic Convention or the Protocols. It provides that any Party may propose amendments to the Convention or any Protocol. Amendments must be adopted by a diplomatic conference which is convened by the Organization at the request of 3/5 of the Contracting Parties. Amendments must be adopted to both the Convention and any Protocol by a 3/5[32] majority vote of the Contracting Parties to the Convention which are represented at the diplomatic conference and submitted by the Depositary for acceptance by all Contracting Parties to the Convention. Acceptance of amendments must be notified to the Depositary in writing and enter into force between Contracting Parties having accepted such amendments on the 30th day following the receipt by the Depositary of notification of their acceptance by at least 3/5 of the Contracting Parties to this Convention or to the Protocol.

Article 17 deals with Annexes and amendments to Annexes. It provides that Annexes to the Convention or to any Protocol form an integral part of the Convention or such Protocol. Except as otherwise provided in any Protocol, the following procedure applies to the adoption and entry into force of any amendments to Annexes to this Convention or to any Protocol. Any Party may propose amendments to the Annexes to this Convention or to Protocols at the meetings referred to in Article 14. Such amendments must be adopted by a 3/5 majority vote of the Contracting Parties to the instrument in question. The Depositary must without delay communicate the amendments so adopted to all Contracting Parties. Any Contracting Party unable to approve an amendment must so notify in writing the Depositary within a period determined by the Contracting Parties concerned when adopting the amendment. The Depositary without delay must notify all Contracting Parties of any notification received. On expiry of this period the amendment to the Annex becomes effective for all Contracting Parties to this Convention or to the Protocol concerned which have not submitted a notification. The adoption and entry into force of a new Annex to this Convention or to any Protocol is subject to the same procedure as for the adoption and entry into force provided that, if any amendment to the Convention or the Protocol concerned is involved the new Annex must not enter into force until such time as the amendment to the Convention or the Protocol concerned enters into force.

Article 18 deals with procedural and financial rules and requires the Parties to adopt rules of procedures for their meetings and conferences envisaged in Articles 14, 15 and 16. The Parties must adopt financial rules, prepared in consultation with the Organization to determine their financial participation.

Article 19 deals with reports and provides that the Parties transmit to the Organization reports on the measures adopted to implement the Arctic Convention and the Protocols to which they are Parties in such form and at such intervals as the meetings of the Parties may determine.

Article 20 provides that the Parties undertake to cooperate in the development of procedures enabling them to control the application of this Convention and the Protocols.

Article 21 deals with settlement of disputes and provides that all disputes between the parties as to the interpretation or application of this Convention or the Protocols, which they have not been above to settle peacefully between themselves, will be settled according to the procedures established in LOSC, Part XV Section 1, Articles 279 - 285; Section 2, Articles 286 - 296; Section 3, Articles 297 - 299; Annex V; and Annex VII.[33]

Article 22 deals with the relationship between the Arctic Convention and the Protocols and provides that no State may become a Contracting Party to this Convention unless it becomes at the same time a Contracting Party to at least 1 of the Protocols.[34] Any Protocol to this Convention is binding only on the Contracting Parties to the Protocol in question. Decisions concerning any

Protocol pursuant to Articles 14, 16 and 17 of this Convention are to be taken only by the Parties to the Protocol concerned.

Article 23 deals with entry into force and provides that this Convention enters into force on the same date as the Protocol first entering into force. The Convention shall also enter into force with regard to the States if they have complied with the formal requirements for becoming Contracting Parties to any other Protocol not yet entered into force. Any Protocol to this Convention, except as otherwise provided in such Protocol, enters into force on the 30th day following the date of deposit of at least 3 instruments of ratification, acceptance, or approval of to such Protocol by the signatory Parties. Thereafter this Convention and any Protocol enters into force with respect to any Arctic rim State on the 30th day following the date of deposit of the instruments of ratification, acceptance or approval.

Article 24 deals with accession and provides that as from (date) the present Convention, the Arctic Oil Rig Protocol, the Arctic Counter-Pollution Protocol, the Arctic Vessel-Source Protocol, the Arctic Dumping Protocol, the Arctic Land-Based Protocol and the Arctic Liability Protocol shall be open for accession by the States enumerated in Article 2 of this Convention. After entry into force of the Convention and of any Protocol, any Arctic rim State may accede to this Convention and to any Protocol. Instruments of accession shall be deposited with the Organization.

Articles which deal with signature, ratification, acceptance or approval, withdrawal and responsibilities of the Depository are generally standard and will not be covered here.[35]

Notes - Chapter 11

1 Convention for the Protection of the Arctic Region against Pollution.

2 The Arctic Convention and all the Arctic Protocols proposed in Part II are modeled upon existing legal documents. This was done in order to benefit from the experience and research existing behind the establishment of these environmental instruments and to present a realistic picture of what composes such a document. At the same time changes were made to reflect ideas for betterment of these documents.

3 UNEP Draft, prepared by the Secretariat of the International Juridical Organization, was used as a model. Upon possible adoption UNEP Draft would become the fifth Protocol to the Barcelona Convention and would immediately be open for ratification. See UNEP, Issue No. 10/III, 1987 (Medwaves) p. 4.

4 Convention for the Protection of the Arctic Region against Pollution (Arctic Convention). See Part II Chapter 11.

5 Protocol for the Protection of the Arctic Region against Pollution resulting from Exploration and Exploitation of the Continental Shelf and the Sea-Bed and its Subsoil (Arctic Oil Rig Protocol). Full document presented in Part II Chapter 12.

6 Protocol for Counter-Pollution Operations in the Arctic Region with Sub Bilateral Agreements (Counter Pollution Protocol). Full bilateral USSR - Norway AGREEMENT presented in Part II Chapter 13.

7 Protocol on Protection of the Arctic Region from Vessel-Source Pollution (Arctic Vessel-Source Protocol). See Part II Chapter 14.

8 Protocol on Protection of the Arctic Region from Dumping of Wastes at Sea (Arctic Dumping Protocol). See Part II Chapter 15.

9 Protocol for the Prevention of Marine Pollution in the Arctic Region from Land-Based Sources (Arctic Land-Based Protocol). See Part II Chapter 16.

10 Protocol for Civil Liability for Oil Pollution Damage and Damage from Dumping in the Arctic Region Resulting from Exploration from and Exploitation of Sea-Bed Mineral Resources (Arctic Liability Protocol). See Part II Chapter 17.

11 See Part II Sections 2.1., 3.2., 4.2. and 5.2. for the legal basis afforded by customary law and the environmental conventions including OILPOL, MARPOL 73/78, the 1958 Geneva Conventions, the LDC, ODC and the Paris Convention.

12 Article 194 (1) (2) (3)(a), (b), (c) and (5). In the following discussion of LOSC Articles, emphasis has been added.

13 Parentheses added. Articles 199, 207, 208, 210-214, 216-220, 222, 234 and 235 are also dealt with respectively in more detail under Part II Chapters 12 - 17 where the legal basis for *each* proposed Protocol is discussed. They are mentioned here as background for the proposed comprehensive Arctic Convention.

14 Hakapää p. 118 notes that it can be questioned whether the internationally agreed standards include only those conventions which have entered in force. Those this question may not have much meaning here due to the requirement 'take into account', it might below where national regulations are required to be 'no less effective' than 'global rules and standards'.

15 Additionally under Articles 207 (3) and 208 (4) States are required to endeavour to *harmonize* their policies at the appropriate regional level.

16 See Part I Section 2.1.4. and Part II Section 10.3.3. for discussion of LOSC marine environmental provisions reflecting international customary law and recommendations for norms.

17 It is understandable that legal authors look to the LOSC semi-enclosed sea regulations for use in the Arctic noted in Part II Section 10.3.3. with the higher standards of State obligations.

18 The State obligations under Articles 208, 199, 211, 210, 207, 212 and 235 respectively are noted above and underlined for emphasis. It must also be noted that for at least one of the pollution sources, dumping, though the LOSC obligation is less than mandatory, in practice developments have already probably exceeded this standard. See Part II Section 3.2.1. Though the State obligation seems somewhat mandatory in Article 235 dealing with State liability and compensation, as noted in Part I Chapter 6, existing international law is at a low stage of development, and it is unclear the extent to which States are liable and responsible for compensating for pollution damage under international law.

19 See Part II Chapters 12 and 13.

20 See Part II Chapter 14.

21 See Part II Chapter 15.

22 See Part II Chapter 16.

23 See Part II Chapter 17.

24 The latter 2 phrases were added to show an existing precedent to environmental cooperation and to promote further State cooperation in the Arctic based upon a flexible consideration for national differences.

25 See Figure 1 Part II Section 9.1.

26 This definition, including the phrase 'and marine ecosystems' is derived from LOSC Article 1 and Article I of the Paris Convention, 2 of the best definitions of pollution according to Birnie pp. 223-224. Article 2 of the Barcelona Convention defines 'pollution' to mean the introduction by man, directly or indirectly, of substances or energy into the marine environment resulting in such deleterious effects as harm to living resources, hazards to human health, hindrance to marine activities including fishing, impairment of quality for use of sea water and reduction of amenities.

27 Pollution monitoring systems could be combined for instance, and uniform standards used for the entire Arctic.

28 As seen below in the Protocols the phrases 'shall endeavour' and 'shall undertake' were consistently changed to 'shall' to make the State obligations of a more mandatory nature. This is not done in the Arctic Convention since it is meant to be general and of a lessor obligatory nature.

29 As noted to an extent this is already being done in bilateral technical agreements noted in Part II Section 9.2.6.

30 The Barcelona Convention uses 2/3 which is impossible here since the Arctic rim States number 5.

31 This is an important clause allowing early adoption of the Protocols should it be desired by the States.

32 The Barcelona Convention uses a 3/4 majority vote.

33 The Barcelona Convention Article 22 and Annex A dealing with the settlement of disputes is not followed here. Basically it requires that Parties seek a settlement through negotiation or any other peaceful means in a dispute as to the interpretation or application of the Convention. If the dispute cannot be thus settled, upon common agreement it should be submitted to arbitration under Annex A. Dispute settlement under the LOSC is probably one of the most comprehensive and complex with its 'dispute-resolution mechanisms', and 'frameworks for arbitration', of any international agreements. See Mounteer, Thomas R., 'The General Agreement on Tariffs and Trade (GATT) and the Law of the Sea (LOS Convention: A Critical Comparison of Arbitration Provisions,' *The International Lawyer*, Fall 1987, Vol. 21, No. 4, (Mounteer). Generally, Part XV, Section 1, Articles 279-285 require parties to resolve disputes peaceably; Section 2, Articles 286-296 provide for compulsory procedures involving binding decisions; and Section 3, Articles 297-299 limit and provide exceptions to the compulsory procedures. Additionally Annex V provides for conciliation procedures and Annex VII for arbitration procedures. This system allows States the choice of settlement procedures ranging from the non-compulsory and nonjudicial to the compulsory judicial. Initially the LOSC imposes a duty on States to exhaust the informal and peaceful means of settling disputes provided for in Part XV Section 1, Article 279; in Annex V conciliation; and in Article 33 of the Charter of the United Nations (U.N. Charter), see Møse p. 61 at 64. Failing this, under Article 286, the disputing parties are subject to the compulsory procedure involving binding decisions set forth in Section 2. Exceptions to the compulsory procedures, set forth in Section 3, include certain disputes that arise out of a coastal State's discretionary exercise of sovereignty within its exclusive economic zone. These exceptions have been criticized as favouring less developed countries. The four forums exercising jurisdiction over LOSC disputes are the ICJ, the International Tribunal for the Law of the Sea, a regional organization, or an Annex VII arbitral tribunal, and States have a choice of all the various forums. They all apply the same law, the LOSC and other rules of international law not incompatible with the LOSC, or if the parties agree, by *ex aqueo et bono* (Article 293). Remedies are left uncertain, since no specific sanctions are

provided for. Whether punitive awards are allowed is as well uncertain. Arbitral decisions are final and shall be complied with by all the parties to the dispute (Article 296 (1)). The arbitral panel's award will be 'final and without appeal unless the parties have agreed in advance to an appellate procedure', (Annex VII Article 11). Enforcement regarding the implementation of the award is provided for by either Party submitting controversies to the arbitral tribunal which made the award, or to another court or tribunal upon agreement by parties to the dispute, (Annex VII Article 12). This does not include national courts but provides for enforcement by the International Tribunal for the Law of the Sea and the ICJ. The LOSC does not allow the U.N. Security Council to enforce ICJ Decisions as does the U.N. Charter. If arbitrators are to be chosen, each State is permitted to nominate 4 qualified arbitrators from a list maintained by the U.N. Secretary-General, (Annex VII Article 1). Each disputing Party nominates one arbitrator who together select 3 arbitrators from the Secretary-General's list, and in all cases there are to be 5 arbitrators, (Annex VII Article 3 (a)-(d)). The arbitral tribunal determines the procedure to be used, unless otherwise agreed, (Annex VII Article 4).

34 This number could be varied to reflect the actual political situation, and if possible the number '2' substituted to include the Arctic Oil Rig Protocol and the Arctic Counter-Pollution Protocol with AGREEMENT and PLAN, leaving the Arctic Vessel-Source Protocol, the Arctic Dumping Protocol, the Arctic Land-based Protocol and the Arctic Liability Protocol for later ratification.

35 For examples see Barcelona Convention Articles 24, 25, 28 and 29.

12. Discussion of Arctic Oil Rig Protocol[1]

12.1. Background

12.1.1. Summary of Existing Sea-bed Regime with Application to the Arctic

As seen in Part I Section 4.2.2. the Paris Convention covering offshore installations is applicable in the Barents Sea, sections of the Greenland Sea, the Arctic Ocean and the Norwegian Sea which is only a small part of the Arctic. If the Soviets accede to the Convention, only the Barents Sea will be covered, and thus most of the Arctic is unregulated. As also seen even under this Convention there lack mandatory obligations required from State Parties, and political problems exist as well involving States which are environmentally conservative and not bordering on the Arctic.[2] Upon this background of *inadequate* international control of pollution from offshore platforms, offshore oil activities exist in the Beaufort Sea, and offshore oil exploration is being conducted or planned in the Barents Sea, the Bering Sea and the Chukchi Sea.[3] Upon this setting the Arctic Oil Rig Protocol is proposed.

12.1.2. Legal Basis for Arctic Oil Rig Protocol under LOSC[4]

As under the Arctic Convention the general obligations Articles 192, 197 and 194 would apply especially the latter under which States are required to take *joint* measures to minimize fully marine pollution from offshore installations under their jurisdiction, particularly to prevent accidents, meet emergencies, ensure operational safety, and regulate the design, construction, equipment, operation and manning of such installations.

As noted under Article 208 States shall establish, acting especially through competent international organizations or diplomatic conference, global and *regional* rules, standards and recommended practices and procedures to prevent, reduce and control marine pollution arising from sea-bed activities under their jurisdiction.[5] This also includes marine pollution arising from artificial islands, installations and structures in that State's exclusive economic zone or on its continental shelf.[6] Such rules, standards and recommended practices and procedures must be re-examined as necessary.[7] Additionally coastal States must adopt national laws and regulations, and take other measures as necessary, to prevent, reduce and control marine pollution arising from sea-bed activities under their jurisdiction and from artificial islands, installations and structures in that State's exclusive economic zone or on its continental shelf.[8] Such laws, regulations and measures shall be no less effective than international rules, standards and recommended practices and procedures, and States must endeavour to harmonize such policies at the appropriate regional level.[9]

Under Article 214 States shall enforce their national laws and regulations adopted in accordance with Article 208 and must adopt national laws and regulations and take other measures necessary to implement applicable[10] international rules and standards established through competent international organizations or diplomatic conference to prevent, reduce and control marine pollution arising from sea-bed activities under their jurisdiction and from artificial islands, installations and structures in that State's exclusive economic zone or on its continental shelf.[11]

Thus, to the extent these rules can be said to be international customary law and recommendations for norms, the Arctic rim States have an obligation to *establish* and enforce an Arctic regional agreement to prevent, reduce and control pollution emanating from offshore platforms under their jurisdiction, which would act as legal support to the formation of the Arctic Oil Rig Protocol.[12]

12.1.3. Use of UNEP Document as Pattern

As a basis for an Arctic Oil Rig Protocol, the Protocol for the Protection of the Mediterranean Sea against Pollution Resulting from Exploration and Exploitation of the Continental Shelf and its Seabed and its Subsoil (UNEP Draft)[13] was used. Though covering an area vastly different than the Arctic, there are several advantages of using this document as a model.

The first is that it comprehensively and exclusively covers marine pollution from offshore oil rigs. It does not as does the Paris Convention encompass pollution from offshore platforms as a section of land-based pollution. It was thought beneficial for the Arctic Region to have a legal document which was up-to-date as well as devoted to *one* pollution source, and hopefully this will result in more uniformity and effectiveness concerning anti pollution regulations for the Arctic region.

Secondly, the UNEP Draft was used because it was realistically formed with the idea of *balancing* oil and environmental interests. 'The strict and sound management of the environment in accordance with the technological development has been chosen'.[14] This is very important in order to obtain a realistic agreement. The countries engaged in exploring and producing oil in the Arctic can only be persuaded to take the environment into account if at the same time the oil interests are given weight. It would even be counter productive to form a document which preserves the pristine beauty of the Arctic environment, while at the same time being completely impossible to implement, since no country would consider ratifying it.[15] If a realistic approach is taken in the original document, this in turn ensures more compliance by the Parties involved. This by no means implies the proposed Arctic Oil Rig Protocol is a anti environmental agreement. As can be noted below Articles have been made more stringent than the UNEP Draft, however an attempt has been made to balance strict environmental management with emphasis on the best available oil technology.

Finally, the UNEP Draft was chosen because of the advantages gained from the instrument itself. In drafting it, legal and technical experts, and organizations including the IMO, The Oil Industry International Exploration and Production Forum, Petroconsultant, Universities and insurance companies were consulted.[16] Additionally, background for its development was obtained from legal texts, treaties, conventions, documents and publications, which necessarily provides a depth of experience beyond any suggestions that could be made solely by the author.[17] However, alternations will be made where it is felt other regulations and specifications including provisions developed under Paris Convention programs covering the North Sea would be more applicable to the Arctic than those set forth by the UNEP Draft.

The Arctic Oil Rig Protocol is structured into 6 Sections, 6 Annexes and 1 Appendix, following the UNEP document.[18]

Sections

I	General Provisions
II	Authorization System
III	Wastes and Harmful or Noxious Substances and Materials
IV	Safeguards
V	Co-operation
VI	Final Provisions

Annexes

I	Harmful or Noxious Substances and Materials Whose Disposal Is Prohibited
II	Harmful or Noxious Substances and Materials Whose Disposal Is Subject to a Special Permit
III	Factors To Be Considered for the Issue of the Permits

IV	Environmental Impact Assessments
V	Oil and Oily Mixtures and Drilling Fluids
VI	Safety Measures
Appendix	List of Oils

12.2. Discussion of Arctic Oil Rig Protocol

The presentation below will consist of introducing the Articles and Annexes appearing in the Arctic Oil Rig Protocol with discussion where relevant appearing in brief footnotes. This is done so as not to interrupt unnecessarily the continuity of the legal instrument. In addition any notable points made in the Arctic Oil Rig Protocol will be commented upon in the footnotes. For clarity the presentation will be divided up into Sections following the divisions of the Arctic Oil Rig Protocol.

12.2.1. Section I - General Provisions

Preamble[19]

The Contracting Parties to the present Protocol

Being Parties to the Convention for the Protection of the Arctic Region against Pollution adopted at ... on ..., and with reference to Article 7 of the said Convention,

Considering the rapid increase in the activities concerning exploration and exploitation of the Arctic region sea-bed and its subsoil,

Recognizing that the pollution which may result therefrom represents a grave danger to both human health and the environment,

Taking particularly into account the specially vulnerable Arctic environment including ice covered areas,

Desirous of protecting and preserving the Arctic region from this source of pollution,

Bearing in mind the United Nations Convention on the Law of the Sea, opened for signature at Montego Bay on 10 December 1982,

Recognizing the differences in legal, political, socio-economic, systems between the Arctic littoral States, and taking account of such distinctions,

Have agreed as follows:

Article 1 - Definitions

(a) 'Convention' means the Convention for the Protection of the Arctic region against Pollution, adopted on (date);

(b) 'Organization' means the body referred to in Articles 2 and 13 of the Arctic Convention;[20]

(c) 'Resources' means all mineral resources, whether solid, liquid or gaseous;

(d) 'Activities concerning exploration and/or exploitation in the Protocol Area' (hereinafter referred to as 'activities') means:

  (i) activities of scientific research concerning the sea-bed and its subsoil;

  (ii) activities conducted for the purpose of recovering resources, including preliminary activities, any treatment before transport to shore, and transport by pipeline to shore; also included is construction, repair, maintenance or similar operations incidental to the main purpose of recovering resources;

(e) 'Pollution' means every act referred to in Article 2 of the Arctic Convention and every other act which results or is likely to result in deleterious effects to human health and to the environment, caused directly or indirectly by activities;[21]

(f) 'Installation' means any fixed or floating structure and any integral part thereof engaged in activities including, in particular:

  (i) fixed, floating or sub sea drilling or production platforms;

  (ii) offshore storage facilities;

  (iii) offshore loading terminals and transport systems for the extracted products, such as submarine pipelines;

  (iv) equipment, and attached apparatus, for the reloading, processing, storage and disposal of substances removed from the sea-bed or its subsoil;

  (v) ships, when they are engaged in the following activities:

    - when being used as offshore storage facilities;

    - when towing the installations referred to in the present definition, during the towing activity;

(g) 'Operator' means:

  (i) the person, authorized by the Controlling State in accordance with this Protocol to carry out activities; or

  (ii) any person who does not hold a valid authorization according to this Protocol, but who is *de facto*, in overall control of the activities;

(h) 'Controlling State' means the Party or Parties exercising jurisdiction over the activities carried out within the area where the installation is located; when an installation is subject to the jurisdictions of more than one Party, these Parties may appoint, by common agreement, the Party which will be the only Controlling State;

(i) 'Safety zone' means a zone established, according to the provisions of general international law, around installations with appropriate markings to ensure the safety of navigation, fishing and the installations;[22]

(j) 'Wastes and harmful or noxious substances and materials' means substances and materials of any kind, form, or description, intended to be abandoned or abandoned into the Protocol Area and/or produced or utilized during or resulting from activities, which might create pollution. It includes, in particular, oil and oily mixtures and drilling fluids, garbage and sewage;

(k) 'Oil' means petroleum in any form including crude oil, fuel oil, sludge, oil refuse and refined products (other than petrochemicals which are subject to the provisions of Annexes I and II to the present Protocol) and, without limiting the generality of the foregoing, includes the substances listed in the Appendix to the present Protocol;

(l) 'Oily mixture' means a mixture with any oil content;

(m) 'Sewage' means:
  (i) drainage and other wastes from any form of toilets, urinals and water-closet scuppers;
  (ii) drainage from medical premises (dispensary, sick bay, etc.) via wash basins, wash tubs and scuppers located in such premises;
  (iii) other waste waters when mixed with the drainages defined above;

(n) 'Garbage' means all kinds of victual, domestic and operational waste excluding fresh fish and parts thereof, generated during the normal operation of the installation and liable to be disposed of continuously or periodically, except those substances which are defined or listed elsewhere in the present Protocol;

(o) 'Freshwater limit' means the place in watercourses where at low tides and in a period of low freshwater flow, there is an appreciable increase in salinity due to the presence of sea water.

Article 2 - Geographical Coverage

1. The area to which this Protocol applies (hereinafter referred to as the 'Protocol Area') shall be:
  (a) The Arctic region as defined in Article 1 of the Convention for the Protection of the Arctic Region against Pollution;[23]
  (b) Waters on the landward side of the baselines from which the breadth of the territorial sea is measured and extending, in the case of watercourses, up to the freshwater limit.

2. The Protocol Area can also include wetlands or coastal areas, as may be designated by each of the Parties.

Article 3 - General Undertakings

1. The Contracting Parties to this Protocol (hereinafter referred to as 'the Parties') shall take individually or jointly all appropriate measures, utilizing the best available technology,[24] to prevent, abate, combat and control pollution of the Protocol Area resulting from activities concerning exploration and exploitation of the sea-bed and its subsoil.
2. The Parties shall ensure that all necessary measures are taken so that activities do not endanger human health, marine fauna and flora, do not cause adverse effects on the environment, and do not cause unjustifiable interference with navigation, fishing or other legitimate use of the Protocol Area according to international law.
3. No provision of the present Protocol shall be interpreted as preventing the Contracting Parties from taking more stringent measures to combat pollution resulting from the exploration and exploitation of the continental shelf and the sea-bed and its subsoil.

12.2.2. Section II - Authorization System

Article 4 - General Principles

1. All activities in the Protocol Area, including erection on site of installations, shall be subject to a prior written authorization from the competent national authority of the Controlling State. Such authority, before granting the authorization, shall be satisfied that the installation has been constructed according to international standards and that the operator has the technical competence and the financial capacity to carry out the activities. Such authorization shall be granted in accordance with the appropriate procedure, as defined by the competent national authority.
2. Authorization shall be refused if there are clear indications that the proposed activities are likely to cause detrimental effects on the environment which could not be avoided by compliance with the conditions in the authorization, referred to in Article 6, paragraph 3 of this Protocol.
3. When considering approval of the siting of an installation, the Controlling State shall ensure that no detrimental effects be caused by such siting to existing installations, including but not limited to,[25] sub sea installations, pipelines, and cables.
4. Additionally, when considering approval of the siting of an installation, the Controlling Party shall ensure that no detrimental effects be caused by

such siting to fishing activities, and commercial or military vessel traffic.[26]

Article 5 - Requirements for Authorizations

The Controlling State requires that requests for authorizations and renewal of authorizations be subject to a presentation of the project by the candidate operator to its national competent authority including but not limited to, the following elements:

(a) A survey concerning the effects of the proposed activities on the environment. The competent national authority shall require, taking into account the nature, scope, duration, technical methods employed in the activities and characteristics of the area, that an environmental impact assessment be prepared in accordance with Annex IV to this Protocol;[27]
(b) The precise geographic areas where the activity is being considered, including safety zones;
(c) The professional and technical qualifications of the candidate operator and personnel on the installation, as well as the composition of the crew;
(d) The safety measures according to Article 15;
(e) The contingency plan according to Article 16;
(f) The monitoring procedures according to Article 17;
(g) The plans for removal of installations according to Article 18;
(h) Precautions for specially protected areas according to Article 19;
(i) The insurance and/or other financial security to cover liability in accordance with Article 26, paragraph 3.

Article 6 - Granting of Authorizations

1. The authorizations, referred to in Article 4, are only to be granted after an examination, by the national competent authority of the Controlling State, of each element listed in Article 5 and Annex IV. Elements additionally required by the Controlling State authorities may be included in this examination.[28]
2. Each authorization shall determine the activities, the duration of the authorization, establish the geographical limits of the area and the safety zones subject to the authorization, and determine the technical exigencies and the authorized installations.
3. The authorization may impose conditions regarding measures, techniques or methods designed to reduce to the minimum risks and damage of pollution resulting from the activities.
4. The Parties shall promptly notify the Organization of granted or renewed authorizations. The Organization shall keep an up-to-date register of all the authorized installations in the Protocol Area.

Article 7 - Sanctions

1. Each Party shall establish sanctions to be imposed for breach of obligations arising out of this Protocol, national laws or regulations implementing this Protocol or the specific conditions accompanying the authorization. These sanctions shall be consistent with international law and as much as possible consistent with those imposed by the other Arctic littoral States.
2. In the conduct of State proceedings in respect of a breach of obligations which may result in the imposition of sanctions, recognized rights of the accused shall be observed.[29]

12.2.3. Section III - Wastes and Harmful or Noxious Substances and Materials

Article 8 - General Obligation

Notwithstanding other standards or obligations of this Section, Parties shall impose a general obligation upon operators at all times to use the best available technology and to observe, as a minimum allowable requirement, internationally accepted standards regarding wastes and harmful or noxious substances and materials, to minimize the risk of pollution.

Article 9 - Harmful or Noxious Substances and Materials[30]

1. The Parties shall ensure that each substance and material used in the Protocol Area, including those to construct and protect the installation, is accompanied by a compound description provided by the entity producing such substance or material.
2. The use and disposal into the Protocol Area of harmful or noxious substances and materials listed in Annex I to this Protocol is strictly prohibited.
3. The use and disposal into the Protocol Area of harmful or noxious substances and materials listed in Annex II to this Protocol require, in each case, a prior special permit from the competent national authority of the Controlling State.
4. The use and disposal into the Protocol Area of all other substances and materials which might cause pollution require a prior general permit from the competent national authority of the Controlling State.
5. The permits referred to in paragraphs 3 and 4 above shall be issued only after careful consideration of all the factors set forth in Annex III to this Protocol.

6. The use and disposal into the Protocol Area of substances and materials of whatever form and compound not previously used are subject to prior temporary permission by the competent national authority of the Controlling State.
7. Each Party shall notify the Organization promptly of temporary permissions granted or refused so that the Parties can take appropriate measures concerning the future use of the substances and materials referred to in paragraph 6 above, in accordance with Article 29.

Article 10 - Oil and Oily Mixtures, Drilling Fluids and Discharges containing Heavy Metals[31]

1. The Parties shall formulate and adopt common minimum standards for the disposal of oil and oily mixtures, drilling fluids and drilling fluids containing heavy metals from installations into the Protocol Area:
   (a) Such common minimum standards shall be formulated in accordance with the provisions of Annex V;
   (b) Such common minimum standards shall not be less restrictive than, in particular, the following:[32]
      (i) for machinery space drainage, a maximum oil content of 15 mg. per litre whilst undiluted;
      (ii) for production water, a maximum oil content of 40 mg. per litre as an average in any 24 hour period.
2. Each Party shall take measures to enforce the common minimum standards adopted pursuant to this Article or to enforce more restrictive standards where such standards have been adopted by that Party.

Article 11 - Sewage[33]

1. The Controlling State shall prohibit the discharge of sewage from installations into the Protocol Area except when:
   (a) The installation is discharging comminuted and disinfected sewage using a system approved by its competent national authority; or
   (b) It has passed through an approved sewage treatment plant certified by its competent national authority.
2. The Controlling State shall impose stricter provisions, as appropriate, where deemed necessary because of, *inter alia*, the regime of the currents in the area or the proximity to any area referred to in Article 19.
3. The exceptions of paragraph 1 shall not apply if the discharge produces visible floating solids or discoloration of the surrounding water.
4. When the sewage is mixed with wastes and harmful or noxious substances and materials having different disposal requirements, the more stringent requirements shall apply.

Article 12 - Garbage

1. Disposal into the Protocol Area of the following is prohibited:
   (a) All plastics, including but not limited to, synthetic ropes, synthetic fishing nets and plastic garbage bags;
   (b) All other garbage, including but not limited to, paper products, rags, glass, metal, bottles, crockery, dunnage, lining and packing materials.
2. Disposal into the Protocol Area of food wastes shall be made as far away as possible from land, but in any case not less than twelve nautical miles from the nearest land.
3. Disposal into the Protocol Area of food wastes consistent with Section 2 shall be made in open water and not in ocean areas covered predominantly by ice.[34]
4. When garbage is mixed with other discharges having different disposal or discharge requirements the more stringent requirements shall apply.

Article 13 - Reception Facilities, Instructions and Sanctions

Parties shall ensure that:

(a) Operators provide adequate and easily accessible receptions facilities for the proper disposal of all wastes and harmful or noxious substances and materials;
(b) Instructions be given to all personnel on proper means of disposal;
(c) Sanctions be imposed on illegal disposals.

Article 14 - Exceptions

1. The provisions of this Section shall not apply in case of:
   (a) *Force majeure* or distress as recognized by rules of international law or when human life or the safety of the installation is threatened;[35] or
   (b) The discharge into the sea of substances containing oil or harmful or noxious substances or materials shall be subject to prior approval by the competent national authority of the Controlling State when being used for the purpose of combating specific pollution incidents in order to minimize the damage from pollution.
2. Disposal effectuated according to paragraph 1 of this Article shall immediately be reported to the Organization and, either through the Organization or directly, to any party or Parties likely to be affected, together with full details of the circumstances and the nature and quantities of wastes or harmful or noxious substances or materials discharged.

12.2.4. Section IV - Safeguards[36]

Article 15 - Safety Measures

1. The Controlling State within whose jurisdiction activities are being considered or are being carried out shall ensure that safety measures are taken with regard to elements, including but not limited to, the design, construction, placement, equipment, marking, operation and maintenance of installations.
2. The Controlling State shall ensure that the operator has at all times on the installations, the most technologically advanced equipment and devices, maintained in good working order, to protect human life, to prevent accidental pollution and to facilitate prompt response to an emergency, in accordance with the best available technology and the provisions of the operator's contingency plan referred to in Article 16.
3. The competent national authority of the Controlling State shall require a certificate of safety and fitness for purpose (hereinafter referred to as 'certificate') recognized by the competent international body[37] for production platforms, sub-sea installations, mobile offshore drilling units, offshore storage facilities, offshore loading systems, pipelines and for such other installations as decided by the Controlling State.
4. The Parties shall formulate and adopt rules and standards in accordance with international practices and procedures to ensure the provisions of Annex VI.
5. The Parties shall ensure through inspection that the activities are conducted by the operators in accordance with this Article.

Article 16 - Contingency Planning

1. The Parties shall promote and maintain a plan for combating pollution or other adverse effects to the environment or to save human life in the Protocol Area in case of emergencies resulting from activities (hereinafter referred to as 'national contingency plan'). The national contingency plan shall establish and co-ordinate in particular, the equipment, ship, aircraft and manpower prepared for operations in cases of emergencies. Parties agree to ratify as soon as possible the Protocol for Counter-Pollution Operations in the Arctic Region.[38]
2. The Controlling State, before it grants an authorization, shall require that the operator has developed a satisfactory contingency plan approved by the competent national authority (hereinafter referred to as the 'operator's contingency plan'), referred to in Article 5(e). Such plan shall be prepared in accordance with the provisions of the Protocol for Counter-Pollution Operations in the Arctic Region.

3. Each Controlling State shall establish national co-ordination and direction in accordance with the Protocol for Counter-Pollution Operations in the Arctic Region.

Article 17 - Monitoring

1. The operator shall be required to measure the effects of the activities on the environment and to report on them periodically or upon request by the competent national authority of the Controlling State for an evaluation by such national authority according to a procedure established by the Controlling State in its authorization system.
2. The Controlling States shall establish progressively and operate within the Protocol area, a permanent monitoring system allowing:
   - the earliest possible assessment of the existing level of marine pollution;
   - the assessment of the effectiveness of measures for the reduction of marine pollution from installations, taken under the terms of the present Protocol and in accordance with which the authorization was granted.

   For this purpose the Contracting Parties shall lay down the ways and means of pursuing individually or jointly systematic and ad hoc monitoring programs. These programs shall take into account the deployment of research vessels and other facilities in the monitoring area. The programs shall also take into account similar programs pursued in accordance with Conventions already in force and by the appropriate international organizations and Agencies.[39]

Article 18 - Removal of Installations

1. Operators shall be required by the Controlling States to remove any installation which is abandoned or disused, to ensure safety of navigation, taking into account any international standards accepted in this regard by the Parties.[40] Such removal shall also have due regard to other legitimate uses of the sea, in particular fishing, the protection of the marine environment and the rights and duties of other States. All necessary measures shall previously have been taken by the operator and under his responsibility to prevent spillage or leakage from the site of the activities.
2. The Controlling State shall require that abandoned or disused pipelines are either removed in accordance with paragraph 1 of this Article or are buried and cleaned inside so that they neither endanger navigation, hinder fishing, threaten the marine environment, nor interfere with other legitimate uses of the sea nor the rights and duties of other States. Its competent national authority shall ensure that appropriate publicity is

given to the depth, position, and dimensions of any buried pipeline and that such information is indicated on charts and notified to the Organization and other competent international organizations and the Parties.

3. The provisions of this Article also apply to the operator whose authorization may have been revoked or suspended pursuant to Article 7.
4. Where the operator fails to comply with the provisions of this Article, the competent national authority of the Controlling State shall undertake, at the operator's expense, such action or actions as shall be necessary to remedy the operator's failure to act.

Article 19 - Specially Protected Areas[41]

For the protection of the areas defined in Article 1 of the Convention for the Protection of the Arctic Region against Pollution, including ice-covered areas, and in furtherance of the goals stated therein, Parties shall take particular measures, either individually or jointly, to prevent, abate, combat and control pollution arising from activities in these areas. These measures shall be non-discriminatory.

In addition to the measures enunciated in said Convention including Articles 4 and 7, such measures may include but are not limited to:

(a) Special restrictions or conditions in such areas for:
  (i) the preparation and evaluation of environmental impact assessments and the granting of authorizations;
  (ii) the elaboration of safety measures, contingency planning, monitoring and removal of installations;
(b) Intensified exchange of information between operators, the competent national authorities, Parties and the Organization regarding matters which may affect such areas.

12.2.5. Section V - Co-operation

Article 20 - Studies and Research Programs

In conformity with Article 11 of the Convention for the Protection of the Arctic Region against Pollution, the parties shall co- operate in promoting studies and undertaking programs of scientific and technological research to develop new methods including but not limited to:

(a) Carrying out activities in a way which minimizes the risk of pollution;
(b) Preventing, abating, combating and controlling pollution, especially in cases of emergency.

Article 21 - Request for Assistance or Information

A Party requiring assistance or information to prevent, abate or combat pollution resulting from activities may call upon the help of the Organization and of the other Parties, who shall do their utmost to provide the assistance or information requested, especially in cases of emergency.

Article 22 - International Rules, Standards and Recommended Practices and Procedures

1. The Parties shall co-operate, either directly or through the Organization or other competent international organizations, in order to:
   (a) Establish appropriate scientific criteria for the formulation and elaboration of international rules, standards and recommended practices and procedures for achieving the aims of this Protocol;
   (b) Formulate and elaborate such international rules, standards and recommended practices and procedures.
2. The Parties shall, as soon as possible, harmonize their laws and regulations with the international rules, standards and recommended practices and procedures referred to in paragraph 1 above.[42]
3. Parties shall exchange information relevant to their national policies, laws and regulations and the harmonization referred to in paragraph 2 above.[43]

Article 23 - Exchange of Scientific and Technical Assistance

1. The Parties shall, if necessary, directly or with the assistance of competent regional or global international organizations, co-operate with a view to formulating and, as far as possible, implementing programs of assistance, particularly in the fields of science, law, education and technology, in order to prevent, abate, combat and control pollution from activities in the Protocol Area.
2. Technical assistance includes but is not limited to, the training of scientific, legal and technical personnel, as well as the acquisition, utilization and production of appropriate equipment on terms to be agreed upon among the Parties concerned.

Article 24 - Mutual Information

The Parties shall inform one another directly or through the Organization of measures taken, of results achieved and, if the case arises, of difficulties encountered in the application of this Protocol. This information shall be public information.[44] Procedures for the collection and submission of such information shall be determined at the meetings of the Parties.

Article 25 - Transfrontier Pollution

1. Each Party shall take all measures necessary to ensure that activities under its jurisdiction are conducted so as not to cause pollution beyond the limits of its jurisdiction.
2. A Party within whose jurisdiction activities are being considered or carried out shall take into account any adverse environmental effects without discrimination as to whether such effects are likely to occur within the limits of its jurisdiction or beyond such limits, *inter alia*, such non-discrimination shall be observed in national laws and regulations.
3. When a Party becomes aware of cases in which the marine environment is in imminent danger of being damaged or has been damaged by pollution, it shall immediately notify other Parties it deems likely to be affected by such damage as well as the Organization and provide them with timely information that would enable them, where necessary, to take appropriate measures.
4. Parties shall, in accordance with their legal systems and, where appropriate, on the basis of an agreement, grant equal access to and treatment in administrative proceedings to persons in other States who may be affected by pollution or other adverse effects resulting from proposed or existing operations.[45]
5. A Party shall not be responsible for any pollution originating in the territory of a non-contracting state. However, the said Party shall co-operate reasonably with the said State so as to make full application of the Protocol possible.

Article 26 - Liability and Compensation

1. The Parties agree to ratify as soon as possible the Protocol on Civil Liability for Oil Pollution Damage and Damage from Dumping resulting from Exploration for and Exploitation of Seabed Mineral Resources, as well as establish appropriate procedures for the determination of liability and compensation for damage resulting from the activities considered in this Protocol.[46]
2. Operators are liable for damage suffered as a result of activities, and they shall pay prompt and adequate compensation for such damage. Compensation should be determined on the basis of strict and unlimited liability.
3. To cover the liability under this Protocol, the operator shall be required to have and maintain insurance or other financial security to such amount, of such type and terms as the Controlling State shall specify.[47]

12.2.6. Section VI - Final Provisions[48]

Article 27 - Appointment of Competent National Authorities

Each Controlling State shall appoint one or more competent national authority to:

(a) Grant, renew and register the authorizations as provided in Section II of this Protocol;

(b) Issue and register the special and general permits and temporary permissions referred to in Article 9 of this Protocol;

(c) Issue the permits referred to in Annex V to this Protocol;

(d) Approve the system and certify the sewage treatment plant referred to in Article 11, paragraph 1 of this Protocol;

(e) Issue the prior approval. for exceptional discharges referred to in Article 14, paragraph 1 of this Protocol;

(f) Carry out the duties regarding safety measures referred to in Article 15, paragraphs 3 and 5 of this Protocol;

(g) Perform the functions relating to contingency planning according to Article 16 and upon ratification the Protocol for Counter-Pollution Operations in the Arctic Region;

(h) Provide for monitoring procedures as provided in Article 17 of this Protocol;

(i) Control the removal operations of the installations as provided in Article 18 of this Protocol.

Article 28 - Transitory Measures

Each Party shall elaborate procedures and regulations regarding activities, whether authorized or not, initiated before the entry into force of this Protocol, to ensure conformity, as far as practicable, with the provisions of this Protocol.

Article 29 - Meetings

1. Ordinary meetings and extraordinary meetings of the Parties shall take place in conjunction with ordinary meetings and extraordinary meetings of the Contracting Parties to the Convention held pursuant to Articles 2 and 14 of the Convention for the Protection of the Arctic Region against Pollution.

2. The functions of the meetings of the Parties to this Protocol shall be, *inter alia*:

    (a) To keep under review the implementation of this Protocol and to consider the efficacy of the measures adopted and the advisability of any other measures, in particular in the form of Annexes and Appendices;

(b) To revise and amend any Annex or Appendix to this Protocol in accordance with Articles 16 and 17 of the Convention for the Protection of the Arctic Region against Pollution;

(c) To consider the records of the authorizations granted or renewed in accordance with Section II of this Protocol;

(d) To consider the records of the permits, permissions and approvals issued in accordance with Section III of this Protocol;

(e) To consider the records of the contingency plans and means of intervention in emergencies adopted in accordance with Article 16 of this Protocol;

(f) To establish criteria and formulate international rules, standards and recommended practices and procedures in accordance with Article 22 of this Protocol, in whatever form the Parties may agree;

(g) To facilitate the implementation of the policies and objectives of Section V, in particular the harmonization of national laws in accordance with Article 22 of this Protocol;

(h) To discharge such other function as may be appropriate for the application of this Protocol.

Article 30 - Final Clause

1. The provisions of the Convention for the Protection of the Arctic Region against Pollution relating to any Protocol shall apply with respect to this Protocol.
2. The rules of procedures and the financial rules adopted pursuant to Article 18 of the Convention for the Protection of the Arctic Region against Pollution shall apply with respect to this Protocol, unless the Parties to this Protocol agree otherwise.
3. This Protocol shall be open for signature, at (place) from (date) to (date), and at (place) from (date) to (date), by any State invited to the Conference of Plenipotentiaries of the Littoral States of the Arctic for the Protection of the Arctic Region against Pollution Resulting from Exploration and Exploitation of the Sea-bed and its Subsoil held at (place) from (date) to (date).
4. This Protocol shall be subject to ratification, acceptance or approval. Instruments of ratification, acceptance or approval shall be deposited with the Government of (specified Arctic State), which will assume the functions of Depository.
5. As from (date), this Protocol shall be open for accession by the States referred to in paragraph 3 above and by any grouping referred to in that paragraph in accordance with Article 24 of the Convention for the Protection of the Arctic Region against Pollution.

6. This Protocol shall enter into force on the thirtieth day following the deposit of at least three instruments of ratification, acceptance or approval of, or accession to, the Protocol by the Parties referred to in paragraph 3 of this Article.
7. In the event of dispute between the Parties and the dispute cannot be settled by the Parties themselves, then it shall be settled according to the provisions outlined in Article 21 of the Convention for the Protection of the Arctic Region against Pollution following LOSC provisions.
8. The Arctic States shall designate a competent Organization to be responsible for secretariat duties in relation to this Protocol in accordance with Articles 2 and 13 of the Convention for the Protection of the Arctic Region against Pollution. The duties of the Organization shall include preparing and assisting, in consultation with the Parties and appropriate international organizations, in the development and implementation of procedures dealing with consideration of any actions required outlined in this Protocol and with emergency situations.[49]

IN WITNESS WHEREOF the undersigned, being duly authorized by their respective Governments, have signed this Protocol.

DONE at (place) on this (date) in a single copy in the Danish, English, Norwegian, and Russian languages, the four texts being equally authoritative.

12.2.7. Annex I

Harmful or Noxious Substances and Materials Whose Disposal Is Prohibited[50]

The following substances, families and groups of substances are listed, not in order of priority, for the purposes of Article 9 (2) of this Protocol. They have been selected mainly on the basis of their persistence, toxicity or other noxious properties,[51] and/or bioaccumulation.

1. Organohalogen compounds and substances which may form such compounds in the marine environment, excluding those which are biologically harmless, or which are rapidly converted in the sea into substances which are biological harmless.[52]
2. Organophosphorus compounds and substances which may form such compounds in the marine environment, excluding those which are biologically harmless, or which are rapidly converted in the sea into substances which are biological harmless.
3. Organotin compounds and substances which may form such compounds in the marine environment, excluding those which are biologically

harmless, or which are rapidly converted in the sea into substances which are biological harmless.

4. Mercury and mercury compounds.
5. Cadmium and cadmium compounds.
6. Used lubricating oils.
7. Persistent synthetic materials which may float, remain in suspension or sink, and which may interfere with any legitimate uses of the sea.
8. Substances having proven carcinogenic, teratogenic or mutagenic properties in the marine environment.
9. Radioactive substances, including their wastes, when their discharges do not comply with the principles of radiation protection as defined by the competent international organizations, taking into account the protection of the marine environment.[53]

12.2.8. Annex II

Harmful or Noxious Substances and Materials Whose Disposal Is Subject to a Special Permit

The following substances, families and groups of substances, or sources of pollution, listed not in order of priority for the purposes of Article 9 (3) of this Protocol, have been selected mainly on the basis of criteria used for Annex I, requiring strict control, while taking into account the fact that they are generally less noxious and are more readily rendered harmless by natural processes.[54]

1. The following elements and their compounds: zinc, copper, nickel, chromium, lead, selenium, arsenic, antimony, molybdenum, titanium, tin, barium, beryllium, boron, uranium, vanadium, cobalt, thallium, tellurium, silver.
2. Biocides and their derivatives not covered in Annex I.
3. Organosilicon compounds and substances which may form such compounds in the marine environment, excluding those which are biologically harmless or are rapidly converted into biologically harmless substances.[55]
4. Cyanides and fluorides.
5. Non-biodegradable detergents and other surface active substances.
6. Pathogenic micro organisms.
7. Thermal discharges.
8. Substances which have a deleterious effect on the taste and/or smell of products for human consumption derived from the aquatic environment, and compounds liable to give rise to such substances in the marine environment.

9. Acid or alkaline compounds of such composition and in such quantity that they may impair the equality of sea water.
10. Substances which, though of a non-toxic nature, may become harmful to the marine environment or may interfere with any legitimate use of the sea owing to the quantities in which they are discharged.

The control and strict limitation of the discharge of substances referred to above must be implemented in accordance with Annex III.[56]

12.2.9. Annex III

Factors To Be Considered for the Issue of the Permits

With a view to the issue of an authorization for the discharge of harmful or noxious substances and materials containing substances referred to in Annex II to this Protocol, particular account shall be taken of but not limited to the following factors:[57]

A. Characteristics and composition of the waste
   1. Type and size of waste source (e.g. industrial process).
   2. Type of waste (origin, average composition).
   3. Form of waste (solid, liquid, sludge, slurry, gaseous).
   4. Total amount (volume discharged, e.g. per year).
   5. Discharge pattern (continuous, intermittent, seasonally variable, etc.).
   6. Concentrations with respect to major constituents, substances listed in Annex I, substances listed in Annex II, and other substances as appropriate.
   7. Physical, chemical and biochemical properties of the waste.

B. Characteristics of waste constituents with respect to their harmfulness
   1. Persistence (physical, chemical, biological) in the marine environment.
   2. Toxicity and other harmful effects.
   3. Accumulation in biological materials or sediments.
   4. Biochemical transformation producing harmful compounds.
   5. Adverse effects on the oxygen content and balance.
   6. Susceptibility to physical, chemical and biochemical changes and interaction in the aquatic environment with other sea-water constituents which may produce harmful biological or other effects on any of the uses listed in section E below.

C. Characteristics of discharge site and receiving marine environment
   1. Hydrographic, meteorological, geological and topographical characteristics of the area.

2. Location and type of the discharge (outfall, canal, outlet, etc.) and its relation to other areas (such as spawning, nursery, and fishing areas, shellfish grounds) and other discharges.
3. Initial dilution achieved at the point of discharge into the receiving marine environment.
4. Dispersion characteristics such as effects of currents, tides and wind on horizontal transport and vertical mixing.
5. Receiving water characteristics with respect to physical, chemical, biological and ecological conditions into the discharge area.
6. Capacity of the receiving marine environment to receive waste discharges without undesirable effects.

D. Availability of waste technologies

The methods of waste reduction and discharge for industrial effluents as well as domestic sewage shall be selected taking into account the availability and feasibility of:

(a) Alternative treatment processes;
(b) Re-use or elimination methods;
(c) On-land disposal alternatives;
(d) Appropriate low-waste technologies.

E. Potential impairment of marine ecosystem and sea-water uses

1. Effects on human life through pollution impact on:
   (a) Edible marine organisms;
   (b) Aesthetics.
2. Effects on marine ecosystems, in particular living resources, endangered species and critical habitats.
3. Effects on other legitimate uses of the sea.

12.2.10. Annex IV

Environmental Impact Assessments

1. Each Party shall require that the environment impact assessment contains at least the following:
   (a) A description of the geographical boundaries of the area within which the activities are to be carried out, including safety zones where applicable;
   (b) A description of the initial state of the environment of the area;
   (c) An statement of the nature, aims, scope and duration of the proposed activities;
   (d) A description of the methods, installations and other means to be used;

(e) A description of the foreseeable direct or indirect short and long-term effects of the proposed activities on the environment, including fauna, flora and the ecological balance;
(f) A description of the expected interferences of the proposed activities with fishing activities and ship navigation including both military and commercial vessels as well as surface and submarine craft;
(g) A statement setting out the measures proposed to reduce to the minimum the risk of damage to the environment from carrying out the proposed activities, in addition, possible alternatives to such measures;
(h) A statement setting out the measures proposed to reduce to the minimum interference from carrying out the proposed activities with fishing activities and ship navigation described in (f) above, in addition, possible alternatives to such measures;
(i) A statement of the measures to be taken for the protection of the environment from pollution and other adverse effects during and after the proposed activities.
(j) A statement of the measures to be taken for the protection of fishing activities and navigation described in (f) above during and after the proposed activities.

2. Each Party shall promulgate standards taking into account the international rules, standards and recommended practices and procedures adopted in accordance with Article 22 by which impact assessments are to be evaluated.[58]

12.2.11. Annex V[59]

Oil and Oily Mixtures and Drilling Fluids

The following provisions shall be ensured by the Parties in accordance with Article 10:

A. Oil and Oily Mixtures
   1. All effluents with a petroleum content, including formation water, ballast water and processing drainage shall be treated by the best available technology to a level of 40 mg per litre (average) before discharge. Machinery spaces drainage shall be treated by the best available technology to a level of 15 mg per litre (average) before discharge. These standards apply as well to effluents from gas/condensate platforms. Testing is to be performed daily to ensure compliance with these standards.[60]

2. Oily waste and sludges from separation processes shall be transported to shore;
3. Oil and condensates from well testing shall be flared, taking all precautions necessary to prevent leakages of oil into the sea.

B. Drilling Fluids

1. Water based drilling fluids are subject to the following requirements:
   (a) The use of such drilling fluids is subject to a prior permit by the competent national authority of the Controlling State;
   (b) The disposal of such drilling fluids and the drill cuttings shall be made as far away as possible from the nearest land, specially protected area, specially designated fisheries area, or fixed fisheries installation, but in any case at a minimum distance established in the permit;
   (c) Where the drilling fluids contain harmful or noxious substances and/or materials, the provisions of Article 9 of the present Protocol shall apply.
2. Oil based drilling fluids are subject to the following requirements:[61]

   Definitions

   (a) Oil based drilling muds are defined as drilling muds containing oil or an emulsion of water and oil.
   (b) Cuttings include drilled formation as well as solids form mud additives.

   Rules and Principles

   (a) In order to control the use of oil based muds, national authorities shall require prior authorization wherever oil based muds are used offshore. Prior permit shall be given only where the national authorities regard the use of oil based mud as essential for geological, safety or economic reasons.
   (b) Necessary economic reasons for the use of oil based muds refer to situations where deviated drilling is necessary, or where the costs of using water-based muds would be prohibitive for the development of marginal fields. In these cases, a full report shall be presented to the Organization by the national authorities.
   (c) In the upper part of the well, the use of oil based muds is prohibited. Exemptions may be granted for strictly geological or safety reasons. In these cases, a full report shall be presented to the Organization by the national authorities.
   (d) Whole oil based muds shall not be dumped or discharged at sea.

(e) All steps shall be taken to prevent leakages and spillages of muds.

(f) The oil content on cuttings discharged into the sea shall not exceed the standard of an average 100 grams of oil per kilogram dry cuttings.[62]

(g) To achieve this standard the following principles shall be applied:

  (i) the optimal use of the best available solids control and cuttings cleaning technology;

  (ii) the use of new mud formulation which of themselves will reduce residual oil on cuttings;

  (iii) new drilling techniques to reduce the quantity of oil contaminated cuttings;

  or combinations of (i), (ii) and (iii) above.

(h) Paragraphs (f) and (g) will apply to all exploration and appraisal drilling operations with oil based muds starting after the end of (1990), and for platforms where the detailed design has started after the end of (1988).[63]

(i) The oil content of cuttings shall be measured as an average over that part of the well drilled with oil based mud, using a 50 ml retort with the heating element outside the test cylinder. Representative samples shall be taken at least every day oil based mud is used.[64]

(j) The use of diesel oil based muds is prohibited. However, diesel oil may exceptionally be added to drilling muds in the following circumstances and after the competent national authorities have given prior authorization:

  (i) in work-over operations:
  producing wells sometime require remedial measures, eg. additional preparation of the casing or modifications to the lining or casing, for which oils are necessary. These operations do not involve drilling or production of cuttings.

  (ii) in well stimulation and completion techniques:
  when a well has been drilled it is necessary to perforate the casing into the reservoir and sometimes to pen up fissures within the reservoir. These operations are carried out at pressure and solid-free fuel oils are often necessary;

  (iii) in emergency drilling operations with water based muds:[65]

if stuck pipe conditions occur with water based muds then diesel oils may be used to free the drill string.

The authorization referred to in subparagraph (i) above shall only be given if the national competent authority is satisfied that the use is justified because of the above circumstances. In cases where such fluids are used, the disposal of such drilling fluids and drill cuttings shall be prohibited.[66]

C. Drilling Fluids containing Heavy Metals[67]

1. All drilling fluids containing heavy metals, including but not limited to mercury, cadmium, zinc, copper, nickel, chromium, lead, selenium, arsenic, antimony, molybdenum, titanium, tin, barium, beryllium, boron, uranium, vanadium, cobalt, thallium, tellurium, and silver shall be used observing the requirements set forth in Annexes I and II of this Protocol. If not appearing in these lists, internationally accepted standards must be observed as a minimum.
2. The Contracting Parties shall as soon as possible develop, formulate and adopt common standards resulting in the entry of these substances, if lacking, onto Annexes I or II to limit the risk of pollution from this source.

12.2.12. Annex VI

Safety Measures

The following provisions shall be ensured by the Parties in accordance with Article 15:

(a) That the installation is safe and fit for the purpose for which it is to be used, including but not limited to, that it is designed and constructed so as to withstand, together with its maximum load, any natural condition, including maximum wind, wave and ice conditions[68] as established by historical weather patterns, earthquake possibilities, sea-bed conditions and stability, and water depth.

(b) That all phases of the activities, including but not limited to, storage and transport of recovered resources, are properly prepared, that the whole activity can be controlled for safety reasons and is conducted in the safest possible way and that the operator applies a monitoring system for all its activities;

(c) That the most advanced safety systems are used and periodically tested in order to minimize the dangers of leakages, spillages, accidental discharges, fire, explosions, blow-outs or any other threat to human safety or the environment, that a trained specialized crew to operate and maintain these systems is present and that this crew undertakes periodic exercises;

(d) That the installation, and the established safety zone, is sufficiently marked so as to give adequate warning of its presence and sufficient details for its identification using appropriate and internationally recognized warning signals;

(e) That the installations are, in accordance with international maritime practice, indicated on charts and notified to those concerned;

(f) That in order to secure the above provisions, the person and/or persons who have the responsibility concerning the installation and/or the activities, including the person responsible for the blowout preventer, have the qualifications required by the Controlling State and that there is permanently sufficient, qualified staff present at the installation. Such qualifications should include, in particular, training, on a continued basis, as regards safety and environmental matters.

12.2.13. Appendix[69]

List of Oils

Asphalt solutions

- Blending Stocks
- Roofers Flux
- Straight Run Residue

Oils

- Clarified
- Crude Oil
- Mixtures containing crude oil
- Diesel Oil
- Fuel Oil No. 4
- Fuel Oil No. 5
- Fuel Oil No. 6
- Residual Fuel Oil
- Road Oil
- Transformer Oil
- Aromatic Oil (excl. vegetable oil)
- Lubricating Oils and Blending Stocks
- Mineral Oil
- Motor Oil
- Penetrating Oil
- Spindle Oil
- Turbine Oil

Distillates

- Straight Run
- Flashed Feed Stocks

Gas Oil

- Cracked

Jet Fuels

- JP-1 (Kerosene)
- JP-3
- JP-4
- JP-5 (Kerosene, Heavy)
- Turbo Fuel
- Kerosene
- Mineral Spirit

Naphtha
Solvent
Petroleum
Heartcut Distillate Oil
Gasoline Blending Stocks
Aklylates - fuel
Reformates
Polymer - fuel
Gasolines
Casinghead (natural)
Automotive
Aviation
Straight Run
Fuel Oil No. 1 (Kerosene)
Fuel Oil No. 1-D
Fuel Oil No. 2
Fuel Oil No. 2-D

Notes - Chapter 12

1 Protocol for the Protection of the Arctic Region against Pollution resulting from Exploration and Exploitation of the Continental Shelf and the Sea-Bed and its Subsoil.

2 See Part II Sections 9.2.2. and 9.3.1.

3 See Part II Sections 8.2. and 10.2.1., and Stokke pp. 8-9.

4 See Part II Sections 1 and 3.2. for convention and customary law regulating State jurisdiction over sea bed pollution lending additional legal support to the formation of the Arctic Oil Rig Protocol.

5 Article 208 (5) and (1).

6 Article 208 (1) and Articles 60 and 80. Articles 60 and 80 give a State exclusive jurisdiction over artificial islands, installations and structures in the State's exclusive economic zone and on its continental shelf respectively. Under Articles 56 and 77 coastal States are given sovereign rights to explore and exploit the natural resources in the exclusive economic zone (including the sea bed and sub sea bed) and over the continental shelf respectively. Article 56 additionally provides a coastal State with jurisdiction for the protection and preservation of the marine environment in its economic zone. The exclusive economic zone is defined in Article 57 as not extending beyond 200 nautical miles from the baselines from which the territorial sea is measured. Article 76 (1) defines the continental shelf generally as comprising the sea bed and subsoil of the submarine areas that extend beyond its territorial sea throughout the natural prolongation of its land territory to the outer edge of the continental margin, or a distance of 200 nautical miles from the baselines from which the territorial sea is measured where the outer continental margin does not extend to this distance. Specifically, Article 76 (2)-(6) limits the continental shelf through a series of complicated equations involving the thickness of sedimentary rocks or straight lines not exceeding 60 nautical miles in length; not exceeding 350 nautical miles from the territorial sea baselines or not exceeding 100 nautical miles from the line connecting depth of 2,500 meters, with the exception of submarine elevations that are natural components of the continental margin.

7 Article 208 (5).

8 Articles 208 (1) (2) and 60 and 80.

9 Article 208 (3) and (4). See Part II Sections 11.2. and 11.3. for a short discussion of the interplay between the national legislation dealing with sea-bed pollution and international regulations.

10 Hakapää p. 118 notes that the word 'applicable' as regards international rules and standards not only obliges the State to enforce conventions and agreements but relevant customary law as well.

11 Articles 214, 60 and 80.

12 See Part I Section 2.1.4. and Part II Section 10.3.3.

13 UNEP/IG.74/Inf.9.

14 UNEP Draft p. 1.

15 An example can be taken in Norway from St.meld.14 s. 10, where key blocks (strategic blocks) for oil exploration in the Barents Sea were dealt out despite objections made by organizations including the Ministries of the Environment and of Fisheries. Even in an environmentally progressive country as Norway, oil interests predominate.

16 UNEP Draft p. 2.

17 *Ibid.* Annex I, pp. 1-5. It must also be noted that various socio-economic, legal, political and philosophical aspects were incorporated, UNEP Draft pp. 2, 4. These differences are clearly not the same as in the Arctic region, and adaptions to reflect this will be made. Though the differences in the Arctic cannot be dealt with here to any great extent, as noted in Part II Section 10.1. the U.S. and the USSR until recently represented two extremes of the political, socio-economic and philosophical spectrum with Canada, Denmark and Norway falling at varying degrees between the two. Greenland, separate from Denmark, represents a indigenous population. The U.S., Canada, Denmark and Norway are considered industrialized countries. The Soviet Union with its economic problems has elements both of being industrialized and developing.

18 This structure was chosen for smoothness and fluency and to avoid frequent amendments to the main body, UNEP Draft pp. 2-3. It is additionally noted that the Annexes might require amendments occasionally to keep up with the development of technology. Annex VII of the UNEP Draft, a contingency plan, was deleted since the Arctic Counter-Pollution Protocol is proposed and discussed in Part II Chapter 13.

19 The Preamble sets the stage by giving various declarations of the Arctic rim States as related to the Arctic and assertions of State desire to preserve this environment. The declaration dealing with the differences between the legal, political and socio-economic systems of the States and taking these into account is an altered version of one similar in the UNEP Draft. See UNEP Draft p. 2. It is included in order to emphasize the necessity for *flexibility* in the Arctic Oil Rig Protocol, both as regards its drafting and its operation, because of these national variations. Under the UNEP Draft the intention is to emphasize the differences in levels of country development and the necessity to take into account the social and economic necessities, and recommendations rather than mandatory provisions for the States appear. These were removed in the Arctic Draft due to the technological developments achieved in the Arctic States.

20 See Part II Section 11.4.2. for comments regarding an Organization established under the Convention for the Protection of the Arctic Region against Pollution. A sub organization to deal specifically with administration of the Arctic Oil Rig Protocol could be formed by the Arctic States if desired.

21 The definition of 'pollution' likewise refers back to the definition given in the Arctic Convention, Article 2. This is done in order to have, as much as possible, a uniform broad definition under the Arctic Convention and Protocols.

22 The definition of 'safety zone' is much the same as that used in the UNEP Draft. Besides ensuring the safety of navigation and the installations, safety to *fishing* is also added. To a great extent it is this activity which takes place in Arctic waters and can be expected at times to operate in the vicinity of the offshore platforms.

23 A definition of the geographical coverage of the Protocol is given corresponding with the Arctic Region defined in the Arctic Convention, Article 1.

24 The phrase 'best available technology' is noteworthy in that it requires the use of the best anti pollution equipment available. Since the world's leading industrialized countries are solidly represented among the Arctic rim States, it can be expected that the equipment required available to be simply the best, which makes the Arctic Oil Rig Protocol definitely environmentally 'friendly'. Note that the same phrase if applied to developing countries could require *less* than the best technology, if 'available' is interpreted narrowly. Regarding the Soviets, in any negotiations the best available technology by Western standards should be emphasized. This should prove attractive to all Parties, the Soviets because it means the supply of sophisticated technology, and the West because it might mean large technology contracts. See Part II Section 8.2.1. regarding Western companies' interest in Soviet oil and gas in the Barents Sea and the Soviet need for oil, technology and expertise. Combining the technology with the proper education of the personnel would benefit the Arctic environment.

25 The phrase 'including but not limited to' was at various times added throughout the Arctic Protocols. This was done in order to allow the States to take into account more elements than listed in establishing their various anti pollution procedures, by which it was hoped to allow the States to take a more 'environmentally friendly' stance if desired.

26 Article 4 (4) was added covering the necessity for ensuring that the placement of an installation does not interfere with known fishing banks and commercial and navy vessel traffic zones. This is both for the preservation of the fishing activities and for platform safety. It seems possible that Norway might be against this paragraph if it considers it detrimental to Norwegian sovereignty over its sea areas, especially as related to the Soviet warships and submarines trafficking through the Barents Sea. See Part II Section 8.3.

27 Though not mentioned in Article 5, under Annex IV a statement is added concerning the inclusion of consideration to *fishing* activities and commercial and military vessel *traffic* zones into this environmental impact statement.

28 The latter sentence was added to allow the States to consider other elements in evaluating an authorization to drill.

29 This phase was added concerning States respecting the rights of the *accused* during legal proceedings for two reasons. First, as can be seen in Butler Sections F.2., F.3., G.1.a., G.3., and J.2. Supplement to J.4. generally the Soviet anti pollution provisions appear heavily in favour of State enforcement rights at the expense of the rights of the accused. This provision might theoretically be of some help to an accused polluter in the Soviet Courts. It was additionally thought that the acknowledgement of general rights of an accused polluter might also allow a somewhat more flexible, positive attitude to the Protocol by the Arctic States with their strong oil interests.

30 See Section 12.2.7. for discussion of toxic chemical lists making up Annexes I and II. Annex I contains a list of harmful or noxious substances the use and disposal of which is prohibited in the Protocol Area. Annex II contains a list of harmful or noxious substances the use or disposal of which in the Protocol Area requires a special permit from the Controlling State. An Article adapted from Article 6 of the Paris Convention could also be added as a *catch-all* to generally reduce pollution from other chemicals not appearing on the lists. This could state for example, '(W)ith a view to preserving and enhancing the quality of the marine environment, the Parties, without prejudice to the provisions of Section III, shall endeavour to reduce the existing pollution from offshore platforms, and to forestall any new pollution from offshore platforms, including that which derives from new substances.'

31 It was considered necessary to add the phrase, 'drilling fluids containing heavy metals', both in Article 10 and to Annex V, based upon comments from Chief Engineer Jens Koefoed of SFT October 7, 1988 mentioned in Part I Section 1.3.6. Other chemicals appearing in drilling fluids determined by experts as noxious should as well be added in the future.

32 The UNEP Draft proved more lenient than the Paris Convention; so Paris Convention and MARPOL 73/78 figures are used. It is possible that for the Arctic marine environment these mixture limits are too lenient. This is dealt with in more detail in discussion covering Annex V, however the two figures in question appear in Article 10 (1)(b)(i) and (ii) and are respectively a maximum oil content of *15 mg. per litre* while undiluted for machinery space drainage; and a maximum oil content of *40 mg. per litre* as an average in any calendar month for production water. As noted, in the past there has been uncertainty concerning MARPOL 73/78's application to offshore platforms, and it has been recently recommended by the IMO that of the four types of discharge from offshore platforms, machinery space discharge, offshore processing drainage, production water and displacement, only machinery space drainage should be subject to MARPOL 73/78. See Part I Section 4.2.2.4. and UNEP Report p. 15. The figure presented here, 15 mg. per litre, is similar to the MARPOL 73/78 figure for discharge of oil or oily mixtures from oil rigs in a *special area.* MARPOL 73/78, Annex I, Regulation 21. Here is used a figure of *15 parts per million.* The UNEP mg per litre will be followed here, especially since it is obviously more stringent than the 40 mg per litre for the other types of platform discharges, followed both by the UNEP Draft and the Paris Commission. Though the Arctic is not declared a special area under MARPOL 73/78, this is a proposal forwarded in Part II Chapter 14. It is possible that even this figure is not restrictive enough for the Arctic environment and needs to be evaluated by marine experts. The 40 mg. per litre average per month is the same as recommended in the North Sea under the Paris Convention and used by Norway. See Part I Section 4.2.2.4. and SFT Utslipp 1988 s. 6.

33 A sewage treatment system on the offshore installations was not thought to be too great an environmental requirement to add for the Arctic, though under the UNEP Draft discharges without treatment are allowed at fixed distances from fisheries installations and land.

34 Disposal of garbage under Article 12 was allowed as long as it was reserved to food wastes, and disposal was not conducted in oceans areas covered with ice.

35 Under UNEP Draft the conditions of threats to human life or the *safety of installation* are added to force majeure. It was at first thought beneficial to drop the platform safety clause just in case an oil company could conceivable unload massive amounts of pollutants to save their platform. It was decided however to allow it. If a blowout or anything drastic occurs which threatens platform safety, then human safety will also be threatened and oil is going conceivably into the sea anyway. Even if no human lives are at stake, in such situations it seems questionable whether the oil companies have much control over the situation, and they are not realistically going to sacrifice a multi million dollar platform to avoid pollution. The environmental benefits seem negligible.

36 In the Articles appearing in Section IV it should be noted that phrases containing the words 'the Controlling State should ensure' were changed to 'the Controlling State *shall* ensure'. Though most of the UNEP Draft Articles used the 'shall' form, Article 15 for example stated that the '(C)ontrolling State ... should ensure that safety measures are taken with regard to the design, construction, placement, equipment, marking, operation and maintenance of installations.'

37 'Competent international body' also appears in the UNEP Draft and has apparently developed standards of safety and fitness for purpose for offshore installations and equipment, but a definition of the term is not given. The IMO is often referred to as the competent international body, but in relation to shipping. Though as noted the IMO has developed IMO Mobile, this applies *only* to marine structures not permanently fixed into

the sea bed, and would thus apply to only floating installations and equipment in this case. As can be seen Article 15 (3) refers to production platforms, sub sea installations, mobile offshore drilling units, offshore storage facilities, offshore loading systems, pipelines and other installations. It is perhaps intended that the term apply to UNEP itself, but this in uncertain. In spite of the vagueness, 'competent international body' was included in the Arctic Oil Rig Protocol to allow the Parties to choose or establish a body to their liking, whether it be the IMO, Paris Commission or other. Regardless, the idea behind this Article is important, that of using established and uniform international standards for safety and fitness for purpose for offshore platforms and equipment.

38 The latter phrase has been added to Article 16 (1). Additionally the latter phrase in 16 (2) has been added as has the phrase appearing in 16 (3). Though somewhat redundant this reference provides a link between the Arctic Oil Rig Protocol and the Arctic Counter Pollution Protocol, while at the same time allowing the Parties time to ratify the different Protocols as they politically see fit. It is hoped by this wording to achieve a balance between a definite linkage and at the same time a flexibility. See Part II Chapter 13 for discussion regarding contingency plans under Article 16 and the Protocol for Counter-Pollution Operations in the Arctic Region.

39 As above 'shall' is substituted for 'should' requiring the operator to measure the effects of the offshore activities on the environment and report them to the Controlling State. The State is additionally required to establish a permanent monitoring system under Article 17 (2). Here the UNEP Draft clause has been substituted with the more specific Article II of the Paris Convention, and 'shall' substituted for 'agree' making the monitoring as mandatory as possible. As seen Article 17 (2) sets up a groundwork for joint monitoring efforts among the Arctic States and also requires the Parties to take into account similar monitoring programs already in operation. This was done so that the Parties would benefit from and utilize the monitoring work and experience being conducted under the Paris Convention of which Norway is a leading participant. See Part I Section 4.2.2.

40 The Arctic Oil Rig Protocol is harmonized with other international regulations through the latter phase. Thus the norms under the IMO, the Paris Convention and the ODC and the LDC, if accepted by the Arctic rim States, can be considered, negotiated and used here. See Part I Section 3.2.1.3. Since this area is still in the developing stage, it was felt that language used should be flexible enough to allow new removal standards to be applied, if consistent with the Arctic conditions.

41 It is proposed in Part II Chapter 14 that the Arctic be declared a Special Area under MARPOL 73/78. Article 19 expands upon this allowing *localized* specially protected areas within the Arctic as related to offshore activities. The idea of localized special areas seems beneficial in order to allow the Coastal States flexibility to require even more stringent measures regarding environmental impact statements, authorizations, platform placement, safety measures, contingency planing, monitoring and the removal of platforms in areas deemed especially sensitive due to fish spawning, bird roosting and the presence of whale and seal congregations, etc. The phrase, '(T)hese measures shall be non-discriminatory', was added since it appears in LOSC Article 234, applied to a Coastal State's regulations to prevent, reduce and control marine pollution from shipping in ice-covered areas. It was thought equally beneficial and necessary in the Arctic, due to the many oil companies from many different States involved in a State's offshore activities.

42 As seen in Part II Section 11.2. this requirement goes further than the LOSC Article 208 (4) recommendation for States to endeavour to harmonize their policies at the regional level. Note that a requirement concerning national policies and measures in the Arctic is added throughout the Arctic Protocols where relevant to promote uniformity and effectiveness.

43 Article 22 (3) has been changed to *require* the Parties to exchange rather than just endeavour to exchange information regarding their national policies, laws and regulations and this harmonization. This has been added in view of the difficulties obtaining legislative information from especially the Soviet Union. See Butler Sections F.2., F.3., G.1.a., G.3., and J.2. Supplement to J.4. for a general survey of Soviet legislation covering offshore installations.

44 This sentence has been added that this information shall be made public to attempt to alleviate the information problems mentioned under the Paris Convention. See Part II Section 9.3.1.

45 Whereas the UNEP Draft requires the States to merely 'endeavour' to provide this access to Courts, the Arctic Oil Rig Protocol requires that States 'shall grant equal access.' This was deemed necessary to achieve cooperation in the Arctic regarding initiating proceedings to receive compensation for pollution damage. If one Party's Courts are particularly inaccessible for these types of proceedings, then this could in turn throw the entire system represented by the Arctic Convention and Sub Protocols into an unbalance and thus affect the success of these agreements in dealing with marine pollution.

46 The phrase was added that the Parties agree to ratify as soon as possible the Arctic Liability Protocol (see Part II Chapter 17) the provisions 26 (1), (2) and (3) dealing respectively and generally with State compensation procedures, operator liability and operator insurance coverage were left in tact. Though redundant if the States ratify the Arctic Liability Protocol, they would fill the gap pending ratification, and as well give emphasis to these points covered in the Protocol. This is to tie in this liability scheme with the pollution source covered by the Arctic Oil Rig Protocol, as well as attempt to give a boost to ratifying the Liability Protocol since States are reluctant to be bound to such schemes. See Part II Section 9.3.3. This linkage might however also prevent the Arctic Oil Rig Protocol from ratification, something which would have to be evaluated. It is left loose, by using the phrase 'agree to ratify as soon as possible' rather than 'shall ratify' to provide for this eventuality.

47 Article 26 (2) and (3) respectively hold operators to strict and unlimited liability for pollution damage and require adequate insurance coverage or other financial security. The UNEP Draft limits the liability. It is noted however in UNEP Draft p. 3 that, 'neither the international organizations nor the insurance companies consulted by the IJO were able to give a realistic minimum figure' for sufficient insurance financial coverage, and noted further is that the minimum amount of insurance established by the 1976 Liability Convention of 35 million Special Drawing Rights is '*much too low*,' emphasis added. The private operator plan OPOL has approximately one-half of the 1976 Liability Convention coverage. See Part I Section 6.3.3. It was therefore felt more environmentally responsible to propose *unlimited* liability for the Arctic region, following Norwegian legislation, even though this will probably be thought unreasonable by several of the Arctic States. See Part II Section 9.2.5.

48 This Section, dealing with among other things the appointment of national authorities, meetings and functions of the meetings was left as in the UNEP Draft, with the exception that when mention is made of contingency plans, reference is made to the Arctic Counter Pollution Protocol.

49 Points 7 and 8 were added and are general for all the Arctic Protocols.

50 UNEP Draft p. 4 notes that the IJO contacted various scientists, Universities, operators and petroleum companies in an effort to compile lists of harmful and noxious chemicals, but no lists were drafted because even the oil companies 'do not know the exact composition of some products where such information is covered by industrial secrecy.' The IJO solution was to study the United Kingdom's legislation covering these chemicals, 'which appears to be both complete and accepted by the operators, in order to provide a solid basis for discussing this matter,' *ibid.* An attempt was made to contact UNEP to obtain more information regarding these lists but a reply was never received, and it was consequently decided to use the Barcelona Protocol III's lists for

land-based pollution as a starting point. See Part I Section 4.2.3.2. where it is noted that this Protocol's 'black list' and 'grey list' as well as procedures for granting discharge permits are more extensive than the corresponding lists from the Paris Convention (which have been in use in the North Sea for over the past 15 years). In the Arctic these lists would have to be updated should it be shown other chemicals are used in offshore activities which are harmful to the Arctic marine environment. Annex II substances could be moved to the Annex I, 'black list', thereby prohibiting the discharge of some or even all of these substances should it be deemed necessary, which would make the Arctic Oil Rig Protocol much more stringent for Arctic conditions.

51 The phrase 'or other noxious properties' was added from Paris Convention, Annex A, Part I.

52 The exception 'biologically harmless substances or rapidly converted into such' could be deleted for chemical families 1., 2., and 3. to make this 'black list' even more stringent.

53 The Barcelona Protocol III's Annex I contains point B. which makes an exception from Annex I for 'discharges which contain substances listed ... which are below the limits defined jointly by the Parties.' This was deleted.

54 The phrase 'requiring strict control' was added from the Paris Convention, Annex A Part II. The phrase 'and therefore generally affect more limited coastal areas' appearing in the Barcelona Protocol III's Annex II, was deleted to ensure a stricter interpretation of this 'grey list.'

55 As in Annex I the phrase 'biologically harmless substances or rapidly converted into such' could be deleted to make the 'grey list' stricter.

56 Crude oils and hydrocarbons of any origin from the Barcelona Protocol III's Annex II was deleted in order to be consistent with Annex V and Appendix.

57 The factors listed are the same as in the UNEP Draft, however the phrase 'but not limited to' was added to ensure that the evaluation concerning disposal authorization should not be limited to these. This list should be viewed by experts to add any necessary points peculiar to the Arctic.

58 As noted previously phrases (f), (h) and (j) have been added to the elements making up the environmental impact statement.

59 This Section was made more stringent based upon recommendations of the Paris Commission and the SFT. See Part I Section 4.2.2.4.

60 The UNEP Draft paragraph 1 dealing with the same reads, 'Spills of high oil content in processing drainage and platform drainage shall be contained, diverted and then treated as part of the product, but the remainder shall be treated to an acceptable level before discharge, in accordance with good oil field practice.'

61 Part 2 of Drilling Fluids is adapted from Par.Comm.10, Annex 12, pp. 1-2 with an addition noted below for the national authorities giving authorization for the use of diesel based drilling fluids. See point j (iii).

62 The word 'standard' was used rather than 'target standard' used in Par.Comm.10. in order to be more mandatory.

63 The dates from Par.Comm.10 were left to give an idea of the expected dates in the North Sea and Barents Sea.

64 UNEP Draft, Annex II pp. 30-31 reads as follows: 'Oil based drilling fluids are subject to the following requirements: (a) Such fluids shall only be used if they are of a sufficiently low toxicity and only after the operator has been issued a permit by the competent national authority of the Controlling State when it has verified such low toxicity; (b) In the cases where such fluids are used, the permit shall contain conditions concerning the disposal of such fluids and the drill cuttings resulting from its use; (c)

In cases where oil based drilling fluids are used, they shall not be disposed of into the sea; and, the drill cuttings shall be effectively washed before discharged into the sea, so that the oil content is less than ten percent in dry weight.

65 This authorization curiously was neglected in Par.Comm.10 Annex 12, and since the idea seems beneficial it was adopted from the less specific UNEP Draft paragraph 3, Annex II, p. 31.

66 UNEP Draft, Annex II p. 31. dealing with diesel based drilling fluids reads as follows: 'Diesel based drilling fluids are subject to the following requirements: (a) The use of diesel based drilling fluids is prohibited except in those exceptional circumstances where the use is necessary because of exceptional technical exigencies and after the operator has been issued a special permit by the competent national authority of the Controlling State; (b) The special permit referred to in subparagraph (a) above shall only be issued if the national competent authority is satisfied that the use is justified because of exceptional technical exigencies; (c) In the cases where such fluids are used, the disposal of such drilling fluids and drill cuttings shall be subject to the conditions in the special permit but in any case, disposal into the Protocol Area shall be prohibited.'

67 This Section dealing with drilling fluids containing heavy metals was added. Discharges containing these pollutants would be covered by Annexes I and II.

68 An addition was made in paragraph (a) for 'ice'. The phrase, 'including but not limited to' is also substituted for 'in particular.'

69 The List of Oils is not necessarily comprehensive. It was compiled from the UNEP Draft list which is identical with MARPOL 73/78 Annex I, Appendix I. For use in the Arctic Oil Rig Protocol the oils need evaluation by Arctic experts and additions made accordingly.

13. Discussion of Arctic Counter-Pollution Protocol with Sub Bilateral Agreements[1]

13.1. Background and Proposals

As noted there is presently no in force bilateral Norwegian - Soviet contingency plan for dealing with emergency oil or hazardous chemical spills in the Barents Sea in spite of such plans being formed by all the Arctic rim States in areas of oil activities.[2] Because of this situation *two* proposals will be made. The first is that a *Protocol* for Counter-Pollution Operations in the Arctic Region be established, similar to the Bonn Agreement[3] with additions from the UNEP Draft Annex VII.[4] This is to be a general Arctic contingency agreement to bring conformity and a uniform minimum level of national competence and cooperation into the different Arctic bilateral contingency plans.[5] The second proposal is that Norway establish a joint contingency plan under the Protocol with the Soviet Union in the Barents Sea.

13.1.1. Protocol for Counter-Pollution Operations in the Arctic

Basically the general points that need to be covered on the *international* level in an Arctic Counter-Pollution Protocol include briefly,[6] a geographical definition of the Arctic in conjunction with Article 1 of the Arctic Convention; statements of desire of State Parties to cooperate in the Arctic in combating pollution;[7] a definition of a pollution incident;[8] statements of promise by the Parties to develop joint operations and bilateral agreements and to harmonize national policies in this regard; ratification of the other Arctic Protocols; and promises to develop routines for notice of incidents to other Parties, consultation, zones of responsibility, exchange of technical information including scientific research, calls for assistance, provision of assistance, economic arrangements, periodic meetings, amendment and voting procedures, settlement of disputes in line with LOSC provisions, accession and duties of the Organization.

There should also be included specific State promises for ensuring at a *minimum* on the *national* level,

(a) co-ordination of the national contingency plan and the operator's contingency plan, and control of the emergency actions;
(b) directions to the operator to take any action the State may specify in the course of preventing, abating or combating pollution or in the preparation of further anti pollution action;
(c) collection and ready availability of all necessary information;
(d) an up-to-date list of the person and entities to be alerted and informed of an emergency;
(e) specified actions, if necessary, including interventions by technical experts and trained personnel with the necessary equipment and materials; and
(f) immediate communication to the competent international organizations with a view to avoiding danger to shipping and other interests.[9]

As seen what is of importance with the proposed Arctic Counter-Pollution Protocol is that the Arctic rim States promise to set up guidelines in such areas as joint clean up operations, exchange of technical information,[10] notification of incidents and operator competence. Under the Protocol these States would meet periodically to discuss their Arctic programs. This would hopefully establish an international milieu which would enhance the overall uniformity and thus effectiveness of Arctic contingency plans.

13.1.2. Sub Bilateral Agreements

The *second* proposal is that Norway establish a specific contingency plan with the Soviet Union in adjoining areas of offshore oil or mineral activities or adjoining areas with moderate to heavy ship traffic in the Barents Sea. As mentioned the form of an Arctic regional agreement encompassing separate bilateral agreements would have the advantage of allowing specific plans to cover the seas in question, while at the same time ensuring uniformity. Thus the proposed Barents Sea contingency plan as well as the already existing Arctic bilateral contingency plans would have to be made consistent with the Protocol points outlined above.

It was thought proposing one contingency plan for all 5 Arctic rim States covering the entire Arctic region to be unrealistic due to the large areas involved, as well as the political and bureaucratic difficulties in practically binding all the countries together. It should be noted however that in practice these difficulties might be less than imagined,[11] and the avenue is of course always open for the States to expand their cooperation.[12]

A model specific for the Norwegian - Soviet contingency plan is proposed and presented in Section 13.4. in which an attempt is made to avoid the boundary conflict. A discussion of the problems involved in arriving at this model is presented below, as well as a brief discussion of legal theory requiring these two countries to establish such a plan in spite of their jurisdictional conflict. In Section 13.3. appears an *alternative* proposal which argues for Norway to enter a regular contingency plan which *ignores* the jurisdictional conflict, similar to what the other Arctic States have done. This is based upon recent positive political developments in the Soviet Union with correspondingly decreased Soviet pressure upon Norway in the North.

13.2. AGREEMENT and PLAN between Norway and the Soviet Union - Background and Supporting Legal Theory

13.2.1. Background

A proposal for a bilateral AGREEMENT is modeled upon the recent US - USSR Agreement.[13] A proposal for a co-operative plan between Norway and the Soviet Union, 'Joint Contingency Plan for Counter-Pollution Operations in the Barents Sea' (PLAN) is based upon NORBRITPLAN.[14]

In a conversation with the Norwegian Ministry of the Environment, it was stated that any eventual contingency agreement between Norway and the Soviet Union, covering offshore oil pollution in the Barents Sea, would probably be fashioned after NORBRITPLAN.[15] A bilateral agreement was also included since an agreement coupled with a contingency plan was the legal arrangement used by the Soviet Union and the US.[16] As noted though negotiations seem to be progressing, the Norwegians until just recently have required any such plan being contingent upon the establishment of a maritime boundary line. At first this view seemed somewhat questionable given the fact that Norway has already entered into the bilateral Rescue at Sea Agreement covering the Barents Sea, in spite of the boundary conflict. It is difficult to see a difference between a Joint Contingency Plan for Counter-Pollution Operations at Sea and a Joint Rescue at Sea Agreement, requiring a different approach to the Soviets.

13.2.2. Supporting Legal Theory for AGREEMENT and PLAN under LOSC[17]

The problem arises under the issue of *jurisdiction*. The control of pollution is tied to a State exercising jurisdiction over a definite area, either the territorial sea, the continental shelf or the exclusive economic zone. As seen a coastal State has jurisdiction over the protection and preservation of the marine environment and at the same time sovereign rights for exploiting natural resources in its exclusive economic zone and on its continental shelf.[18] A coastal State also has exclusive jurisdiction over artificial islands, installations and structures in the State's exclusive zone and on its continental shelf.[19] As seen States are obliged to adopt and enforce anti pollution regulations for sea bed activities and artificial islands under their jurisdiction, must adopt and may enforce to a certain extent anti pollution regulations against ships in their territorial seas and exclusive economic zones, and must adopt and enforce anti pollution regulations for dumping activities under their jurisdiction.[20] What this means is that with the exceptions noted other States cannot take anti pollution measures over areas, vessels or installations where they do not have jurisdiction.

In contrast the duty to render assistance is independent of jurisdiction. Article 98 requires States to demand of ships under their jurisdiction 'to render assistance to any person found *at sea* in danger of being lost', if it can be done safely, and to promote setting up and operating search and rescue services covering safety on and over the sea, including through regional arrangements.[21] Though the use of the word 'jurisdiction' may be somewhat confusing here, in seeing the difference, an example using the Barents Sea may help. Under the Articles mentioned a Soviet ship steaming through the Norwegian territorial sea would be required to assist Norwegian fishermen in distress even though the area is under complete Norwegian jurisdiction. On the contrary if a Soviet ship collides with another ship within the Norwegian territorial sea, it is Norway which has jurisdiction over the anti pollution control operations.[22] Thus, it is perhaps more clear why the Joint Norwegian - Soviet Rescue at Sea Agreement is possible, while a bilateral Agreement and Joint Contingency Plan for Counter-Pollution Operations at Sea is more difficult.

Note that the word 'impossible' is *not* used here. As noted also appearing in LOSC is Article 199 which states in particular,

> '(I)n the cases referred to in Article 198, States in the area affected, in accordance with their capabilities, and the competent international organizations shall cooperate, to the extent possible, in eliminating the effects of pollution and preventing or minimizing the damage. To this end, States *shall jointly develop* and promote contingency plans for responding to pollution incidents in the marine environment.'[23]

Article 198 requires States, when they become aware of cases in which the marine environment is in imminent danger of being damaged or has been damaged by pollution, to notify the competent international organizations and the other States deemed likely to be affected. These Articles would seem to require both Norway and the Soviet Union to set up a joint contingency plan for the Barents Sea in light of the on going oil exploration taking place in the region plus expected Northeast Passage shipping activity, in spite of their jurisdictional differences.[24]

Both Norway and the Soviet Union have made efforts to settle the boundary dispute.[25] Norway has made it clear that it is willing to compromise its legal standpoint in the boundary conflict in order to come to a solution, as has the Soviet Union recently with its proposal for a 'joint economic zone'. Future meetings for negotiations are planned after the respective Foreign Ministers, Stoltenberg and Sjevardnadse, have talked in New York at the U.N.[26]

Thus, though it appears that things are moving in the right direction both practically and politically, and eventually a boundary solution and subsequent joint contingency plan covering a clearly jurisdictionally divided Barents Sea might emerge, still it is a question of how long this will take. With this in mind the proposed AGREEMENT and the PLAN were developed, trying to *circumvent* the jurisdictional problem caused for Norway by the lack of a maritime boundary. The difficulties encountered will be illustrated in the following Section showing why a more comprehensive environment plan was not possible to achieve.

13.2.3. Problems

As noted it is assumed most beneficial that a Norwegian - Soviet bilateral agreement and contingency plan follow the arrangement recently used by the US and the Soviet Union.

Article II of US - USSR Agreement defines 'Area of Responsibility of a Party' as,

> 'the waters within the Bering Sea and Chukchi Seas which are the respective Party's international waters or territorial sea, and the sea area beyond the territorial sea in which that Party exercise its sovereign rights and jurisdiction in accordance with international law. Areas of responsibility of the Parties where they are adjacent will be separated by the *maritime boundary* between the two countries'.[27]

NORBRITPLAN in Article 2 defines the Plan Area as within the North Sea (which lies south of latitude 62 degrees North),

> 'contained within a band following the median line between Norway and the United Kingdom and extending to a distance of 50 nautical miles on either side of that line'.

In order to avoid these definitions based around boundary lines, yet maintain the structure of both US - USSR Agreement and NORBRITPLAN, it was decided to alter and use both definitions yet merged with one another.

As can be seen, in Section 13.4.2. the AGREEMENT's Article II, the 'area of responsibility' of a Party includes within the Barents Sea the Party's internal waters or territorial sea and the sea area containing that respective Party's PLAN Area. 'PLAN Area' is defined in Article II as that area surrounding offshore installations subject to authorization and control of either one of the Parties with a radius of *50 nautical miles*. Thus a delimitation is achieved by withdrawing the boundary line reference yet referring to an area surrounding a platform under either of the Parties' jurisdiction. This postpones the broad issue of jurisdiction in the Barents Sea, because it is based upon the practical fact that in order to be doing the oil drilling in the first place either Party must claim jurisdiction over the spot the drilling is taking place. So far at least neither has drilled in a disputed area, though the Soviets have come close.[28]

The advantage with having a definite geographical area is that fits nicely into the US - USSR Agreement and NORBRITPLAN for use as models. Using a circular area around each offshore installation resulted in *two* problems however, one with respect to broad environmental considerations; the other with respect to environmental considerations concerning two activities with special jurisdictional characteristics, shipping and pipelines.

Environmentally, it is obvious that only small ocean areas are covered by this 'circular' scheme.[29] From the above a Party in the Barents Sea is responsible only for its internal waters, territorial sea and these circular areas. As noted in the AGREEMENT the 50 mile figure can be varied, yet this still leaves large sea areas that are not encompassed by the AGREEMENT and PLAN. This was thought to be a necessary evil.

At first it was believed that disputed areas in the Barents Sea could merely be excluded from an agreement. However it was discovered that there is not consensus of what area is disputed. One Norwegian diplomat notes one area from a Norwegian standpoint,[30] while a Soviet diplomat notes another area from a Soviet standpoint.[31] The 'Grey Zone' Agreement involving fishing regulations covers yet another area.[32]

A solution thought possible was to summate all the areas negotiated by the Norwegians and the Soviets in the past 20 years and using their geographical definitions, exclude these areas from an agreement. However, the Soviets at least once have claimed there is *no* disputed area,[33] and considering this view it seems clear that any reference in an agreement to an area as disputed which is unfavourable to one of the Parties would not be accepted as a viable solution by that Party. Any acceptance of such an agreement could be considered as an admission of the other Party's sovereignty over the area or part of the area in question. However if an agreement could be reached as to an exact area which is disputed, an AGREEMENT and PLAN covering broad areas of the Barents Sea would be a straightforward matter. The disputed area could merely be excluded, pending further negotiations, and an AGREEMENT and PLAN could be fashioned almost directly following US - USSR Agreement and NOR-BRITPLAN, using the disputed area's coordinates rather than a boundary line.

Given this ambiguity of disputed areas in order to arrive at a realistic proposal, any mention of the terms 'boundary lines' and 'disputed areas' had to be avoided, even if this resulted in a less than desired environmental solution.

The second problem, involving shipping and pipelines, again results in a less than adequate environmental solution. It should be fairly obvious that these activities are adequately covered in each Parties' internal waters, territorial seas or PLAN Areas.[34] If a ship collides with an oil platform or a pipeline bursts within these areas, either a contingency plan of the Party with jurisdiction over these areas or the joint PLAN come into action. However, as noted above there are large expanses of sea and sea bed over which ships travel and pipelines can run. Normally these present no problems since as seen the coastal State can take environmental measures concerning foreign ships sailing in their internal waters, territorial sea or exclusive economic zones, or concerning foreign pipelines running across their continental shelf.[35]

In this case however it is exactly the delimitation of the continental shelf and the exclusive economic zone which is in issue, and as seen above no reference can be made to these disputed boundary areas because of the jurisdictional question. To fill in the gaps in the AGREEMENT and the PLAN to attempt to cover ships and pipelines outside the mentioned areas, it was first thought that the same definition of 'areas of responsibility' in Article II of US - USSR Agreement could be used, but without reference to the marine boundary. This would include reference made to the respective Party's internal waters or territorial sea,

> 'and the sea area beyond the territorial sea in which that Party exercises its sovereign rights and jurisdiction in accordance with international law'.[36]

Thus it was hoped that a *vague* reference would be made to the Parties' responsibility for cleaning up pollution in areas outside the PLAN Area from ships and pipelines, and blowouts which have drifted outside the 50 mile radius circles.

The problem with this however is that since the second phrase must be deleted,[37] this might be seen as 'opening the door' and acceding to the other Party's jurisdictional claim and allowing that Party rights in a disputed area, which neither Party would agree to. Thus the above phrase was deleted from the definition and the phrase added covering only the PLAN area as noted above.[38]

In practice this was felt to be no problem, since in *undisputed* areas, Norway in the western Barents Sea and the Soviet Union in the eastern Barents Sea, clearly have jurisdiction over their continental shelves and exclusive economic zones under international law to deal with pollution problems emanating from platforms, ships or pipelines. This jurisdiction is not in any way compromised by the AGREEMENT or PLAN, since the definitions of 'Area of Responsibility' and 'PLAN Area' are specifically 'For the purposes of this AGREEMENT and PLAN'[39] and as well 'The Parties will implement this AGREEMENT in accordance with rules and principles of general international law and the respective laws and regulations.'[40]. Both countries can take anti pollution measures in these areas without any problem under the Articles mentioned in Part II Section 13.2.2. In addition should they desire, they could even agree with the other Party for joint clean up assistance in these areas or even disputed areas under Article 2.2. of the PLAN.[41]

Thus the *unstated* seems in fact to be the best here since no jurisdictional ambiguity is caused by referring to sea areas beyond the territorial sea over which the Parties exercise their sovereign rights and jurisdiction in accordance with international law, yet parts of which are undefined and in dispute.[42]

In summary, though not environmentally the most desirable since large sea areas are not specifically addressed, the AGREEMENT and PLAN offer a bilateral contingency plan which is better than nothing and which could be put into force in the Barents Sea pending the resolution of the boundary conflict. Should a definite marine boundary for the continental shelf and the exclusive economic zone come into being, the AGREEMENT and PLAN should be amended, the phrases dealing with the circular PLAN Area being replaced by phrases dealing with a boundary line.

13.3. Alternative Proposal

The alternative proposal centres around the idea that as noted most of the Arctic States have formed joint contingency plans in spite of maritime boundary disputes, and in two of the cases the plans encompass the disputed areas themselves. In linking the formation of a joint contingency plan in the Barents Sea to the establishment of a definite maritime boundary, Norway is putting the Barents Sea in a less than acceptable environmental situation, since the boundary negotiations have been in progress over 20 years, and the prospect of a settlement being negotiated can be questioned. As also noted Norway's reluctance to enter a bilateral agreement is based upon fear of Soviet influence and domination in the North,[43] yet at the same time Norway has recently entered into 5 bilateral agreements, most of which include provisions encompassing the northern areas, including the Barents Sea.[44]

Because of this unclarity, it is deemed possible for Norway to re evaluate their position in light of the new political developments in the Soviet Union. The legal basis of such a joint contingency plan would be those LOSC and other provisions indicated in Section 13.2.2., specifically Articles 198 and 199. Although Norway certainly has no obligation to enter into a bilateral agreement with a State exerting unacceptable political pressure, still the combination of political developments, Arctic State practice and LOSC provisions would indicate that the formation of a contingency plan along the lines of an unaltered US - USSR Agreement or the US - Canadian Agreement in the Barents Sea would be the better practice.

For the first it appears that on Svalbard and in the Barents Sea the Soviets are now respecting Norwegian policy and administration and have been doing so for the past few years.[45] This of course is not conclusive, however generally it can be said that the Soviets seem to be very aware of their international obligations and are willing to respect them in accordance with international law.[46]

For the second, the other Arctic States' boundary conflicts have not been problem free, yet this has not hindered development of joint contingency plans.[47] In the disputed areas in both the Bering and Beaufort Seas, the US has already sold leases to oil companies. Canada has done the same. Soviet Bear H bombers have 'buzzed' American oil rigs in the Bering Sea, and Soviet ships as well have disturbed both seismic explorations as well as scientific activities. Canada has claimed the entire archipelago in the North as inner waters and subject to full Canadian sovereignty, a claim the US rejects. The US consequently has twice sailed its icebreakers through the archipelago without Canadian permission, resulting in Canadian protests of violations of territorial sovereignty. Norway is clearly not the only State subject to pressure in the North from its neighbours.

Though the Arctic maritime boundary conflicts are different and must be evaluated each on their own merits both legally and politically, an element which was found to be specific to the conflict in the Barents Sea was the *unclarity* as to what was the disputed area.[48] In the information made public concerning the disputes in the Bering, Chukchi and Beaufort Seas, nowhere was there evidence that either Canada, the US or the USSR attempted to enlarge the disputed area over the original claim.[49] In addition the Soviets have their largest navy base on the Kola Peninsula in the near vicinity of Northern Norway, a factor having no parallel in the other Arctic boundary conflicts. These 2 factors would thus indicate that Norway the smallest Arctic rim State has been subjected to the *most* pressure in its Arctic areas.

In spite of this however the arguments for establishing a joint contingency plan seem to outweigh those against. The Soviets seem so intent on finding a place in the international community and solving their internal economic and political problems, it seems unlikely they would try to expand their influence in the Barents Sea.[50] Thus if the boundary negotiations fail to produce any result, as they characteristically have in the past, Norway should evaluate establishing and entering a joint contingency plan in spite of the history of Soviet domination. Especially if the Northeast Passage is opened and large oil finds are made, a concrete plan with effective routines and technology is needed. The Barents Seas has enough problems in addition to that which a blow out or a large oil tanker catastrophe can add, and an effective plan will be of help.[51] On the international level it must also be to Norway's advantage that an environmentally progressive country as Norway be a Party to a joint contingency plan covering its Arctic areas the same as the more environmentally conservative countries, the US and the Soviet Union.[52]

13.4. AGREEMENT[53] and PLAN[54]

13.4.1. Introduction

In the following presentation of the proposed AGREEMENT and PLAN any reference to a boundary line, median line or indefinite areas of jurisdiction will be altered in an attempt to circumvent the Norwegian - Soviet boundary dispute in the Barents Sea and will be indicated by a (*), with the original phrase from the US - USSR Agreement or NORBRITPLAN appearing in a footnote. The original is shown in order to indicate how the AGREEMENT could appear in the event a definite boundary line was negotiated between the Norwegians and the Soviets. Any comments reflecting these passages will be placed in parentheses. As before this form is chosen so as not to interrupt the continuity of the document.

13.4.2. AGREEMENT

The Government of the Kingdom of Norway and the Government of the Union of Soviet Socialist Republics, hereinafter referred to as the 'Parties',

Conscious that exploration, exploitation, and production of natural resources, as well as related marine transport, pose a threat of significant pollution by oil and other hazardous substances in the Barents Sea,

Recognizing, that, in the event of a pollution incident or the threat thereof, prompt and effective action should be taken, to organize and coordinate prevention and pollution combating activities,

Taking into account the Agreement on Environmental Cooperation between the Government of the Kingdom of Norway and Government of the Union of Soviet Socialist Republics, signed 15 January 1988, and desiring to develop further the principles of mutually beneficial cooperation, and

Desiring to avert, through the adoption of measures to prevent and combat pollution resulting from oil and other hazardous substance spills, damage to the marine environment of the Barents Sea, including coastal areas,

Have agreed as follows:

Article I

The Parties undertake to render assistance to each other in combating pollution incidents which may affect the areas of responsibility of the Parties (*).[55] Such assistance shall be rendered consistent with the provisions of this AGREEMENT. To such end the competent authorities of the Parties shall develop the Joint Contingency Plan Against Pollution in the Barents Sea, hereinafter referred to as the 'PLAN', which shall enter into force upon their written agreement.

Article II

For the purposes of this AGREEMENT and the PLAN:

'Pollution Incident' means a discharge or an imminent threat of discharge of oil or other hazardous substances[56] from any source of such a magnitude or significance as to require an immediate response to prevent such a discharge or to contain, clean-up or dispose of the substance to eliminate the threat to or to minimize its harmful effects on living resources and marine life, public health or welfare.

'Area of responsibility of a Party' means the waters within the Barents Sea defined by [COORDINATES][57] which are the respective Party's internal waters or territorial sea, and the sea area (*)[58] containing that respective Party's PLAN Area.

'PLAN Area' means that area contained within the sea area beyond the territorial sea, surrounding offshore installations subject to authorization and control of either one of the Parties with a radius of 50 nautical miles.[59] The area such defined shall be known as the 'PLAN Area'.[60]

'Oil' means oil of any form, including crude oil, fuel oil, sludge, oil wastes and refined products.

'Hazardous substances' means elements and compounds which when discharged into the marine environment could present an imminent and substantial danger to the public health or welfare, harm living resources and marine life, damage amenities or interfere with other legitimate uses of the sea.

'Ship' means a vessel of any type whatsoever operating in the marine environment including hydrofoil boats, air-cushion vehicles, submersibles and floating craft but excluding fixed or floating offshore installations.

'Offshore Installation' means fixed or floating artificial islands and structures used for the exploration, exploitation and associated offshore processing of sea bed mineral resources.

'Competent Authority' means with respect to Norway ..., and with respect to the Union of Soviet Socialist Republics ...[61]

'Response resources' means the personnel, vessels, equipment and other means for combating pollution.

Article III

The Parties, consistent with their means, commit themselves to the development of national systems that permit detection and prompt notification of the existence or the imminent possibility of the occurrence of pollution incidents, as well as providing adequate means within their power to eliminate the threat posed by such incidents and to minimize the adverse effects to the marine environment and the public health and welfare.

Article IV

The Parties shall routinely exchange up-to-date information and consult to guarantee adequate cooperation between their competent authorities, with regard to activities pertaining to this AGREEMENT and the PLAN.

Article V

The implementation of the PLAN shall be the primary responsibility of the competent authorities of the respective Parties, and of other authorities of the Parties to the extent of such other authorities' competence under applicable law. The PLAN may be amended from time to time, consistent with this AGREEMENT and the procedures set forth in the PLAN, by the competent authorities.

Article VI

The competent authority of the Party in whose (*)[62] PLAN Area a pollution incident occurs, or whose area of responsibility is affected by such an incident, shall direct response operations within that area.

Article VII

The PLAN may be invoked whenever a pollution incident occurs that affects or threatens to affect the areas of responsibility of both Parties or, although only directly affecting the area of responsibility of one Party, is of such a magnitude as to justify a request of the other Party's assistance.

Article VIII

The joint response provided for under the PLAN can only be undertaken when the competent authorities of the Parties agree. The competent authorities of the Parties will determine the appropriate response action required for each pollution incident.

Article IX

Requests for assistance will be communicated between the competent authorities of both Parties. Request for assistance by telephone shall be confirmed by telex, telegraph or facsimile.

A Party shall endeavour to promptly provide requested assistance as soon as possible to the extent that the Party determines the resources are available. The availability of response resources for a specific pollution incident is understood to be dependent upon funding and the requirements of other missions.

The requesting Party shall provide all possible support to the response resources of the assisting Party.

Article X

The assisting Party may fully or partly terminate its assistance if that Party determines that it is necessary to do so. Notice of the termination shall be communicated to the competent authority of the requesting Party. The requesting Party shall release the response resource made available as soon as possible after the termination.

The requesting Party shall promptly inform the assisting Party when the need for assistance no longer exists, and release as soon as possible the response resources made available by the assisting Party.

Article XI

The Parties will periodically conduct joint pollution response exercises and meetings in accordance with the provisions of the PLAN. The competent authorities of the Parties shall alternate in the supervision of the exercises.

Article XII

The requesting Party shall, to the greatest extent possible, facilitate the arrival and departure of response resources made available by the assisting Party for response activities pertaining to this AGREEMENT. The Parties shall cooperate in implementing the provisions of this Article.

Article XIII

The requesting Party shall reimburse the assisting Party for expenses associated with response resources. The amount to be reimbursed will be determined by the rates of the assisting Party.

In all other cases and circumstances unless otherwise agreed, each Party will bear the expense of its own activities pertaining to this AGREEMENT.

The expenses involved in conducting joint exercises shall be borne by each Party respectively.

Article XIV

Nothing in this AGREEMENT shall affect in any way the rights an obligations of either Party resulting from other bilateral and multilateral international agreements.

The Parties will implement this AGREEMENT in accordance with rules and principles of general international law and their respective laws and regulations.

Article XV

This AGREEMENT shall enter into force upon the date the Parties notify each other in writing that necessary international procedures have been completed, and shall remain in force unless terminated by either party upon six months advance written notice to the other Party of its intention to terminate this AGREEMENT.

Termination of this AGREEMENT shall not affect response operations which have been taken hereunder and are not yet completed at the time of termination unless otherwise agreed by the Parties.

This AGREEMENT may be amended by written agreement between the Parties.

DONE at (place), in duplicate, this (date), in (English), Norwegian, and Russian languages, all three texts being equally authentic.

FOR THE GOVERNMENT OF THE KINGDOM OF NORWAY

FOR THE GOVERNMENT OF THE UNION OF SOVIET SOCIALIST REPUBLICS

13.4.3. PLAN

Index

Contents

Annexes

Article 1 - Introduction

1. Operations of any kind which involves international co-operation can be defined as joint operations. In joint oil and hazardous substances pollution clearance operations in the marine area covered by the PLAN, the Norwegian and Soviet authorities have coordination tasks to perform in accordance with their own national arrangements. Since, however, divided national control may easily result in operational problems, the authorities are agreed that coordination of oil and hazardous substances pollution clearance action should not be divided, and that the identity of the Action Coordinating Authority (ACA) should be decided and promulgated at an early stage.
2. In order to avoid operational problems and to increase the effectiveness of the measures taken, the Norwegian and Soviet authorities have agreed to act in concert in accordance with the principles of the present PLAN.
3. Whichever part they play in a joint operation, national authorities will operate in accordance with their own instructions and procedures. The success of the joint operation will thus depend on:
   (a) adequate knowledge and understanding of the instructions and procedures of the other country;
   (b) a full understanding of the techniques used by, and equipment available to, the other country, and how the best use can be made of these resources.
   (c) prepared procedures for joint operation within the framework of national plans.
4. The object of the present PLAN is to establish such procedures to be followed in joint Norway-Soviet counter-pollution operations at sea. Insofar as the PLAN aims primarily to cater for major incidents (*)[63] in the PLAN Area, it is oriented largely towards oil or hazardous substances spills resulting from major incidents at off-shore installations, such as blow-outs and damage to submarine pipelines. However, it is not confined to such events, and will apply as necessary and appropriate to any spills (of whatever origin) within the PLAN Area which are of sufficient severity to warrant joint action.
5. Guidelines should be established for co-operation at working level between the Rescue Coordination Centre at .. in maritime Search and Rescue operations.[64] The PLAN is not meant to encompass these arrangements in any way.
6. (*).[65]
7. *Definitions and abbreviations*
   For definitions of terms and abbreviations used in PLAN, see Annex 5 and AGREEMENT Article II.

8. *Amendments*
   (a) Amendments to the PLAN must be jointly agreed to by the Norwegian and Soviet authorities and addressed as follows:
      (i) Norway, The Ministry of Environment, Myntgata 2, Oslo 1.
      (ii) Soviet Union, ...
   (b) Annexes to the PLAN will be updated as necessary.

Article 2 - The Geographic Region

1. The geographic region of the PLAN is within the Barents Sea as defined by [COORDINATES], and is contained within the sea area beyond the territorial sea, (*)[66] surrounding offshore installations subject to authorization and control of either one of the Parties with a radius of 50 nautical miles. The area such defined shall be known as the 'PLAN Area'.[67]
2. The PLAN may also be activated in the case of oil or other hazardous substances spills outside this Area by agreement between the two countries.[68]

Article 3 - Types of Incidents and Action Required

1. *Incident involving an offshore installation.*
   The respective approaches of the two Governments to such incidents have broadly similar characteristics in that in respect of each country:
   (a) It is the policy to seek to protect against damage from oil or hazardous substances pollution:
      (i) sea fisheries;
      (ii) wildlife, including important concentrations of sea birds, and also aquatic mammals;
      (iii) areas of amenity value, and
      (iv) ecologically sensitive areas.
   (b) The operator of an offshore installation has the primary obligation to react and respond to any spillage of oil or hazardous substances from that installation, at the same time notifying the respective Government of the occurrence and the measures taken and proposed. It is required by the 'parent' Government to have for this purpose:
      (i) a contingency plan, and
      (ii) access to the resources necessary to implement that plan.
   (c) Should the resources of the operator prove insufficient for the task, it should seek additional resources and/or other assistance.
   (d) The role of Government is basically to ensure that the response of the operator is adequate, speedy, likely to be effective and in accordance with Government policy. In fulfilling this role, Govern-

ment's own action will consist as appropriate of monitoring, tendering advice, rendering practical assistance, giving positive instructions or taking over control. This might be done by virtue of some statutory or regulatory power, or by agreement or other understanding between Government and operator.

2. *Incident Involving a Shipping Casualty*

The respective approaches of the two Governments are broadly similar in that in respect of each country:

(a) The Government is forced to accept the impracticality for the owner or master of the casualty vessel to have the capability to respond to actual or threatened spillage of oil or hazardous substances.

(b) The Government must therefore take any action required to prevent, eliminate or mitigate any oil or hazardous substances pollution or the threat of same.

(c) The International Convention Relating to Intervention on the High Seas in Cases of Oil Pollution Casualties 1969 and Protocol (1973), both of which both Governments are Parties, and the provisions of which are implemented in the national legislation of both countries, lay down the principles governing intervention on the high sea in shipping casualties which cause or threaten to cause oil or hazardous substances pollution likely to result in major harmful consequences to a national coastline or related interests.[69]

3. *Action Common to Both Types of Incident*

Action to be taken to restrict the movements of aircraft and shipping above or near the incident area is described in Article 7.

Article 4 - Principles of Co-operation

1. *Implementation of the PLAN*

(a) The authority of the country (*)[70] with authority and control over the PLAN Area in which a major oil or hazardous substances spill occurs or is threatened, that affects or threatens to affect the areas of responsibility of both Parties or is of such a magnitude as to justify a request of the other Party's assistance, shall immediately notify the authority of the other country.[71] This notification shall be transmitted both between the RCC .. and the RCC .. and the respective national authorities .. (see Annex 1). It may be sent either in plain language or in the POLREP form set out in Annex 2.

(b) Following this notification, the authority of either country may request the implementation of the PLAN, using the same two channels of communication. Such a request shall be given immediate consideration.

(c) The PLAN will become operative when agreement to the request is transmitted.

2. *Determination of Lead and Supporting Roles*

(a) Where an incident is deemed to be of such severity and/or magnitude as to justify joint action by both Governments - which fact assumes that one or other Government is in control of operations or is in a position to give directions to the operator concerned - the responsibility for initiating such joint action will rest in principle with the country in whose PLAN Area the incident occurs. This lead country will designate an Action Coordinating Authority who will thereafter assume responsibility, and will request and coordinate any assistance which may be necessary and forthcoming. The other country will provide such support as is possible and appropriate, and will similarly designate an Action Liaison Authority (ALA).

(b) (*)[72]

(c) (*)[73]

(d) (*)[74]

(e) (*)[75]

3. *Command and Control*

(a) Resources deployed by either Government, although coordinated by the ACA, remain under national command.

(b) An offshore operator concerned in an incident remains subject to control and direction by its 'parent' Government. Should the ACA role be with the other country, any instructions from the ACA to the operator should be passed through the 'parent' Government.

4. *Counter-Pollution Measures*

(a) The preferred technique in appropriate sea conditions is the physical recovery of oil or hazardous substances from the sea surface. However, it may be necessary in appropriate circumstances to make use of chemical dispersants.

(b) In controlling counter-pollution operations in the PLAN Area (*)[76] that affects or threatens to affect the areas of responsibility of both Parties, the ACA should as a general rule consult the ALA as to the proposed clean-up methods, the equipment available and the magnitude of the operation.[77] Questions related to priorities and the proposed concentration on special tasks or geographical areas should also be mutually discussed.

(c) The use of dispersants on a large scale is unlikely to be necessary save in the event of a prolonged spillage (e.g. an uncontrolled blowout) where there is a significant possibility of it becoming a threat to one of the interests cited in 3.1(a). The Norwegian and Soviet guidelines to operators (*)[78] under their respective jurisdictions regarding the use of dispersants are listed in Annex III.[79]

(d) Where the large-scale use of dispersants is contemplated, the ACA should consult the ALA, as well as organizations representing fisheries and wild life interests as to:
   (i) the respective merits of using dispersants or relying on natural forces;
   (ii) the sensitivity of particular areas; and
   (iii) other aspects liable to cause anxiety.

(e) The use of chemical dispersants to protect human life, or to reduce the risk of damage to offshore installations, is within the discretion of the ACA.

(f) It is open to either country, following consultation between ACA and ALA, to seek assistance from other State Parties under the Protocol for Counter-Pollution Operations in the Arctic Region, or otherwise.

(g) Whether or not such assistance is requested, the lead country should ensure that information on the occurrence of the incident, and on the progress of operations to deal with it, is passed to the other State Parties to the Protocol.

5. *Liaison Requirements*

(a) In the pollution combating phase of an incident, when the emphasis is on practical measures to stop or limit the spill and to minimize its effects, company liaison personnel of sufficient standing and with delegated authority should be available to the ACA.

(b) Additionally, in prolonged joint oil-combating operations, each country should provide liaison personnel to assist the ACA or ALA of the other country on questions arising out of the joint operations.

Article 5. - Emergency Organizations

In the event of a 'blow-out', or other serious pollution incident at sea, both the Norwegian and the Soviet Governments have special organizations - either in permanent existence or in reserve - to monitor, advise, assist or carry out such operations as may be required. A Blow-out Emergency Team (BET) will be established within 24 hours and will be located in .. The functions of the BET are to monitor, advise and assist the operator concerned in action to deal with the continued escape of oil. They do not, however, include supervision of the oil clearance operations which are the particular province of the ... The team has the specific responsibility of conveying any special instructions from the Government to the operator.

1. Norway[80]
   (a) The Governmental Action Control Group (GACG) will be established by the Norwegian Government whenever an accident occurs in Norwegian coastal waters or offshore areas that has led to, or may lead to, situations involving large scale oil or hazardous substances pollution and/or material damage.
   (b) The Governmental Action Control Group (GACG) will be supported by a staff element and will normally be located in the RCC at ...
   (c) It will be headed by a high-ranking official from the Ministry of Environment/SFT, and will have representatives from the following authorities:
      (i) the Ministry of Justice and Police;
      (ii) the Ministry of Defence (Navy);
      (iii) the Ministry of Local Government and Labour (Petroleum Directorate); and
      (iv) the Ministry of Trade (Maritime Directorate).
   (d) The Governmental Action Control Group (GACG) will coordinate initial measures and the surveillance of the accident, and may at any time take over, in full or in part, the responsibility for remedial action in connection with the emergency situation. The time and extent of take over is to be decided by the Action Command.
   (e) The Governmental Action Control Group (GACG) will normally undertake on behalf of the Norwegian Government the functions of ACA or ALA, as the case may be.
   (f) A team of Press Officers will be located alongside the Action Command, and will handle contact with the news media on its behalf.
2. *U.S.S.R.*

   ...
3. *Offshore operators*
   (a) The respective national associations of offshore operators provide links:
      (i) between individual operators within the association concerned;
      (ii) between those operators and the 'parent' Government; and
      (iii) with other national associations and their member operators.
   (b) The link with the 'parent' Government is particularly important as providing a means of promulgating statutory regulations, emergency and contingency measures, and of implementing Government policy. Special sub-committees of the associations deal with oil spill related matters.
   (c) The Norwegian Organization

The Norwegian Offshore Clean Seas Association (NOCSA) of the Norwegian Industry Association for Operating Companies (NIFO) has been assigned the administration of the commonly owned counter-pollution resources on behalf of the operators. The establishment of these resources has been carried out in close co-operation with the Norwegian authorities, and they are based on requirements given by the SFT.

(d) In an emergency the NOCSA will supply the equipment and install it on board the assigned vessels. The NOCSA will also provide the key operating personnel, while the operator itself will have the operational responsibility for the system.

(e) The NIFO has established a stock of counter-pollution equipment at supply bases in ...

(f) USSR

...

(g) The Sector Club Organization

In order to respond more speedily and effectively to emergencies involving installations - particularly those arising from e.g. blow-outs, fires and other disasters - the operators active in the Arctic Region oil and gas fields have formed themselves into groups within geographical sectors, identified by colours.

(k) Each such 'sector club' has certain identified emergency vessels and resources available to it, together with an efficient communications network. Some of the vessels are of a multi-purpose type and incorporate a dispersant spraying capability, although this function is secondary to their main safety role. A member operator is designated to coordinate the resources of each group and their deployment.

Article 6 - Decisions at Higher Level

1. Where a pollution emergency requires decision or action at a higher administrative or possibly a Ministerial level, operation authorities report as follows:

(a) The Norwegian RCC reports to the Duty Officer of the SFT in Oslo on oil clearance matters, and to the Duty Officer of the Norwegian Petroleum Directorate on technical or other emergency matters concerning offshore installations.

(b) USSR

...

(c) At Headquarters level, information regarding the pollution situation can be exchanged, consultation carried out, problems resolved and decisions taken respectively between the SFT in Norway, and the .. in the U.S.S.R.

(d) Dependent on the nature and magnitude of the emergency, both Governments may establish the Action Command to coordinate national measures. In this case, any such exchanges regarding pollution should be made directly between the Action Command, normally located at ...

Article 7 - Establishment of Danger Zones

1. A danger area is to be established by the ACA if it is considered that vessels and/or aircraft not concerned with the operation should be warned to keep clear.
2. Broadcasts to Shipping
   The need for simplicity in the designation of the danger area is emphasized. If possible one circle of suitable radius and centred on a position given in longitude and latitude or on the position of a charted offshore installation should be designated. The reason should be briefly given.
3. Control of Airspace
   The procedures which are to be followed for obtaining flight restrictions over a danger area are given in Annex 4.

Article 8 - Communications

1. Communications between the two countries may be required at any level. Initial notification of an emergency, or a request to implement the PLAN, is to be passed through RCC ..., and also between the SFT in Norway and the .. in the U.S.S.R.
2. The operational control functions including those of ACA and ALA, will normally be undertaken:
   (a) In Norway by the Governmental Action Control Group (GACG) which will normally be located in RCC ...
   (b) In the USSR by ...

Article 9 - Resources

1. Arrangements are established under the Agreement between the Governments of the Kingdom of Norway and the Government of the Union of Soviet Socialist Republics concerning cooperation in combating Pollution in the Barents Sea in Emergency Situations to provide on a regular basis details of the resources likely to be available in the PLAN Region for dealing with emergencies covered by the PLAN.

Annex I

Arrangements for Communication

1. Norway

   RCC .... Telephone:
   (24 hours) .... Telex:
   Government Action Control Group .... Telephone:
   (GACG) .... Telex:
   State Pollution Control Authority .... Telephone:
   (SFT) .... Telex:
   (out of office hours, contact via RCC ...)

2. USSR

   ...

   There is also a permanent direct telephone link between the RCC in .. and ..

CONTENTS OF MESSAGE AND CODE LETTERS

Part I - INFORMATION WHICH SHOULD BE PROVIDED IN AN INITIAL POLREP

A. CLASSIFICATION of report - (i) Doubtful, (ii) Probable, (iii) Confirmed.

B. DATE and TIME pollution observed/reported, and identity of observer/reporter.

C. POSITION and EXTENT of pollution. (If possible state range and bearing of some prominent land mark or Decca position, and estimated amount of pollution (e.g. size of polluted area, number of tons of oil or hazardous substances spilled, or number of containers, drums etc. lost). When appropriate, give position of observer relative to pollution.)

D. TIDE, WIND speed and direction.

E. WEATHER conditions and SEA state.

F. CHARACTERISTICS of pollution. (Give type of pollution, e.g. oil (crude or otherwise), packaged or bulk chemicals, sewage. For chemicals give proper name or United Nations number if known. For all, give also appearance, e.g. liquid, floating solid, liquid oil, semi-liquid sludge, tarry lumps, weathered oil, discoloration of sea, visible vapour. Any markings on drums, containers etc. should be given.)

G. SOURCE and CAUSE of pollution, (e.g. from ship or other undertaking. If from ship say whether as a result of a deliberate discharge or a casualty. If the latter, give brief description. Where possible, give name, type, size nationality and port of registry of polluting vessel. If vessel is proceeding on its way, give course, speed and destination.)

H. Details of SHIPS IN THE AREA. (To be given if the polluter cannot be identified, and the spill is considered to be of recent origin.)

J. Whether PHOTOGRAPHS have been taken, and/or SAMPLES for analysis.
K. REMEDIAL ACTION taken or intended, to deal with the spillage.
L. Forecast of likely effect of pollution, (e.g. arrived on beach), with estimated timing.
M. Names of other States and organizations informed.
N. Any OTHER relevant information, (e.g. names of other witnesses, reference to other instances of pollution pointing to source.)

Part II - SUPPLEMENTARY INFORMATION TO BE PROVIDED LATER
O. RESULTS of SAMPLE analysis.
P. RESULTS of PHOTOGRAPHIC analysis.
Q. RESULTS of SUPPLEMENTARY ENQUIRIES (e.g. inspections by surveyors, statements of ship's personnel, etc. if applicable).
R. RESULTS OF MATHEMATICAL MODELS.

Annex II
POLREP (POLLUTION REPORT)[81]
NOTE:
1. The form is intended for use to bring a major spillage or threat thereof to the attention of another country.
2. It is arranged for the rapid transmission of information (e.g. by Telex) using the alphabetical code.
3. The report should be made as comprehensive and informative as possible, but its dispatch must not be unduly delayed on this account.
4. For updating reports, the appropriate letters only need be used.
5. The message should be prefixed by the code-work POLREP, followed in turn by:
   (i) the country of origin
   (ii) a number to identify the incident
   (iii) a serial number of the reports relating to that incident.

The final message relating to a particular incident should be marked 'Final' in addition to the serial number.

Annex III
NORWAY: RESPONSE BY OFFSHORE OPERATORS TO OIL AND HAZARDOUS SUBSTANCE SPILLS FROM NORWEGIAN OFFSHORE INSTALLATIONS[82]

Regulations relating to Safe Practices for the Production etc. of Submarine Petroleum Resources were issued by Royal Decree on July 9, 1976. Similar

Regulations with respect to drilling operations were issued by Royal Decree on August 29, 1975.

By letter dated September 17, 1976 from the SFT to operators on the Norwegian Continental Shelf, the operators were instructed to establish a system for mechanical recovery of oil from a blow-out of up to 8000 tons per day under favourable weather conditions. The equipment shall operate in up to 2.5 meters significant wave height and currents of 1.5 knots.

By Royal Decree dated April 26, 1977 a Governmental Action Control Group (GACG) was established. This Group will meet in case of large scale pollution incident, or when danger of such incidents exists. The Group will have delegated authority to initiate necessary measures to limit or stop the pollution, and will on behalf of the government coordinate the national combating measures. The Group is also empowered, wholly or in part, to take over the combating operations if this is considered necessary.

ORGANIZATION

In case of a large scale pollution incident, or in case of danger of such incident, the GACG will meet in their operations headquarters at ... They will immediately establish contact with the operating company concerned in order to be fully briefed on the situation. Thereafter GACG will continue to monitor the company's actions, and as appropriate give advice, render assistance, give positive instruction or, if necessary, take over control. Frequent meetings with the operator will be required in this respect. Since the operator by regulation is responsible for all combating measures offshore, GACG will normally not be engaged in recovery or spraying operations in these areas. Governmental anti-pollution operations will be confined to coastal and inshore waters, and take place in close co-operation between GACG, the SFT and the coastal municipalities. GACG will coordinate pollution combating measures with neighbouring countries, and will under the PLAN act in the role of Action Command Authority (ACA) or Action Liaison Authority (ALA) when assembled. If the pollution incident takes place in an area where it does not represent a direct threat to Norwegian interests, or if it is of minor magnitude and importance, GACG will not be mobilized. In such cases the SFT will act in the role of ALA, and will also coordinate all activities with respect to rendering assistance outside the country.

OPERATOR ACTIONS

A basic principle of Norwegian oil pollution control policy is that oil spills as far as possible be combated through the use of mechanical recovery equipment. In heavy seas or under high wind conditions it is, however, difficult with today's technology to recover oil from the sea. If it proves impossible to do so with the help of mechanical equipment one is faced with the choice of either leaving the oil to itself, in the hope that as much as possible will evaporate, or be absorbed in the water masses and be diluted; or of using chemical dispersants. In the case of a comprehensive and long-lasting oil spill accident on the continental shelf it is probable that a combination of both methods would be used.

Use of Chemical dispersants is regulated in pursuance of the 'Regulations concerning the Composition and Use of Dispersants to Combat Oil Spills', issued on February 2, 1980.

The operators North of .. have plans to join in a co-operative (Norsk Oljevernforening for Operatører - NOFO) in order to comply with the national requirements.

NOFO now has in stock equipment to contain and remove oil from a blow-out, and the necessary ships to handle the equipment have been modified and will be made available in an emergency. The operator concerned will be responsible for making available qualified and trained personnel to handle the equipment.

In case of a major oil spill the operator will immediately take the necessary steps to mobilize the NOFO equipment, and at the same time alert the SFT. During continuous consultations between operator and SFT, and later with GACG is established, the magnitude and urgency of the response action will be agreed on the basis of the situation and the expected development.

In smaller spill situations the operator is also obliged to report to the SFT, but may otherwise act in accordance with the above mentioned Dispersant Regulations.

Decisions with respect to application of dispersants in excess of 1000 litres or leaving the oil to degrade will be taken in consultation with the SFT.

REGULATIONS CONCERNING THE COMPOSITION AND USE OF DISPERSANTS TO COMBAT OIL AND HAZARDOUS SUBSTANCE SPILLS

Issued by the Ministry of Environment in pursuance of the Act date 11 June 1976 No. 79 concerning Product Control, paragraph 4. Given 2 February 1980.

1. *The scope of the regulations*

   These Regulations apply to the composition and use of dispersants in the combat of oil spills.

2. *Definitions*

   (a) Dispersant: A chemical product to be spread on oil spills, and which causes the oil to be finely split up/spread i the water masses.

   (b) Emulsifier: The part of the dispersant which causes the oil to be combined with the water molecules and spread in the water masses.

(c) Solvent: The part of the dispersant which keeps the emulsifiers dissolved and ensures contact between the emulsifiers and the oil.

3. *Requirements for a dispersant*

The dispersant shall fulfil the following requirements:

(a) Prohibited components.

The dispersant shall not contain components which are toxic to the environment such as carbon-tetrachloride and other chlorinated organic compounds. Neither shall it contain benzene or other carcinogenic and toxic aromatic hydrocarbons, phenols and cresols. Further, the dispersant shall not contain components which make it difficult to store or to use, including strong acids and alkalis.

(b) Effectiveness.

The dispersant shall fulfil the requirements in force at the time in question concerning effectiveness, as defined in the Appendix to the Regulations..

(c) Acute environmental toxicity.

The dispersant shall fulfil the requirements in force at the time in question concerning acute toxicity, as defined in the Appendix to the Regulations.

(d) Degradation.

The dispersant shall be biologically degradable.

(e) Other properties.

The properties of the dispersant from the point of view of viscosity, and its other physical properties, shall be such that the dispersant may be stored and employed at air temperatures down to -15 degrees C.

As regards inflammability, the dispersant shall not belong to a higher danger class than class C, see paragraph 3 in the Act dated 21 May 1971 concerning Inflammable Products. The storage lifetime of the dispersant shall be at least five years.

4. *Requirements for the solvent*

The solvent utilized in the dispersant shall contain a maximum of 3% by volume of aromatic hydrocarbons.

5. *Area of use*

The use of dispersants in water courses is prohibited.

Dispersants may be employed to limit the harmful effects of oil spills at sea and in the beach zone bordering on the sea.

Dispersants may be employed only when the use of mechanical recovery equipment has been shown to be ineffective, and if on the basis of a total assessment it is assumed that the use of dispersants will reduce the harmful effects on the environment.

It is permitted to use only that amount of dispersant which is necessary to disperse the oil.
All spraying equipment shall be able to spray (atomize) the dispersant in a satisfactory manner.
If necessary, the oil/water masses shall be agitated (stirred up) after the dispersant has been sprayed on.
The dispersant shall not normally be used in the following cases:
(a) When the oil slick is thicker than approximately 1/2 cm.
(b) When the oil slick is composed of fluid oil components such as petrol, diesel, etc.
(c) At low temperatures when the oil slick is composed of heavy oil components.
(d) After the oil has been strongly emulsified.
(e) In the case of smaller oil spills in the open sea.
(f) When the production of ice makes it difficult to spread the dispersant effectively.

6. *Approval*

Dispersants shall be approved by SFT. Anyone importing or manufacturing such dispersants is responsible for ensuring that the dispersants which are sold have been approved. The application for approval shall include information concerning the composition of the dispersant. It must be documented that the requirements in paragraphs 3 and 4 are complied with.
The packaging shall indicate that the dispersant has been approved.

7. *Special license*

The use shall obtain special permission from the SFT in each individual case involving:
(a) The use of more than 1000 litres of dispersant (ready-to-use solution).
(b) The use of dispersants in the beach zone.
If an oil spill results in danger of fire or of explosion, it is permitted to use dispersants independent of the limitations to their use defined in these Regulations. Notification of the. use of dispersants in such cases shall be sent as soon as possible to the SFT.

8. *Dispensation*

The SFT may made exceptions to the rules given in these Regulations.

9. *Complaint*

A decision made by the SFT pursuant to these Regulations may be appealed to the Ministry of Environment. The appeal shall be forwarded through the SFT. Otherwise the rules stipulated in the Public Administration Act apply to cases dealt with in pursuance of the Regulations.

10. *Control and Enforcement*
The SFT shall undertake control and enforcement action to see that these Regulations are observed.
11. *Civil Penalty*
In case of violation of these Regulations or of a decision made pursuant to these Regulations, the SFT may impose a compulsory fine pursuant to paragraph 13 of the Product Control Act.
12. *Punishment*
Violation of these regulations or of a decision made pursuant to these Regulations will be punished pursuant to paragraph 12 of the Product Control Act unless the situation is such that more severe penal provisions apply.
13. *Entry into force*
These regulations enter into force immediately.
Dispersants which have been bought by the user before these Regulations entered into force may nevertheless be utilized until up to one year after the date on which the Regulations enter into force.

USSR: RESPONSE BY OFFSHORE OPERATORS TO OIL AND HAZARDOUS SUBSTANCE SPILLS FROM SOVIET OFFSHORE INSTALLATIONS

APPENDIX TO THE REGULATIONS CONCERNING THE COMPOSITION AND USE OF DISPERSANTS TO COMBAT OIL AND HAZARDOUS SUBSTANCE SPILLS

1. *Effectiveness*
The dispersants shall satisfy Warren Spring Laboratory's[83] requirements for efficiency.
2. *Acute toxicity*
   (a) The dispersant mixed with oil shall not be more toxic towards brown shrimps and the common limpet than the pure oil tested in accordance with RAA Blackman *et.al's* methods described in Fisheries Research Technical Report No. 39 (1977).
   (b) The dispersant mixed with oil shall have a LC50 - value of minimum 200 ppm towards the unicellular alga *Chlamydomonas reinhardti*, tested in accordance with Norland *et.al's* method described in Chemosphere No. 3 (1978):. Toxicity Testing with Synchronized Cultures of the Green Alga Chlamydomonas. Concentrates will be tested in dilution 1:10.
   (c) (Standards of dispersants with different hazardous chemicals).

Annex IV

CONTROL OF AIRSPACE

1. *General Provisions*
   (a) It may be necessary to advise aircraft not taking part in an operation to deep clear of the incident area in order to avoid interference with aircraft which are engaged. Outside territorial waters, a danger area may be established by the Civil Aviation Authority (CAA) in Norway and the U.S.S.R.
   (b) The establishment of a danger area does not mean that aircraft are obliged not to enter it. The air traffic service unit responsible for the airspace in question may, however, assist in diverting traffic by informing aircraft in flight.
   (c) The establishment of a danger area is notified to users by NOTAM Class I within 24 hours and repeated by NOTAM Class II distributed once weekly in Norway and the USSR. Users are notified of cancellation by NOTAM Class I.
2. *Application to the PLAN*
   (a) The decision to request establishment of a danger area is the responsibility of the ACA which determines the boundaries of the area to be established. (*)[84]
3. *Procedures*
   (a) Norwegian procedures
      (i) The ACA and ALA may ask for the establishment of a temporary danger area by telephone to ...
   (b) USSR procedures

      ...
   (c) In both countries the request must contain the following information:
      (i) Nature of incident and reason for request.
      (ii) Position where the incident has occurred in latitude and longitude coordinates.
      (iii) The horizontal and vertical dimensions of the airspace which is to be established as a danger area.
      (iv) Representative of the authority (ACA or ALA) and telephone number where information regarding possible entry into the area may be sought.
   (d) Subsequently, the ACA is to advise:
      (i) Any changes required in the dimensions or of the location of the temporary danger area.
      (ii) When the established danger area may be terminated.

Annex V
Definitions and Abbreviations

Action Coordinating Authority (ACA) The authority designated by the government of the lead country to coordinate the joint counter-pollution operations.

Action Liaison Authority (ALA) The authority designated by the government of the supporting country to maintain contact with the ACA, to control its own resources and if necessary to provide a link between the ACA and the responsible operator.

Blow-out Emergency Team (BET) See Article 5.

Bonn Agreement The 1969 Agreement for Co-operation in Dealing with Pollution of the North Sea by Oil

Governmental Action Control Group (GACG) See Article 5.1.

Lead Country Generally speaking the country (*)[85] in whose PLAN Area the incident occurs or the main action is taking place. (See Article 4 of the PLAN.) (*).[86]

(*)[87]

NIFO Norwegian Industry Association for Operating Companies (See Article 5.3 (c) and (e)).

NOCSA Norwegian Offshore Clean Seas Association (See Article 5.3 (c) and (d)).

Protocol Protocol for Counter-Pollution Operations in the Arctic Region

Parent Government The government by which an operator is licensed and controlled.

RCC Rescue Coordination Centre

POLREP Pollution Report in alphabetically coded form (See Annex II).

Sector Clubs See Article 5.3 (g) and (k).

SFT State Pollution Control Authority (Norway).

Notes - Chapter 13

1 Protocol for Counter-Pollution Operations in the Arctic Region with Sub Bilateral Agreements.

2 See Part II Sections 9.3.4.

3 See Part I Section 7.4.

4 See Part II Section 12.1.3.

5 The legal support for forming such an agreement under LOSC and other provisions would be the same as presented below in Section 13.2.2., supporting the establishment of the AGREEMENT and PLAN.

6 Adapted from Bonn Agreement Articles 2-12, 16 and 17.

7 This cooperation would include combating pollution from oil and other hazardous substances which affects the Arctic environment and State Parties' interests.

8 This would be consistent with Article 2 of the Arctic Convention and include a discharge of oil or other hazardous substances representing a serious and imminent danger to one or several of the Parties' coastal or other interests.

9 Other necessary State promises include availability of the most advanced alarm system and communication system at the installation in good working order; communication of any emergency immediately to the competent national authority, and reception of the alarm and appropriate assistance and co-ordination of assistance organized and supervised without delay; immediate information about the nature and extent of the emergency given to the crew; constant information to the national authority about the progress of combating the emergency; availability, at all times, of sufficient and most advanced materials and equipment, including stand-by boats and aircraft; knowledge of the most advanced methods and techniques by specialized crews in order to combat leakages, spillages, accidental discharges, fire, explosions, blowouts and any other threats and in order to deal with the reduction and prevention of long term adverse environmental effects; knowledge by the crews of the operator's contingency plan, and periodic emergency exercises held so that the crew has a thorough working knowledge of the equipment and procedures; cooperation by the operator on an institutional basis with other operators or entities so as to assure that, when required, proper assistance can be rendered. Adapted from UNEP Draft Annex VII. The list is not inclusive. This proposal is to ensure that a uniform minimum level of State and operator competence is provided in the national Arctic contingency programs, since it will be these which will ultimately be of most importance. A quick look at Figure 1 Part I Section 9.1. shows that roughly 60% of the total marine area around the North Pole is either the Soviet or the Canadian, most of which are not in the vicinity of a boundary line where a joint contingency plan would be most applicable. Additionally the national plans will be that which is used in any joint contingency plans.

10 A good example of exchange of technical information is given by SFT who are interested in films made of the recent ESSO VALDEZ oil spill incident in Alaska. These films have valuable research value showing first hand the effects large quantities of oil spilled had upon Arctic animals, and effects of the oil slick's depth at places unexpectedly extending *90 feet* in depth. Unfortunately, these films are Court documents and cannot be released. SFT therefore sent their own film team to attempt to record this information. Interview with Chief Engineer Jens Koefoed, August 1, 1989.

11 Chief Engineer Jens Koefoed SFT August 1, 1989, notes that in the wake of the ESSO VALDEZ accident Norway gave assistance, and Norway is probably the Arctic country furthest from Alaska. See Figure 1 Section 9.1. Additionally it was noted that requests for cooperative efforts have been forwarded to Norway by Canada and the Soviet Union; so it appears that in spite of the distance and political and bureaucratic problems, Arctic cooperation is developing in the absence of a Regional Arctic Protocol. The argument for uniformity and effectiveness still applies however.

12 See Section 13.4.3. PLAN Article 2.2. regarding special agreement to operate PLAN outside specified regions, and Section 13.4.2. AGREEMENT Articles V and XV regarding amendment of PLAN and AGREEMENT.

13 See Part I Section 7.8.1.

14 See Part I Section 7.6.

15 Conversation with Chief Consultant, Per Antonsen, Norwegian Ministry of the Environment, September 26, 1988.

16 This was thought to be more realistic than solely a plan based upon NORBRIT-PLAN, which was entered into by letters of exchange. Because of the differences between Norway and the Soviet Union, regarding size and politics, a more formal agreement negotiated and ratified seems more likely.

17 See Part I Sections 2, 3.2., 4.2. and 5.2. for customary law and the conventions regulating State jurisdiction over sea-bed and vessel-source pollution and dumping, lending additional legal support to the formation of the Arctic Counter-Pollution Protocol with Sub Bilateral Agreements. As legal support the use of the LOSC provisions are limited to the extent they can be said to be international customary law and recommendations for norms. See Part I Section 2.1.4. and Part II Section 10.3.3.

18 Articles 56, 60, 77 and 80. See Part II Section 12.1.2.

19 *Ibid.*

20 Articles 208, 214, 211, 217, 218, 220, 210 and 216 respectively. See Part II Sections 11.2., 12.1.2., 14.1.2. and 15.1.2. for further discussion of LOSC provisions. Under LOSC Articles 92, 94, 211 and 217 on the high seas, outside the exclusive economic zone or the territorial waters, generally only the flag State has exclusive jurisdiction over its ships. Under LOSC Article 209 in the Area, the sea bed beyond national jurisdiction, anti pollution measures for installations are required from the States with authority over them. See Part I Section 4.2.1. for 1958 Continental Shelf Convention Article 5. See also Part I Section 5.2.1. for 1958 High Seas Convention, Articles 6, 10, and 11. Thus anti pollution measures are to be taken with respect to ships only by the State with jurisdiction over them. The Intervention Convention and Protocol discussed in Part I Section 5.3.1. is an exception.

21 Article 98 (1) and (2). See also Part I Section 5.2.1. 1958 High Seas Convention, Article 12.

22 Some exceptions are made under the sovereign immunity principle for war ships or other State owned non commercial ships. See Articles 30-32 and 17-26.

23 Emphasis and parentheses added. In addition as noted previously the general Articles 192, 197, and 194 give legal support, especially the latter which requires States to take all *joint* measures necessary to minimize fully accidents and *meet emergencies* from any pollution sources including oil platform and shipping, emphasis added. See Part II Section 11.2.

24 See Part II Section 9.2.2. for the legal basis for both the Soviet and Norwegian claims to different marine boundaries in the Barents Sea under the 1958 Continental Shelf Convention Article 6.

25 See generally Tresselt s. 75-93, especially s. 81 where it is stated, '(E)ven though one of the Parties in the delimitation negotiations, or for that matter both, has expressed a willingness to reach a pragmatic compromise, their legal positions continue to play a major role.' (Translated by author). It is noted that international law *limits* some of the Parties' activities in the disputed area, for example positively exercising their jurisdiction over the continental shelf or carrying on activities which are dependent upon permission from the coastal State. Yet both Parties can carry on activities which all States can conduct on the high seas for example navigation and overflight. See LOSC Article 87. Fishing, constructing artificial islands, laying submarine cables and pipelines and scientific research also appearing in Article 87 are not listed here, since they are subject to other rules including those regarding the continental shelf and exclusive economic zone which are again what are disputed. Note additionally that LOSC Articles 74 and 83 make the requirement that States, pending delimitation agreement, 'in a spirit of understanding and cooperation shall make every effort to enter into provisional arrangements of a practical nature and, during this transitional period, not to jeopardize or hamper the reaching of the final agreement.'

26 Two official meetings of experts are planned, one in Autumn 1989 and the other in the first half of 1990, after which discussions will take place between the Foreign Ministers, Aftenposten, 4 oktober 1989, Uke 40, Nr.458, 130 årg., s. 4. An additional meeting between legal experts is planned for September 1990. Conversation Norwegian Ministry of Foreign Affairs, August 9, 1990.

27 Emphasis added.

28 See Tresselt s. 82. It is noted both Parties however have conducted seismic examinations in disputed areas.

29 The area of such a circle is approximately 7,850 square nautical miles.

30 See Tresselt specifically s. 79-82 where he refers to 'a disputed area', and 'the disputed area' as something defined.

31 *Ibid.* Additionally former Prime Minister Ryzhkov's proposed joint oil operations in the Barents Sea in January 1988, according to the Norwegian National Television, covered an area apparently not identical to the disputed area outlined by Tresselt.

32 Tresselt s. 84. See also Ulfstein s. 67-68.

33 Tresselt s. 84.

34 See Section 13.4.2. AGREEMENT Articles I - II and Section 13.4.3. PLAN Articles 1.4., 2.1. and 3.

35 See Part I Sections 4.2.1. and 5.2.1. and Part II Sections 12.1.2. and 14.1.2. Under LOSC Articles 58 and 79 all States have the right, subject to the rights and duties of the coastal State, to lay and operate submarine pipelines in the coastal State's exclusive economic zone and on the continental shelf respectively. Article 4 of the 1958 Continental Shelf Convention and Article 26 of the 1958 High Seas Convention give much the same right.

36 See Section 13.4.2. AGREEMENT Article II where the deleted phrases appear as a footnote for reference. For reference the deleted phrases are indicated throughout the AGREEMENT and PLAN by (*) and reproduced in the footnotes.

37 ''Areas of responsibility' of the Parties where they are adjacent will be separated by the maritime boundary between the two.'. See *ibid.* footnote to AGREEMENT Article II.

38 *Ibid.* Article II. Another scheme was considered to cover pollution from ships and pipelines outside the PLAN Area, but inside a *defined* disputed area, if this eventuality ever occurs. This was based upon respective Party environmental jurisdiction over and thus responsibility for Norwegian and Soviet flag ships and ships trading in their respective harbours, plus any ships involved in trafficking the Soviet Northeast Passage. The same could be used for pipelines crossing a disputed area, having the same jurisdiction as the platform from which they emanated. This scheme of course is not feasible in the Barents Sea as it would give both Parties environmental jurisdiction over vessels and pipelines in areas which are clearly under the other Parties control.

39 *Ibid.* Article II.

40 *Ibid.* Article XIV.

41 See Section 13.4.3. PLAN Article 2.2.

42 Note that the reference to territorial sea is left in tact since Norway and the Soviet Union have agreed upon a boundary in this sea area. See Tresselt pp. 75 and 85.

43 See Part II Section 8.3.

44 See Part II Section 9.2.6.

45 There have been no clear incidents recently to the author's knowledge in which Norway has sent a formal protest to Soviet officials regarding a Soviet violation of Norwegian sovereignty. In addition according to Norwegian State Television the Soviet postal system on Svalbard is now linked to the Norwegian which would indicate the two countries are less than adversaries.

46 See Aftenposten 26 september, 1990, Uke 39., Nr.438., 131.årg. s. 9. regarding Soviet interest in broad international cooperation through the U.N. in international dispute settlement.

47 See generally Antinori pp. 1-10.

48 As noted the Soviets have claimed at one point that there was no disputed area. See Section 13.2.3.

49 Pharand pp. 753-754 and Antinori pp. 10-23. Antinori p. 9 notes however that very little information has been made public concerning the Soviet - US dispute.

50 See however Aftenposten 26 september, 1990, Uke 39. Nr.438, 131.årg. s. 10, an interview with the Norwegian political experts Holst (Norwegian Institute for International Affairs), Eide (Vice Minister) and Østerud (University of Oslo) regarding world instability and disintegration of the Soviet Union. What disintegration means in terms of Soviet - Norwegian relations in the Barents Sea, specifically a contingency plan, is not known, but the article at the same time calls for Norwegian political forces to set to work to settle the Barents Sea boundary conflict.

51 See Part I Sections 1.2. and 1.3.

52 As noted in Part II Sections 7.1. and 7.8.4. it appears that the second alternative is being followed by Norwegian officials. An Agreement is presently in the evaluation stage but a Plan has yet to be developed.

53 Agreement between the Government of the Kingdom of Norway and the Government of the Union of Soviet Socialist Republics concerning Cooperation in Combating Pollution in the Barents Sea in Emergency Situations (AGREEMENT). Adapted from the US - USSR Agreement concerning Cooperation in Combatting Pollution in the Bering and Chukchi Seas in Emergency Situations, signed May 11, 1989.

54 Norway - Soviet Union Joint Contingency Plan for Counter-Pollution Operations at Sea (PLAN). Adapted from NORBRITPLAN, signed August 25, 1983.

55 (*), regardless of where such incidents may occur. (This phrase was removed and 'from incidents occurring within the Parties' respective PLAN Areas' added to narrow down the scope of the AGREEMENT and PLAN to incidents occurring within the PLAN Area noted below).

56 'Hazardous substances' does not appear in NORBRITPLAN but does in the US - USSR Agreement. The phrase was added where relevant throughout the PLAN to enlarge the scope both because of the precedence set by the US - USSR Agreement and because of the environmental benefits achieved.

57 See Part II Section 8.2.

58 (*) beyond the territorial sea in which that Party exercises its sovereign rights and jurisdiction in accordance with international law. Areas of responsibility of the Parties where they are adjacent will be separated by the maritime boundary between the two. (The territorial sea was left intact as it appears from Tresselt p. 85 that it is the boundaries of the continental shelves and the 200 mile economic zones which are in question in the Barents Sea, not the territorial seas).

59 The 50 nautical mile figure was chosen since it was the figure used by NORBRITPLAN on both sides of the boundary line and is believed to be the figure used as a radius for 'danger zones' established to surround several offshore accidents in the North Sea. It is of course variable, and the Parties could negotiate it larger or smaller as the case may be.

60 It was attempted with this definition to avoid any reference to and dependency upon boundary lines or disputed areas, which is the major hinderance to a joint Norwegian Soviet Contingency Plan. This system is based upon Party jurisdiction over offshore installations. Should either Party begin drilling in the disputed areas, a contingency plan would follow whatever jurisdictional arrangement has been established allowing this drilling. Both Norway and the Soviet Union in the event of a shipping or pipeline pollution incident occurring outside the PLAN Area but in uncontested areas of the Barents Sea over which they have clear jurisdiction in accordance with international law

could call the PLAN into action, if specially agreed to under PLAN Article 2.2. The same is true for contested areas though this seems unlikely before the jurisdiction issue is settled. A jurisdictional system covering ships and pipelines not specifying boundary lines or disputed areas yet respecting uncontested Norwegian or Soviet exclusive economic zones would have been difficult to achieve. See Part I Section 5.2.4. for further discussion.

61 In the US - USSR Agreement the Soviet agency cited is the Marine Pollution Control and Salvage Administration attached to the USSR Ministry of Merchant Marine. In Norway, under NORBRITPLAN Article 1.8(a), Amendments, the Norwegian authority cited is the Ministry of the Environment.

62 (*) area of responsibility. (This again is removed to narrow down the scope of the AGREEMENT and the PLAN to incidents occurring within the PLAN Area).

63 (*) close to the boundary line

64 Norway and the Soviet Union have entered the bilateral Rescue at Sea Agreement for use in the Barents Sea.

65 (*) Guidelines should also be established between the Norwegian Oil Directorate and the Soviet ... concerning the drilling of relief wells close to the respective boundary lines. (This section was omitted due to its dependence upon geographic location. As long as both the Soviets and the Norwegians do not drill in disputed areas, the drilling of relief wells a short distance from the original well probably should present no problem. Should either Party begin drilling in the disputed areas, the drilling of relief wells would follow whatever jurisdictional arrangement that has been established allowing the original drilling).

66 (*) a band following the boundary line between Norway and the Soviet Union and extending to a distance of 50 nautical miles on either side of that line.

67 This definition though appearing in the AGREEMENT Article II appears in the PLAN also for sake of clarity and emphasis.

68 This paragraph allows the Parties flexibility to enlarge the scope of the PLAN to include combating oil or hazardous substances pollution in areas outside the PLAN Area including disputed areas.

69 The Protocol is added here to cover hazardous substances.

70 (*) on whose side of the boundary line

71 The phrase, 'that affects or threatens to affect the areas of responsibility of both Parties or is of such a magnitude as to justify a request of the other Party's assistance' was added to *limit* the number of necessary notifications. Without this the Norwegians would have to notify the Soviets of a blowout or spill occurring in the western Barents sea regardless of whether they could affect Soviet interests or not.. Due to the Norwegian reluctance to participate too much in any cooperative arrangement with the Soviets, it is attempted to hold the cooperation as narrow as possible.. See Part II Section 8.3. Notifications in a system based upon boundary lines are required only for incidents occurring within 50 nautical miles on either side of the boundary line.

72 (*) Where an incident involves an offshore oil field which extends across the jurisdictional boundary (median line), the country from which the operator concerned is regulated shall assume the lead role and designate the ACA, regardless of which side of the boundary the incident occurs. (In the event this happens the jurisdiction issue will be established and both Norway's and the Soviet Union's 'environmental jurisdiction' will follow that regime.)

73 (*) Where desired and appropriate, the lead and supporting roles of the two countries may be reversed by agreement between them. This may, for instance, be appropriate when the main area of pollution or of activity moves from one side of the boundary line to the other, or when the major part of the resources required and deployed are provided by the supporting country. Such a change should be preceded by

prior consultation and agreement. (Paragraphs (c), (d), and (e) were removed both due to the sensitivity of the Norwegians to Soviets exercising too much control in a bilateral agreement and to the boundary line problem. It seems unlikely that either the Norwegians or the Soviets would be too positive to the other Party leading clean up activities in their respective economic zones.)

74 (*) Should there be a major and separate area of pollution on each side of the median line, it may be appropriate for each country to assume the lead for operations on its own side of the line.

75 (*) Where spillage is from a submarine pipeline, the initiation of action and assumption of lead role are similarly determined by the position of the incident in relation to the median line, regardless of the ownership of the installation from which it runs, or the jurisdictional control of the pipeline. In appropriate circumstances, the lead role may pass to the other country as provided for in (c) above.

76 (*) close to or beyond the boundary line

77 The phrase, 'that affects or threatens to affect the areas of responsibility of both Parties' was added as above to limit the number of necessary consultations.

78 (*) on their respective continental shelves. (The phrase, 'under their respective jurisdictions' was added to avoid the continental shelf boundary problem.)

79 Norway's guidelines are briefly listed to give an example.

80 Norway is presented as is under NORBRITPLAN. The Soviet Union is left blank.

81 A form for reporting pollution incidents and omitting irrelevant details was adopted by the Contracting Parties to the Bonn Agreement. A similar form may be used for the rapid notification of incidents under the PLAN arrangements. It is not duplicated here, but can be obtained from the Norwegian Ministry of the Environment.

82 The legal regulations and technical limits are reproduced from NORBRITPLAN in order to give an idea of the essential elements of such a contingency plan. Note that the phrase 'oil spills' was changed to 'oil and hazardous substance spills' in the title to be consistent with the AGREEMENT and PLAN. It was *not* done in the text since it would give a mistaken representation of the Norwegian legislation. In addition the title of both the Norwegian and the Soviet Response by Offshore Operators was changed from, 'RESPONSE BY OFFSHORE OPERATORS IN THE NORWEGIAN SECTOR TO OIL SPILLS FROM OFFSHORE INSTALLATIONS', to the titles shown in order to de emphasize the areal idea of sector and emphasize each country's jurisdiction over the respective platforms.

83 These standards could be based upon more recent tests if necessary.

84 (*) If this area is divided by the boundary of the FIR, it is the responsibility of each national authority to submit a request in respect of the relevant part to the air navigation services in accordance with the appropriate procedures for each country.

85 (*) on whose side of the boundary line

86 (*) The roles of the countries may however be reversed, by agreement.

87 (*) Median line:. The line dividing the continental shelf for the purposes of exploration and exploitation of the sea bed and its resources.

14. Discussion of Arctic Vessel-Source Protocol[1]

14.1. Background

This Protocol is somewhat different from the other Arctic Protocols in that it is generally meant to *supplement* existing conventions covering vessel-source pollution rather than provide environmental legal coverage for the Arctic. As has been noted previously the area of vessel-source pollution is the one most regulated at the international level, but in spite of these regulations there have been problems, both of a technical nature and with State compliance.[2] At the same time shipping is increasing in the Arctic.[3] This Protocol attempts to focus upon these problems and require from the Arctic States explicit promises of compliance with provisions found in the shipping conventions and the development of appropriate technology. In addition, where beneficial, new provisions are added based generally upon LOSC rules. It is hoped by this combination to achieve a regime built upon the existing international regulations, but carrying them further to meet the special environmental conditions required of shipping in the Arctic. A balance was attempted in increasing environmental demands while at the same time respecting shipping interests. Where these provisions might result in non ratification by the Arctic States, note is made. The format used is *informal*, where only the ideas which are considered pertinent to each Article are presented, together with a short discussion.

14.1.1. Summary of Main Existing Vessel-Source Pollution Regimes with Application to the Arctic

As noted MARPOL 73/78[4] encompasses the Arctic region covering the prevention of pollution from oil, hazardous chemicals, hazardous chemicals in containers, sewage and garbage. The Intervention Convention and Protocol[5] encompass the Arctic and gives coastal State Parties the right to take necessary measures on the high seas to prevent danger to the coastlines or other interests for marine pollution. The SOLAS Convention and Protocol[6] apply to the Arctic and generally cover international safety standards. INMARSAT[7] applies to the

Arctic and contributes through satellites to improving safety communications and management of ships. STCWS[8] applies to the Arctic, and establishes basic standards of crew training and ship operation including in conditions involving ice. The Load Lines Convention,[9] applies in the Arctic and deals with the overloading of vessels. The Collisions Convention[10] with application in the Arctic is a safety convention dealing with vessel traffic separation zones. The Containers Convention[11] applies in the Arctic and covers the safe handling and shipping of containers.

14.1.2. Legal Basis for Arctic Vessel-Source Protocol under LOSC[12]

14.1.2.1. Legislative Jurisdiction

It must be noted that of the 74 States which have claimed an exclusive economic zone, *few* have any specific legislation governing pollution in the zone and *fewer* have provided for port State jurisdiction in their legislation.[13] These States could take a more offensive attitude in practice under these Articles following the LOSC, and therefore they will be presented comprehensively as legal support for the Arctic Vessel-Source Protocol. It is the emphasis upon more offensive State practice which is attempted in this Protocol's provisions.

As noted in Part II Section 11.2. the general obligations set forth by Articles 192, 197 and 194 apply, especially the latter under which States must jointly take all measures necessary to minimize fully pollution from ships, particularly preventing accidents and dealing with emergencies and safety, preventing intentional and unintentional discharges, and regulating the design, construction, equipment, operation and manning of their flag ships.

Generally as noted flag States must establish, acting through the competent international organization[14] or general diplomatic conference, international rules and standards to prevent, reduce and control pollution of the marine environment from vessels, and establish routing systems minimizing the threat of polluting accidents.[15] Flag States must adopt laws and regulations for the prevention, reduction and control of vessel-source pollution for ships of their registry at least as effective as that of generally accepted international rules established through the IMO or diplomatic conferences.[16]

Coastal States are given further jurisdictional powers in 12 mile territorial seas, 24 mile contiguous zones, 200 mile exclusive economic zones, archipelagic waters and especially environmentally endangered waters.[17] Coastal States may adopt laws and regulations for the prevention, reduction and control of marine pollution from foreign ships in their territorial sea, which do not hamper vessels' innocent passage.[18] Additionally, the coastal State may adopt laws and regulations which are in conformity with LOSC provisions and other rules of international law regarding a ship's innocent passage through the territorial sea with respect to the preservation and prevention, reduction and control of pollution of the coastal State's environment.[19] These anti pollution laws and regulations do *not* 'apply to the design, construction, manning or equipment of foreign ships unless they are giving effect to generally accepted international rules or standards,' and must be publicized.[20] Additionally such laws and regulations must not practically deny or impair the right of innocent passage, discriminate in form or fact or generally hamper the innocent passage of foreign ships.[21]

Coastal States may in accordance with their enforcement rights under LOSC adopt laws and regulations for the prevention, reduction and control of pollution from ships in their exclusive economic zones which give effect to generally accepted international rules and standards established through the IMO or diplomatic conferences.[22] Coastal States may upon certain conditions establish *special areas* within their exclusive economic zone and adopt additional laws and regulations for these areas.[23] The laws and regulations may relate to discharges or navigational practices but not to established design, construction manning or equipment standards.[24]

As noted previously Article 234 gives coastal States the right to adopt and enforce non-discriminatory laws and regulations for the prevention, reduction and control of marine pollution from vessels in *ice-covered areas* within the exclusive economic zone, where severe climatic conditions and the presence of ice covering such areas for most of the year create obstructions or exceptional hazards to navigation, and pollution of the marine environment could cause major harm to or irreversible disturbance of the ecological balance. Such laws and regulations must take consideration to navigation and the protection and preservation of the marine environment based on the best available scientific evidence. Thus coastal States have the right to regulate the design, construction, manning and equipment of foreign ships in ice-covered areas under its jurisdiction as well as control the entry, passage and conditions of the ship's operation in these areas. In addition a coastal State is *unrestricted* in enforcing its anti pollution rules for ice covered areas in its territorial sea and/or exclusive economic zone, since the usual restrictions covering coastal State *enforcement* jurisdiction Articles 220, 228-233 noted below are not mentioned in Article 234.[25]

The LOSC makes no changes in the legislative competence of port States except that consistent with the enforcement jurisdiction given port States noted below. States must also publicize and file with the IMO their particular anti pollution measures regarding prevention, reduction and control required as a condition for foreign vessels to enter their ports, internal waters or off-shore terminals.[26]

14.1.2.2. Enforcement Jurisdiction

LOSC Section 6 deals with enforcement. Enforcement, said to be the 'crucible of law, the test of its reality', includes surveillance, stopping and boarding, search or inspection, arrest and detention, and trial and sanctions.[27] It is classified under the LOSC according whether the State is a flag State, a coastal State or a port State.[28]

Flag States must provide for the effective enforcement of applicable international anti pollution rules and standards, established though the IMO or general diplomatic conference, and of State anti pollution laws and regulations adopted in accordance with LOSC, irrespective of where a violation concerning vessel source pollution occurs.[29] In particular flag States must take measures to ensure that ships under their registry are prohibited from sailing until compliance with the applicable international rules and standards, including design, construction, equipment and manning standards, are met.[30] Additionally flag States must ensure that their ships carry on board the proper certificates required by the applicable international rules and standards and must ensure that their ships are periodically inspected for verification of certificate conformity with actual ship condition.[31] If a violation occurs of the anti pollution rules and standards established through the IMO or general diplomatic conference, the flag State must provide for immediate investigation and where appropriate institute proceedings irrespective of where the violation occurred.[32] Assistance may be requested by the flag State of other States for clarification of the polluting incident, and other States must endeavour to meet the flag State requests.[33] Upon written request of any State, flag States must investigate any violation alleged to have been committed by ships under their registry, and upon sufficient evidence institute legal proceedings without delay.[34] The IMO and the requesting State must be informed of legal action taken, and the result and such information made available to all States.[35] Penalties administered by flag States must be adequate in severity to discourage violations.[36]

When a ship is at a coastal State port or offshore terminal, subject to the safeguards noted below, the coastal State may prevent the ship from sailing if it is in violation of applicable international rules and standards relating to seaworthiness and hence threatens the marine environment.[37] Port States must, as far as practicable and in accordance with the Section 7 safeguards noted below, take administrative measures to prevent unseaworthy[38] ships from sailing from their ports or offshore terminals.[39] When a ship is voluntarily within a port or at an offshore terminal of a coastal State, that State may, subject to the safeguards noted, institute proceedings for violation of its anti pollution laws and regulations adopted in accordance with LOSC or applicable international rules and standards where such violation has occurred within the State's territorial sea or its exclusive economic zone.[40]

A coastal State may, where there are *clear grounds* for believing that a ship navigating in its territorial sea, has *illegally* discharged in that State's territorial sea, without prejudice to relevant provisions concerning *innocent passage*, inspect the ship and where the evidence so warrants, take legal proceedings including detaining the ship, subject to the safeguards noted.[41] A coastal State, where there are clear grounds for believing that a ship navigating in its exclusive economic zone or territorial sea has illegally discharged in that State's exclusive economic zone, may require the ship to give information covering its identity, port of registry, its last and next port of call and other relevant information required to determine whether a violation has occurred.[42] A coastal State, where there are clear grounds for believing that a ship navigating in its exclusive economic zone or territorial sea has illegally released a *substantial discharge* causing or threatening significant marine pollution in that State's exclusive economic zone, may physically inspect the ship if the ship has refused to give information or the information given is manifestly incorrect and the circumstances justify such inspection.[43] A coastal State, where there is clear objective evidence that a ship navigating in its exclusive economic zone or territorial sea has illegally released a discharge causing *major damage* or threat of major damage to the coastline or related interests of the coastal State or to any resources in that State's territorial sea or exclusive economic zone, may in accordance with the safeguards noted and upon sufficient evidence institute legal proceedings including detaining the ship.[44] The detained ship must be allowed to proceed if bonding or other appropriate financial security has been assured through IMO procedures or otherwise agreed and the coastal State is bound by such procedures.[45] These coastal State rules[46] apply to national laws and regulations adopted by a coastal State with respect to special areas established by a coastal State in its exclusive economic zone, mentioned above.[47]

As noted a coastal State may take proceedings against a foreign ship in one of its ports which illegally discharged in that State's territorial sea or exclusive economic zone or if the violation occurred outside these areas when it has caused or is likely to cause pollution in its own internal waters, territorial sea or exclusive economic zone.[48] Article 218 additionally allows port States, when a ship is voluntarily within a port or at an off shore terminal of that State, to investigate and where the evidence so warrants to take proceedings against that vessel for any illegal discharge *outside* that State's internal water, territorial sea or exclusive economic zone.[49] This strong enforcement power is limited, however, for illegal discharges in the internal waters, territorial seas or exclusive economic zones of other States unless *requested* by those States, the flag State, a State damaged or threatened by damage.[50] The port State must as far as practicable comply with requests for investigations of ships, voluntarily with its ports or offshore terminals, from any State covering illegal discharges occurring in their internal waters, territorial seas or exclusive economic zones or for investigations from flag States covering illegal discharges occurring anywhere.[51] Records of the investigation must be transmitted upon request to the flag State or the coastal State, and any proceedings instituted by the port State, in accordance with the safeguards noted below, must be suspended upon request of the coastal State when the violation has occurred within the coastal State's internal waters, territorial sea or exclusive economic zone.[52]

Coastal States also have enforcement powers over ships polluting on the high seas in case of maritime casualties, ship collisions and strandings resulting in material damage, which allows a coastal State beyond the territorial sea to take and enforce measures to protect their coastline or related interests from pollution or the threat thereof with expected major harmful consequences.[53]

The Section 7 *safeguards*, mentioned several times above, are set forth in Articles 223 through 233.[54] They include measures to facilitate proceedings including the hearing of witnesses and the admission of evidence, the attendance at such proceedings by IMO officials, the flag State and the polluted State; the exercise of enforcement only by State officials and vessels and aircraft; a duty by States to avoid adverse consequences while exercising enforcement including endangering navigation or the marine environment; a duty not to delay foreign vessels unnecessarily;[55] a duty not to discriminate against foreign ships; a duty to suspend proceedings upon the taking of proceedings by the flag State for the same incident; a prohibition against preclusion of civil proceedings; limitation to only monetary penalties; observance of the rights of the accused in all proceedings which might result in penalties; notification of the flag State and other States concerned of enforcement proceedings and of official reports; and the attachment of liability for unlawful enforcement proceedings or proceedings exceeding those reasonably required.

An additional exception is provided by Article 236 giving sovereign immunity for warships, naval auxiliary, other vessels or aircraft owned or operated by a State and used for government non-commercial service. However a State must provide, by regulations 'not impairing operations or operational capabilities of such vessels,' that its vessels or aircraft 'act in a manner consistent, so far as is reasonable and practicable,' with the Convention. For non compliance of a coastal State's regulations governing in its territorial sea including those related to marine pollution and innocent passage, under Article 30 a warship may be required to leave the territorial sea immediately.

14.1.2.3. Expected Problems[56]

Jurisdictional and enforcement problems may result under the LOSC regime. As seen above Article 211 (2) requires flag States to adopt anti pollution laws and regulations for their ships at least as effective as that of 'generally accepted international rules and standards'. The scope of the 'generally accepted international rules and standards' is not clear however, and Article 211 (2) thus may require flag States to follow provisions of conventions to which they are not parties, which may discourage ratification of LOSC.[57]

Under Article 211 (6) a coastal State may adopt anti pollution legislation in a special area, upon certain conditions noted above, but as in the territorial sea and exclusive economic zone covering only discharges or navigational practices and *not* the design, construction, manning or equipment of foreign ships, unless they are in accordance with 'generally accepted international rules or standards'. The same comments regarding this phrase in Article 211 (2) apply equally here. Additionally Article 211 has increased the geographical areas of application for coastal State legislative jurisdiction into the exclusive zone, but the competence to adopt different kinds of pollution regulations has been reduced, excluding design, construction, manning or equipment standards, unless in accordance with accepted international provisions.[58]

Under Article 217 (1) States must provide effective enforcement of applicable international anti pollution rules and standards and State anti pollution laws and regulations, which was not required under MARPOL 73/78, however this is based upon use of MARPOL 73/78 flag State enforcement which as seen has not been too effective.[59] Article 217 (4) provides for flag State enforcement for violations by its ships of anti pollution rules and standards no matter where the violations occur, however it is *vague* whether a flag State may arrest its own ships in the exclusive economic zone of another State.[60]

Article 228 of the Section 7 Safeguards, which allows for the *suspension* of coastal and port State proceedings (pre-emption) in favour of flag State proceedings, is often complained of as undermining more stringent coastal and port State enforcement procedures occurring under Articles 220 and 218 respectively.[61]

Under Article 220 (2) the coastal State is allowed only to make laws and regulations conforming to international regulations and standards except those established under the Article 211 (6) special area noted above. Thus the term 'illegal' is limited to violations of international regulations and standards.[62] There are several additional coastal State enforcement problems under Article 220 (2) including that of the coastal State's requirement of proving that its activities do not violate Article 24 (1)(a) regarding innocent passage, and foreign ships using this to resist coastal State enforcement measures.[63] Also regarding innocent passage foreign ships could claim under Article 21 (2) that the coastal State cannot adopt or enforce laws and regulations relating to the design, construction, manning or equipment of foreign ships more strict than generally accepted international rules or standards. A circular 'no win' situation is met regarding stopping a ship in a coastal State's territorial sea under Article 220 (2). Foreign ships must not be hampered nor unreasonably delayed under Article 211 (4), Article 24 (1) and the Section 7 safeguards, and the reason for the stopping must be lawful. Under Article 19 (2)(h) the discharge must have been 'wilful and serious', but the coastal State to obtain evidence of a 'wilful' violation which gives it the right to stop the ship and deny innocent passage must stop the ship for interrogation and inspection.[64] There are similar problems with the term 'clear grounds' used in Article 220 (2). It is doubtful a State has anything more than prima facie evidence before making a physical investigation, and the only way of obtaining sufficient evidence amounting to 'clear grounds' would be through inspection, which can be made only if 'clear grounds' have already been established.[65]

Under Article 220(3) problems are similar to above. The wording 'applicable international rules and standards' may allow coastal States to enforce conventions to which they and possibly the flag ship are not parties. Regarding 'clear grounds' the coastal States are limited to requiring only the ship's identity, port or registry, its last and next port of call and other relevant information in establishing a violation. However it can be questioned whether the information may be obtained by physical inspection as in the territorial sea or only by radio or visual signals. One author believes that physical inspection may be considered more limiting to the ship's right of free navigation and therefore *not* permitted in the exclusive economic zone, though permissible in the territorial sea.[66]

Under Article 220 (5) the same arguments regarding the scope and the meaning of the terms 'illegal' and 'clear grounds' apply here. Additionally what is 'substantial' discharge allowing physical inspection under Article 220 (5) and 'slight' under Article 220 (3) above is not defined, either by LOSC or MARPOL 73/78 and is left to the coastal State's *discretion*.[67] The substantial discharge does allow a further inspection under Article 226 (1) by the coastal State than the usual inspection of certificates, records or other ship documents under Article 220 (3).[68]

Under Article 220 (6) the difference between 'major damage' allowing arrest and 'substantial discharge' of Article 220 (5) is unclear and left to coastal State discretion.[69] It seems possible for coastal States to use Article 220 (6) rather than Article 220 (5) since they probably will prefer arrest over inspection if given a choice, and they have this with the unclarity in terminology.[70] An additional problem is that if a ship violates a discharge law in the exclusive economic zone and later sails into the territorial sea, the ship is liable for the violations in the zone, but enforcement procedures can be carried out only as if the ship were still in the exclusive economic zone.[71]

Under Article 220 (8) the type of enforcement procedures by the coastal State outlined above in Articles 220 (3) (5) and (6) apply as well to special areas established by the coastal State in its exclusive economic zone under Article 211 (6). Thus, because of the ambiguity noted regarding 'major damage', 'substantial discharge' and 'slight', a coastal State could arrest a ship in the exclusive economic zone as if the area had special status unless the pollution was too slight.[72]

As seen a coastal State may take proceedings under Article 220 (1) against a foreign ship in one of its ports which illegally discharged in that State's territorial sea or exclusive economic zone. Also under Article 218 when a ship is voluntarily within a port or at an off shore terminal of that State, the port State may investigate and upon sufficient evidence take proceedings against that ship for any illegal[73] discharges *outside* that State's internal waters, territorial sea or exclusive economic zone.[74] This power is *not* mandatory, however and this can be considered a weakness in port State enforcement.[75]

Additional limitations upon port State enforcement appear in Article 218 (2) and include proceedings may not be brought for illegal discharges in the internal waters, territorial seas or exclusive economic zones of other States unless *requested* by those States, the flag State, or a State damage or threatened by damage. Additionally under Article 218 (4) records of the investigation must be transmitted upon request to the flag State or the coastal State, and any proceedings instituted by the port State, in accordance with the Section 7 safeguards, must be *suspended* upon request of the coastal State when the violation has occurred with the coastal State's internal waters, territorial sea or exclusive economic zone.[76] Thus port State enforcement is subject to *both* a coastal State preemption under Article 218 (4) and a flag State preemption under Article 228.

Finally the usual sovereign immunity exception for military or State non commercial vessels under Article 236 reiterating both customary law and MARPOL 73/78 provisions, allows all the LOSC anti pollution regulations enumerated above to be non applicable for such vessels. This can be especially unfortunate for the Arctic since it is chiefly these kinds of ships operating there.

In a broader scope, in spite of the possible problems, to the extent that LOSC environmental provisions can be said to be international customary law and recommendations for norms, the Arctic States have an obligation, acting through the competent international organization or diplomatic conference, to establish and to enforce within the proper jurisdictional limits, provisions to prevent, reduce and control vessel-source pollution, which would act as support to the formation of the Arctic Vessel-Source Protocol.[77]

14.2. Discussion of Proposed Arctic Vessel-Source Protocol

14.2.1. Introduction

Based upon the difficulties noted in Part II Section 9.3.2. and with legal foundation in the LOSC provisions noted in Section 14.1.2. proposals will be made for the Arctic Vessel-Source Protocol. As noted since the Articles proposed often deal with controversial subjects, there is a good possibility that these would not be acceptable to the Arctic rim States for Protocol ratification since they are too environmentally 'friendly' and cut too deeply into traditional shipping sovereignty. Several of these provisions provide some of the reasons behind State reluctance in ratifying the LOSC, and *none* of the Arctic States have ratified this Convention. Still for Arctic environmental protection, perhaps the Arctic States would be willing to establish a more stringent regime.

14.2.2. Articles

Appearing first is the *Preamble*, modeled upon the Preambles for MARPOL 73 and MARPOL 78, and as in the other Arctic Protocols the Preamble consists of general State proclamations.

The Parties to the Arctic Vessel-Source Protocol:

• Being Conscious of the need to preserve the Arctic human environment in general and the Arctic marine environment in particular,

• Recognizing the deliberate, negligent or accidental release of oil and other harmful substances from ships constitutes a serious source of pollution,

• Recognizing Also the need to improve further the prevention and control of marine pollution from ships, particularly oil and chemical tankers,[78]

• Recognizing the significant contribution which can be made by the LOSC, MARPOL 73/78, SOLAS, INMARSAT, STCWS, Intervention Convention and Protocol, Load Lines Convention, Collisions Convention and Container Convention among others, to the protection of the Arctic marine environment from pollution by ships,

• Recognizing Further the need for implementing the regulations of the prevention of pollution by oil contained in MARPOL 73/78 Annex I, the prevention of pollution by noxious liquid substances in bulk contained in Annex

II, the prevention of pollution by harmful substances in packaged forms and freight containers contained in Annex III, the prevention of pollution by sewage from ships contained in Annex IV and the prevention of pollution by garbage from ships contained in Annex V,

• Desiring to achieve the complete elimination of intentional pollution of the marine environment by oil and other harmful substances and the minimization of accidental discharge of such substances,

• Considering these objectives to best be achieved by conclusion of a Protocol drafted by the Arctic rim States in accordance with the Arctic Convention and the above Conventions,

Agree as follows:

Article 1 gives the Arctic States the right to regulate foreign ships in ice-covered areas of their exclusive economic zones, in accordance with Article 234 of LOSC. This is done for emphasis and because LOSC is not in force. Thus under Article 1 the Arctic States shall adopt and enforce[79] regulations covering illegal discharges and design, construction, manning and equipment of foreign ships in ice covered areas in their exclusive economic zones as well as control the entry, passage and conditions of the ship's operation in these areas. The States shall harmonize their policies in this connection in order to provide for uniform requirements for illegal discharges and design and equipment, cargo limits and crewing standards.[80]

Article 2 deals with State anti pollution jurisdiction and enforcement over ships in accordance with LOSC provisions. As noted in Part II Section 9.3.2. jurisdictional limits of the coastal State in its territorial sea and of the flag State on the high seas remains traditional under MARPOL 73/78 and therefore the more liberal and stringent provisions of LOSC are followed and emphasized in the Arctic Vessel-Source Protocol. These provisions could present a problem for ratification of the Arctic Vessel-Source Protocol since as noted the broadening of coastal and port State authority under LOSC at the expense of the traditional shipping sovereignty is what has kept many countries from ratifying LOSC. Facing such problems, Article 2 might have to be reformed more in line with traditional MARPOL 73/78 provisions. However, Article 228 of the Part 7 Safeguards, dealing with the suspension of port or coastal State proceedings if the flag State initiates similar proceedings is a concession to flag State jurisdiction, which is somewhat of a safety value and might help in ratification of the Arctic Vessel-Source Protocol by the Arctic States.

Article 2 thus provides that the Arctic States in exercising their jurisdiction and enforcement powers over ships suspected of polluting or with deficiencies in design, construction, manning and equipment shall do so in accordance and in full compliance with the provisions of LOSC including particularly but not limited to Articles 19, 21 and 24; Article 211; Articles 217-220; Articles 223-233; and Article 234.

Additionally under Article 2 the Arctic States shall give effect to the LOSC jurisdiction and enforcement provisions noted above through specific legislation governing marine pollution in their territorial seas and exclusive economic zones.[81] In putting these provisions into practice the Arctic States shall vigorously enforce illegal discharges and violations of design, construction equipment and manning standards, including surveillance, stopping and boarding, search and inspection, arrest and detention, prosecution and sanctions in accordance with LOSC provisions. States shall harmonize national policies for the Arctic in this regard.[82]

In addition to and supplementing the legal measures outlined in this Protocol the Arctic States under Article 2 shall endeavour to jointly extend and develop political measures to assist controlling vessel-source pollution including but not limited to, exerting political pressure by international organizations such as the IMO; bilateral and regional enforcement systems among States against each other's ships; increasing domestic political and public opinion for giving oil and chemical pollution a higher priority among States; screening by the oil and chemical companies of tankers before they charter; and giving foreign aid to developing countries to help them economically and incentive wise to bring tanker fleets within the international standards.[83]

Article 3 deals generally with compliance of MARPOL 73/78 provisions and states that the Arctic States shall legislate and establish measures and procedures to fully implement the provisions of MARPOL 73/78, with the exception of those State jurisdictional and enforcement measures set out in this Protocol incorporating LOSC provisions.[84] The States shall harmonize national policies for the Arctic in this regard. In addition the Arctic States shall, if not already realized, as soon as possible ratify MARPOL 73/78 and Annexes I - V.[85]

Article 4 requires the States to take all measures to amend MARPOL 73/78 to include the Arctic as a 'special area' with corresponding restrictions for oil, chemical and garbage discharges from ships. In these special areas generally discharges except clean ballast and segregated ballast are prohibited, and pollutants must be retained on board and reception facilities developed to receive these. Thus under Article 4 the Arctic States jointly shall take, under the proper procedures and in conjunction with the competent international organizations all appropriate measures towards amending MARPOL 73/78, Annex I Regulation 10(1), Annex II Regulation 1(7), and Annex V Regulation 5(1) declaring the Arctic a 'special area'.[86]

Article 5 deals with a special area in the respective exclusive economic zones vulnerable to vessel-source pollution under LOSC Article 211 (6), which could be suggested for the Arctic.[87] Thus Article 5 states that the Arctic States shall in addition to or in the alternative to declaring the Arctic a special area under MARPOL 73/78 declare the Arctic a special area with special binding anti pollution measures under LOSC Article 211 (6).[88] The Arctic States shall harmonize their policies in this regard.

Article 6 covers the obligation to provide adequate oil reception facilities.[89] It states the Arctic States shall ensure that not later than January 1, 199_[90] all oil loading terminals and repair ports with the Arctic special area are provided with facilities adequate for the reception and treatment of all the dirty ballast and tank washing water from oil tankers. In addition all ports within the Arctic special area designated in Article 4 shall be provided with adequate reception facilities for other residues and oily mixtures from all ships. Such facilities shall have adequate capacity to meet the needs of the ships using them without causing undue delay.

Article 6 further provides that the Arctic States having under its jurisdiction entrances to seawater courses with low depth contour which might require a reduction of draught by the discharge of ballast shall ensure the provision of facilitates referred to above with the proviso that ships required to discharge slops or dirty ballast could be subject to some delay.

Article 6 also provides that following the date of January 1, 199_ the Organization[91] shall make an examination of all the Arctic State reception facilities and notify the IMO for transmission to Contracting Parties of MARPOL 73/78 of all cases where the facilities are alleged to be inadequate. The Arctic States shall ensure that such alleged inadequate facilities meet this Protocol's standards not more than 6 months following the report to the IMO. After this date ships calling at Arctic ports where such facilities are not yet available, must fully comply with the requirements of Article 9.[92]

Article 7 covers the provision of chemical reception facilities. Although it at first was considered beneficial to follow the same provisions as Article 6, the less stringent MARPOL 73/78 Regulation 7 of Annex II was used as a model instead. This was because of State reluctance to follow Annex II generally. Should it be shown that States would be interested in more stringent provisions, then Article 7 could be modeled after MARPOL 73/78 Annex I Regulation 10(7) 'Reception facilitates within special areas' the same as Article 6.

Thus Article 7 provides that Arctic States shall ensure[93] the provision of reception facilities according to the needs of ships using its Arctic ports, terminals or repair ports. Cargo loading and unloading ports and terminals shall have facilities adequate for reception without undue delay to ships of such residues and mixtures containing noxious liquid substances as would remain for disposal from ships carrying them as a consequence of the application of MARPOL 73/78 Annex II; and ship repair ports undertaking repairs to chemical tankers shall have facilities adequate for the reception of residues and mixtures containing noxious liquid substances.

Further, each Arctic State shall determine the types of facilities required to cover shipping demands in accordance with this Article at each cargo loading and unloading Arctic port, terminal and ship repair port in its territories and shall notify the Organization. The Organization must examine all the Arctic State reception facilities and notify the IMO for transmission to Contracting Parties of MARPOL 73/78 of all cases where the facilities are alleged to be inadequate.[94] The Arctic States shall ensure that such alleged inadequate facilities meet this Protocol's standards not more than 6 months following the report to the IMO. After this date ships calling at Arctic ports where such facilities are not yet available, must fully comply with the requirements of Article 9.[95]

Article 8 deals with the reception facilities within special areas for garbage. It is modeled upon MARPOL 73/78 Annex V Regulation 5(4) and Regulation 7 and has been upgraded as noted. This Article requires that each Arctic State shall ensure[96] that not later than January 1, 199_[97] all its Arctic ports and terminals are provided with reception facilities for garbage, without causing undue delay to ships.[98]

Article 8 also provides that following the date of January 1, 199_ the Organization shall make an examination of all the Arctic State reception facilities and notify the IMO for transmission to Contracting Parties of MARPOL 73/78 of all cases where the facilities are alleged to be inadequate.[99] The Arctic States shall ensure that such alleged inadequate facilities meet this Protocol's standards not more than 6 months following the report to the IMO.[100] After this date ships calling at Arctic ports where such facilities are not yet available, must fully comply with the requirements of Article 9.

Article 9 covers discharge standards in the Arctic special area. There will not be proposed here any tightening of the discharge standards in this Article under the Arctic Vessel-Source Protocol, but the possibility exists should the Arctic States agree to more stringent standards. As noted previously Canada's standards regarding ship discharges in its Arctic areas are stricter than those of MARPOL 73/78.[101]

Under MARPOL 73/78 in special areas the discharge of oil, oil mixtures, and noxious chemicals and garbage, is allowed upon meeting special conditions including vessel size and velocity and low effluent concentrations.[102] Briefly the discharge of oil or oil mixtures, including all oil drainage and sludge, dirty ballast and tank washing water, is prohibited except for ships other than oil tankers under 400 grt. For these, discharges of oil or oil mixtures is allowed if the content does not exceed 15 ppm or when the ship is proceeding en route, the oil content is less than 100 ppm and the discharge is made at least 12 nautical miles from land. For noxious chemicals the standards are more lenient, allowing discharge of *all* chemicals if the concentrations are low enough and the ship is travelling at least 7 knots, the discharge is made below the water line and the discharge is made more than 12 nautical miles from land and in a depth of water not less than 25 meters. For garbage the disposal of foodwastes[103] is allowed if the disposal is made more than 12 nautical miles from land.[104] These discharge 'loopholes', especially those involving noxious chemicals and oil, could be closed for the Arctic if the Arctic States were willing.[105]

Thus Article 9 states that the Arctic States shall ensure complete compliance through legislative measures with MARPOL 73/78 provisions for discharges in special areas set forth in Annexes I, II and V. The States shall harmonize their policies in this regard. In the event it is found though documented scientific research that the discharge standards for special areas set forth in MARPOL 73/78 Annexes are inconsistent with sound principles of Arctic marine environmental protection, than the Arctic States shall prohibit the discharge of all oil, oil mixtures, noxious chemicals, noxious chemical mixtures and garbage in the Arctic by 199_. The States shall harmonize their policies in this regard.

Article 10 deals with the development of reasonably priced and effective monitoring equipment for use on ships, which can register and give witness to illegal discharges under MARPOL 73/78.[106] It provides that the Arctic States shall contribute to the joint research and development of reasonably priced and effective discharge monitoring equipment for use on vessels. In accordance with the development of such equipment the Arctic States shall legislate provisions to implement the use of such equipment by ships of their registry. Additionally the Arctic States shall legislate provisions to allow the use of such monitoring equipment in enforcement procedures, by the flag States, the coastal States and the Arctic port authorities in accordance with LOSC provisions.[107] The States shall harmonize national policies for the Arctic in this regard.

Article 11 deals with port, coastal and flag State enforcement of certificates and oil record book and cargo record book regulations under MARPOL 73/78, Annexes I, II and IV in accordance with LOSC provisions.[108] As noted the 'Oil Pollution Prevention Certificate' is probably the most important element in detecting MARPOL 73/78 violations of construction, design, equipment and manning standards, which are of special importance in the Arctic. The same could probably be said for the 'Noxious Liquid Substance Prevention Certificate' and the 'International Sewage Pollution Prevention Certificate' as well.[109]

Thus this Article states that the Arctic States shall legislate and establish measures in accordance with LOSC provisions[110] to ensure the implementation of MARPOL 73/78 survey and inspection requirements by flag States regarding a ship's 'Oil Pollution Prevention Certificate', the 'Noxious Liquid Substance Prevention Certificate', the 'International Sewage Pollution Prevention certificate', the oil record book and the cargo record book. Additionally the Arctic States shall legislate and establish measures to ensure the implementation of MARPOL 73/78 ad hoc spot survey and inspection requirements by coastal and port States of said Certificates and oil and cargo record books. Flag, coastal and port States shall ensure that the quality and reliability of the inspections made and the respective certificates issued by the ship classification societies under their jurisdiction are of the highest quality.[111] The States shall harmonize national policies for the Arctic in this regard.

The Arctic States shall legislate and establish measures that prevent and control the fraudulent entry of discharges in oil and cargo record books. States shall undertake to establish systems encompassing both the oil and cargo record books and the oil and noxious chemicals discharge monitoring devices under Article 10 following the availability of such devices.[112] The States shall harmonize national policies for the Arctic in this regard.

Article 12 deals with surveillance of illegal discharges.[113] This is an especially difficult problem in the Arctic due to the large ocean spaces involved, polar darkness, currents and weather conditions. This Article consequently states that the Arctic States shall jointly develop and establish effective surveillance systems for detecting illegal vessel-source discharges. The systems shall use the best available technology to trace as much as is technically possible the discharged oil, chemical, containers, sewage or garbage discharges to those vessels responsible. The States shall harmonize national policies for the Arctic in this regard.

In addition the Arctic States shall establish measures in their respective legal systems to facilitate the use of circumstantial evidence gathered in accordance with this Article in bringing legal proceedings against vessels allegedly responsible for illegal discharges within the State's respective jurisdictions.[114] The States shall cooperate in the detection of violations and the enforcement of the provisions[115] and shall harmonize national policies for the Arctic in this regard.

Article 13 deals with the difficulty in arresting a moving ship unwilling to comply with coastal State demands. This Article states that the Arctic States shall establish legislation, measures and procedures which facilitate the inspection and arrest of ships alleged to have engaged in illegal discharges, including those not stopping and complying with coastal State requests and demands made in accordance with LOSC provisions.[116]

Article 14 attempts to deal with upgrading the technical standard of oil and chemical tankers in the Arctic in ice-covered areas.[117] This Article states that the Arctic States jointly shall implement legislation and measures to require that by the year 199_ all oil and chemical tankers used in the Arctic shall be constructed with double sides and double bottoms.[118] The Arctic States shall harmonize national policies for the Arctic in this regard.

Article 15 attempts to deal with the problem of flags of convenience ships.[119] This Article states that Arctic port State authorities with knowledge of ship equipment or operations deficiencies shall refuse clearance for port entry until the deficiencies are corrected or take other measures ensuring correction of these deficiencies. The Arctic States shall implement legislation and measures requiring vessels operating in the Arctic ice-covered areas to be restricted to those 15 years and younger, unless there is 50% national ownership.[120] The Arctic States shall harmonize national policies for the Arctic in this regard.

Article 16 deals with ratification of the Arctic Convention and Arctic Protocols, which is meant to provide a link to these agreements. The Article states that the Arctic States agree to ratify as soon as possible, if not already realized, the Arctic Convention with Arctic Protocols, especially the Protocol for Counter-Pollution Operations in the Arctic Region, and to enter into a bilateral contingency plans with their Arctic neighbours in areas characterized by moderate to heavy shipping activities. The link is left somewhat loose by the phrase 'agree to ratify as soon as possible' rather than 'shall ratify' in order to leave the States some flexibility under the Convention and Protocol 'umbrella' system discussed in Part II Chapter 11.

Article 17[121] deals with liability for vessel-source pollution and calls on all Arctic States to ratify as soon as possible, if not already realized, the International Convention on Civil Liability for Oil Pollution Damage (1969), the International Convention on the Establishment of an International Fund for Compensation for Oil Pollution Damage (1971), the Protocol of 1984 to Amend the International Convention on Civil Liability for Oil Pollution Damage 1969,[122] and the Protocol of 1984 to Amend the International Convention on the Establishment of an International Fund for Compensation for Oil Pollution Damage.[123]

Additionally under Article 17 the Arctic States shall directly and with the assistance of the competent international organizations draft a Convention on Liability and Compensation in Connection with the Carriage of Noxious and Hazardous Substances by Sea for use in the Arctic and ratify this document as quickly as possible.[124]

Also under Article 17 States must cooperate directly and with the assistance of competent regional or global international organizations to formulate and implement programs and procedures for the determination of liability and compensation for damage resulting from pollution from vessels in accordance with the Conventions and Protocols listed in this Article. These programs and procedures must specifically ensure that those requirements involved in determining liability and compensation for pollution damage, including vessel identification, proof of causality, proof of fault, jurisdiction, enforcement and compensation can be determined as rapidly and adequately as possible.[125]

Article 18 deals with Arctic weather forecasting and provides that the Arctic States shall on a regional basis and as soon as possible establish joint programs and exchange data in the fields of Arctic weather and ice forecasting.

Article 19 deals with Arctic research regarding ships. This Article provides that the Arctic States shall on a regional basis and as soon as possible establish joint research programs and exchange data in the fields of ship design, construction and testing in the Arctic and the effects of vessel-source pollution on Arctic marine ecosystems.

Article 20 deals with personnel training in connection with Arctic shipping. This Article provides that the Arctic States shall as soon as possible if not realized already ratify the International Convention on the Standards of Training Certification, and Watchkeeping for Seafarers[126] (STCWS) (1978). In addition the Arctic States shall establish on a regional basis joint training and proficiency testing of Arctic ship personnel, including navigation and position determination in ice, ship manoeuvring and handling in ice or ice accumulation, and establishing adequate and safe watchkeeping arrangements under special operating conditions including ice. The States shall harmonize national policies in this regard.

Article 21 deals with the International Convention for the Safety of Life at Sea (1974) (SOLAS) and its 1978 Protocol,[127] including standards covering the intact stability of ships, fire safety requirements, life-saving appliances, main propulsion machinery, periodically unattended machinery spaces, nuclear ships, steering systems, radar systems and safety measures for ships including chemical and gas tankers. This Article states that the Arctic States shall as soon as possible, if not already realized, ratify the 1974 SOLAS Convention and its 1978 Protocol and legislate and take other measures to give these instruments full effect. The States shall harmonize national policies for the Arctic in this regard.

Article 22 deals with joint use of satellite systems of navigation, communication and search and rescue in the Arctic. This Article states that the Arctic States shall as soon as possible, if not already realized, ratify, legislate and take other measures, including joint coordination, to implement the satellite communications system INMARSAT, the satellite navigation systems TRANSIT and NAVISTAR, and the satellite search and rescue system

COSPAS - SARSAT, the universal coverage communications systems LORAN-C - TSJAJKA and OMEGA; the position fixing systems very low frequency (VLF)OMEGA system; and the ice and ocean condition surveillance and monitoring Landsat and NOAA National Environmental Satellite Services, giving these systems full effect. The States shall harmonize national polices for the Arctic in this regard.[128]

Article 23 deals with intervention in cases of marine casualties on the high seas and provides that the Arctic States shall as soon as possible, if not already realized, ratify, legislate and take other measures to implement the Intervention Convention and Protocol,[129] including the right to take measures on the high seas as may be necessary to prevent, mitigate or eliminate grave danger to their coastline or related interests from threats of or oil pollution itself. The States shall harmonize national policies for the Arctic in this regard.

Article 24 deals with overloading of vessels, which can result in destabilizing the ship as well as increasing the danger of running aground. This Article provides that the Arctic States shall as soon as possible, if not already realized, ratify, legislate and take other measures to implement the International Convention on Load Lines and its Protocol.[130] The States shall harmonize national policies for the Arctic in this regard.

Article 25 deals with the prevention of ship collisions and ship traffic separation zones and in the future could be of use in both the Arctic Northwest and the Northeast Passages. This Article provides that the Arctic States shall as soon as possible, if not already realized, ratify, legislate and take other measures to implement the Convention on the International Regulations for Preventing Collisions at Sea and its amendments.[131] The States shall harmonize national policies for the Arctic in this regard.

Article 26 deals with ship containers carrying dangerous substances. This Article provides that the Arctic States shall as soon as possible, if not already realized, ratify, legislate and take other measures to implement the Convention on the International Convention for Safe Containers.[132] The States shall harmonize national policies for the Arctic in this regard.

Article 27 deals with the military sovereign immunity exception to anti pollution regulations and reiterates the MARPOL 73/78 Article 3(3) and the LOSC Article 236 provisions somewhat limiting this exception.[133] This is done for *emphasis*, because of the high number of military vessels present in the Arctic. Thus, States shall ensure by the adoption of appropriate measures not impairing the operations or operational capabilities of military or non commercial ships owned or operated by it, that such ships act in a manner consistent, so far as is reasonable and practicable, with the present Protocol.

Article 28 deals with dispute settlement and makes a general statement that in the event of dispute between the Parties and the dispute cannot be settled by the Parties themselves, then the dispute must be settled according to the provisions outlined in Article 21 of the Arctic Convention following LOSC provisions.[134]

Article 29 deals with the establishment and duties of the Organization. Similar to the other Protocols a central organization is proposed with sub organizations dealing with the different Protocols. Thus Article 29 provides that the Arctic States must designate a competent Organization to be responsible for Secretariat duties in relation to this Protocol in accordance with Articles 2 and 13 of the Arctic Convention. The duties of the Organization include but are not limited to convening consultative meetings of the Arctic States supplying information to the public regarding the functioning of the Arctic Vessel-Source Protocol in practice,[135] and preparing and assisting in the development and implementation of procedures dealing with emergency situations, and consideration of any additional actions required as necessary. The Parties must establish procedural rules as necessary.

Article 30 deals with amendments to the Protocol and amendments to Annexes. This Article states that any amendments to the present Protocol must be carried out consistent with and by the procedures outlined in Article 16 of the Arctic Convention. Any amendments to any Annexes must be carried out consistent with and by the procedures outlined in Article 17 of the Arctic Convention.

Article 31 deals with Arctic State accession and requires States which may accede to the Arctic Vessel-Source Protocol to do so in accordance with the provisions outlined in Article 24 of the Arctic Convention.

General Articles which deal with among other things the place and date of signature, where the instrument may be deposited upon ratification, when the Protocol comes into force, methods of withdrawal, and authenticity of language are basically standard and will not be covered here.[136]

Notes - Chapter 14

1 Protocol on Protection of the Arctic Region from Vessel-Source Pollution.

2 See Part II Sections 9.2.3. and 9.3.2.

3 See Part II Sections 8.2.1. and 10.2.1.

4 See Part I Section 5.2.2.

5 See Part I Section 5.3.1.

6 See Part I Section 5.3.2.

7 See Part I Section 5.3.3.

8 See Part I Section 5.3.4.

9 See Part I Section 5.4.2.

10 See Part I Section 5.4.3.

11 See Part I Section 5.4.4.

12 See Part I Sections 2, 5.2., 5.3. and 5.4. for IMO and ILO Conventions, and customary law lending additional legal support to the formation of this Protocol.

13 Churchill p. 260 and Curtis p. 709.

14 Churchill p. 256 notes that it is generally recognized that the IMO is the 'competent international organization'. Gold p. 37 notes that through LOSC's clear designation of the IMO as the 'competent international organization' that IMO treaties are thus further legitimized.

15 Article 211 (1).

16 Article 211 (2). See Part II Sections 11.2. and 11.3. for a short discussion of the interplay between national legislation dealing with vessel-source pollution and international regulations

17 Article 211 (4) (5) (6). Territorial sea is defined in Article 3 as not exceeding *12 nautical miles*, measured from baselines, defined in Article 5 as being the low-water line.

18 Article 211 (4). Innocent passage by definition in Article 19 does not include 'any act of wilful and serious pollution contrary to LOSC provisions,' as well as any passage prejudicial to the peace, good order or security of the coastal State, including any threat of use of force against the sovereignty of the State; any exercise or practice with weapons; any act aimed at collecting information to the prejudice of the State; any act of propaganda aimed at affecting the defence or security of the State; the launching, landing or taking on board of any aircraft; the launching, landing or taking on board of any military device; the loading or unloading of any commodity, currency or person contrary to the customs, fiscal, immigration or sanitary laws of the State; any fishing activities; the carrying out of research or survey activities; any act aimed at interfering with any systems of communication or any other facilities or installations of the State; and any other activity not having a direct bearing on passage.

19 Article 21 (1)(f).

20 Article 21 (2) and (3).

21 Articles 24 and 211 (4). Regarding laws and regulations covering straits in the territorial sea, Article 42 is relevant where similar requirements of non discrimination, non hampering and publication appear. The only anti pollution legislation possible is that 'giving effect to applicable international regulations regarding the discharge of oil, oily wastes and other noxious substances in the strait, Article 42 (b).

22 Article 211 (5).

23 Article 211 (6). The conditions include that the usual international rules and standards must be ecologically inadequate for such an area and the area must be clearly defined and within the state's respective exclusive economic zone. The other concerned States must be consulted through the IMO and the designation accepted by the IMO. 15 months notice of the entry into force of these laws and regulations must be made, and they may be made in accordance with international rules and standards made applicable for *special areas* through the IMO or may be additional laws and regulations made by the coastal State on its own.

24 Article 211 (6)(c). Article 211 (7) notes that the international rules and standards referred to in Article 211 include *inter alia* prompt notification to affected coastal States or discharges or probable discharges.

25 See Section 14.1.2.2.

26 Article 211 (3).

27 Wang pp. 309 and 311.

28 *Ibid.* p. 311 notes that other players in the enforcement interaction between the coastal State, the port State and the flag State include the oil, shipping and insurance industries (oil, where it is the polluting cargo) and national or international environmental lobbyists, whose influence is 'not negligible'.

29 Article 217 (1). The language used is States 'shall ensure', 'shall adopt', 'shall take' and 'shall provide', indicating an obligation to take positive action. The flag State enforces rules and standards on the high seas.

30 Article 217 (2).

31 Article 217 (3).

32 Article 217 (4). This must be without prejudice to regulations covering enforcement by port States, coastal States or the regulations concerning the suspension and restrictions on institution of proceedings.

33 Article 217 (5).

34 Article 217 (6).

35 Article 217 (7).

36 Article 217 (8).

37 Article 219. Such action can be taken upon request or upon the coastal State's initiative. The vessel may be permitted to proceed to the nearest repair yard and must be permitted to sail immediately upon removal of the causes of the violation.

38 'Unseaworthy' under Article 219 means, 'in violation of applicable international rules and standards relating to seaworthiness of vessels and thereby threatens damage to the marine environment.'

39 Article 219.

40 Article 220 (1). Churchill p. 259 notes that 220 (1) follows customary international law with the addition of covering enforcement powers in the exclusive economic zone.

41 Article 220 (2). Emphasis added. 'Illegal' in Article 220 (2) means 'violated laws and regulations of that State adopted in accordance with this Convention or applicable *international rules and standards* for the prevention, reduction and control of pollution from vessels', (emphasis added).

42 Article 220 (3). 'Illegal' in Article 220 (3) means 'a violation of applicable international rules and standards for the prevention, reduction and control of pollution from vessels or laws and regulations of that State conforming and giving effect to such rules and standards'. Flag States under Article 220 (4) must adopt laws and regulations and take other measures to ensure that their flag ships comply with such requests for information.

43 Article 220 (5). Emphasis added. 'Illegal' in Article 220 (5) has the same definition as Article 220 (3).

44 Article 220 (6). Emphasis added. 'Illegal' in Article 220 (6) has the same definition as Article 220 (3).

45 Article 220 (7).

46 This is excepting Article 220 (1) regarding a vessel voluntarily at a port or offshore terminal; and Article 220 (2) a vessel navigating in the territorial sea.

47 Article 220 (8).

48 Article 220 (1) and Article 218 (2).

49 Article 218 (1). Emphasis added. 'Illegal' is defined in Article 218 (1) as 'in violation of applicable international rules and standards established through the IMO or general diplomatic conference'. This includes the high seas area where there have been no claims of preemption under Articles 218 (4) or 228.

50 Article 218 (2). Emphasis added. 'Illegal' means the same as above in Article 218 (1).

51 Article 218 (3). 'Illegal' has the same definition as in Article 218 (1).

52 Article 218 (4). In such a case any evidence gathered and any financial security must be transmitted to the coastal State, and such transferal precludes the continuation of the port State proceedings. Port State proceedings could proceed in the event the coastal State does not request a suspension. In any event no proceedings could be instituted by the port State regarding a violation in another State's internal waters, territorial sea or exclusive economic zone unless meeting the conditions noted above under Article 218 (2).

53 Article 221.

54 Wang p. 329 adds that enforcement measures are subject to the basic values of customary international law such as 'reasonableness' and 'proportionality'.

55 Article 226 establishes further conditions for investigations under Articles 218 and 220, including requirements that physical inspections of the ship following ship document inspection be made only upon clear grounds of non correspondence between the ship's condition and the stated condition of the documents; upon invalid documents and upon insufficient documental contents.

56 See Sections 14.1.2.1. and 14.1.2.2. for specific LOSC Articles discussed below.

57 Churchill pp. 255-256. The author as noted believes that OILPOL provisions are included but probably not MARPOL 73/78, especially including all 5 Annexes.

58 *Ibid.* Also, in the Article 211 (6) special areas a ship's right of navigation is probably not any different than its innocent passage right in the territorial sea. See also Wang p. 325.

59 See Part II Section 9.3.2. and Wang p. 329.

60 Some authors believe it can, due to the coastal State having no general right to arrest foreign ships for violation of LOSC regulations in its own exclusive economic zone which can limit flag State exclusive jurisdiction. See Churchill p. 257.

61 Emphasis added. See Curtis p. 709. Churchill pp. 259-260 counters this point however noting that this acts as a double jeopardy safeguard, and additionally under Article 229 there is no obligation to suspend proceedings where the pollution occurred in the territorial sea; where the pollution caused major damage to the coastal State; where the flag State has not instituted proceedings within six months of the coastal or port State taking action; and where the flag State has repeatedly disregarded its duty to enforce effectively violations of applicable international anti pollution rules and standards. Also where the flag State does not bring its legal proceedings to a conclusion, the port or coastal State may lift the suspension and continue its own proceedings.

62 Churchill p. 258 notes however as previously that the wording appearing in Articles 220 (2) and (3) 'applicable international rules and standards' may allow coastal States, if using a broad interpretation, to enforce conventions to which they and possibly the flag ship are not parties.

63 See Wang pp. 321-326.

64 *Ibid.*

65 *Ibid.*

66 *Ibid.*

67 *Ibid.*

68 *Ibid.*

69 *Ibid.* p. 327 notes that arbitrary enforcement by different coastal States could be clarified in the future through State practice and third party dispute settlement procedures.

70 See Churchill p. 258. See also Curtis p. 709 where it is noted LOSC has thus allowed port and coastal States to initiate proceedings against ships from flag States which are hesitant to prosecute for illegal discharges. Just this point however has kept many countries from ratifying or signing LOSC since it bites too deeply into traditional shipping sovereignty, and it thus can be questioned how much international interest there really is in increasing port and coastal State jurisdiction in favour of marine environmental protection. This is especially interesting when seen in light of the legislative histories governing pollution in the exclusive economic zones noted in Section 14.1.2.1.

71 See Wang pp. 327-328.

72 *Ibid.*

73 'Illegal' is defined in Article 218 (1) as 'in violation of applicable international rules and standards established through the IMO or general diplomatic conference.'

74 This power includes the high seas where there have been no claims for preemption under Articles 218 (4) or 228.

75 See Wang p. 329.

76 Emphasis added.

77 See Part I Section 2.1.4. and Part II Section 10.3.3.

78 See Part I Section 5.2.3.1. for a brief history of State opposition to MARPOL 73 Annex II dealing with pollution by noxious chemicals.

79 In this Article as well as those following, if necessary, the mandatory 'shall' could be substituted for 'agree to' where too controversial for State ratification. Also as is noted in Section 14.1.2.1. coastal States are *unrestricted* in enforcing its anti pollution rules for ice covered areas in its exclusive economic zone since the usual restrictions including LOSC Section 7 safeguards do not appear in Article 234.

80 This sentence was added to the basic Article 234 requirement.

81 This provision is added due to the problems noted in Part II Section 9.3.2. related to lack of specific State anti pollution legislation in the exclusive economic zone.

82 This was added in order to attempt to tighten up lax State legislative and enforcement practices in their exclusive economic zones or port State reluctance to prosecute ships which have illegally discharged far from the State's waters. As noted in Part II Section 9.2.3.1. and Butler Sections F.2., F.3. and J.2., Supplement to J.4., of the Arctic States it appears that the Soviet Union and Canada conduct the most stringent enforcement of their anti pollution regulations in its exclusive economic zone.

83 See Part II Section 9.3.2. and Wang pp. 334-335.

84 This measure was added to require States to ensure MARPOL 73/78 technical standards be met. See Part II Section 9.3.2.

85 See Part I Section 5.2.3. Canada has not ratified. Denmark reserves its accession with respect to Greenland. Norway and the US lack ratification of Annex IV.

86 See Part I Section 5.2.3.1. and Part II Section 9.3.2. A weaker phrase could be used such as 'shall endeavour to amend' depending upon the possibility of amendment under the IMO system and the Arctic State's receptivity to the more stringent provisions governing a special area. As noted a special area under MARPOL 73/78 Annex I Regulation 1 (10) and Annex II Regulation 1(7) is defined as an area 'where for recognized technical reasons in relation to its oceanographical and ecological condition and to the particular character of its traffic the adoption of special mandatory methods for the prevention of sea pollution by oil (and noxious liquid substances) is required.'

87 Birnie p. 230 notes that no such area has been designated for the North Sea, but once LOSC comes into force those North Sea States which are Parties to the Convention plus the European Community may consider making use of this provision.

88 This alternative might allow the Arctic States to pass even more stringent legislation modeled upon Canada's anti pollution legislation concerning vessel-source pollution. See however Section 14.1.2.3. for comments regarding LOSC Article 211 (6) special areas.

89 This Article is taken from MARPOL 73/78 Annex I Regulation 10(7) 'Reception facilities within special areas' and is reproduced in the Arctic Vessel-Source Protocol to emphasize and require specific State compliance by a certain date to provide adequate reception facilities for oily water ballast and tank washings from oil tankers. As noted previously State compliance with requirements under MARPOL 73/78 has been lacking. See Birnie pp. 230 and 234. In the event the Arctic is not designated a special area the usual MARPOL 73/78 Annex Regulation 12 reception facilities provisions could be substituted.

90 This date could be negotiated.

91 'Organization' is meant to signify an Organization established under the Arctic Convention Articles 2 and 13 with sub organizations formed for the administration of each Arctic Protocol. See Part II Section 11.4.2.

92 The last 2 sentences were added. The 6 months could be negotiated.

93 The MARPOL 73/78 Annex II Regulation 7 provision states, 'undertakes to ensure'.

94 *Ibid.* provides each State shall notify the IMO of any cases where facilities are alleged to be inadequate.

95 The last 2 sentences were added. The 6 months could be negotiated.

96 MARPOL 73/78 Annex V Regulation 5(4) states, 'undertakes to ensure that as soon as possible'.

97 This date could be negotiated. The provision is included in order to place a mandatory duty upon the Arctic States to provide adequate reception facilities for garbage.

98 MARPOL 73/78 Annex V Regulation 5 (4) and Regulation 7 (1) add, 'taking into account the special needs of ships operating in these areas', which is deleted here to make the provision more stringent. Article 8 as with Article 7 could be modeled upon the more stringent MARPOL 73/78 Annex I Regulation 10(7) similar to Article 6 above if the Arctic States were willing.

99 This provision is added taken from MARPOL 73/78 Annex I Regulation 7. Annex V Regulation 5 (4) requires only that the States notify the IMO of measures taken regarding reception facilities for garbage and the IMO will set a date for compliance. Regulation 7 (2) requires only that each Party must notify the IMO of all cases where the facilities provided for garbage are inadequate.

100 This sentence was added. The 6 months could be negotiated.

101 See Part II Section 9.3.2.

102 See Part I Section 5.2.3.1. See MARPOL 73/78 Annex I Regulation 10 (3), Annex II Regulation 5 (7)(8)(9) and Annex V Regulation 5 (2).

103 The disposal of garbage not including plastics, paper products, rags, glass, metal, bottles, crockery, dunnage, lining and packing materials is possibly allowed. See MARPOL 73/78 Annex V Regulation 5 (2).

104 See MARPOL 73/78 Annex I Regulation 10 (3), Annex II Regulation 5 (7)(8)(9) and Annex V Regulation 5 (2).

105 Roginko, Alexei, 'Environmental Conservation in the Soviet Arctic Seas,' Chapter 7, to be published in the *Soviet Maritime Arctic*, (1989), (Roginko Paper) p. 14.

106 As noted in Part II Section 9.3.2. though MARPOL 73/78's strongest enforcement is through certification of construction standards, reliable monitoring if possible would be of great help in combatting vessel-source pollution.

107 See *ibid.* where it is noted that both flag State and port State enforcement is needed of ships for discharge violations as well as violations related to construction, design, equipment and manning standards. This Article attempts to give assistance in this direction.

108 *Ibid.* and Wang p. 328 where it is noted that in order to understand coastal State enforcement competence in its own ports, LOSC must be read together with relevant MARPOL 73/78 provisions.

109 There are no general certificates required for Annexes III and V.

110 By emphasis upon the LOSC provisions plus the harmonious development of State policies in this regard it is hoped this would allow some joint 'pressure' to increase State enforcement practice in the Arctic.

111 See Part II Section 9.3.2. where it is noted that the quality and reliability of the ship classification societies has varied, with shipping interests sometimes running the inspections.

112 An additional provision could provide the States shall ensure that to the fullest extent possible the control measures and technology devised for MARPOL 73/78 Annex I (oil pollution) shall be transferred into use where relevant for Annex II (noxious substances pollution), Annex III (containers), Annex IV (sewage) and Annex V (garbage).

113 See Part II Sections 9.3.2. and 14.1.2.3.

114 As noted in Section 14.1.2.3. alleged violations must be supported by 'clear grounds' or 'clear objective evidence' under Article 220 (3) (5) and (6), the burden of which the claimant must carry. If circumstantial evidence were of a high enough technological quality and more acceptable in terms of State evidence law, this difficulty could be alleviated, allowing more aggressive coastal and port State enforcement measures.

115 This clause is taken from MARPOL 73/78 Article 6 for emphasis.

116 As seen in Section 14.1.2.3. a problem in the exercise of coastal State jurisdiction is arresting moving ships while at the same time respecting 'innocent passage' provisions.

117 This Article and those following addressing 'ice-covered areas' could be dropped if it were shown that design, construction and equipment standards mentioned were consistent with IMO environmental standards. See *ibid.*

118 The date could be extended. This construction is already being introduced into shipping. See Aftenposten 9 mars 1990, Uke 10., Nr.115., 131 årg., s. 26. where it is noted that 'environmental ships' with double bottoms and double sides are being ordered by Norwegian shipping interests. Additionally the US Congress recently passed

legislation with more stringent environmental requirements for ships, something with which the new ships should comply. See OPA.

119 See Part II Section 9.3.2.

120 The age could be negotiated even lower. See *ibid.*

121 This Article also appears in the Arctic Liability Protocol, Part II Section 17.2.2.

122 See Part I Section 6.4.1. As noted Canada and the US have not ratified the CLC, and none of the Arctic States has ratified the Protocol.

123 See Part I Section 6.4.2. As noted Canada and the US have not ratified the Fund Convention, and none of the Arctic States has ratified this Protocol.

124 See Part I Section 6.4.6.2. regarding the HNS Draft Convention. The IMO and many States are interested in establishing this Convention despite widespread disagreement, and therefore the HNS Draft Convention is emphasized here. This clause also appears in the Arctic Liability Protocol and could be deleted failing State ratification.

125 This Article could be deleted failing State ratification.

126 See Part I Section 5.3.4. All the Arctic States have ratified, etc. with the US recently becoming a State Party.

127 See Part I Section 5.3.2. All of the Arctic States have ratified, etc. the Convention, and all have ratified, etc. the Protocol with the exception of Canada.

128 See Part I Section 5.3.3. for further information and also noting a recent agreement standardizing the US 'LORAN C' and the Soviet 'TSJAJKA' radio navigation systems with the land based stations connected in the Arctic Pacific.

129 See Part I Section 5.3.1. As noted all the Arctic States but Canada have ratified, etc. both the Convention and its Protocol.

130 See Part I Section 5.4.2. All the Arctic States have ratified, etc. the Convention.

131 See Part I Section 5.4.3. All the Arctic States have ratified, etc. the Convention.

132 See Part I Section 5.4.4. All the Arctic States have ratified, etc. the Convention.

133 See Part I Section 5.2.3.1. and Section 14.1.2.2.

134 As noted in Part II Section 11.4.2. Article 21 of the Arctic Convention contains provisions incorporating LOSC dispute settlement provisions, Articles 279-299 and Annex V, which are meant to apply to all the Arctic Protocols.

135 This provision is deemed necessary due to the informational problems noted in Part II Section 9.3.2.

136 See for example Articles 18, 19 and 20 of MARPOL 73 and Articles IV, V and VII, VIII and IX of MARPOL 78.

15. Discussion of Arctic Dumping Protocol[1]

15.1. Background

15.1.1. Summary of Existing Dumping Regimes with Application in the Arctic

The LDC and the ODC cover dumping in the Arctic, the former globally and the latter regionally in the Barents Sea, sections of the Greenland Sea, the Arctic Ocean and the Norwegian Sea.[2] However as seen there exist problems with these regimes related to insufficient definitional coverage, lack of mandatory obligations, environmentally conservative non Arctic Parties and a want of public information.[3] Because of these deficiencies and the possible Arctic dumping activities already being conducted,[4] the Arctic Dumping Protocol is proposed.[5]

15.1.2. Legal Basis for Arctic Dumping Protocol under LOSC[6]

Similar to other Protocols already mentioned, the general obligations set forth by Articles 192, 197 and 194 apply, especially the latter under which States are required to jointly take all measures to minimize fully the release of toxic, harmful or noxious substances, especially persistent from *dumping*.[7]

As noted under Article 210 States, acting especially through international organizations or diplomatic conference, shall *endeavour* to establish global and *regional* rules, standards and recommended practices and procedures to prevent, reduce and control dumping pollution. Such provisions must be examined as necessary.[8] Additionally States *shall* adopt national laws and regulations to prevent, reduce and control marine pollution from dumping and shall take other measures necessary to prevent, reduce and control such pollution.[9] Such national laws, regulations and measures must ensure that dumping is not carried out without the permission of the State authorities, and the national laws, regulations and measures must be *no less effective* in preventing, reducing and controlling dumping than the global rules and standards.[10] Dumping in the territorial sea, in the exclusive economic zone or onto the continental shelf must not be carried out without the express prior approval of the coastal State, and this State has the right to permit, regulate and control such dumping after due consideration of this matter with other States, which may be adversely affected.[11]

Under Article 216 States, depending upon the jurisdiction exercised in accordance with (a), (b) and (c), shall enforce laws and regulations relating to dumping adopted in accordance with the LOSC, as well as applicable international rules and standards established through international organizations or diplomatic conference,

(a) the coastal State shall enforce with regard to dumping in its territorial sea, in its exclusive economic zone or onto its continental shelf;

(b) the flag State shall enforce with regard to its flag ships or ships or aircraft under its registry;

(c) any State shall enforce with regard to acts of loading of wastes or other matter occurring within its territory or at its off shore terminals.[12]

No proceedings are required when another State has already brought proceedings in accordance with this Article.[13]

The result of this is that a coastal State can stop a ship that has violated a dumping regulation in its exclusive economic zone and if necessary detain it for bringing legal proceedings against it at a coastal port. The same would apply to ships sailing in the exclusive economic zone following unlawful dumping in the territorial sea or in some cases even on the high seas if waste settles onto the continental shelf.[14]

Thus to the extent these LOSC provisions can be said to be international customary law and recommendations for norms, the Arctic States have an obligation to *endeavour* to establish and thereafter to enforce an Arctic regional agreement to prevent, reduce and control marine pollution from dumping, which would act as legal support to the formation of the Arctic Dumping Protocol.[15]

15.1.3. Use of LDC as Pattern for Arctic Dumping Convention with Inputs - Model of Helsinki Convention as Alternative

It is proposed to use the LDC with additions as a basis for forming an Arctic Dumping Protocol. Although the ODC could also have been chosen, the LDC was considered because it appears formulated more extensively.[16] Since many authors consider the LDC and the ODC to be similar, it is probably not of importance which is used as a model as long as the additions noted below are made.

Importantly, an alternative for the Arctic Dumping Protocol based upon the Helsinki Convention is also briefly presented in Section 15.2.4., since as noted it takes a much stricter approach than either the ODC and the LDC by prohibiting *all* dumping with the exception of dredged spoils.[17] This is given as an alternative model should the Arctic States agree that dumping should be more strictly regulated in the Arctic. The Helsinki Convention's dumping provisions were not chosen as model for the Arctic Dumping Protocol, since realistically it seems doubtful that all dumping will be prohibited in the Arctic, at least in the near future.[18] Thus the special and general permit system from the LDC and the ODC for the dumping of various wastes and other materials is worth viewing. The alternative posed by the Helsinki model can rather be thought of as goal towards which the Arctic Dumping Protocol may proceed.

15.2. Discussion of Proposed Arctic Dumping Protocol

15.2.1. Introduction

The discussion of this Protocol encompasses 14 Articles and 3 Annexes. Formal Articles are *not* presented, but rather an idea of the desired content of each Article along with a brief discussion.

15.2.2. Articles

The *Preamble* begins by giving general proclamations by the Arctic States recognizing the importance of the Arctic marine environment and its importance to humanity, the limit of its capacity to assimilate wastes and regenerate natural resources, and the responsibility of the Arctic States to ensure that activities under their jurisdiction do not damage the environments of other States or areas outside of State jurisdiction. Additionally the States recall Resolution 2749 (XXV) of the U.N.'s General Assembly on the principles governing the sea bed and the subsoil beyond the limits of national jurisdiction. The Arctic States note that it is important that they use the best practicable means to prevent pollution from dumping and develop products and processes which will reduce the amount of harmful wastes to be disposed of. The States are convinced that regional action in the Arctic is necessary to control marine pollution by dumping and must be taken without delay, but that this action should not preclude ratification as soon as possible of the other Arctic Protocols, which take measures to control other sources of marine pollution.[19] From the above declarations the State agree to the following Articles.

Under *Article 1* the Arctic States individually and collectively promote the effective control of all marine pollution through the prompt ratification of the Arctic Convention and the Arctic Protocols. They pledge themselves especially to take all steps to prevent marine pollution from the dumping of wastes and other matter hazardous to human health and harmful to living resources and marine life or to interfere with other legitimate uses of the sea.[20]

Article 2 provides that the Arctic States must take effective measures individually and collectively according to their best scientific and technical capabilities[21] to prevent marine pollution caused by dumping and must harmonize their policies in this respect.

Article 3 gives definitions used in the Protocol. This Article is very important for it is here as noted the LDC avoids covering dumping from offshore oil operations. Consequently under the Arctic Dumping Protocol the following definitions are proposed.

(a) 'Dumping' means:

(i) the disposal of wastes or other matter directly arising from, or related to the exploration, exploitation and associated offshore processing of seabed mineral resources;[22]

(ii) any deliberate disposal at sea of wastes or other matter from vessels, aircraft, platforms or other man made structures at sea;

(iii) any deliberate disposal at sea of vessels, aircraft, platforms or other man made structures at sea.

(b) 'Dumping' does not include:

(i) the disposal at sea of wastes or other matter incidental to, or derived from the normal operations of vessels, aircraft, platforms or other man made structures at sea and their equipment, other than wastes or other matter transported by or to vessels, aircraft, platforms or other man made structures at sea, operating for the purpose of disposal of such matter or derived from the treatment of such wastes or other matter on such vessels, aircraft platforms or structures;

(ii) placement of matter for a purpose other than the mere disposal therefor, provided that such placement is not contrary to the aims of this convention.

(c) Under no circumstances shall it be considered incidental to, or derived from normal operations to dispose at sea of wastes or other matter related to the exploration, exploitation and associated offshore processing of sea bed mineral resources.[23]

Additionally defined are 'vessels and aircraft' meaning waterborne or airborne craft of any type, including air cushioned craft and floating craft, whether self propelled or not; 'sea' as meaning all marine waters including the States' internal waters;[24] 'wastes or other matter' as meaning material and substance of any kind, form or description; 'special permit' as meaning permission granted specifically on advance application according to provisions of Annexes II and III; 'general permit' as meaning permission granted on advance application in accordance with Annex III; and finally, 'Organization' as meaning the organization designated by the Parties in accordance with Articles 2 and 13 of the Arctic Convention and Article 14 in this Protocol.

Article 4 sets the regulatory boundaries for the dumping of wastes. Under its provisions the Arctic States *prohibit* the dumping of any wastes or other matter in whatever form or condition except as permitted by special or general permit below. The dumping of wastes or other matter listed in Annex I is prohibited. The dumping of wastes or other matter listed in Annex II requires a prior special permit.[25] The dumping of all other wastes or matter requires a prior general permit. A permit can be issued only upon careful consideration of all the factors listed in Annex III. Any Party may for itself, without violating the Protocol, prohibit the dumping of wastes or other matter not listed in Annex I, and notify such measures to the Organization. The States shall harmonize national policies for the Arctic in this regard.

Under *Article 5* exceptions to the Article 4 dumping requirements are listed for special situations. These include, when necessary to secure the safety of human life or ships, aircraft, or offshore installations in cases of *force majeure* caused by weather, or in any case where there is a threat to the same. This is allowed if dumping appears to be the only way of averting the threat and if the damage from the dumping will be less than the damage that otherwise would occur. In these situations the dumping must be carried out so as to minimize damage to human or marine life and must be reported to the Organization.

Additionally under Article 5 a Party may issue a special permit as an exception to Article 4 prohibiting the dumping of Annex I substances if there is an unacceptable risk to human health and there is no other feasible solution. The Party must consult other States likely to be affected and the Organization. The Organization after consulting other Parties and appropriate international organizations must in accordance with Article 11 promptly recommend to the Party the most appropriate procedures to adopt. The Party must follow these recommendations to the maximum extent possible. The Party shall inform the Organization of the action taken, and the Parties pledge themselves to assist one another in such situations. Any Arctic State may waive these rights related to emergencies of Annex I substances at the time of or subsequent to ratification or accession to the Protocol.[26]

Article 6 deals with State authorities issuing special permits, required prior to and for the dumping of Annex II substances in accordance with Article 4; and general permits required prior to and for the dumping of all other substances. In addition, the authorities must keep public records[27] of dumpings allowed, including the nature, quantities, location, time and methods. The authorities must also monitor individually or jointly with other Parties and international organizations the marine environment with respect to this Protocol.[28] All records of dumpings allowed as well as monitoring of the marine environment must be kept on file with the Organization where they will be considered public information.

The State authorities must issue special or general permits for matter intended for dumping, loaded in its territory or loaded by a ship or aircraft registered in a State's territory or flying its flag, when the loading occurs in the territory of a State not a Party to this Protocol. In issuing these permits the State authorities must comply with the provisions of Annex III together with additional measures and requirements considered relevant by the individual States. Each Party must report directly to the Organization[29] and where appropriate other Parties the information in the records, the monitoring and the permit requirements adopted specified above. The States shall harmonize national policies for the Arctic in this regard.

Article 7 deals with State jurisdiction over dumping activities and states that each Party must apply the provisions of this Protocol to all ships and aircraft registered in its territory or flying its flag; ships and aircraft loading in its territory, territorial seas or exclusive economic zone[30] matter which is to be dumped; and ships and aircraft and fixed or floating platforms under its jurisdiction believed to be dumping. Each Party must take in its territory, territorial seas and exclusive economic zone[31] measures to prevent and punish conduct which is in violation with this Protocol's provisions. The Parties must cooperate in developing procedures for the effective application of this Protocol particularly on the high seas, including procedures for reporting ships and aircraft observed dumping in violation of this Protocol.[32]

The Parties must cooperate within the Organization and the competent international organizations in developing procedures for the effective application of this Protocol and international law, regarding the disposal of abandoned or disused offshore platforms, installations, storage facilities, submarine pipelines and other offshore equipment, taking into account other legitimate uses of the sea, including but not limited to fishing, the protection of the marine environment, navigational safety and the rights and duties of other States.[33] The States shall harmonize national policies for the Arctic in this regard.

Under Article 7 this Protocol does not apply to ships and aircraft protected by *sovereign immunity* under international law.[34] Each Party however must ensure[35] that such ships and aircraft protected by sovereign immunity owned or operated by it, act in a manner consistent with the object and purpose of this Protocol and must inform the Organization accordingly. Finally, nothing in this

Protocol affects the right of each Party to adopt other measures, in accordance with the principles of international law, to prevent marine dumping.

Article 8[36] deals with State support regarding training, equipment and waste disposal. This would promote closer cooperate efforts among the Arctic States. The Arctic States must promote through collaboration within the Organization and other international bodies, support for those Parties which request it for the training of scientific and technical personnel; the supply of necessary equipment and facilities for research and monitoring; and the disposal and treatment of waste and other measures to prevent or mitigate pollution caused by dumping.

Article 9 makes a general statement regarding State responsibility and dispute settlement. It states that in accordance with the principles of international law regarding State responsibility for environmental damage caused by dumping, the Arctic States *undertake* to promote measures toward ratification of the Arctic Liability Protocol, which includes measures regarding liability for general dumping and dumping related to oil field activities.[37] In the event of dispute between the Parties and the dispute cannot be settled by the Parties themselves, then the dispute shall be settled according to the provisions outlined in Article 21 of the Arctic Convention following LOSC provisions. The States shall harmonize national policies for the Arctic in this regard.

Article 10 is meant as somewhat of a catchall, covering other types of pollution including radioactive. Under Article 10 the Parties pledge themselves to promote within the competent specialized agencies and other international bodies environmental measures against pollution by radioactive pollutants from all sources, including ships, and agents of chemical and biological warfare. The Parties pledge also to promote within the appropriate international organization the codification of signals to be used by vessels engaged in dumping.

Article 11 deals with the duties of the Organization. Similar to the other Protocols a central organization is proposed under the Arctic Convention Articles 2 and 13 with sub organizations dealing with the different Protocols. Any reference however is made to the central 'Organization'. Under Article 11 the Arctic States must designate a competent Organization to be responsible for secretariat duties in relation to this Protocol in accordance with Articles 2 and 13 of the Arctic Convention.

The Secretariat duties of the Organization include those consistent with Article 13 of the Arctic Convention including convening of consultative meetings of the Arctic States and of special meetings of the Parties; preparing and assisting, in consultation with the Parties and appropriate international organizations, in the development and implementation of procedures referred to in Article 5 dealing with emergency situations;[38] and consideration of any additional actions required. The Parties must establish procedural rules as necessary.

Article 12 deals with amendments to the Protocol. At meetings called in accordance with Article 11, amendments may be adopted in accordance with Article 16 of the Arctic Convention.

Under *Article 13* amendments to the Annexes are to be based upon scientific or technical considerations. Recommendations for amendments to the Annexes must be made in accordance with Article 17 of the Arctic Convention.

Article 14 deals with Arctic State accession and requires States which may accede to the Arctic Dumping Protocol to do so in accordance with the provisions outlined in Article 24 of the Arctic Convention.

General Articles which deal with among other things the place and date of signature, where instruments may be deposited upon ratification, acceptance or approval, when the Protocol comes into force, withdrawal from the Protocol, and authenticity of language are basically standard and will not be covered here.[39]

15.2.3. Annexes

Annex I as noted contains the 'black list' substances and are compiled from both the LDC and the ODC Annex I 'black lists'.[40] As seen in Part II Section 9.3.1. this hazardous chemical list will have to be *upgraded* and *updated* as more knowledge is obtained regarding the chemicals and their effects upon Arctic waters and life. Measures regarding the determination of new substance toxicity should be harmonized nationally by the Arctic States.

The present 'black list' substances include organohalogen compounds and compounds which may form such substances in the marine environment;[41] organosilicon compounds and compounds which may form such substances in the marine environment;[42] substances which have been agreed between the Arctic States as likely to be carcinogenic under the conditions of disposal;[43] mercury and mercury compounds; cadmium and cadmium compounds; persistent plastics and other persistent synthetic materials, such as netting and ropes;[44] crude oil, fuel oil, heavy diesel oil, and lubricating oils, hydraulic fluids, and any mixtures containing any of these;[45] high level radioactive wastes or other high level radioactive matter defined as unsuitable for dumping at sea;[46] and materials in whatever form produced for biological and chemical warfare.[47] Annex I applies to wastes or other material containing the matters referred to above as trace contaminants.[48] This list does *not* apply to substances which are rapidly rendered harmless by physical, chemical or biological processes in the sea provided they do not make edible marine organisms unpalatable, or endanger human or domestic animal health.[49]

Annex II contains the 'grey list' substances requiring special care and a special permit under Article 6. These are compiled from both the LDC and the ODC, Annex II 'grey lists'. As above this list must also be updated and upgraded as more knowledge is obtained regarding toxicity in the Arctic, if not moving Annex II 'grey list' substances to Annex I's 'black list', allowing only general permit substances to be dumped in the Arctic. Measures regarding the determination of new substance toxicity should be harmonized nationally by the Arctic States.

The 'grey list' substances include,

1) Wastes containing significant amounts of arsenic, lead, copper, zinc (and compounds of theses substances), cyanides, fluorides, pesticides and their by products not covered in Annex I.[50]
2) In the issue of permits for the dumping of large quantities of acids and alkalis, consideration must be given to the possible presence in such wastes of the paragraph 1) substances and to beryllium, chromium, nickel and vanadium (and compounds of these substances);
3) Containers, scrap metal, tar like substances liable to sink to the sea bottom[51] and other bulky wastes liable to sink to the sea bed and present a serious obstacle to fishing or navigation;
4) Radioactive wastes or other radioactive matter not included in Annex I. In the issue of permits for the dumping of this matter, the Arctic States must take full account of the recommendations of the competent international body in this field, at present the International Atomic Energy Agency;
5) Substances which though of a non toxic nature may become harmful due to the quantities in which they are dumped or which are liable to seriously reduce amenities;
6) When in the application of the provisions of Annexes II and III, it is considered necessary to deposit waste in deep water, this should be done only when the depth is not less than 2000 meters and the distance from the nearest land is not less than 150 nautical miles.[52]

Finally *Annex III* lists the *criteria* to be considered in the issue of permits for dumping in accordance with Article 4. These are compiled from both the LDC and the ODC Annex III provisions. As above these must also be updated and upgraded as more knowledge is obtained regarding toxicity in the Arctic. Measures regarding the determination of new substance toxicity should be harmonized nationally by the Arctic States. The criteria include characteristics and composition of the matter,[53] characteristics of dumping site and method of deposit,[54] and general considerations and conditions.[55]

15.2.4. Alternative Arctic Dumping Protocol based upon Helsinki Convention (Alternative)

The Alternative will be only briefly discussed to give an idea of its contents, since the Helsinki Convention's Article 9 and Annexes I, II, IV, V and VI are discussed in Part I Section 3.2.2.

The provisions of the Alternative require Arctic States to prohibit dumping in the Arctic Area subject to exceptions noted. The exceptions include the dumping of dredged spoils which are subject to a prior special permit by the appropriate national authority in accordance with Annex V. Each Contracting Party must ensure compliance[56] with the provisions of this Article by vessels and aircraft registered in its territory or flying its flag; loading, within its territory or territorial sea or exclusive economic zone, matter which is to be dumped; or believed to be engaged in dumping within its territorial sea and exclusive economic zone.[57]

These provisions do not apply when the safety of human life or of a vessel or aircraft at sea is threatened in much the same circumstances as noted previously in Article 5 of the Arctic Dumping Protocol in Section 15.2.2. In cases of dumping which is suspected to be in violation of this Article, the Arctic States must cooperate in investigating the matter in accordance with Annex IV.

Annex V consists of 4 regulations. Regulation 1 includes exceptions to the prohibition of dumping and provides that prohibition of dumping does not apply to the disposal at sea of dredged spoils provided that they do not contain significant quantities and concentrations of substances to be defined by the Organization and listed in Annexes I and II; and provided that the dumping is carried out under a prior special permit issued by the proper national authority, either within the area of the exclusive economic zone of the Contracting Party or outside the area of the exclusive economic zone, whenever necessary, after prior consultations with the Organization.[58] Regulation 2 deals with the issuance of permits and is essentially the same as appears in Article 6 of the Arctic Dumping Protocol but is limited to permits for dumping dredge spoils. Regulation 3 deals with factors the national authorities must take into account when issuing Regulation 1 permits and include much the same as noted above in Annex III of the Arctic Dumping Protocol, but as related to only dredged spoils. Regulation 4 deals with information which must be included in the reports of dumpings required in accordance with the Article and includes much the same as noted above under the Arctic Dumping Protocol Annex III.

Annex I lists the hazardous substances DDT (1,1,1-trichloro-2,2-bis-(chlorophenyl)-ethane) and its derivatives DDE and DDD; and PCB's (polychlorinated biphenyls). *Annex II* lists the noxious substances including mercury, cadmium, and their compounds; antimony, arsenic, beryllium, chromium, copper, lead, molybdenum, nickel, selenium, tin, vanadium, zinc, and their compounds, as well as elemental phosphorus.[59] *Annex IV* provides for appropriate State assistance, taking into account immunity for warships or ships used on government non commercial service, in investigating violations of existing legislation on anti pollution measures which have occurred or are suspected to have occurred within the Arctic region.[60] *Annex VI* includes State cooperation in combatting marine pollution and covers State contingency plans and bilateral or multilateral agreements, surveillance, salvage and recovery.

Notes - Chapter 15

1 Protocol on Protection of the Arctic Region from Dumping of Wastes at Sea.

2 See Part I Section 3.2.1.

3 See Part II Section 9.3.1.

4 See Part II Section 10.2.1. and Sovjet Nytt, 21 juli 1989 Nr. 20 s. 6.

5 An alternative would be the amendment of the LDC to include dumping associated with offshore oil activities, but since this has not occurred since 1972, it is doubtful it is a viable alternative.

6 See Part I Sections 2 and 3.2.1. for conventions and customary law regulating State jurisdiction over dumping lending additional support to the formation of this Protocol.

7 Emphasis added.

8 Article 210 (4). Emphasis added.

9 Article 210 (1) and (2). Emphasis added.

10 Article 210 (3) and (6). Emphasis added. See Part II Sections 11.2. and 11.3. for a short discussion of the interplay between the national legislation dealing with dumping and international regulations. Churchill p. 272 notes that Article 210, by requiring States to adopt legislation no less effective than global rules and standards, thus imposes by reference obligations contained in other international dumping conventions on States which may not be parties to those conventions. This may have undesirable effects on State ratification of both LOSC and international dumping conventions, since States do not want to be bound to something to which they do not agree.

11 Article 210 (5). Hakapää p. 247 notes that though the meaning of the word 'consideration' is vague, providing information is not enough, and thus a coastal State in order to permit dumping must consult surrounding States.

12 Article 216 (1).

13 Article 216 (2).

14 Hakapää 248 and 279. It is noted however that it probably does not encompass all dumping on the high seas which are carried by waves and currents to end up on a State's continental shelf. Also noted is that coastal State jurisdiction does not preempt concurrent and good faith flag State jurisdiction, and a flag State could interfere with its vessels for dumping violations in a foreign State's exclusive economic zone. At the same time the concurrent jurisdiction does *not* allow flag State pre emption and limitation of penalties, as under vessel-source pollution, which 'is well justified in view of the particularly offensive character of dumping activities.' If a coastal State has started proceedings against a ship which had illegally dumped in its economic zone, the

flag State is no longer obliged to do so, but it would still have the necessary legal competence.

15 See Part I Section 2.1.4. and Part II Section 10.3.3.

16 Churchill pp. 271-272 notes otherwise stating, '(A)part from their geographical scope, the differences between the London Convention and these other instruments (including the ODC) are relatively minor. Thus, for example there are small variations in the contents of the black and grey lists, and the institutional arrangements differ in each case' (parentheses added). See also Hakapää p. 98.

17 See Part I Section 3.2.2.

18 The positive statements of the British Parliamentary Committee and the Norwegian Ministry of the Environment noted in Part II Section 9.3.1. relate only to lack of pollution damage. As confirmed by the newspaper articles, they do not indicate that dumpings have ceased.

19 Phrases were added to the Preamble concentrating on the Arctic region. The Preamble of the LDC additionally includes a declaration by which the States wish to improve protection of the marine environment through encouraging States with a common interest in particular geographical areas to enter into appropriate agreements supplementary to this Convention. This provision which is repeated as well in Article 8 of the LDC, would in fact give legal support for the Arctic rim States to enter a regional Arctic Dumping Protocol supplementary to the LDC.

20 This Article was made more stringent than the LDC, where States pledge to take all 'practicable' steps to prevent marine pollution. The first sentence establishes a link to the other Arctic Protocols.

21 The LDC uses 'take effective measures individually, according to their scientific, technical and economic capabilities, and collectively'. The Arctic States are all industrialized nations, and their best scientific and technical capabilities independent of cost both individually and collectively should be required for the best environmental protection in the Arctic.

22 Under the LDC a similar subsection appears under '(b), 'Dumping' does not include'. See Part I Section 3.2.1. and Part II Section 9.3.1. for discussion of these definitions and resulting problems.

23 This paragraph was added to prevent any vagueness arising from the phrase, 'incidental to, or derived form the normal operations' as applied to dumping occurring in offshore oil activities. This phrase as noted in Part II Section 9.3.1. appears both in the LDC Article III and the ODC Article 19, and under the latter allows a possible loophole for offshore oil field dumping.

24 See Part I Section 3.2. regarding possible inclusion of 'internal waters' within the geographical coverage of the ODC.

25 Note that an important option for the Arctic Dumping Protocol could be to list the Annex II 'grey list' substances under Annex I 'black list' substances. This option would move in the direction of the total dumping prohibition in the Baltic under the Helsinki Convention but with more flexibility, since the dumping of general permit substances would still be allowed. This is also mentioned under Annex II. See Part II Section 9.3.1.

26 This second paragraph of Article 5 might be deleted so as to totally prohibit the dumping of 'black list' substances by special permit as suggested in Part II Section 9.3.1.

27 The word 'public' as well as the last sentence of this paragraph were added to ensure that information concerning the dumping practice under the Arctic Dumping Protocol is more available than under the LDC, noted in Part II Section 9.3.1.

28 This last requirement could be included in a joint monitoring program of all the Arctic States covering all the pollution sources as mentioned in Part II Section 11.4.2. Article 10 of the Arctic Convention.

29 The LDC uses additionally 'or through a regional organization'. See discussion under Article 8 regarding 'regional agreement'.

30 The phrase 'exclusive economic zone' was added to be consistent with LOSC provisions.

31 The phrase, 'territorial seas and exclusive economic zone' was added.

32 The LDC uses, 'agree to cooperate in the development of procedures'.

33 This paragraph was added to cover recent developments in the IMO and the Paris-Oslo Commissions regarding this form for dumping. See Part I Section 3.2.1.3.

34 It was considered better to allow this exception to remain in spite of its anti environmental quality and in spite of the fact that dumping possibly exists from the Soviet military ships based on Kola Peninsula and from other Arctic State military vessels as well. This is because of the traditional importance of the sovereign immunity exception accorded in international law, and the fact that two of the Arctic States, the US and the Soviet Union are the world's naval powers. Any provision abolishing sovereign immunity would ensure non ratification of this Protocol. The weaker requirement for the military vessels appearing in the second sentence however is 'beefed up' from the LDC provision, based upon LOSC Article 236. If the Arctic States would agree, military dumping should be stopped.

35 The LDC also includes the phrase, 'by the adoption of appropriate measures' which was considered to weaken the requirement.

36 Under the LDC Article 8 deals with *regional* agreements, which is briefly mentioned here due to its relevance as legal support to the Arctic Dumping Protocol, especially since all the Arctic States are Parties. Article 8 states that to further the objectives of the LDC the Contracting Parties with common interests in a given geographical area *shall endeavour*, taking into account characteristic regional features, to enter into regional agreements consistent with the LDC for the prevent of marine pollution, especially from dumping (emphasis added). The Contracting Parties to the LDC shall endeavour to act consistently with the objectives and provisions of such regional agreements, which shall be notified to them by the Organization. The Contracting Parties shall seek to cooperate with the Parties to regional agreements in order to develop harmonized procedures to be followed by Contracting Parties to the different conventions concerned. Special attention shall be given to cooperation in monitoring and scientific research. As can be seen this Article would support the formation of an Arctic Dumping Protocol and as well require other LDC Parties to endeavour to act consistently with its objectives and to cooperate to develop harmonized procedures.

37 See Part II Section 17.2.2. The phrase 'undertake to promote' was used to provide a weak linkage to the Arctic Liability Protocol. Since State ratification of a liability regime is unlikely, then any stronger requirement in this Protocol would result in its not being ratified. The LDC Article X as noted in Part II Section 6.2. uses the phrase 'undertake to develop procedures for the assessment of liability and the settlement of disputes regarding dumping.'

38 These include criteria for determining exceptional and emergency situations, procedures for consultative advice and the safe disposal of matter in such circumstances, including the designation and recommendation of appropriate dumping areas.

39 See LDC Articles XIV, XVI - XXII.

40 As also noted under the LDC, amendments to the Annexes were made in 1978 regarding incineration, and 1980 regarding the lists of substances. See Part I Section 3.2.

41 The phrase, 'and compounds which may form such substances in the marine environment' was added from the ODC.

42 This phrase was added from the ODC.

43 This phrase was added from the ODC.

44 The LDC Annex I.4. states further, 'netting and ropes, which may float or may remain in suspension in the sea in such a manner as to interfere materially with fishing, navigation or other legitimate uses of the sea'. This was deleted as too lenient.

45 This list could be enlarged if necessary to one similar to the Arctic Oil Rig Protocol's List of Oils, Part II Section 12.2.13. This would for example prohibit dumping of engine oil from fishing boats. See Nordlys, 27 november 1989, Nr.275, 88 årgang, s. 7.

46 The LDC ANNEX I.6. states further, 'radioactive matter, defined on public health, biological or other grounds, by the competent international body in this field, at present the International Atomic Energy Agency, as unsuitable for dumping at sea.' This was deleted as too lenient.

47 The LDC Annex I.7. specifies 'e.g. solids, liquids, semi-liquids, gases or in a living state.'

48 The LDC Annex I.9. states that trace elements of the above substances such as sewage sludges and dredged spoils are not Annex I substances and such wastes are subject to the provisions of Annexes II and III as appropriate. This was considered too lenient.

49 The LDC Annex I.8. notes further that the consultative procedure provided for (under Article XIV) should be followed by a Party if there is doubt about the harmlessness of the substance. The last sentence could be deleted if the Arctic States would agree.

50 Organosilicon compounds also appear in Annex II of the LDC but as noted were moved up to Annex I of the Arctic Dumping Protocol since this was done in the ODC.

51 The ODC additionally notes in Annex II.2. that substances of this nature should always be deposited in deep water. The 'tar like substances' were added from Annex II of the ODC.

52 This paragraph was added from the ODC Annex II.4.

53 Specifically these include total amount and average composition of matter dumped (per year); form (solid, sludge, liquid, or gaseous); properties, physical (solubility and density), chemical and biochemical (oxygen demand, nutrients) and biological (presence of viruses, bacteria, yeasts, parasites); toxicity; persistence (physical, chemical and biological); accumulation and biotransformation in biological materials or sediments; susceptibility to physical, chemical and biochemical changes and interaction in the marine environment with other dissolved organic and inorganic materials, including the possible formation of new compounds ('(I)ncluding the possible formation of new compounds' was added from the ODC Annex III.1. (h)); probability of production of taints or other changes reducing marketability of resources (fish and shellfish).

54 Specifically these include location (coordinates of the dumping area, depth and distance from the coast), location in relation to other areas (spawning, nursery and fishing areas, living resources in adult or juvenile phases ('(L)iving resources in adult or juvenile phases' was included from the ODC Annex III.2. (b) to include living resources other than fish), exploitable resources and amenity areas); rate of disposal per specific period (quantity per time period); methods of packaging and containment, if any; initial dilution achieved by proposed method of release; dispersal characteristics (effects of currents, tides and wind on horizontal transport and vertical mixing); water characteristics (temperature, pH, salinity, stratification, oxygen indices of pollution - dissolved oxygen (DO), chemical oxygen demand (COD), biochemical oxygen demand (BOD) - nitrogen present in organic and mineral form including ammonia, suspended

matter, other nutrients and productivity); bottom characteristics (topography, geochemical and geological characteristics and biological productivity); existence and effects of other dumpings which have been made in the dumping area (heavy metal background reading and organic carbon content), including accumulative effects ('(I)ncluding accumulative effects' was added from the ODC Annex III.2. (g)); a consideration of whether an adequate scientific basis exists for assessing the consequences of such dumping according to Annex III criteria, taking into account seasonal variations.

55 Specifically these include possible effects on amenities (presence of floating or stranded material, turbidity, objectionable odour, discoloration and foaming); possible effects on marine life, fish and shellfish culture, fish stocks and fisheries; possible effects on other uses of the sea (impairment of water quality for industrial use, underwater corrosion of structures, interference with ship operations from floating materials, interference with fishing or navigation through deposits of waste or solid objects on the sea bed and protection of areas of special importance for scientific or conservation purposes); the practical availability of alternative land based methods of treatment, disposal or elimination, or of treatment to render the matter less harmful for marine dumping.

56 Article 9 of the Helsinki Convention uses 'undertakes to ensure compliance', which seems weak.

57 The phrase 'exclusive economic zone' was added to be consistent with LOSC provisions.

58 The Helsinki Convention Annex V uses the phrase 'territorial sea'. This exception could be deleted to make the Alternative even more stringent.

59 See Part I Section 3.2.2. for additional Annex II noxious substances from the Helsinki Convention Annex II.

60 This assistance may include but is not limited to inspection by authorities of oil record books, cargo record books, log books and engine log books and taking oil samples for analytical identification purposes and in respect of the system of tagging oil residues.

16. Discussion of Arctic Land-Based Protocol[1]

16.1. Background

16.1.1. *Summary of Existing Land-based Pollution Regime with Application to the Arctic*

In relation to the Arctic the Paris Convention is applicable in about 25% of the area (with Soviet accession) covering the Barents Sea, sections of the Greenland Sea, the Arctic Ocean and the Norwegian Sea.[2] The Convention does not cover the problematic Soviet rivers Yenisey, Lena or Ob and their areas of discharge into the sea.[3] Although experience dealing with pollution from land-based sources can be gained from this Convention in practice, there are substantial problems involving lack of mandatory State obligation, political problems, insufficient 'black list' prohibitions as well as the obvious insufficient geographical scope.[4]

Although provision is made for airborne pollution in the Paris Convention, the Protocol covering this pollution source was not in force as of June 1987. Additionally developments related to the Arctic seem to be progressing mostly under the 1979 ECE of which all the Arctic rim States and the Eastern and Western European Countries are Parties.[5] Problems with the ECE include however that the largest polluters have not consented to reduce their admissions, and there has been conducted insufficient monitoring.[6]

Upon this background of probable Arctic pollution from Soviet rivers with presently less than 25% of the Arctic covered by international regulation, (including however an elementary control of Arctic airborne pollution), the Arctic Land-Based Protocol is proposed.

16.1.2. *Legal Basis for Arctic Land-Based Protocol under LOSC*[7]

As with the preceding Arctic Protocols the general obligations set forth by Articles 192, 197 and 194 apply, especially the latter under which States are required to jointly take all measures to minimize fully the release of toxic, harmful or noxious substances, especially persistent, from *land-based sources* and from or through the atmosphere.

As noted under Article 207 States shall *endeavour*, acting especially through competent international organizations or diplomatic conference, to establish global and *regional* rules, standards and recommended practices and procedures to prevent, reduce and control marine land-based pollution taking into account characteristic regional features. Such provisions must be re-examined as necessary.[8] Additionally States shall adopt national laws and regulations to prevent, reduce and control marine pollution from land-based sources[9] taking into account internationally agreed rules, standards and recommended practices and procedures.[10] States must also take other necessary measures to prevent, reduce and control such pollution and must endeavour to harmonize such policies at the appropriate regional level.[11]

Under Article 213 States must enforce their national laws adopted in accordance with Article 207 and adopt national laws and regulations and take other measures necessary to implement applicable[12] international rules and standards established through competent international organizations or diplomatic conference to prevent, reduce and control marine land-based pollution.

To the extent these LOSC provisions can be said to be international customary law and recommendations for norms, the Arctic rim States have an obligation to *endeavour* to establish and thereafter to enforce an Arctic regional agreement to prevent, reduce and control marine pollution emanating from land-based sources under their jurisdiction, which would act as legal support to the formation of the Arctic Land-based Protocol.[13]

16.1.3. Use of the Paris Convention as a Pattern with Inputs from the Barcelona Protocol III and the Helsinki Convention

In spite of the Paris Convention's shortcomings, somewhat paradoxically it is proposed to use the Paris Convention as a basis for forming the upgraded Arctic Land-Based Protocol. This is because of this Convention's coverage of heavily industrialized Europe and the experience gained therefrom, as well as its direct relation to the Arctic.

As noted the Paris Convention upon ratification by the Soviet Union will include as Parties 3 of the 5 Arctic rim States. It has had most to do with non Arctic areas, mainly Europe, however Europe is heavily industrialized. Though there is not found the variety of industry with pollutants in the Arctic, still because of the large variety of pollutants found in Europe, there would be a chance that should industrial pollution crop up in the Arctic, the pollutants would appear on the Paris Convention's lists.[14] Additionally as a look at a Paris Commission Annual Report will show, the programs initiated under the Paris Convention are extensive and based upon some 15 years experience. There is thus a possibility to study and better these programs for implementation and use in combatting land-based pollution in the Arctic.[15] Inputs from the Barcelona Protocol III are used in the Arctic Land-Based Protocol, since this Protocol's Annex I 'black list' and Annex II 'grey list' are even more comprehensive than Paris Convention's, and the issuance of permits is more extensively covered in the Barcelona Protocol III's Annex III than under the Paris Convention's Part II Annex A.[16]

A point of consideration is the relationship of the Arctic Land-Based Protocol with the Paris Convention, the former which will overlap the latter. This actually would not be detrimental to environmental protection of the Arctic because the Arctic Land-Based Protocol with its higher proposed standard of environmental protection, would effectively prevent the lesser effective application of the Paris Convention over the common area. By this is meant that many of the political problems regarding EEC stalling and State inactivity would be avoided.[17] Only Norway, the Soviet Union and Denmark would be members of both agreements and could agree to higher standards of environmental protection without violating any provision of the Paris Convention.[18]

Inputs could also be included as well from the Helsinki Convention which lists among other things pollutants used in agriculture and waste water from industrial plants.[19] This is probably very relevant to the Arctic Land-Based Protocol considering the facts noted concerning the Yenisey and 51% of the Soviet territory draining through this river into the Arctic Ocean. Another positive feature of the Helsinki Convention is that 2 of its Parties are Arctic rim States, the USSR and Denmark, and experience gained in the Baltic, including that of an east - west environmental cooperative effort is relevant to the Arctic Land-Based Protocol.

In preparing the Arctic Land-Based Protocol additions have been made where thought relevant, and these will be generally noted. The most important deal with *upgrading* State Party obligations from ambiguous terms such as 'shall undertake', 'agree to', 'shall endeavour' and 'shall take account of', to 'shall'. Though as noted above the duty required under LOSC provisions generally is only to *endeavour* to establish regional rules, standards, recommended practices and procedures to combat pollution from this source, the Arctic Land-Based Protocol goes beyond this by making it *mandatory*. The Arctic Land-based Protocol is thus more stringent than both the LOSC and the Paris Convention provisions, and this can cause difficulties for State ratification. If it is discovered that this Protocol is *not* being ratified by the Arctic rim States because of the increased duties, then the mandatory language could be softened by reintroducing 'shall endeavour' and 'shall undertake'. This could be done in order to achieve State ratification of some regional agreement and thus some degree pollution control. An idea concerning these areas of international and national control less popular with States would be for the States to enter a vague Protocol which would be capable of modification towards a stricter regime in the future.[20]

16.2. Discussion of Proposed Arctic Land-Based Protocol

16.2.1. Introduction

The discussion covers 25 Articles and 2 Annexes. A Protocol text is *not* presented, but rather an idea of the desired content of each Article along with a brief discussion.

16.2.2. Articles

A *Preamble* sets the stage by giving various declarations of the Arctic rim States of a desire to preserve the sensitive Arctic environment, combat marine pollution from land-based sources and promote cooperative international action among the Arctic States to attain these goals.

Under *Article 1* the Arctic States pledge themselves to take all possible steps to prevent marine pollution, which is defined as every act referred to in Article 2 of the Arctic Convention and every other act which results or is likely to result in deleterious effects to human health and to the environment, caused directly or indirectly by such activities.[21] Also included is the provision requiring that the Arctic States adopt individually and jointly measures to combat marine pollution from land-based sources in accordance with the provisions of the present Protocol and harmonize their policies in this regard.

Article 2 deals with area of Protocol application. This is to be the same as outlined in the Arctic Convention Article 1.[22]

Definitions are covered in *Article 3.* Most important is the definition of 'Pollution from land-based sources', which is the same as under the Paris Convention with reference to offshore platforms *dropped.* Thus, 'Pollution from land-based sources' means the pollution of the maritime area (i) through watercourses and (ii) from the coast, including introduction through underwater or other pipelines. 'Maritime area', 'Freshwater limit' and 'Organization' also appear and are as follows. 'Maritime area' means the high seas, economic zones and the territorial seas of the Contracting Parties and waters on the landward side of the base lines from which the breadth of the territorial sea is measured, extending, in the case of watercourses up to the freshwater limit unless otherwise decided under Article 16 of the present Protocol. 'Freshwater limit' means the place in the water course where, at low tide and in a period of low freshwater flow, there is an appreciable increase in salinity due to the present of sea water. 'Organization' means the body referred to in Articles 2 and 13 of the Arctic Convention. This is done to make a link with the Arctic Convention.

At first it was thought advantageous to include other definitions such as 'Activities', 'Wastes and harmful or noxious substances and materials', 'Oil' and 'Oily mixture' similar to those appearing in the Arctic Oil Rig Protocol in Appendix I and the Agreement and Contingency Plan in Appendix II. However due to the complexity and variety of pollution from the land-based source, it is probably advantageous to follow the Paris Convention having just 3 definitions. By trying to be too definite, some land-based pollution will bound to be omitted and thus not subject to control.

Article 4 is a very *central* and important and contains provisions to the effect that the Contracting Parties shall eliminate,[23] if necessary by stages, through joint or individual implementation of measures and programs, marine pollution from land-based sources of substances listed in Annex I, and strictly limit marine pollution from land-based sources of substances listed in Annex II.[24] These latter substances are to be discharged only upon approval by the appropriate State Authorities, and such approval must be periodically reviewed in accordance with considerations set forth in Annex III.[25] Additionally, the programs and measures implemented shall include[26] specific regulations or standards covering environmental quality, discharges into the maritime area or watercourses affecting the maritime area, and the composition and use of substances and products, and are to be based[27] upon the latest technical developments. The programs must contain time limits for their completion. Finally the Contracting Parties shall, jointly or individually as appropriate, implement[28] programs or measures to eliminate or reduce marine pollution[29] from land-based sources by a substance not then listed in Annexes I and II of the present Protocol, if scientific evidence has established that a serious hazard may be created in the marine area by that substance.[30] The States shall harmonize national policies for the Arctic in this regard.

Article 5 deals with radioactive substances. The Arctic Land-Based Protocol while respecting other International Agencies and Organizations dealing with these substances at the same time takes a stronger stance than the Paris Convention, by making the provisions related to radioactive substances *mandatory*. Thus, the Contracting Parties shall adopt measures to eliminate pollution of the maritime area from land-based sources by radioactive substances listed in Annex I of the present Protocol.[31] Additionally, without prejudice to their obligations under other Treaties and Conventions, in implementing this Article[32] the Contracting Parties must take full account of the recommendations of the appropriate International Organizations and Agencies; take account of the monitoring procedures recommended by these international Organizations and Agencies; and, coordinate their monitoring and study of radioactive substances in accordance with Articles 10 and 11 of the present Protocol.[33] The States shall harmonize national policies for the Arctic in this regard.

Article 6 deals with general unclassified polluting substances from land-based sources. Since as noted, there are many different varieties of such pollutants, it seems probable that at least a few such substances are presently being emptied into Arctic waters. This Article is therefore important since it is a 'catch-all'. As with the above Articles however it is made mandatory. Thus it provides that in order to preserve and enhance the quality of the marine environment, the Contracting Parties, without prejudice to the provisions of Article 4, shall[34] reduce existing pollution from land-based sources and forestall any new pollution from land-based sources, including that which derives from new substances. In implementing this program[35] the Arctic States shall at a minimum include an evaluation of the nature and quantities of the pollutants under consideration; the level of existent pollution; the quality and absorptive capacity of the receiving waters of the maritime area; and the need for an integrated planning policy, consistent with the requirement of environmental protection.[36] The States shall harmonize national policies for the Arctic in this regard.

Article 7 attempts to limit pollution from other sources and to other areas through application of the Arctic Land-Based Protocol. This Article provides that the Contracting Parties shall apply[37] the adopted measures in such a way as to avoid increasing marine pollution outside the area to which the present Protocol applies; and in the maritime area covered by the present Protocol, originating otherwise than from land-based sources. The States shall harmonize national policies for the Arctic in this regard.

Article 8 provides that no provision of the present Protocol may be interpreted as preventing the Contracting Parties from taking more stringent measures to combat marine pollution from land-based sources.

Article 9 refers to the substances not appearing in Annex I, that is the less noxious and unclassified substances. It is intended as a settlement dispute arrangement. It provides that when pollution from these substances from land-based sources is likely to prejudice the interests of one or more Arctic States, the Parties concerned undertake to enter into consultation, at the request of any one of them, in order to negotiate a cooperation agreement. The Organization referred to in Article 3 and 15, at the request of any Party, must consider the dispute and may make recommendations. The cooperation agreements referred to above may among other things, define the areas of application, the quality objectives to be achieved, and the methods for achieving these objectives, including methods for the application of appropriate standards, and the scientific and technical information to be collected. The Arctic States forming and signing these agreements must through the Organization inform the other Contracting Parties of the agreements and the progress of putting them into effect.[38]

Dispute settlement relating to the interpretation or application of the present Protocol is also included in this Article. Thus provisions include that in the event the dispute cannot be settled by the Arctic States themselves, then the dispute must be settled according to the provisions outlined in Article 21 of the Arctic Convention following LOSC provisions. The States shall harmonize national policies for the Arctic in this regard.

Article 10 deals with complementary or joint programs by Arctic States of scientific and technical research, including the best methods of eliminating or replacing noxious substances so as to reduce marine pollution. The information so obtained must be transferred to one another and the work carried out should take into account the work carried out by the relevant International Organizations and Agencies. The States shall harmonize national policies for the Arctic in this regard.

Article 11 deals with establishing monitoring programs, and under it the Arctic States shall[39] establish progressively, within the Protocol Area a permanent monitoring system covering the earliest possible assessment of the existing level of marine pollution and the assessment of the effectiveness of measures for the reduction of marine pollution under the terms of the Arctic Land-Based Protocol. To achieve this the Arctic States must establish the means for pursuing individually or jointly systematic and ad hoc monitoring programs, which shall take into account the deployment of research vessels and other facilities in the monitoring areas. Additionally the programs must take into account[40] similar programs followed in accordance with Arctic Protocols and Conventions already in force and by the appropriate International Organizations and Agencies. The States shall harmonize national policies for the Arctic in this regard.

Article 12 deals with enforcement provisions and upgrades the duty required of the Arctic States to mandatory compliance with and enforcement of the Protocol. Under it each Party shall[41] ensure compliance with the provisions of this Protocol and must take in its territory appropriate measure to prevent and punish conduct in contravention of its provisions. The Arctic States must inform the Organization of the legislative and administrative measures taken to implement the above provision. The States shall harmonize national policies for the Arctic in this regard.

Article 13 deals with Party cooperation and the duty is upgraded from 'undertake to assist one another' to 'shall assist one another'. Thus the Contracting Parties shall assist one another as appropriate to prevent incidents which may result in pollution from land-based sources, to minimize and eliminate the consequences of such incidents and to exchange information to that end. The States shall harmonize national policies for the Arctic in this regard.

Article 14 covers situations where the pollution has its origin in a non Contracting State. It states that the provisions of the present Protocol may not be used against a Contracting Party to the extent that the Party is prevented, as a result of pollution having its origin in the territory of a non Contracting State, from ensuring the full application of those provisions. However the Contracting Party must endeavour to cooperate with the non Contracting State in order to make possible the full application of the present Protocol.[42]

Under *Article 15* an Organization made up of representatives of each of the Contracting Parties is to be established in accordance with Articles 2 and 13 of the Arctic Convention. It is to meet regularly and at any time due to special circumstances when decided in accordance with the rules of procedure. This Organization is mentioned in Article 3 above and is meant to be established under the umbrella Arctic Convention.

Article 16 covers the duties of the Organization in accordance with Articles 2 and 13 of the Arctic Convention and includes the exercise of supervision over the implementation of the Protocol; the convening of consultative meetings of the Arctic States and of special meetings of the Parties; the review of the condition of the seas within the area of Protocol application, the effectiveness of the control measures being adopted and the need for additional or different measures; the establishment in accordance with Article 3 of the extent of the maritime area into watercourses; the establishment in accordance with Article 4 of programs and measures for the elimination or reduction of pollution from land-based sources; the formation of recommendations in accordance with Article 9 regarding disputes; the receipt and review of information and distribution of it to the Arctic States and the public[43] in accordance with Articles 11 and 17 (monitoring) and Article 12 (compliance and enforcement); and the formation of recommendations in accordance with Article 13 (assistance to prevent polluting incidents) regarding amendment of Annex A substances lists; and finally the discharge of other functions appropriate under the Protocol terms.

Article 17 concerns the required transmission by the Arctic States to the Organization of monitoring results obtained under Article 11, and of the most detailed information available on the Annex substances. The Arctic States additionally are to endeavour to improve progressively sampling techniques which can contribute to the revision of the pollution reduction programs adopted under Article 4.[44] The States shall harmonize national policies for the Arctic in this regard.

Article 18 deals with airborne pollution and requires the Arctic States to do their utmost to promote measures eliminating, reducing and containing airborne pollution of all types, currently under progress in International Organizations and Agencies, including but not limited to the ECE and the Paris Commission. The States shall harmonize national policies for the Arctic in this regard.[45]

Article 19 deals with liability for land-based pollution and states that the Arctic States must cooperate directly and if necessary with the assistance of competent regional or global international organizations to formulate and implement programs and procedures for the determination of liability and compensation for damage resulting from pollution from land-based sources under their jurisdiction.[46] The States shall harmonize national policies for the Arctic in this regard.

Article 20 deals with rules of procedure to be followed by the Organization which requires the Arctic States to adopt rules of procedure for their meetings and conferences in accordance with Articles 14, 15 and 16 of the Arctic Convention.

Article 21 deals with amendments to the Protocol and states that at meetings called in accordance with Article 16, amendments may be adopted in accordance with Article 16 of the Arctic Convention.

Article 22 deals with amendments to the Annexes which are to be based upon scientific or technical considerations. At meetings called in accordance with Article 16 recommendations for amendments to the Annexes must be made in accordance with Article 17 of the Arctic Convention.

Article 23 deals with Arctic State accession and requires States which may accede to the Arctic Land-Based Protocol to do so in accordance with provisions outlined in Article 24 of the Arctic Convention.

Article 24 which duplicates a Paris Conference Resolution[47] is useful for the interim period between the signing of the Protocol and its coming into force, which as noted might be some time. Article 24 states that the participants to the Conference culminating in the adoption of the text to the Arctic Land-Based Protocol, shall endeavour[48] to facilitate the later implementation of the Arctic Land-Based Protocol; to further the objectives and undertakings contained in the Convention and in doing so to give particular weight to taking early action with respect to Annex I substances, including PCB's and DDT.

Article 25 is meant as a linking Article to the other Arctic Protocols and the Arctic Convention. Under Article 25 the Arctic States agree to promote measures towards ratification of the Arctic Protocols not already ratified and to promote within the competent specialized agencies and other international bodies environmental measures against pollution from all sources.

General Articles which deal with among other things, the place and date of signature, where instruments may be deposited upon ratification, acceptance or approval, when the Protocol comes into force, withdrawal from the Protocol, information dispatched to all Arctic States concerning the above and authenticity of language are basically standard and will not be discussed here.[49]

16.2.3. Annexes

Annex I, Harmful or Noxious Substances and Material Whose Disposal Is Prohibited, of the Arctic Land-Based Protocol is derived as noted from Annex I of the Barcelona Protocol III.[50] As noted in PART II Section 9.3.1. these lists will have to be *updated* and *upgraded* as knowledge is obtained regarding chemicals' effect on Arctic waters and life. The list seems suspiciously small.

Annex II, Harmful or Noxious Substances and Material Whose Disposal Is Subject to a Special Permit is correspondingly derived from the Barcelona Protocol III's Annex II.[51] A possible stringent alternative, as noted previously involves deleting all 'grey list' substances and moving them onto the 'black list' for use in the Arctic.[52] Due however to State reluctance to be bound to stringent agreements covering the land-based pollution sources, this alternative seems of little feasibility under present political conditions.

Annex III, Factors to Be Considered for the Issue of the Permits is also adopted from the Barcelona Protocol III's Annex III due to its comprehensiveness.[53]

Notes - Chapter 16

1 Protocol for the Prevention of Marine Pollution in the Arctic Region from Land-based Sources.

2 See PART I Section 4.2.2.

3 See PART II Section 10.2.1.

4 See PART II Section 9.3.1.

5 See PART I Section 4.2.5.

6 See PART II Section 9.3.1.

7 See PART I Sections 2 and 4.2.2. for conventions and customary law regulating State jurisdiction over land-based pollution lending additional legal support to the formation of the Arctic Land-based Protocol.

8 Article 207 (4).

9 These sources include 'rivers, estuaries, pipelines and outfall structures'.

10 Article 207 (1). See PART II Sections 11.2 and 11.3. for a short discussion of the interplay between the national legislation dealing with land based pollution and international regulations.

11 Article 207 (2) and (3). Laws regulations, measures, rules, standards and recommended practices and procedures referred to in the above provisions must include those designed to minimize as much as possible the release into the marine environment of toxic, harmful or noxious substances, especially those which are persistent., Article 207 (5).

12 See PART II Section 12.1.2. for discussion of the term 'applicable' obliging the State to enforce relevant customary law.

13 See PART I Section 2.1.4. and PART II Section 10.3.3. However, even if these provisions could be said to be international law, as Churchill p. 278 notes this obligation is so imprecisely and broadly formulated, it is unlikely to have much effect in practice.

14 As an example could be mentioned the mining and metal processing activities, mentioned in PART II Section 10.2.1., polluting the Soviet river Yenisey. Though pollutants are not mentioned specifically, there is a good chance that the mining and processing of similar metals are found in Europe and thus covered by the Paris Convention.

15 See Par.Comm.10.

16 See PART I Section 4.2.3.2. The UNEP instrument was not used solely as a model in spite of its more rigorous Annexes because it was thought better to have the Arctic Land-based Protocol based upon an instrument where 2 and possible 3 of the Arctic States are Parties.

17 See PART II Section 9.3.1.

18 See Article 8 of the Paris Convention in which it is stated nothing prevents any Party from taking more stringent measures to combat marine pollution from land-based sources.

19 See PART I Section 4.2.3.1.

20 See also PART II Section 17.1.1. for the same recommendation concerning the Arctic Liability Protocol.

21 A definition of pollution is given referring back to the Arctic Convention in order to have a uniform broad definition. The definition suggested is more encompassing than the Paris Convention Article 1 definition.

22 See PART II Section 11.4.2.

23 'Undertake to eliminate' from the Paris Convention was upgraded as shown.

24 The Paris Convention contains a paragraph dealing with the mandatory implementation of programs and measures to carry out these *undertakings*. This was omitted here, since the entire Article was made mandatory. A simple implementation phrase is however included as noted.

25 This allows the State Authorities wide discretion in their application.

26 The phrase 'as appropriate' in the Paris Convention is excluded here for introducing vagueness.

27 'Take into account' was upgraded as shown.

28 The phrase 'may, jointly or individually as appropriate, implement' was made a degree more mandatory by the wording shown. 'As appropriate' was retained to give a degree of flexibility.

29 'To forestall, reduce or eliminate pollution' was upgraded as shown.

30 'And if urgent action is necessary' is dropped here in order to require the States to take action in this situation.

31 Article 5 of the Paris Convention uses the phrase 'undertake to adopt measures to forestall and, as appropriate, eliminate pollution ...'

32 The phrase 'in implementing this undertaking' was upgraded as shown.

33 It is unclear the extent of control being exercised over discharges of radioactive substances in the Arctic by the relevant International Organizations and Agencies. See PART II Section 9.3.1. However, if it be shown lacking, this Article could be upgraded to be more stringent in order to better contain this problem.

34 The Paris Convention uses the phrase, 'shall endeavour'.

35 The phrase, '(I)n implementing this undertaking' was upgraded as shown in order to better emphasize this Article.

36 The phrase 'at a minimum include an evaluation of' was added here to make this requirement more specific and allow the States flexibility to add more elements to the list of the evaluation. The Paris Convention uses the phrase 'shall take account of'.

37 The Paris Convention uses the phrase 'agree to apply'.

38 This Article is left vague and indefinite, in order to give State Parties flexibility regarding these non 'black list' substances and provide counter balance to the mandatory provisions noted above. Though environmentally it might be advisable to change the classification of the substance to the next more stringent listing if it causes *prejudice* to another State Party, still Articles 4 and 6 will cover situations where the substances appear especially harmful. The dispute settlement provision which follows is added.

39 The Paris Convention uses the term 'agree to'.

40 The vague phrase 'take into account' was deliberately retained in order to allow the States flexibility in establishing the means of such a monitoring program.

41 The Paris Convention uses the term 'undertakes'.

42 The phrase 'shall endeavour' is not changed since the cooperation is also dependent upon the non Contracting Party.

43 'And the public' was added to alleviate the information problems noted under the Paris Convention in PART II Section 9.3.1.

44 The phrase 'to endeavour' is not upgraded since improving sampling techniques can only be attempted.

45 Articles 18-25 are added.

46 This Article which also appears in the Arctic Liability Protocol could be dropped if shown that the States had no interest in considering the question of liability for damage from this pollution source.

47 Resolution N V, Paris Convention.

48 Under Resolution N V, the Conference 'strongly urges' Conference participants to take the steps noted.

49 See Articles 19-29 of the Paris Convention.

50 This is the same as for *Annex I* for the Arctic Oil Rig Protocol, and a glance at PART II Section 12.2.7. Annex I will indicate the substances included on this list.

51 See PART II Section 12.2.8. for ANNEX II of the Arctic Oil Rig Protocol which is the same as that proposed here.

52 See PART II Section 9.3.1.

53 See PART II Section 12.2.9. for ANNEX III of the Arctic Oil Rig Protocol which is the same as that proposed here.

17. Discussion of Arctic Liability Protocol[1]

17.1. Background

17.1.1. Brief Summary of Applicable International Law or Agreements Covering Liability in the Arctic[2]

The international shipping liability conventions and agreements applicable in the Arctic include the CLC,[3] the Fund Convention,[4] and for instances where neither the CLC nor the Fund Convention apply, the private regimes established by tanker owners or oil companies to limit their liability, TOVALOP, CRISTAL and PLATO.[5] As noted in Part I Section 6.4.6. there are 2 shipping liability Conventions in force covering pollution damage caused by substances *other* than oil or radioactive matter, the LLMC and the 1957 Liability Convention. However it is apparent that the IMO and many States are interested in establishing a HNS Draft Convention despite widespread disagreement and despite the existence of these 2 Conventions. Thus it is the HNS Draft Convention which is emphasized, and it would be applicable in the Arctic *if* it were in force.

As seen in Part I Section 6.3.2. oil companies, operating platforms chiefly on the British continental shelf, have formed a *private* liability regime, OPOL.[6] This agreement parallels the 1976 Liability Convention by using strict but limited liability, but OPOL is more oriented to industry interests than the Convention. Although nothing similar is currently present among oil companies operating in the Norwegian or Soviet Arctic,[7] such agreements are under development in the US and Canadian Arctic areas, due to their generally more 'industry friendly' legal provisions.[8]

Liability for pollution damage from dumping in the Arctic is limited to the general provisions of Article X of the LDC, under which offshore oil activities are excluded, however some developments may be occurring under the LDC which may in the future materialize liability for pollution damage from other dumping activities.[9]

There is no regime applicable in the Arctic covering liability for damage caused by land-based pollution.[10]

Finally as noted in Part I Section 2.1.3. and Part II Section 6.5. although it is generally acknowledged that a State has responsibility for an act or omission which results in a breach of a legal obligation resting upon a treaty, custom or other, still at this point States seem reluctant to either make or allow claims for pollution damage based upon solely custom.

As seen there are problems associated with the existing regimes consisting of non Arctic State ratification, insufficient liability coverage, limited geographic coverage, lapse and lack of application.[11] Because of the inadequacies of the liability regimes it is a relatively simple job to propose a legal document obliging States to provide a system for adequate compensation for marine pollution damage. However drafting a document is not the problem. The problem is trying to convince reluctant States to enter into such regimes, and for this reason alternatives are noted trying also to make the Arctic Liability Protocol more palatable to State interests. Although this might not always be in the best environmental interests, the strategy as under the Arctic Land-based Protocol, Part II Section 16.1.3., is that perhaps after State ratification of a less stringent document, a better environmental alternative could be achieved through gradual amendments over time.

17.1.2. Legal Basis for Arctic Liability Protocol under LOSC[12]

As in all the Arctic Protocols discussed previously, the basic LOSC Articles could be mentioned as providing a legal groundwork for establishing a liability regime in the Arctic.[13] Although liability is not specifically referred to in these Articles, the argument exists that a good liability regime, to ensure satisfactory compensation for damages suffered from pollution, would go far in aiding in the carrying out of principles in these Articles, the prevention, reduction and control of pollution of the marine environment.[14] Thus there would be reciprocal support afforded these regimes and that of liability.

In addition to this, as noted, LOSC Article 235 declares specifically that States are responsible for fulfilling their international obligations concerning marine environmental protection and preservation and shall be liable in accordance with international law.[15] Under this obligation States must ensure that legal recourse is available for prompt and adequate compensation or other relief for damage caused by marine environmental pollution by natural or juridical persons under their jurisdiction.[16] In order to ensure prompt and adequate compensation for all marine environmental pollution, States shall cooperate in implementing existing international law and further developing international law related to responsibility and liability for the assessment of and compensation for damage and the settlement of related disputes. Where appropriate, the development must also include the formation of criteria and procedures for the payment of adequate compensation such as compulsory insurance or compensation funds.[17]

Thus to the extent the LOSC provisions can be said to be international customary law and recommendations for norms, the Arctic rim States have an obligation to cooperate and develop international law covering liability for pollution damage and the payment of compensation for such, which would act as legal support for the formation of the Arctic Liability Protocol.[18]

17.2. Discussion of Proposed Arctic Liability Protocol

17.2.1. Introduction

In spite of its unpopularity among States, it is proposed to use the 1976 Liability Convention as a model for forming the Arctic Liability Protocol. This is chiefly because it was the most extensive document found covering this subject. The Barcelona Convention contains as noted only 1 paragraph in Article 12 briefly and generally mentioning civil liability for pollution damage, while as the 1976 Liability Convention covers the subject in 26 comprehensive Articles. This can perhaps be the reason why it is *not* in force. It may be too extensive for most States to be willing to bind themselves to it.[19]

Though the use of the 1976 Liability Convention may seem unwise since this Convention has not been ratified, it was considered worthwhile to present a document truly oriented towards compensating a person damaged by pollution. However in order to also deal with reality, suggestions are noted for 'toning down' the Arctic Liability Protocol to be more compatible with State interests and less beneficial to damaged persons. Changes could even be made to make the Arctic Liability Protocol less 'friendly' to damaged persons than the 1976 Liability Convention, which include graduated limits of liability, which if low enough might induce State ratification, but which hopefully over a period of time would be reasonably increased to attempt to reach satisfactory compensation levels.[20]

The Arctic Liability Protocol encompasses chiefly liability for damage from offshore *oil field pollution* and *dumping* of wastes associated with offshore oil activities.[21] Liability for vessel source pollution is already covered by the IMO Conventions noted above, though insufficiently. A general Article 16 is added to the Arctic Liability Protocol to address this insufficiency and assist in the process of providing better coverage over shipping. Liability for damage from dumping *not* related to offshore oil activities is briefly and generally addressed in Article 17. Liability for land based pollution is covered very briefly and generally in Article 18 and is not covered extensively because of the difficulties noted in Part II Section 4.2. concerning agreements dealing with land-based pollution. Additionally problems of causality,[22] and extreme State reluctance to allow compensation for damages from this pollution source[23] require a sober approach to this subject. As noted, these latter Articles could be dropped in order to made the Arctic Liability Protocol more 'State friendly'.

17.2.2. Articles

The *Preamble* is of the same general content as in previous Protocols wherein the States declare that they are conscious of the dangers of pollution from oil and harmful or noxious substances and materials[24] posed by exploration and exploitation activities of certain seabed mineral resources; that they are convinced of the need for ensuring adequate compensation for persons damaged by such pollution; and that they desire to adopt uniform rules and procedures for determining liability questions and providing adequate compensation in such cases.

Article 1 covers definitions which are somewhat detailed. Under the 1976 Liability Convention this is probably due to the large amounts of monetary compensation which possibly could be in issue here. Where large sums are involved, then more care seems taken with clear definitions to set limits to liability.

Under Article 1 'oil' means crude oil and natural gas liquids, whether or not such oil or liquids are mixed with or present in other substances. 'Crude oil' includes crude oil treated to render it suitable for transmission (such as by adding or removing certain fractions).

'Harmful or Noxious Substances and Materials' means those substances listed in Annex I and Annex II of the Arctic Oil Rig Protocol.[25]

'Installation' means (a) any well or other fixed or mobile facility which is used for exploring for, producing, treating, storing, transmitting or regaining control of the flow of crude oil from the sea bed or its subsoil; (b) any well which has been used for the purpose of exploring for, producing or regaining control of the flow of crude oil from the sea bed or its subsoil and which has been abandoned after the entry into force of this Protocol for the State concerned; (c) any well which is used for the purpose of exploring for, producing or regaining control of the flow of gas or natural gas liquids from the sea bed or its subsoil during the period that any such well is being drilled, including completion, or worked upon except for normal maintenance operations; (d) any well which is used for the purpose of exploring for any mineral resources other than crude oil, gas or natural gas liquids, which such exploration involves the deep penetration of the subsoil of the sea bed; and (e) any facility which is normally used for storing crude oil from the sea bed or its subsoil. These wells must be located seaward of the low water line of States as marked on official charts. Wells directly connected to a platform or similar facility are considered part of that platform and the entire unit is considered one installation. A ship shall not be considered to be an installation.[26]

'Operator' means the person, whether licensee or not, designated as operator for the purpose of this Protocol by the Controlling State, or lacking such designation, the person who is in overall control of the installation's activities. For purposes of channelizing liability 'operator' is meant to include the operator, licensees and all their servants and agents.[27]

'Controlling State' means the Arctic State with sovereign rights over exploration and exploitation of the resources of the sea bed and its subsoil in the area in or above which the installation is situated. When an installation extends over areas in which 2 or more Arctic States have such rights, these States may agree which shall be the Controlling State.

'Person' means any individual or partnership or any public or private body, corporate or not, including an Arctic State or any of its constituent subdivisions.

'Pollution damage' means

(1) loss or damage outside the installation caused by contamination resulting from the escape or discharge of oil and/or harmful or noxious substances and materials[28] from the installation and includes the cost of preventive measures and further loss or damage outside the installation caused by preventive measures, and/or

(2) loss or damage outside the installation caused by contamination resulting from the disposal of wastes or other matter listed in Annexes I and II of the Arctic Dumping Protocol directly arising from, or related to the exploration, exploitation and associated offshore processing of seabed mineral resources from the installation or any vessels associated with the activities carried out at the installation and includes the cost of preventative measures and further loss or damage outside the installation caused by preventative measures.[29]

'Preventative measures' means any reasonable measures taken by any person in relation to a particular incident to prevent or minimize pollution damage with the exception of well control measures and measures taken to protect, repair or replace an installation.[30] 'Incident' means any occurrence, or series of occurrences having the same origin, which causes pollution damage. 'Special Drawing Right' means Special Drawing Right as defined by the International Monetary Fund and used in its operations and transactions.

Article 2 deals with the geographic coverage of the Protocol. It states that this Protocol shall apply exclusively to pollution damage, resulting from an incident which occurred beyond the coastal low water line[31] under the jurisdiction of a Controlling State, and suffered in the territory, including the internal waters and territorial sea, of a State Party or in the areas in which, in accordance with international law, it has sovereign rights over natural resources; and to preventative measures, wherever taken, to prevent or minimize such pollution damage.

Article 3 deals with operator's liability. It states that with certain exceptions noted below the operator of the installation at the time of an incident is liable for any pollution damage resulting from the incident.[32] When the incident consists of a series of occurrences, liability for pollution damage arising out of each occurrence attaches to the operator of the installation at the time of that occurrence. Where an installation has more than one operator they are jointly and severally liable.

Exceptions for liability for pollution damage include act of war, hostilities, civil war, insurrection or a natural phenomenon of an exceptional, inevitable and irresistible character. The burden of proof of rests upon the operator.

Liability for pollution damage will not attach to the operator of an abandoned well if he proves that the incident which caused the damage occurred more than 10 years[33] after the date of authorized abandonment in addition and in accordance to the time limitations in which to raise a claim set forth in Article 10.[34] Where a well has been abandoned in other circumstances, operator liability will be governed by the applicable national law.

If it can be proven by the operator that the pollution damage resulted wholly or partly either from an intentional act or omission to cause damage by the person who suffered the damage or from the negligence of that person, the operator may be exonerated wholly or partly from liability to that person. If it can be proven by the operator that the pollution damage resulted wholly or partly either from an intentional act or omission to cause damage by State licensing authorities or from the negligence of State licensing authorities, the operator may exonerated wholly or partly from liability to persons suffering the damage.[35] The States shall harmonize national policies for the Arctic in this regard.

Article 4 specifies the circumstances under which claims for compensation can be made. No claim for compensation for pollution damage can be made against the operator and others except in accordance with this Protocol. Claims for compensation for pollution damage under this Protocol may be made jointly and severally against the operator, licensees and their servants or agents.[36] Nothing in this Protocol prejudices the question whether the operator liable for damage has a right of recourse. The States shall harmonize national policies for the Arctic in this regard.

Article 5 concerns situations where there are several installations involved or several operators. It states that when oil and/or harmful or noxious substances and material have escaped or have been discharged from 2 or more installations and there is resulting pollution damage, the operators of all the installations are jointly and severally liable subject to the Article 3 exceptions for all damage which is not reasonably separable. Additionally, when oil and/or harmful or noxious substances and material have escaped or has been discharged from one installation as a result of an incident and there is resulting pollution damage and during the incident there has been a change of operators, all operators of the installation are jointly and severally liable subject to the Article 3 exceptions for all such damage which is not reasonably separable.

Article 5 states additionally that when wastes or other matter have been disposed from vessels associated with 2 or more installations and there is resulting pollution damage, the operators of all the installations shall be jointly and severally liable subject to the Article 3 exceptions for all damage which is not reasonably separable. Additionally, when wastes or other matter have been disposed from 1 installation or from vessels associated with 1 installation as a result of an incident and there is resulting pollution damage and during the incident there has been a change of operators, all operators of the installation are jointly and severally liable subject to the Article 3 exceptions for all such damage which is not reasonably separable.[37] The States shall harmonize national policies for the Arctic in this regard.

Article 6 concerns unlimited liability.[38] The operator and shall be subject to unlimited liability under this Protocol for each installation and each incident, regardless that it is proven that the pollution damage occurred as a result of negligence (*culpa*), recklessness (*culpa lata*) or as a result of an act or omission by the operator himself, done deliberately with actual knowledge that pollution damage would result (*dolus*). Where operators of different installations are liable in accordance with Article 5, the liability of the operator of any one installation is unlimited for one incident.[39] Where in the case of any one installation more than one operator is liable under this Protocol, all of them in respect of any one incident are subject to unlimited liability.[40]

The operator must constitute a fund of _____[41] with the Court or other authority of any one of the State Parties in which action is brought under Article 11. A fund established by one of the operators mentioned in Article

3 is deemed to be established by all of them. The fund can be constituted either by depositing the sum or producing a bond guarantee or other acceptable guarantee and considered adequate by the Court or Controlling State authorities. The fund is to be distributed among the claimants in proportion to the amounts of their established claims. If before the fund is distributed, the operator or any of its servants or agents, or insurer or other financial security has paid compensation for pollution damage from an incident, such person up to the amount he has paid acquires by subrogation the rights which the person so compensated would have enjoyed under this Protocol. The right of subrogation may also be exercised by other persons than those mentioned in respect to any amount of compensation for pollution damage paid, but only to the extent that such subrogation is permitted under the applicable national law.

If the operator or other person establishes that they may have to pay compensation at a later date with regard to which such person would have a right of subrogation if the compensation had been paid before the fund was distributed, the Court or State authority where the fund has been established may order that a sufficient sum be set aside to enable such person at such later date to enforce his claim against the fund. An operator who has taken preventive measures has with respect to these measures the same rights against the fund as any claimant.[42]

The fund amount referred to must be converted into the national currency of the State Party in which the fund is established on the basis of the average value of that currency during the 30 days immediately preceding the date on which the fund is constituted, of the Special Drawing Rights as published by the International Monetary Fund.

The insurer or other person providing financial security is entitled to establish a fund following the same conditions mentioned above. This can be done alone or in conjunction with the operator, and this fund has the same effect as if it were established by the operator. Such a fund may be established even in cases of intentional operator pollution, but the establishment of such a fund does not prejudice the rights of any claimant against the operator. The States shall harmonize national policies for the Arctic in this regard.

Article 7 deals with the effects of establishing a fund. Where the operator after an incident has established a fund under Article 6,[43] no person having a claim for pollution damage from that incident can exercise any right against any other assets of the operator with respect to such claim. The Court or other State authority must order the release of any property belonging to the operator which has been seized in respect to a claim for pollution damage from that incident and must similarly release any bail or other security furnished to avoid such seizure. This only applies however if the claimant has access to the Court administering the fund and the fund is actually available to his claim. The States shall harmonize national policies for the Arctic in this regard.

Article 8 deals with operator financial coverage of the liability and requires the operator to have and maintain sufficient insurance or other financial security of such type and terms as the Controlling State specifies, provided that that amount shall not be less than 66 million Special Drawing Rights until 5 years from the date when this Protocol is opened for signature and not less than 100 million Special Drawing Rights afterwards.[44] The Controlling State may exempt the operator wholly or in part from this requirement wholly caused by an act of sabotage or terrorism. Insurance or other financial security does not satisfy the requirements of this Article if it can cease, for reasons other than the expiry of the period of validity of the insurance or security, before 2 months have elapsed from the date on which notice of its termination is given to the Controlling State authority. This provision similarly applies to any modification which results in the insurance or security no longer satisfying the requirements of this Article.

Any claim for compensation for pollution damage may be brought directly against the insurer or other person providing financial coverage for the operator's liability for pollution damage. In such case the liability of the defendant is limited to the amount specified in accordance with the amounts stated above irrespective of the fact that the pollution damage occurred as a result of a deliberate act or omission by the operator himself, done with actual knowledge that pollution damage would result. The defendant may further avail himself of the defenses, which the operator himself would have been entitled to invoke, with the exceptions of bankruptcy or winding up of the operator. The defendant may avail himself of the defense that the pollution damage resulted from the wilful misconduct of the operator himself, but may not avail himself of any other defense which he might have been entitled to invoke in proceedings brought by the operator against him. The defendant has the right to require the operator to be joined in the proceedings.

Any sums provided by insurance or other financial security in accordance with this Article must be available in the first instance for the satisfaction of claims under this Protocol. Where the operator is a State Party, the operator is not required to maintain insurance or other financial security to cover its liability. The States shall harmonize national policies for the Arctic in this regard.

Article 9 deals with an Organization formed under Articles 2 and 13 of the Arctic Convention to evaluate the sums required under Articles 6 and 8 above. An Organization composed of a representative of each State Party is to be established. If a State Party considers that any of the amounts currently applicable under Articles 6 or 8 above is no longer adequate or is otherwise unrealistic, it may convene a meeting of the Organization to consider the matter. The Organization may recommend to the States Parties an amendment to any of the amounts in accordance with Articles 16 and 17 of the Arctic Convention. In making such a recommendation the Organization must take into account any information concerning events causing or likely to cause pollution damage having a bearing on the objects of this Protocol; any information on increases and decreases in the costs of goods and services involved in the treatment and remedying of marine oil and hazardous chemical spillages and disposal of waste and other materials; the availability of reliable insurance coverage against liability for pollution damage.

Under *Article 10* rights of compensation under this Protocol are extinguished unless, within 2 years[45] of the date on which the person suffering the damage knew or ought reasonably to have known of the damage, the claimant has in writing notified the operator of his claim or has brought an action in respect of it. In no case can an action be brought after 4 years from the date of the incident which caused the damage. Where the incident consists of a series of occurrences, the 4 years' period runs from the date of the last occurrence but does not operate to exclude a claim under Article 3 if the first occurrence was before the 10 year period.[46]

Article 11 deals with where compensation actions may be brought and states legal actions for compensation under this Protocol may be brought only in the Courts of any State Party where pollution damage was suffered as a result of the incident, or in the Courts of the Controlling State. For the purpose of determining where the damage was suffered, damage suffered in an area in which in accordance with international law a State has sovereign rights over natural resources is deemed to have been suffered in that State. Each State Party must ensure that its Courts possess the necessary jurisdiction for such actions for compensation. After the fund has been established in accordance with Article 6, the Courts of the State Party in which the fund is established are exclusively competent to determine all matters relating to the apportionment and distribution of the fund. The States shall harmonize national policies for the Arctic in this regard.

Article 12 deals with enforcement of judgments and states that any judgment given by a court with Article 11 jurisdiction, which is enforceable and no longer subject to review, must be recognized by any State Party. Exceptions include where the judgment was obtained by fraud or where the defendant was not given reasonable notice and a fair hearing. A judgment recognized under this Article is enforceable in each State Party as soon as the formalities required in that State have been complied with. This does not permit the merits of the case to be re-opened nor a reconsideration made of the applicable law. The States shall harmonize national policies for the Arctic in this regard.

Under *Article 13*, where a State Party is the operator, it is subject to suit in the jurisdiction specified in Article 11 and must waive all defenses based on sovereign immunity.

Article 14 deals with liability for nuclear incidents and specifies that no liability arises under this Protocol if the operator of a nuclear installation or ship is liable for such damage under other Conventions,[47] or if the operator of a nuclear installation or ship is liable for such damage under national law, provided that this law is as favourable to persons damaged as under the Conventions listed. The States shall harmonize national policies for the Arctic in this regard.

Under *Article 15*[48] a State can convene a Conference of States Parties for the revision or amendment of this Protocol in accordance with Articles 16 and 17 of the Arctic Convention.

Article 16 deals with liability for vessel-source pollution and calls on all Arctic States to ratify as soon as possible, if not already done, the International Convention on Civil Liability for Oil Pollution Damage (1969), the International Convention on the Establishment of an International Fund for Compensation for Oil Pollution Damage (1971), the Protocol of 1984 to Amend the International Convention on Civil Liability for Oil Pollution Damage 1969, and the Protocol of 1984 to Amend the International Convention on the Establishment of an International Fund for Compensation for Oil Pollution Damage.[49] The Arctic States shall directly and with the assistance of the competent international organizations draft a Convention on Liability and Compensation in Connection with the Carriage of Noxious and Hazardous Substances by Sea for use in the Arctic and ratify this document as quickly as possible.[50]

Additionally under Article 16 States must cooperate directly and with the assistance of competent regional or global international organizations to formulate and implement programs and procedures for the determination of liability and compensation for damage resulting from pollution from vessels in accordance with the Conventions and Protocols listed above. These programs and procedures must specifically ensure that those requirements involved in determining liability and compensation for pollution damage, including vessel identification, proof of causality, proof of fault, jurisdiction, enforcement and compensation can be determined as rapidly, adequately and sufficiently as possible. The States shall harmonize national policies for the Arctic in this regard.[51]

Article 17 deals with liability for pollution from dumping *not* associated with offshore oil activities and states that the Arctic States must cooperate directly and if necessary with the assistance of competent regional or global international organizations, especially those under the Intergovernmental Conference on the Convention on the Dumping of Wastes at Sea (LDC), to formulate and implement programs and procedures for the determination of liability and compensation for damage resulting from pollution from dumping other that than associated with offshore oil activities.[52] The States shall harmonize national policies for the Arctic in this regard.

Article 18 deals with liability for land-based pollution and states that the Arctic States must cooperate directly and if necessary with the assistance of competent regional or global international organizations to formulate and implement programs and procedures for the determination of liability and compensation for damage resulting from pollution from land-based sources under their jurisdiction.[53] The States shall harmonize national policies for the Arctic in this regard.

Article 19 makes a general statement regarding State Parties and dispute settlement. It states that in the event of dispute between the Arctic States and the dispute cannot be settled by the Parties themselves, then the dispute shall be settled according to the provisions outlined in Article 21 of the Arctic Convention.

Article 20 deals with State accession to the Protocol and requires Arctic States which may accede to the Arctic Liability Protocols to do so in accordance with the provisions outlined in Article 24 of the Arctic Convention.

General Articles which deal with among other things the place and date of signature, where the instrument may be deposited upon ratification, when the Protocol comes into force, methods of withdrawal, and authenticity of language are basically standard and will not be discussed here.[54]

17.3. Conclusion

In conclusion it is hoped by the discussion and proposals presented in Part II Chapters 8 - 17 that an idea can be obtained of the present Arctic marine environmental regimes, where they exist, and the *deficiencies*. It is also hoped that an understanding can be obtained of why it is deemed necessary to establish new regimes covering the entire Arctic in order to achieve optimum environmental protection for the Barents Sea. It is hoped that an idea is conferred of the possible legal support given by international customary law including the LOSC and general principles towards establishing new regimes. Finally it is hoped that the proposals submitted for the Arctic regimes achieve that delicate *balance* required, being both environmentally 'friendly', yet at the same time economically and politically feasible in order to allow State ratification.

Notes - Chapter 17

1 Protocol for Civil Liability for Oil Pollution Damage and Damage from Dumping Resulting from Exploration for and Exploitation of Seabed Mineral Resources.

2 National liability law which regulates liability and compensation for damage from marine pollution from sources other than shipping in the Arctic will not be covered here as being beyond the scope of the project. However, for a brief discussion of the national liability legislation of the US, Canada and Denmark/Greenland, see Part II Section 9.2.4.1. and Gold pp. 79-89, 91-111; for the Soviet and Norwegian liability legislation see *ibid.* and Butler Sections F.2., J.3., Supplement to J.4.; and Bull Erstatning s. 682-693 respectively.

3 See Part I Section 6.4.1. for State Parties, amounts of liability and other particulars. Also in this Section appears the 1984 Protocol to amend the CLC, which would be applicable in the Arctic *if* it were in force. It raises the limits of liability and hence the amount of compensation available.

4 See Part I Section 6.4.2. Also appearing in this Section is the 1984 Protocol to amend the Fund Convention which raises the liability limits. It would be applicable in the Arctic *if* it were in force.

5 See Part I Sections 6.4.3., 6.4.4. and 6.4.5. respectively for limits of liability and applicability.

6 See Part I Section 6.3. for both OPOL and the 1976 Liability Convention.

7 Although it is difficult to say it seems doubtful that the private Western oil companies if operating on the Soviet continental shelf would be able to limit their liability through an OPOL type agreement any more than they have been able to on the Norwegian continental shelf.

8 See Part II Section 9.2.4.

9 See Part I Section 6.2.

10 See Part I Section 6.3.

11 See Part II Sections 9.3.4.

12 See Part II Chapters 2 and 6 for conventions, customary and soft law regulating State liability and compensation lending additional legal support to the formation of this Protocol.

13 These include Articles 192, 194, 197, and the pollution control series including Articles 207 (land-based), 208 (offshore installations), 210 (dumping), 211 (vessel-source), 212 (airborne and atmospheric) and the enforcement series, Articles 213-222.

14 See Churchill p. 264.

15 Article 235 (1).

16 Article 235 (2).

17 Article 235 (3). Timagenis pp. 63 and 624 notes that in the development of LOSC it was agreed unanimously that LOSC would not contain detailed provisions regarding liability and compensation and that only general principals would be included. It was felt better that a separate convention could cover this subject in the future, and as noted in Part I Section 6.5. work is in progress by the ILC on State responsibility in general, and treaty Articles being drafted.

18 See Part I Section 2.1.4. and Part II Section 10.3.3.

19 See Part II Section 9.3.4.

20 For the sake of discussion changes are also noted oriented towards compensating a person damaged by pollution more than the 1976 Liability Convention. These changes reflect the legal concepts of *joint and several liability* and *unlimited liability*, as well as a broader liability scope than under the 1976 Liability Convention.

21 The provisions related to liability for dumping have been added.

22 As an example consider, could it actually be proven that land-based pollution from specific industries into Soviet rivers really caused damage to fish stocks in the Norwegian exclusive economic zone in the Barents Sea? There are problems of distance, time, dilution as well as location involved.

23 See Part I Section 6.3. Consider in Norway how far lawsuits have progressed (1) demanding compensation for damage from radioactive pollution from Chernobyl (direct causation probably easily proven) to meat industry of reindeer and sheep (amount of damages probably easily quantified); (2) demanding compensation from factories in both the UK and Germany for damages to fish and forests from pollution from acid rain (problems proving causation and amount of damages). To the author's knowledge, no compensation has been awarded in either case.

24 'Pollution from harmful or noxious substances and materials' was added to enlarge the scope of the Arctic Liability Protocol from just oil pollution. These could be dropped in the face of State reluctance to ratify and a phrase built in to add hazardous or noxious substances and materials after a period of time such as 5 to 10 years.

25 The definition of harmful or noxious substances and material was added, since these substances are covered by the Arctic Oil Rig Protocol and consequently as well need be covered in an Arctic Liability Protocol. See Part II Sections 12.2.7. and 12.2.8., Annexes I and II. Both Annexes were listed, even though Annex II substances are less noxious or harmful on the principle that if damage from these substances has taken place, then it should be compensated. These substances could be dropped in the absence of Arctic State ratification to make it more compatible to State interests.

26 Reference is made by the 1976 Liability Convention to a definition of 'ship' given by the International Convention on Civil Liability for Oil Pollution Damage, 29 November 1969, Brussels which is not used here.

27 The last sentence was added so that subcontractors can be joined in claims, something neither the 1976 Liability Convention nor Norwegian legislation primarily allows. See Part I Sections 6.3.3. and 9.2.4. and Bull Erstatning pp. 115-118. Since this sentence is 'claimant friendly' because it allows more parties to be sued, this could be dropped in the interest of State ratification. The channelization of responsibility as under the 1976 Liability Convention and Norwegian legislation seems to be an advantage as long as the licensee is solvent. This is probably no problem due to the economic size of the oil companies, but from a claimant's point of view, joint and several liability covering operators, licensees and subcontractors might be better, since full claims could be brought against companies more easily reached and most solvent. It is noted the groups protected by channelization include drilling and other sub entrepreneurs, equipment suppliers, maintenance companies and employees. The State is not normally included in one of these protected groups.

28 The 1976 Liability Convention uses simply, 'the escape or discharge of oil.' The addition noted is made throughout the Arctic Liability Protocol and could be dropped in order to induce State ratification.

29 Paragraph (2) was added, and the provisions related to damage from *dumping* could be *dropped* in the interest of State ratification. As noted the coverage denoted by this paragraph limits the compensation available only to damage from dumping associated with offshore oil activities.

30 The exclusion of 'well control methods and measures taken to protect, repair or replace an installation' from being 'preventative measures' is so that the operators cannot apply these costs against the compensation fund established under Article 6.

31 The phrase 'at an installation' from the 1976 Liability Convention was deleted so as to include dumping activities associated with offshore oil activities.

32 Note that the 1976 Liability Convention is as broad, stating *any* pollution damage (emphasis added). Using the definition under the 1976 Liability Convention of pollution damage this means any loss or damage outside the installation caused by contamination resulting from the escape or discharge of oil. Conceivably this would also include any demonstrable contamination of *fishing grounds*, even though the UK does not allow such compensation, and it is questionable whether Norway does. See Hjort pp. 210-226, Bull Erstatning s. 123-124, and Churchill Conflict p. 203.

33 The 1976 Liability Convention uses a 5 years limitation. 10 years is used here since the dumping of scrap is included in the Arctic Liability Protocol, and it is conceivable that scrap might not cause problems to fishing until several years have passed. A longer or shorter period of limitation might be considered in an attempt to balance fishing interests with State interests.

34 The latter phrase was added for clarity in order to ensure Article 10 also operates in this case.

35 This sentence was added to cover those situations where the dumping of 'grey list' substances was licensed and permitted by State authorities in areas such as fishing grounds. The operator can still be partly liable in those situations where it knew or should have known of the inapplicability for dumping in these areas.

36 Under the 1976 Liability Convention no claim for compensation for pollution damage, under the Convention or otherwise, may be made against the servants of the operator, since as noted liability is channelized to the operator. Claims against servants of the operator are allowed in the Arctic Liability Protocol in order to allow claimants the broadest possible choice of suits depending upon the damage suffered. Should this be considered too 'claimant friendly', the more restrictive language 'operator' of the 1976 Liability Convention could be substituted.

37 This latter sentence has been added to cover liability for incidents involving dumping resulting in damage.

38 This follows Norwegian legislation. As noted it was felt by Norwegian authorities that the oil companies could take out *several* insurance policies in order to met these requirements for drilling on the Norwegian continental shelf. See discussion of Article 8 below for insurance requirements. The 1976 Liability Convention states that the operator shall be entitled to limit his liability for each installation and each incident to *30 million* SDR's until 5 years from the date the Convention is opened for signature and *40 million* SDR's thereafter. Under the Convention the operator is not entitled to limit his liability if it is proved that the pollution damage occurred as a result of an act or omission by the operator himself, done deliberately with actual knowledge that pollution damage would result. In the interest of State ratification, rather than unlimited liability the limited liability of the 1976 Liability Convention could be reintroduced and these figures could be *reduced* to even less than the 1976 Liability Convention's figures and then increased in a graduated manner over a period of time. Thus for example 15 million SDR's liability limitation could be required for the first 5 years and then 30 million for the next 5 years, 60 million for the next 5 years and finally 100 million

SDR's thereafter. These figures and the time period are of course arbitrary, but the idea is to have several graduations over a longer period of time. This would give the States some leeway in their liability requirement while at the same time allowing for *some* liability coverage with the prospects of sufficient coverage in the not too distant future. Note that Article 9 addresses procedures for amending these amounts, procedures which must be reasonable and useable.

39 Under the 1976 Liability Convention the liability of the operator of any 1 installation shall not for any 1 incident exceed any limit which may be applicable to him.

40 Under the 1976 Liability Convention in such circumstances the aggregate liability of all of them in respect of any one incident shall not exceed the highest amount that could be awarded against any of them, but none of them shall be liable for any amount in excess of the limit applicable to him.

41 With respect to unlimited liability this amount is open for discussion. A proposal is for each installation and each incident 90 million Special Drawing Rights until 5 years have elapsed from the date on which the Protocol is opened for signature and 120 million Special Drawing Rights thereafter. Since UNEP Draft, p. 3 noted that none of the consultants were able to give a realistic figure but considered the original 35 million SDR's, the minimum amount of insurance required by the 1976 Liability Convention, too low, approximately 3 times Article 6 figures (30 million SDR's the first 5 years and then 40 million SDR's), 90 million SDR's and 120 million SDR's respectively were considered reasonable proposals. See Part II Section 12.2.5. and Article 8 discussion below. Under the 1976 Liability Convention in order to qualify for the liability limits, the operator must constitute a fund for the total sum representing the limits of his liability.

42 See discussion above under definition of 'preventative measures' regarding what can be claimed by the operator in such cases.

43 The 1976 Liability Convention continues, 'and is entitled to limit his liability.'

44 These amounts were obtained using the same ratios as used in 1976 Liability Convention. Under this Convention to cover the 30 million SDR's limit, at least 22 million SDR's of insurance is required for the first 5 years, which is 73.3% of the amount. The liability limit of 40 million SDR's is required covered by 35 million SDR's, which is 87.5%. For the figures proposed above for unlimited liability, these percentages calculate to approximately 66 million SDR's and 100 million SDR's respectively. In the interest of State ratification these figures could be reduced to even *less* than the 1976 Liability Convention's insurance figures and then increased in a graduated manner over a period of time. Thus for instance 11 million SDR's insurance could be required for the first 5 years, and then 22 million for the next 5 years, 44 million for the next 5 years, 66 million the next 5 years and finally 100 million SDR's thereafter. These figures and the time period are of course arbitrary, but the idea is to have several graduations of figures over a longer period of time, which would give the States some leeway in their liability requirement. See Article 9 for amendment procedures of these insurance amounts.

45 Though a 1 year Statute of Limitation used in the 1976 Liability Convention is usual in English and American legal systems, it seems reasonable to increase the time within which to raise a claim due to the special Arctic environmental conditions including period of darkness and ice, which might conceal causal effects longer than in other more temperate zones.

46 This phrase was added for clarity to be consistent with Article 3 abandonment time limits.

47 Under the 1976 Liability Convention these include the Paris Convention of 29 July 1960 on Third Party Liability in the Field of Nuclear Energy, or the Vienna Convention of 21 May 1963 on Civil Liability for Nuclear Damage and The Brussels Convention of 25 May 1962 on the Liability of Operators of Nuclear Ships.

48 The 1976 Liability Convention allows a State to provide for unlimited liability or a higher limit of liability than that specified under this Convention. As noted this provision should leave it open for Norway to apply its unlimited liability under the 1976 Liability Convention. This is not applicable in the Arctic Liability Protocol since all the Arctic States are required to provide for unlimited liability.

49 As noted none of the Arctic States has ratified these Protocols which raise the limits of available compensation.

50 See Part I Section 6.4.6.2.

51 This Article could be dropped failing State ratification.

52 This Article could be worded even more vaguely or dropped if shown that the States did not want to consider the question of liability for damage from this pollution source.

53 This Article could be dropped if shown that the States did not want to consider the question of liability for damage from this pollution source.

54 See Articles 16-22, 25 and 26 of the 1976 Liability Convention.

Appendix

F.2.: Edict on the Economic Zone of the USSR; On the Procedure for Applying Articles 19 and 21 of the Edict of the Presidium of the USSR Supreme Soviet "On the Economic Zone of the USSR"; Statute on the Protection of the Economic Zone of the USSR (Butler).

F.3.: Statute on the Protection and Preservation of the Marine Environment in the Economic Zone of the USSR (Butler).

G.1.a.: Edict of the Presidium of the USSR Supreme Soviet Concerning the Continental Shelf (Butler).

G.1.b.: Decree on the Application of the Edict of the Presidium of the USSR Supreme Soviet "On the Continental Shelf of the USSR" (Butler).

G.3.: Decree on the Procedure of Conducting Work on the Continental Shelf of the USSR (Butler).

G.4.: Charter on Discipline for Workers and Employees Engaged in Work Relating to Exploitation of Oil and Gas Resources on the Continental Shelf of the USSR (Butler).

J.2.: Decree on Intensifying Responsibility for Pollution of the Sea by Substances Harmful for the Health of People or for Living Resources of the Sea or by Other Wastes of Materials (Butler).

J.3.: Edict on the Amounts of Compensation by Shipowners for Losses Caused by Pollution of the Sea by Oil and Other

Substances Harmful for the Life of People and for Living Resources of the Sea (Butler).

J.4.: Edict on Intensifying Nature Protection in Areas of the Far North and Marine Areas Adjacent to the Northern Coast of the USSR (Butler).

J.5.: Barabolia, P.D., "Some Questions of the Legal Status and Legal Regime of the Economic Zone of Coastal States", Butler W.E. (ed.), *The Law of the Sea and International Shipping: Anglo-Soviet Post-Unclos Perspectives*, 1985 (Butler).

J.6.: Rusina, N. Ia., "International Legal Principles of Protection of the Marine Environment against Pollution", Butler W.E. (ed.), *ibid.*

J.7.: Brown, E. D., "The International Oil Pollution Compensation Fund: An Analytical Report on Fund Practice", Butler W.E. (ed.), *ibid.*

Bibliography

Adede, A.O., "International Environmental Law from Stockholm to Rio - An Overview of Past Lessons and Future Challenges," *Environmental Policy and Law*, 22/2 (1992), p. 88.

Akehurst, M., *A Modern Introduction to International Law*, 6th Ed. (1989).

Anker-Nilssen, Tyncho, "Metoder til Konsekvensanalyser olje/sjøfugl," Direktoratet for naturforvaltning, 1987.

Antinori, Camille M., "The Bering Sea: A Maritime Delimitation Dispute between the United States and the Soviet Union," *Ocean Development and International Law*, Vol. 18, No. 1, 1987.

Armstrong, T., "The Northern Sea Route," *Polar Research*, Vol, 25 (157).

Atlas *Gyldendals Store Verdensatlas*, Gyldendal Norsk Forlag.

Backer, Inge, *Naturvern og naturinngrep*, Universitetetsforlaget (1986).

Belsky, Martin, "Management of Large Marine Ecosystems: Developing a New Rule of Customary International Law," *San Diego Law Review*, Vol. 22, No. 4, 1985.

Bergsager, Egil, "Hydrocarbon Resources and Energy Politics in Arctic Areas Offshore Norway and USSR." Updated version of paper given at Offshore Northern Seas. Stavanger 20-24/8 1986.

Bergstrand, S. and R. Doganis, "The Impact of Flags of Convenience (Open Registries)," *The Law of the Sea and International Shipping: Anglo-Soviet Post-UNCLOS Perspectives*, (ed.) W.E. Butler, Centre for Socialist Legal Systems, University College, London (1985).

Birnie, Pat W., "Pollution and Fisheries in the North Sea," *The regulation of fisheries: legal, economic and social aspects, Proceedings of a European workshop*, University of Tromsø, Norway, (eds.) Ulfstein, Andersen and Churchill, June 2-4, 1986, p. 219.

Brigham, Lawson, Retired US Coast Guard Officer, Lecture given at FNI, 3 October, 1989.

Brown, E.D., "The International Oil Pollution Compensation Fund; An Analytical Report on Fund Practice, the Law of the Sea and International Shipping: Anglo-Soviet Post-UNCLOS Perspectives" (ed.) Butler, Studies on Socialist Legal Systems, Faculty of Laws, University College, London (1985) pp. 275-304.

Brownlie, Ian, *Principles of Public International Law*, Third Edition, Principles: pp. 431-477.

Brownlie, Ian, *System of the Law of Nations - State Responsibility*, Part I.

Brownlie, Ian "Causes of Action in the Law of Nations," 50 *Brit. Y.B. Int'l L.* 13, (1979)

Brundtland, A., "Sovjetisk utfordring og vestlige svar - Norsk sikkerhets politikk på nye vilkår?" *Internasjonal Politikk*, nr. 3, 1989,s. 135.

Brundtland Environment Report: *Our Common Future*, The World Commission on and Development 1987, Oxford.

Bull, Hans Jacob, "Petroleumsrett," *Knophs Oversikt over Norges Rett*, 9. Utgave, 1987, Universitetsforlaget, s. 682-693.

Bull, Hans Jacob, "Erstatnings ansvar i petroleumsvirksomheten" *Tidsskrift for Rettsvitenskap*, 1-2/1986, s. 84-115.

Butler, William E., *The USSR, Eastern Europe and the Development of the Law of the Sea*, Compiled and Edited, with Translations and Annotations by William E. Butler, Professor of Comparative Law in the University of London, Director, Centre for the Study of Socialist Legal Systems, University College, London.

Calmet, D.P. and J.M. Bewers, "Radioactive waste and ocean dumping, The role of the IAEA", *Marine Policy*, Vol 15, Number 6, November 1991, p. 413

Churchill, R.R. and A.V. Lowe, *The Law of the Sea*, 2nd Edition, 1988.

Churchill, Robin, "The Conflict between Oil and Fisheries: A Survey of Norwegian and United Kingdom Law and Practice," *The regulation of fisheries: legal, economic and social aspects, Proceedings of a European workshop*, University of Tromsø, Norway, (eds.) Ulfstein, Andersen and Churchill, June 2-4, 1986.

Churchill, R.R., A.V. Lowe, *The Law of the Sea*, 1983.

Churchill, R.R., D. Freestone (eds.), International Law and Global Climate Change, Chapter 9, "Controlling Emissions of Greenhouse Gases", (1991), p. 147.

Copenhagen Inom Report: Rapport Från Det Adertonden Nordiska Plenarmotet Køpenhamnsavtalet I Vasa Den 29.8.-1.9.1988 Angående Samarbete I Fråga Om Åtgärder Mot Oljeførurening Av Havet, Utkast 25.10.1988, sider 1-16, obtained from SFT, November 1988.

Curtis, Jeff B., "Vessel-Source Oil Pollution and MARPOL 73/78: An International Success Story?" *Environmental Law*, Vol. 15, (1985).

Dempsey, Paul Stephen and Lisa L. Helling, "Oil Pollution by Ocean Vessels - An Environmental Tragedy: The Legal Regime of Flags of Convenience, Multilateral Conventions, and Coastal States," *Denver Journal of International Policy*, Vol. 1, (1980).

De Cesari, P. et.al. (eds.) *Index of Multilateral Treaties on the Law of the Sea*, Milano, Dott. A. Giuffrè Editore 1985.

Eaton, J., "Oil Spill Liability and Compensation: Time to Clean-Up the Law," *George Washington Journal of International Law and Economics*, Vol. 19, No. 12, 1985.

Engelhardt, F.R. (ed.), *Petroleum Effects in the Arctic Environment*, Ottawa, Canada, (1985).

Euroil, Vol. Issue 5, August 1990.

Fleischer, Carl August, *Folkerett*, 5.utgave, (1986).

Fleischer, Carl August, "Et folkerettslig perspektiv på global ressursforvaltning -Behov for forskning?" NAVF's forskningspolitiske konferanse, 22-23 mars 1988 for oppfølging av Brundtland-kommisjonens rapport, s. 17.

FOH: Final Report from the Norwegian Marine Pollution Research and Monitoring Programme (FOH), "The Fate and Significance of Oil in the Sea," Sponsored by the Ministry of the Environment, 1976-1984.

Franckx, Erik, "The New USSR Legislation on Pollution Prevention in the Exclusive Economic Zone," *International Journal of Estuarine and Coastal Law*, Vol. 1, No. 2, (1986).

Frydenlund, Knut, *Lille land - hva nå*?

Fyhn, Hans Jørgen (ed.) "Fish Larval Physiology and Anatomy, Basic Research and Effects of Oil, Final Report 1983-1985," Zoological Laboratory, University of Bergen, Supported by A/S Norsk Shell.

Ganten, R., "The International System for Compensation for Oil Pollution Damage, An Assessment Based on the Experiences of the International Oil Pollution Compensation Fund," *Marius Nr. 62*, Norwegian Maritime Institute (1981).

Gorbatsjov's Speech given in Murmansk October 1, 1987, by Soviet President Mikhail Gorbatsjov in which an appeal for Arctic cooperation Speech: was made including environmental protection, scientific research, economic development and security issues. Obtained from the Soviet Embassy, Oslo Norway.

Gold, Edgar, *Handbook on Marine Pollution*, Halifax, 1985.

Griffiths, J.J., N.A. Øritsland, T. Øritsland, "Marine Mammals and Petroleum Activities in Norwegian Waters, A Review of the Literature on the Effects of Petroleum on Marine Mammals and Recommendations for Further Research," The Fishing Directorate's Institute of Marine Research, Bergen, 1987.

Hakapää, Kari, *Marine Pollution in International Law, Material Obligations and Jurisdiction, with Special Reference to the Third United Nations Conference on the Law of the Sea*, Dissertationes Humanarum Litterarum 28, (1981).

Handl, Gunther, "International Liability of States for Marine Pollution," *Canadian Yearbook of International Law*, Vol. 21, p. 85.

Harders, J. Enno, "In quest of an Arctic legal regime," *Marine Policy*, October 1987, pp. 285-298.

Harders, J. Enno, "The Arctic Ocean: Environmental Protection and Circumpolar Co-operation - An Assessment and a Proposal for a Legal Regime, Seminar Proceedings, Munich, Federal Republic of Germany, *Energy Law*, 1986.

Helsinki Report: Baltic Sea Environmental Proceedings No. 23, Activities of the Commission 1986, Including the Eighth Meeting of the Commission held in Helsinki 24-27 February 1987.

Henkin, Pugh, Schachter, Smit, *International Law, Cases and Materials*, American Casebook Series, West Publishing Co. (1980).

Hey, Ellen, "The Precautionary Approach. Implications of the revision of the Oslo and Paris Convention," *Marine Policy*, July 1991.

Hjort, Harald, Gunnar Meyer, *Forurensningsansvar i oljevirksomhet*, Nordisk Institutt for sjørett (1981).

Hoel, Alf Håkon, "Ressursforvaltning i nordområdene," *Norsk utenrikspolitisk årbok*, 1989.

Holst, Johan Jørgen, "Petroleum i norsk sikkerhetspolitikk," *Internasjonal Politikk*, nr. 2, 1986.

Hustoft, A.G., M. Østby, E. Næess-Ulseth, "Vurdering av oljevernberedskapen på Tromsøflaket, Rapport nr. 830215-2, 9 mai 1984.

Hydrographic: *Limits of Oceans and Seas*, International Hydrographic Organization Special Publication 23, Draft 4th ed. 1986.

ILC Report: International Law Commission, Report of May 9 - July 29, 1988, General Assembly, United Nations, Official Records: Forty-Third Session, Supplement No. 10 (A/43/10).

IMO Document: *Publications of the International Maritime Organization*, (1989), 938 78.15.E.

IMO Status: International Maritime Organization, "Status of Multilateral Conventions and Instruments in Respect of which the International Maritime Organization or its Secretary General Performs Depositary or Other Functions as of 31 December 1988," J/2735/Rev.3.

Johnston, Douglas M., (ed.) *The Environmental Law of the Sea*, International Union for Conservation of Nature and Natural Resources, (IUCN) Switzerland, 1981.

Karstad, Arne, "Security Aspects of Norwegian Oil and Energy Policy," Address by The Norwegian State Secretary for Defence, to the Conference on Norway, OPEC and the Middle East at SAS Royal Hotel, Bergen, 8 October, 1987.

Kasoulides, George C., "State Responsibility and Assessment of Liability for Damage resulting from Dumping Operations," *San Diego Law Review*, Vol. 26, No. 3, (1989) pp. 497-523.

Kibsgaard, "Militære operasjoner og mulige interessekonflikter i de nordlige havområdene," *Internasjonal Politikk*, Nr. 2-3, 1988, s. 35.

Kindt, "International Environmental Law and Policy; An Overview of Transboundary Pollution" (Law of Sea Symposium), 23 *San Diego Law Review* 583 (1986), p. 600.

Kwiatkowska, Barbara, "New Soviet Legislation on the 200 Mile Economic Zone and Certain Problems of Evolution of Customary Law," *Netherlands International Law Review*, Vol. 33, (1986).

Lamson, Cynthia, "Arctic shipping, marine safety and environmental protection," *Marine Policy*, January 1987.

Lamson, Cynthia and David VanderZwaag, "Arctic Waters: Needs and Options for Canadian-American Cooperation," *Ocean Development and International Law*, Vol. 18, No. 1. 1987, pp. 49-99.

MacRae, "Customary International Law and the United Nations Law of the Sea Treaty,", 13 *Cal.W.Int'l. L.J.* 181.

Malone, "Who Needs the Sea Treaty?", 45 *Foreign Affairs*, Spr. 1984 p. 44.

McIntyre, A.D., "Pollution in the North Sea from Oil Related Industry: An Overview," paper presented at the International Conference on Environmental Protection of the North Sea, March 1987.

Medwaves, UNEP, Issue No. 10/III, 1987.

Miles, Edward, Laura Geselbracht, "Fisheries and Conflicting Uses of the Sea," *Proceedings of a European workshop*, University of Tromsø, Norway, (eds.) Ulfstein, Andersen and Churchill, June 2-4, 1986.

Mounteer, Thomas R., "The General Agreement on Tariffs and Trade (GATT) and the Law of the Sea (LOS) Convention: A Critical Comparison of Arbitration Provisions," *The International Lawyer*, Fall 1987, Vol. 21, No. 4, p. 989.

Mundheim, Ø., Tørhaug, M., Helgaker, P., Ose, T., Westergaard, Rich., H., Sjøen K., Vesterhaug, O., Olshausen, K., Wold, M. (1981), "Offshore Blow out Control". OTTER Group Report No. SFT 88 A81004, Trondheim, Norway.

Møse, Erik, "Folkerettslige tekster," Universitetesforlaget, A/S 1986.

Nelson, J. C. and R. D. Needham, "The Arctic as a Regional Sea," *Environmental Conservation* 12, 1 (Spring 1985).

Norwegian Pollution Act: Forurensningsloven.

Northern Perspectives, Vol, 18, Nr. 3, September - October 1990, Canadian Arctic Resources Committee.

NOU:1989: Norges Offentlige Utredninger 1989: 9, "Norsk Polar Forskning."

O'Connell, D.P., *The International Law of the Sea*, 1982.

Oil and Gas: Norwegian Bureau of Statistics, Oil and Gas Activity, 3rd Quarter 1990, Statistics and Analysis.

Oil Sea: *Oil in the Sea, Inputs, Fates, and Effects*, National Research Council, et al., National Academy Press, 1985.

OPN 1988: Ocean Policy News, *Council on Ocean Law*, The Newsletter of International Ocean Law, 1709 New York Ave., N.W. Washington D.C. 20006, USA, January/February, 1988.

Oppedal, Georg, Fiskeridirektoratet. "Fangstdata Barentshavet Syd," Dokument Nr. 04-01-01, Olje- og Energidepartmentet Konsekvenssutredninger for Barentshavet Syd, April, 1987.

Osherenko, Gail, "Environmental Security in the Arctic: Prospects for Soviet Cooperation," Dartmouth College, Prepared for the International Studies Association Annual Convention, London, March 1989.

Oslo Comm: Thirteenth Annual Report on the Activities of the Oslo Commission 1989, Oslo Commission, New Court, 48 Carey Street, 2A 2JE.

OTTER: Mundheim, Ø., Tørhaug, M., Helgaker, P., Ose, T., Westergaard, Rich., H., Sjøen K., Vesterhaur, O., Olshausen, K., Wold, M. (1981), "Offshore Blowout Control," OTTER Group Report No. STF 88 A81004. Trondheim, Norway.

Oslo-Paris Comm. The Oslo and Paris Commissions, *The First Decade, International Cooperation in Protecting the Marine Environment*, (1984).

Our Future: *Our Common Future, The World Commission on Environment and Development*, Oxford University Press, 1987.

Par.Comm.7: Seventh Annual Report on the Activities of the Paris Commission, Paris Commission 1985, New Court, 48 Carey Street, London WC2A 2JE.

Par.Comm.8: Eighth Annual Report on the Activities of the Paris Commission, Paris Commission 1986, (as above).

Par.Comm.9: Ninth Annual Report on the Activities of the Paris Commission, Paris Commission 1987, (as above).

Par.Comm.10: Tenth Annual Report on the Activities of the Paris Commission, Paris Commission 1988, (as above).

Pharand, Donat, "The legal regime of the Arctic: some outstanding issues," *International Journal*, Vol. XXXXIX, No. 4, Autumn 1984.

Roginko, Alexei, "Arctic Environmental Cooperation: Prospects and Possibilities," Institute of World Economy and International Relations, USSR Academy of Sciences, Prepared for the International Studies Association Annual Convention, London, March 1989.

Roginko, Alexei, "Environmental Conservation in the Soviet Arctic Seas," Chapter 7, to be published in *The Soviet Maritime Arctic* (1989).

SFT Akutt: Statens forurensningstilsyn, "Rapport om akutt forurensning i 1986. Registrerte utslipp til land, kyst og hav." SFT mai 1987.

SFT Miljø 85: Miljøundersøkelser Rundt Oljeplattformer på Norsk Sektor, Årsrapport for 1985, SFT, mai 1987.

SFT Miljø 86: Miljøundersøkelser Rundt Oljeplattformer på Norsk Sektor, Årsrapport for 1986, SFT, januar 1988.

SFT Utslipp 1986 Statens forurensningstilsyn, "Utslipp av olje fra plattformene på norsk kontinentalsokkel i 1986," SFT, mai 1987.

SFT Utslipp 1988 Utslipp av Olje og Andre Kjemikalier Fra Plattformene på Norsk Kontinentalsokkel i 1988, Statens forurensningtilsyn, august 1989.

St.meld.14: Stortingsmelding. nr. 14 (1987-88), Statens samlede engasjement i petroleums virksomheter i 1988. (Storting melding is the Norwegian Parliamentary Report).

St.meld.39: Stortingsmelding nr. 39 (1974-1975), Justis- og politidepartementet vedrørende Svalbard.

St.meld.40: Stortingsmelding nr. 40 (1988-89), Åpening av Barentshavet Syd for letevirksomhet.

St.meld. 79: Stortingsmelding nr. 79 (1984-1985).

St.meld.101: Stortingsmelding nr. 101 (1981-1982), Om sikkerhet og nedrustning, Rustningskontroll og nedrustnings arbeidets plass i sikkerhetspolitikken.

Stokke, Olav Schram, "Environmental Threats in the Arctic: Ignorance, Poverty and Impotence, Fridtjof Nansen Institutt, September 1989.

Sætevik, Sunneva, "Miljøvernsamarbeid mellom Nordsjøstatene, virkelighetsoppfatninger, handlingspreferanser og utfall. En studie av Pariskommisjons arbeid for regulering av landbasert havforurensning." Fridtjof Nansens Institutt, April 1987.

Timagenis, Gr. J., *International Control of Marine Pollution*, Oceana Publications, Inc., Dobbs Ferry, New York, Sitjhoff & Noordhoff International Publishers BV; Alphen aan den Rijn, The Netherlands, Vols. I-II, 1980.

Tresselt, Per, "Norsk-sovjetiske forhandlinger om avgrensning av kontinentalsokler og økonomiske soner," *Internasjonal Politikk - Petroleum og sikkerhet i Barentshavet*, Nr. 2-3. 1988.

Ulfstein, Geir, *Økonomiske Soner - Hva Nå?*, Universitetsforlaget, (1982).

Ulfstein, Geir, "The Conflict between Petroleum Production, Navigation and Fisheries in International Law." *Ocean Development and International Law*, Vol. 19, pp. 229-262.

UNEP Report of the Technical Consultation on Pollution Resulting from Exploration and Exploitation of the Continental Shelf and the Sea-Bed and the Sub-Soil, UNEP/WG.155/1.

VanderZwaag, David and Cynthia Lamson, "Ocean Development and Management in the Arctic: Issues in American and Canadian Relations," Conference paper, *Transit Management in the Northwest Passage: Problems and Prospects*, Dalhousie Ocean Studies Programme, Halifax Canada, 1984.

Van Hanswyk, Beth, "The 1984 Protocols to the International Convention on Civil Liability for Oil Pollution Damages and the International Fund for Compensation for Oil Pollution Damages," *The International Lawyer*, Summer 1988, Vol. 22, No. 2., p. 319.

Veritec Report No. 84-3372, juli 1984, "Sannsynligheten for oljesøl ved tankskipsuhell og ved kontinentalsokkelvirksomhet (blowouts) i norsk farvann i perioden 1985-2000.

Wang, Cheng-Pang, "A Review of the Enforcement Regime for Vessel-Source Oil Pollution Control, *Ocean Development and International Law*, Vol. 16, No. 4.

Westergaard, Rich. H., *All About Blowout*, Senter for Industriforskning, Oslo, Norway, 1987.

Young, Oran R., "The Age of the Arctic," *Foreign Policy*, 1985-86, No. 61, pp. 160-179.

Ørebech, Peter, *Om allemannsrettigheter, Særlig med Henblikk på Rettsvernet for Fiske ved Igangsetting og Utøving av Petroleumsvirksomhet*, Osmundsson Forlag AS, (Ørebech), 1991.

Østreng, Willy., Bergesen, Moe and Østreng, *Soviet Oil and Security Interests in the Barents Sea*, Fridtjof Nansens Institutt (FNI) (1987).

Østreng, Willy, "Delimitation arrangements in the Arctic seas, Cases of precedence or securing of strategic/economic interests?," *Marine Policy*, April 1986, Vol. 10, No. 2, pp. 132-154.

Østreng, Willy, "The Changing Mood of the Kremlin," *International Challenges*, Vol. 9, No. 2, 1989, pp. 10-14.

Index of Treaties

Helsinki Convention: Convention on the Protection of the Marine Environment of the Baltic Sea Area, 13 *International Legal Materials* 544, (1974). 59, 60, 78-80, 82, 83, 90, 92, 100, 105, 107, 135, 136, 150, 151, 182-184, 246, 276, 288, 299, 409, 410, 412, 417, 418, 424, 425

HNS Draft Convention: Draft Convention on Liability and Compensation in Connection with the Carriage of Noxious and Hazardous Substances, (1984), 23 *International Legal Materials* 150. 162, 241, 397, 436

ICJ Statute: Statute of the International Court of Justice, 26 June 1945, Møse, Erik, (ed.), *Folkerettslige tekster*, Universitetetsforlaget A/S, 1986, p. 74. (Attached to U.N. Charter). 61, 63, 64, 232, 284

ILO No. 92: ILO Convention No. 92 of 1949 supplemented by No. 133 (1979) concerning crew Accommodation on Board Ship, 160 *United Nations Treaty Series* 223. 134

ILO No. 109: ILO Convention No. 109 of 1958 concerning Wages, Hours of Work on Board Ship and Manning (not in force), Singh, *International Convention of Merchant Shipping* p. 1203. 134

ILO No. 145: ILO Convention No. 145 of 1976 concerning Continuity of Employment of Seafarers, *International Labour Conference Convention* 145. 134

ILO No. 147: ILO Convention No. 147 of 1976 concerning Minimum (Crewing) Standards in Merchant Ships, VI *New Directions in the Law of the Sea* 449. 134

IMO Contingency: International Convention on Oil Pollution Preparedness, Response and Cooperation, (not in force). Can be obtained from the IMO. 180, 181

IMO Dangerous: IMO International Maritime Dangerous Goods Code of 1988. IMO Document p. 10. 135

IMO Mobile: Draft Convention on Offshore Mobile Craft, International Maritime Organization, LEG XXXXIV/66, 19 December 1977. 37, 106, 152, 154, 155, 320

LDC: Intergovernmental Conference on the Convention on the Dumping of Wastes at Sea, (London), 11 *International Legal Materials*, 1294, (1972). 59, 60, 73-79, 82, 83, 91, 100, 106, 107, 123, 150, 151, 165, 180, 234, 240, 244-248, 256, 295, 298, 321, 408-416, 436, 447

LLMC: Convention on Limitation of Liability for Maritime Claims (1976), IMO Document, 444 77.04.E, p. 21. 161, 162, 241, 436

Load Lines Convention: International Convention on Load Lines (1966), IMO Document, 701 81.01.E, (1989), p. 33. 59, 134, 157, 237, 382, 390

LOSC: United Nations Convention on the Law of the Sea, 21 *International Legal Materials* 1261, (1982). 1, 2, 11, 33, 58, 60, 62, 64, 65, 75-77, 82, 90, 92, 119, 121, 124, 157, 182, 185, 232-234, 245, 246, 250-252, 254, 257, 283-288, 295, 297, 298, 300, 304, 309, 322, 323, 327, 344, 346, 347, 348, 350, 352, 381-385, 387-393, 395, 396, 399, 400, 408, 409, 413, 414, 417, 423-425, 429, 437, 438, 447

MARPOL 73/78: 1973 International Convention for the Prevention of Pollution from Ships, 12 *International Legal Materials* 1319. 122, 123, 125-129. The 1978 Protocol to the 1973 International Convention for the Prevention of Pollution from Ships, 17 *International Legal Materials* 546. 4-6, 59, 77, 95, 100, 119, 121-129, 134-136, 157, 162, 180, 221, 236, 238-240, 244, 249-255, 298, 318, 322, 335, 381, 387, 388-396, 399, 457

Mexico - US Agreement: Mexico - United States: Agreement of Cooperation regarding Pollution of the Marine Environment, 20 *International Legal Materials* 696. 183, 184

MOU: Memorandum of Understanding on Port State Control, 21 *International Legal Materials* 1 (1982). 134, 238-240

NORBRITPLAN: Norway - United Kingdom Joint Contingency Plan for Counter-Pollution Operations at Sea, obtained from the Norwegian Ministry of Foreign Affairs. 184, 186-188, 242, 346, 348-350, 353-355, 363, 367

Nordic Agreement: Agreement between Denmark, Finland, Norway and Sweden concerning Co-operation in Measures to deal with Pollution of the Sea by Oil, II *New Directions in the Law of the Sea* 637. 90

180, 181, 236, 242, 244, 245, 247, 248, 278, 295, 298, 300, 309-311, 317, 318, 321, 323, 327, 328, 423, 424, 425-430, 432, 446, 459

PLAN: (proposed) Norway - Soviet Union Joint Contingency Plan for Counter-Pollution Operations at Sea. 188, 304, 344, 345, 346, 348, 349, 350, 351, 353, 354, 355, 356, 357, 358, 359, 360, 361, 365, 367, 368, 373, 374

PLATO: Pollution Agreement among Tanker Owners (1985). Can be obtained from PLATO Limited, Portland House, 6th Floor, Stag Place, London SW1E 5BH. 6, 160-162, 241, 255, 436

Polar Bear Convention: 1973 Agreement on Conservation of Polar Bears, 17 *International Legal Materials* 13. 243

Rescue at Sea Agreement: Overenskomst mellom Kongeriket Norges Regjering og Regjeringen i Unionen av Sovjetiske Sosialistiske Republikker om Samarbeid ved Ettersøking av Savnede og Redning av Nødstedte Mennesker i Barentshavet, (january 1988). Obtained from the Norwegian Foreign Ministry. 243, 346, 347, 358

SOLAS: International Convention for the Safety of Life at Sea (1974), 14 *International Legal Materials* 959. Protocol to the International Convention for the Safety of Life at Sea (1978), 17 *International Legal Materials* 579. 59, 128, 129, 130, 157, 237, 251, 254, 381, 390, 398

Space Convention: Convention on International Liability for Damage caused by Space Objects, March 1972, 24 *United Stats Treaties and Other international Agreements* 2289. 61

STCWS: International Convention on the Standards of Training, Certification, and Watchkeeping for Seafarers (1978), IMO Document p. 319. 133, 237-240, 251, 382, 390, 398

Stockholm Declaration: Stockholm Declaration and Action Plan on the Human Environment, Report of the United Nations Conference on the Human Environment U.N. Doc. A/CONF. 48/14 and Cor. I, 11 *International Legal Materials* 1416 (1972). 59, 64, 150, 163, 287

TOVALOP: Tankers Owners' Voluntary Agreement Concerning Liability for Oil Pollution (1969), 8 *International Legal Materials* 497. 5, 152, 159, 160, 162, 241, 255, 436

Index